Path Integrals in Physics
Volume I
Stochastic Processes and Quantum Mechanics

Path Integrals in Physics

Volume I
Stochastic Processes and Quantum Mechanics

M Chaichian

Department of Physics, University of Helsinki
and
Helsinki Institute of Physics, Finland

and

A Demichev

Institute of Nuclear Physics, Moscow State University, Russia

IoP

Institute of Physics Publishing
Bristol and Philadelphia

© IOP Publishing Ltd 2001

All rights reserved. No part of this publication may be reproduced, stored in a retrieval system or transmitted in any form or by any means, electronic, mechanical, photocopying, recording or otherwise, without the prior permission of the publisher. Multiple copying is permitted in accordance with the terms of licences issued by the Copyright Licensing Agency under the terms of its agreement with the Committee of Vice-Chancellors and Principals.

British Library Cataloguing-in-Publication Data

A catalogue record for this book is available from the British Library.

ISBN 0 7503 0801 X (Vol. I)
     0 7503 0802 8 (Vol. II)
     0 7503 0713 7 (2 Vol. set)

Library of Congress Cataloging-in-Publication Data are available

Commissioning Editor: James Revill
Production Editor: Simon Laurenson
Production Control: Sarah Plenty
Cover Design: Victoria Le Billon
Marketing Executive: Colin Fenton

Published by Institute of Physics Publishing, wholly owned by The Institute of Physics, London

Institute of Physics Publishing, Dirac House, Temple Back, Bristol BS1 6BE, UK

US Office: Institute of Physics Publishing, The Public Ledger Building, Suite 1035, 150 South Independence Mall West, Philadelphia, PA 19106, USA

Typeset in TEX using the IOP Bookmaker Macros
Printed in the UK by Bookcraft, Midsomer Norton, Bath

Fate has imposed upon our writing this tome the yoke of a foreign tongue in which we were not sung lullabies.

Freely adapted from Hermann Weyl

Contents

	Preface		ix
	Introduction		1
		Notational conventions	9
1	**Path integrals in classical theory**		**12**
	1.1	Brownian motion: introduction to the concept of path integration	12
		1.1.1 Brownian motion of a free particle, diffusion equation and Markov chain	12
		1.1.2 Wiener's treatment of Brownian motion: Wiener path integrals	22
		1.1.3 Wiener's theorem and the integration of functionals	29
		1.1.4 Methods and examples for the calculation of path integrals	36
		1.1.5 Change of variables in path integrals	45
		1.1.6 Problems	49
	1.2	Wiener path integrals and stochastic processes	56
		1.2.1 A short excursion into the theory of stochastic processes	56
		1.2.2 Brownian particles in the field of an external force: treatment by functional change of variables in the path integral	63
		1.2.3 Brownian particles with interactions	66
		1.2.4 Brownian particles with inertia: a Wiener path integral with constraint and in the space of velocities	69
		1.2.5 Brownian motion with absorption and in the field of an external deterministic force: the Bloch equation and Feynman–Kac formula	72
		1.2.6 Variational methods of path-integral calculations: semiclassical and quadratic approximations and the method of hopping paths	78
		1.2.7 More technicalities for path-integral calculations: finite-difference calculus and Fourier decomposition	94
		1.2.8 Generating (or characteristic) functionals for Wiener integrals	101
		1.2.9 Physics of macromolecules: an application of path integration	108
		1.2.10 Problems	111
2	**Path integrals in quantum mechanics**		**122**
	2.1	Feynman path integrals	123
		2.1.1 Some basic facts about quantum mechanics and the Schrödinger equation	123
		2.1.2 Feynman–Kac formula in quantum mechanics	137
		2.1.3 Properties of Hamiltonian operators from the Feynman–Kac formula	141
		2.1.4 Bohr–Sommerfeld (semiclassical) quantization condition from path integrals	144
		2.1.5 Problems	149
	2.2	Path integrals in the Hamiltonian formalism	153

		2.2.1	Derivation of path integrals from operator formalism in quantum mechanics	154
		2.2.2	Calculation of path integrals for the simplest quantum-mechanical systems: a free particle and a harmonic oscillator	161
		2.2.3	Semiclassical (WKB) approximation in quantum mechanics and the stationary-phase method	169
		2.2.4	Derivation of the Bohr–Sommerfeld condition via the phase-space path integral, periodic orbit theory and quantization of systems with chaotic classical dynamics	176
		2.2.5	Particles in a magnetic field: the Ito integral, midpoint prescription and gauge invariance	183
		2.2.6	Applications of path integrals to optical problems based on a formal analogy with quantum mechanics	187
		2.2.7	Problems	190
	2.3	Quantization, the operator ordering problem and path integrals		200
		2.3.1	Symbols of operators and quantization	200
		2.3.2	General concept of path integrals over trajectories in phase space	209
		2.3.3	Normal symbol for the evolution operator, coherent-state path integrals, perturbation expansion and scattering operator	216
		2.3.4	Problems	226
	2.4	Path integrals and quantization in spaces with topological constraints		230
		2.4.1	Point particles in a box and on a half-line	231
		2.4.2	Point particles on a circle and with a torus-shaped phase space	238
		2.4.3	Problems	243
	2.5	Path integrals in curved spaces, spacetime transformations and the Coulomb problem		245
		2.5.1	Path integrals in curved spaces and the ordering problem	245
		2.5.2	Spacetime transformations of Hamiltonians	251
		2.5.3	Path integrals in polar coordinates	258
		2.5.4	Path integral for the hydrogen atom: the Coulomb problem	266
		2.5.5	Path integrals on group manifolds	272
		2.5.6	Problems	282
	2.6	Path integrals over anticommuting variables for fermions and generalizations		286
		2.6.1	Path integrals over anticommuting (Grassmann) variables for fermionic systems	286
		2.6.2	Path integrals with generalized Grassmann variables	298
		2.6.3	Localization techniques for the calculation of a certain class of path integrals	304
		2.6.4	Problems	315

Appendices 318
A General pattern of different ways of construction and applications of path integrals 318
B Proof of the inequality used for the study of the spectra of Hamiltonians 318
C Proof of lemma 2.1 used to derive the Bohr–Sommerfeld quantization condition 322
D Tauberian theorem 326

Bibliography 328

Index 333

Preface

The importance of path-integral methods in theoretical physics can hardly be disputed. Their applications in most branches of modern physics have proved to be extremely fruitful not only for solving already existing problems but also as a guide for the formulation and development of essentially new ideas and approaches in the description of physical phenomena.

This book expounds the fundamentals of path integrals, of both the Wiener and Feynman type, and their numerous applications in different fields of physics. The book has emerged as a result of many courses given by the authors for students in physics and mathematics, as well as for researchers, over more than 25 years and is based on the experience obtained from their lectures.

The mathematical foundations of path integrals are summarized in a number of books. But many results, especially those concerning physical applications, are scattered in a variety of original papers and reviews, often rather difficult for a first reading. In writing this book, the authors' aim was twofold: first, to outline the basic ideas underlying the concept, construction and methods for calculating the Wiener, Feynman and phase-space quantum-mechanical path integrals; and second, to acquaint the reader with different aspects concerning the technique and applications of path integrals.

It is necessary to note that, despite having almost an 80-year history, the theory and applications of path integrals are still a vigorously developing area. In this book we have selected for presentation the more or less traditional and commonly accepted material. At the same time, we have tried to include some major achievements in this area of recent years. However, we are well aware of the fact that many important topics have been either left out or are only briefly mentioned. We hope that this is partially compensated by references in our book to the original papers and appropriate reviews.

The book is intended for those who are familiar with basic facts from classical and quantum mechanics as well as from statistical physics. We would like to stress that the book is not just a linearly ordered set of facts about path integrals and their applications, but the reader may find more effective ways to learn a desired topic. Each chapter is self-contained and can be considered as an independent textbook: it contains general physical background, the concepts of the path-integral approach used, followed by most of the typical and important applications presented in detail. In writing this book, we have endeavored to make it as comprehensive as possible and to avoid statements such as 'it can be shown' or 'it is left as an exercise for the reader', as much as it could be done.

A beginner can start with any of the first two chapters in volume I (which contain the basic concepts of path integrals in the theory of stochastic processes and quantum mechanics together with essential examples considered in full detail) and then switch to his/her field of interest. A more educated user, however, can start directly with his/her preferred field in more advanced areas of quantum field theory and statistical physics (volume II), and eventually return to the early chapters if necessary.

For the reader's convenience, each chapter of the book is preceded by a short introductory section containing some background knowledge of the field. Some sections of the book require also a knowledge of the elements of group theory and differential (mainly Riemann) geometry. To make the reading easier, we have added to the text a few supplements containing some basic concepts and facts from these mathematical subjects. We have tried to use a minimum of mathematical tools. Thus, the proofs of a number of theorems and details of applications are either briefly sketched or omitted, adequate references being given to enable the interested reader to fully grasp the subject. An integral part of the presentation of the material is the problems and their solutions which follow each topic discussed in the book. We do hope that their study will be helpful for self-education, for researchers and teachers supervising exercise sessions for students.

During the preparation of both volumes of this book the authors have benefited from discussions on various physical and mathematical aspects related to path integrals with many of their colleagues. We thank all of them for useful discussions and for their advice. Especially, it is a pleasure to express our gratitude to Alexander Beilinson, Alan Carey, Wen-Feng Chen, Vladimir Fainberg, Dmitri Gitman, Anthony Green, John van der Hoek, Mikhail Ioffe, Petr Kulish, Wolfgang Kummer, Antti Kupiainen, Jorma Louko, the late Mikhail Marinov, Kazuhiko Nishijima, Matti Pitkänen, Dmitri Polyakov, Adam Schwimmer, Konstantin Selivanov and the late Euan Squires, and to acknowledge their stimulating discussions, suggestions and criticism. Over the years, many students have provided us with useful remarks and suggestions concerning the presentation of the material of the book. We thank all of them, in particular Jari Heikkinen and Aleksi Vuorinen. We are deeply grateful to Claus Montonen, Peter Prešnajder and Anca Tureanu for their invaluable contributions and improvements throughout the book. It is also a great pleasure for us to express our gratitude to Jim Revill, Senior Academic Publisher of IOP, for his fruitful cooperation and for his patience.

The financial support of the Academy of Finland under Project No 163394 is greatly acknowledged.

Masud Chaichian, Andrei Demichev
Helsinki, Moscow
December 2000

Introduction

The aim of this book is to present and explain the concept of the path integral which is intensively used nowadays in almost all the branches of theoretical physics.

The notion of *path integral* (sometimes also called *functional integral* or *integral over trajectories* or *integral over histories* or *continuous integral*) was introduced, for the first time, in the 1920s by Norbert Wiener (1921, 1923, 1924, 1930) as a method to solve problems in the theory of diffusion and Brownian motion. This integral, which is now also called the *Wiener integral*, has played a central role in the further development of the subject of path integration.

It was reinvented in a different form by Richard Feynman (1942, 1948) in 1942, for the reformulation of quantum mechanics (the so-called '*third formulation* of quantum mechanics' besides the Schrödinger and Heisenberg ones). The Feynman approach was inspired by Dirac's paper (1933) on the role of the Lagrangian and the least-action principle in quantum mechanics. This eventually led Feynman to represent the propagator of the Schrödinger equation by the *complex-valued* path integral which now bears his name. At the end of the 1940s Feynman (1950, 1951) worked out, on the basis of the path integrals, a new formulation of quantum electrodynamics and developed the well-known *diagram technique* for perturbation theory.

In the 1950s, path integrals were studied intensively for solving functional equations in quantum field theory (*Schwinger equations*). The functional formulation of quantum field theory was considered in the works of Bogoliubov (1954), Gelfand and Minlos (1954), Khalatnikov (1952, 1955), Mathews and Salam (1954), Edwards and Peierls (1954), Symanzik (1954), Fradkin (1954) and others. Other areas of applications of path integrals in theoretical physics discovered in this decade were the study of Brownian motion in an absorbing medium (see Kac (1959), Wiegel (1975, 1986) and references therein) and the development of the theory of superfluidity (Feynman 1953, 1954, ter Haar 1954, Kikuchi 1954, 1955). Starting from these pioneering works, many important applications of path integrals have been found in statistical physics: in the theory of phase transitions, superfluidity, superconductivity, the Ising model, quantum optics, plasma physics. In 1955, Feynman used the path-integral technique for investigating the polaron problem (Feynman 1955) and invented his variational principle for quantum mechanics. This work had an important impact on further applications of path integrals in statistical and solid state physics, as well as in quantum field theory, in general.

At the same time, attempts were initiated to widen the class of exactly solvable path integrals, i.e. to expand it beyond the class of Gaussian-like integrals. In the early 1950s, Ozaki (in unpublished lecture notes, Kyushu University (1955)) started with a short-time action for a free particle written in Cartesian coordinates and transformed it into the polar form. Later, Peak and Inomata (1969) calculated explicitly the radial path integral for the harmonic oscillator. This opened the way for an essential broadening of the class of path-integrable models. Further important steps in this direction were studies of systems on multiply connected spaces (in particular, on Lie group manifolds) (Schulman 1968, Dowker 1972) and the treatment of the quantum-mechanical Coulomb problem by Duru and Kleinert (1979), who applied the so-called Kustaanheimo–Stiefel spacetime transformation to the path integral.

In the 1960s, a new field of path-integral applications appeared, namely the quantization of *gauge fields*, examples of which are the electromagnetic, gravitational and Yang–Mills fields. The specific properties of the action functionals for gauge fields (their invariance with respect to gauge transformations) should be taken into account when quantizing, otherwise wrong results emerge. This was first noticed by Feynman (1963) using the example of Yang–Mills and gravitational fields. He showed that quantization by straightforward use of the Fermi method, in analogy with quantum electrodynamics, violates the unitarity condition. Later, as a result of works by De Witt (1967), Faddeev and Popov (1967), Mandelstam (1968), Fradkin and Tyutin (1969) and 't Hooft (1971), the problem was solved and the path-integral method turned out to be the most suitable one for this aim. In addition, in the mid-1960s, Berezin (1966) took a crucial step which allowed the comprehensive use of path integration: he introduced integration over Grassmann variables to describe fermions. Although this may be considered to be a formal trick, it opened the way for a unified treatment of bosons and fermions in the path-integral approach.

In the 1970s, Wilson (1974) formulated the field theory of quarks and gluons (i.e. quantum chromodynamics) on a Euclidean spacetime *lattice*. This may be considered as the discrete form of the field theoretical path integral. The lattice serves as both an ultraviolet and infrared cut-off which makes the theory well defined. At low energies, it is the most fruitful method to treat the theory of strong interactions (for example, making use of computer simulations). A few years later, Fujikawa (1979) showed how the *quantum anomalies* emerge from the path integral. He realized that it is the 'measure' in the path integral which is not invariant under a certain class of symmetry transformations and this makes the latter anomalous.

All these achievements led to the fact that the path-integral methods have become an indispensable part of any construction and study of field theoretical models, including the realistic theories of unified electromagnetic and weak interactions (Glashow 1961, Weinberg 1967, Salam 1968) and quantum chromodynamics (the theory of strong interactions) (Gross and Wilczek 1973, Politzer 1973). Among other applications of path integrals in quantum field theory and elementary particle physics, it is worth mentioning the derivation of asymptotic formulas for infrared and ultraviolet behaviour of Green functions, the semiclassical approximation, rearrangement and partial summation of perturbation series, calculations in the presence of topologically non-trivial field configurations and extended objects (solitons and instantons), the study of cosmological models and black holes and such an advanced application as the formulation of the first-quantized theory of (super)strings and branes. In addition, the path-integral technique finds newer and newer applications in statistical physics and non-relativistic quantum mechanics, in particular, in solid body physics and the description of critical phenomena (phase transitions), polymer physics and quantum optics, and in many other branches of physics. During the two last decades of the millennium, most works in theoretical and mathematical physics contained some elements of the path-integral technique. We shall, therefore, not pursue the history of the subject past the 1970s, even briefly. Functional integration has proved to be especially useful for the description of collective excitations (for example, quantum vortices), in the theory of critical phenomena, and for systems on topologically non-trivial spaces. In some cases, this technique allows us to provide solid foundations for the results obtained by other methods, to clarify the limits of their applicability and indicate the way of calculating the corrections. If an exact solution is possible, then the path-integral technique gives a simple way to obtain it. In the case of physically realistic problems, which normally are far from being exactly solvable, the use of path integrals helps to build up the qualitative picture of the corresponding phenomenon and to develop approximate methods of calculation. They represent a sufficiently flexible mathematical apparatus which can be suitably adjusted for the extraction of the essential ingredients of a complicated model for its further physical analysis, also suggesting the method for a concrete realization of such an analysis. One can justly say that path integration is an integral calculus adjusted to the needs of contemporary physics.

Universality of the path-integral formalism

The most captivating feature of the path-integral technique is that it provides a unified approach to solving problems in different branches of theoretical physics, such as the theory of stochastic processes, quantum mechanics, quantum field theory, the theory of superstrings and statistical (both classical and quantum) mechanics.

Indeed, the general form of the basic object, namely the transition probability $W(x_f, t_f|x_0, t_0)$, in the theory of stochastic processes, reads

$$W(x_f, t_f|x_0, t_0) \sim \sum_{\substack{\text{all trajectories} \\ \text{from } x_0 \text{ to } x_f}} \exp\left\{-\frac{1}{4D} F[x(\tau)]\right\} \qquad (0.0.1)$$

where x_0 denotes the set of coordinates of the stochastic system under consideration at the initial time t_0 and $W(x_f, t_f|x_0, t_0)$ gives the probability of the system to have the coordinates x_f at the final time t_f. The explicit form of the functional $F[x(\tau)]$, $t_0 \leq \tau \leq t_f$, as well as the value and physical meaning of the constant D, depend on the specific properties of the system and surrounding medium (see chapter 1). The summation sign symbolically denotes summation over all trajectories of the system. Of course, this operation requires further clarification and this is one of the goals of this book.

In quantum mechanics, the basic object is the transition amplitude $K(x_f, t_f|x_0, t_0)$, not a probability, but the path-integral expression for it has a form which is quite similar to (0.0.1):

$$K(x_f, t_f|x_0, t_0) \sim \sum_{\substack{\text{all trajectories} \\ \text{from } x_0 \text{ to } x_f}} \exp\left\{\frac{i}{\hbar} S[x(\tau)]\right\} \qquad (0.0.2)$$

or, in a more general case,

$$K(x_f, t_f|x_0, t_0) \sim \sum_{\substack{\text{all trajectories} \\ \text{in phase space} \\ \text{with fixed } x_0 \text{ and } x_f}} \exp\left\{\frac{i}{\hbar} S[x(\tau), p(\tau)]\right\}. \qquad (0.0.3)$$

Here, $S[x(\tau)]$ is the action of the system in terms of the configuration space variables, while $S[x(\tau), p(\tau)]$ is the action in terms of the phase-space variables (coordinates and momenta). Though now we have purely imaginary exponents in contrast with the case of stochastic processes, the general formal structure of expressions (0.0.1)–(0.0.3) is totally analogous. Moreover, as we shall see later, the path integrals (0.0.2), (0.0.3) can be converted into the form (0.0.1) (i.e. with a purely real exponent) by a transition to purely imaginary time variables: $t \to -it$ and, in many cases, this transformation can be mathematically justified.

In the case of systems with an infinite number of degrees of freedom, it was also realized, even in the 1960s, that an essential similarity between quantum field theory and (classical or quantum) statistical physics exists. In particular, the vacuum expectations (Green functions) in quantum field theory are given by expressions of the type:

$$\langle 0|\widehat{A}(\hat{\varphi})|0\rangle \sim \sum_{\substack{\text{all field} \\ \text{configurations}}} A(\varphi) \exp\left\{\frac{i}{\hbar} S[\varphi]\right\} \qquad (0.0.4)$$

where, on the left-hand side, $\widehat{A}(\hat{\varphi})$ is an operator made of the field operators $\hat{\varphi}$ and on the right-hand side $A(\varphi)$ is the corresponding classical quantity. After the transition to purely imaginary time $t \to -it$

4   *Introduction*

(corresponding to the so-called Euclidean quantum field theory), the vacuum expectation takes the form:

$$\langle 0|\widehat{A}(\hat{\varphi})|0\rangle \sim \sum_{\substack{\text{all field}\\\text{configurations}}} A(\varphi)\exp\left\{-\frac{1}{\hbar}S[\varphi]\right\} \quad (0.0.5)$$

while in classical statistical mechanics, thermal expectation values are computed as

$$\langle A(\varphi)\rangle_{\text{cl.st.}} \sim \sum_{\substack{\text{all}\\\text{configurations}}} A(\varphi)\exp\left\{-\frac{1}{k_{\text{B}}T}E[\varphi]\right\} \quad (0.0.6)$$

(k_{B} is the Boltzmann constant and T is the temperature). The similarity of the two last expressions is obvious.

It is worth noting that historically quantum field theory is intimately linked with the classical field theory of electromagnetism and with particle physics. Experimentally, it is intimately connected to high-energy physics experiments at accelerators. The origins of statistical mechanics are different. Historically, statistical mechanics is linked to the theory of heat, irreversibility and the kinetic theory of gases. Experimentally, it is intimately connected with calorimetry, specific heats, magnetic order parameters, phase transitions and diffusion. However, since equations (0.0.5) and (0.0.6) are formally the same, we can mathematically treat and calculate them in the same way, extending the methods developed in statistical physics to quantum field theory and vice versa.

It is necessary to stress the fact that both statistical mechanics and field theory deal with systems in an infinite volume and hence with an infinite number of degrees of freedom. A major consequence of this is that the formal definitions (0.0.4)–(0.0.6) by themselves have no meaning at all because they, at best, lead to $\frac{\infty}{\infty}$. There is always a further definition needed to make sense of these expressions. In the case of statistical mechanics, that definition is embodied in the thermodynamical limit which first evaluates (0.0.6) in a finite volume and then takes the limit as the size of the box goes to infinity. In the case of quantum field theory, the expressions (0.0.4), (0.0.5) need an additional definition which is provided by a 'renormalization scheme' that usually involves a short-distance cut-off as well as a finite box. Thus, in statistical mechanics, the (infrared) thermodynamical limit is treated explicitly, whereas in quantum field theory, it is the short-distance (ultraviolet) cut-off that is discussed extensively. The difference in focus on infrared cut-offs versus ultraviolet cut-offs is often one of the major barriers of communication between the two fields and seems to constitute a major reason why they are traditionally considered to be completely different subjects.

Thus, path-integral techniques provide a unified approach to different areas of contemporary physics and thereby allow us to extend methods developed for some specific class of problems to other fields. Though different problems require, in general, the use of different types of path integral—Wiener (real), Feynman (complex) or phase space—this does not break down the unified approach due to the well-established relations between different types of path integral. We present a general pattern for different ways of constructing and applying path integrals in a condensed graphical form in appendix A. The reader may use it for a preliminary orientation in the subject and for visualizing the links which exist among various topics discussed in this monograph.

The basic difference between path integrals and multiple finite-dimensional integrals: why the former is not a straightforward generalization of the latter

From the mathematical point of view the phrase 'path integral' simply refers to the generalization of integral calculus to functionals. The general approach for handling a problem which involves functionals was developed by Volterra early in the last century (see in Volterra 1965). Roughly speaking, he considered a functional as a function of infinitely many variables and suggested a recipe consisting of three steps:

(i) replace the functional by a function of a finite number of N variables;
(ii) perform all calculations with this function;
(iii) take the limit in which N tends to infinity.

However, the first attempts to integrate a functional over a space of functions were not very successful. The historical reasons for these failures and the early history of Wiener's works which made it possible to give a mathematically correct definition of path integrals can be found in Kac (1959) and Papadopoulos (1978).

To have an idea why the straightforward generalization of the usual integral calculus to functional spaces does not work, let us remember that the basic object of the integral calculus on $\mathbb{R}^n$ is the Lebesgue measure (see, e.g., Shilov and Gurevich (1966)) and the basic notion for the axiomatic definition of this measure is *a Borel set*: a set obtained by a countable sequence of unions, intersections and complementations of subsets $\mathcal{B}$ of points $x = (x_1, \ldots, x_n) \in \mathbb{R}^n$ of the form

$$\mathcal{B} = \{x \mid a_1 \leq x_1 \leq b_1, \ldots, a_n \leq x_n \leq b_n\}.$$

The Lebesgue measure μ (i.e. a rule ascribing to any subset a number which is equal, loosely speaking, to its 'volume') is uniquely defined, up to a constant factor, by the conditions:

(i) it takes finite values on bounded Borel sets and is positive on non-empty open sets;
(ii) it is invariant with respect to translations in $\mathbb{R}^n$.

A natural question now appears: Does the Lebesgue measure exist for infinite-dimensional spaces? The answer is negative. Indeed, consider the space $\mathbb{R}^\infty$. Let $\{e_1, e_2, \ldots\}$ be some orthonormal basis in $\mathbb{R}^\infty$, B_k the sphere of radius $\frac{1}{2}$ with its centre at e_k and B the sphere of radius 2 with its centre at the origin (see in figure 0.1 a part of this construction related to a three-dimensional subspace of $\mathbb{R}^\infty$). Then, from the property (i) of the Lebesgue measure, we have

$$0 < \mu(B_1) = \mu(B_2) = \mu(B_3) = \cdots < \infty.$$

Note that the spheres B_k have no intersections and, hence, the additivity of any measure gives the inequality $\mu(B) \geq \sum_k \mu(B_k) = \infty$, which contradicts condition (i) for a Lebesgue measure.

Thus the problem of the construction of path integrals can be posed and considered from a purely mathematical point of view as an abstract problem of a self-consistent generalization of the notion of an integral to the case of infinite-dimensional spaces. Investigations along this line represent indeed an important field of mathematical research: see, e.g., Kac (1959), Gelfand and Yaglom (1960), Kuo (1975), Simon (1979), De Witt-Morette *et al* (1979), Berezin (1981), Elliott (1982), Glimm and Jaffe (1987) and references therein. We shall follow, however, another line of exposition, having in mind a corresponding physical problem in all cases where path integrals are utilized. The deep mathematical questions we shall discuss on a rather intuitive level, with the understanding that mathematical rigour can be supplied whenever necessary and that the answers obtained do not differ, in any case, from those obtained after a sound mathematical derivation. However, we do try to provide a flavour of the mathematical elegance in discussing, e.g., the celebrated Wiener theorem, the Bohr–Sommerfeld quantization condition, properties of the spectra of Hamiltonians derived from the path integrals etc.

It is necessary to note that the available level of mathematical rigour is different for different types of path integral. While the (probabilistic) Wiener path integral is based on a well-established mathematical

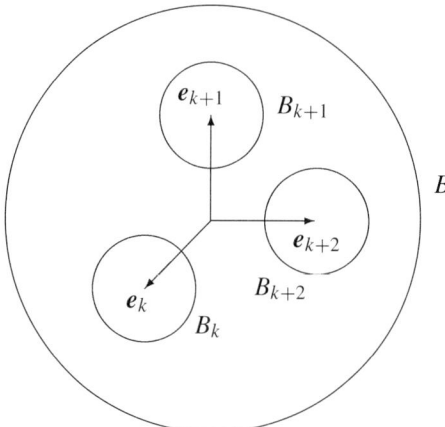

Figure 0.1. A three-dimensional part of the construction in the finite-dimensional space $\mathbb{R}^\infty$, which proves the impossibility of the direct generalization of the Lebesgue measure to the infinite-dimensional case.

background, the complex oscillatory Feynman and phase-space path integrals still meet some analytical difficulties in attempts of rigorous mathematical definition and justification, in spite of the progress achieved in works by Mizrahi (1976), Albeverio and Høegh-Krohn (1976), Albeverio *et al* (1979), De Witt-Morette *et al* (1979) and others. Roughly speaking, the Wiener integral is based on a well-defined functional integral (Gaussian) measure, while the Feynman and phase-space path integrals do not admit any strictly defined measure and should be understood as more or less mathematically justified limits of their finite-dimensional approximation. The absence of a measure in the case of the Feynman or phase-space quantum-mechanical path integrals is not merely a technicality: it means that these in fact are *not* integrals; instead, they are *linear functionals*. In a profound mathematical analysis this difference might be significant, since some analytical tools appropriate for integrals are not applicable to linear functionals.

What this book is about and what it contains

Different aspects concerning path integrals are considered in a number of books, such as those by Kac (1959), Feynman and Hibbs (1965), Simon (1979), Schulman (1981), Langouche *et al* (1982), Popov (1983), Wiegel (1986), Glimm and Jaffe (1987), Rivers (1987), Ranfagni *et al* (1990), Dittrich and Reuter (1992), Mensky (1993), Das (1993), Kleinert (1995), Roepstorff (1996), Grosche (1996), Grosche and Steiner (1998) and Tomé (1998). Among some of the main review articles are those by Feynman (1948), Gelfand and Yaglom (1960), Brush (1961), Garrod (1966), Wiegel (1975, 1983), Neveu (1977), DeWitt-Morette *et al* (1979) and Khandekar and Lawande (1986).

In contrast to many other monographs, in this book the concept of path integral is introduced in a deductive way, starting from the original derivation by Wiener for the motion of a Brownian particle. Besides the fact that the Wiener measure is one whose existence is rigorously proven, the Brownian motion is a transparent way to understand the concept of a path integral as the way by which the Brownian particle moves in space and time. Thus, the representation in terms of Wiener's treatment of Brownian motion will serve as a prototype, whenever we use path integrals in other fields, such as quantum mechanics, quantum field theory and statistical physics.

Approximation methods, such as the semiclassical approximation, are considered in detail and in the subsequent chapters they are used in quantum mechanics and quantum field theory. Special attention is devoted to the change of variables in path integrals; this also provides a necessary experience when dealing with analogous problems in other fields. Some important aspects, like the gauge conditions in quantum field theory, can similarly be met in the case of the Brownian motion of a particle with inertia which involves path integrals with constraints. Several typical examples of how to evaluate such integrals are given.

With the background obtained in chapter 1, chapter 2 continues to the cases of quantum mechanics. We essentially use the similarity between Wiener and Feynman path integrals in the first section of chapter 2 reducing, in fact, some quantum-mechanical problems to consideration of the corresponding Wiener integral. On the other hand, there exists an essential difference between the two types (Wiener and Feynman) of path integral. The origin of this distinction is the appearance of a new fundamental object in quantum mechanics, namely, the probability *amplitude*. Moreover, functional integrals derived from the basic principles of quantum mechanics prove to be over paths in the *phase space* of the system and only in relatively simple (though quite important and realistic) cases can be reduced to Feynman path integrals over trajectories in the *configuration space*. We discuss this topic in sections 2.2 and 2.3. A specific case in which we are strongly confined to work in the framework of phase-space path integrals (or, at least, to start from them) is the study of systems with *curved* phase spaces. The actuality of such a study is confirmed, e.g., by the fact that the important *Coulomb problem* (in fact, any quantum-mechanical description of atoms) can be solved via the path-integral approach only within a formalism including the phase space with curvilinear coordinates (section 2.5).

A natural application of path integrals in quantum mechanics, also considered in chapter 2, is the study of systems with *topological constraints*, e.g., a particle moving in some restricted domain of the entire space $\mathbb{R}^d$ or with non-trivial, say periodic, boundary conditions (e.g., a particle on a circle or torus). Although this kind of problem can, in principle, be considered by operator methods, the path-integral approach makes the solution simpler and much more transparent. The last section of chapter 2 is devoted to the generalization of the path-integral construction to the case of particles described by operators with anticommutative (fermionic) or even more general defining relations (instead of the canonical Heisenberg commutation relations).

In chapter 2 we also present and discuss important technical tools for the construction and calculation of path integrals: operator symbol calculus, stochastic Ito calculus, coherent states, the semiclassical (WKB) approximation, the perturbation expansion, the localization technique and path integration on group manifolds. This chapter also contains some selected applications of path integrals serving to illustrate the diversity and fruitfulness of the path-integral techniques.

In chapter 3 we proceed to discuss systems with an infinite number of degrees of freedom, that is, to consider quantum field theory in the framework of the path-integral approach. Of course, quantum field theory can be considered as the limit of quantum mechanics for systems with an infinite number of degrees of freedom and with an arbitrary or non-conserved number of excitations (particles or quasiparticles). Therefore, the starting point will be the quantum-mechanical phase-space path integrals studied in chapter 2 which we suitably generalize for the quantization of the simplest field theories, at first, including scalar and spinor fields. We derive the path-integral expression for the generating functional of Green functions and develop the perturbation theory for their calculation. In most practical applications in quantum field theory, these path integrals can be reduced to the Feynman path integrals over the corresponding *configuration spaces* by integrating over momenta. This is especially important for relativistic theories where this transition allows us to keep *explicitly* the relativistic invariance of all expressions.

Apparently, the most important result of path-integral applications in quantum field theory is the formulation of the celebrated *Feynman rules* and the invention of the Feynman diagram technique for the perturbation expansion in the case of field theories with constraints, i.e. in the case of *gauge-field theories* which describe all the realistic fundamental interactions of elementary particles. This is one of the central topics of chapter 3. For pedagogical reasons, we start from an introduction to the quantization of quantum-mechanical systems with constraints and then proceed to the path-integral description of gauge theories. We derive the covariant generating functional and covariant perturbation expansion for Yang–Mills theories with exact and spontaneously broken gauge symmetry, including the realistic *standard model* of electroweak interactions and *quantum chromodynamics*, which is the gauge theory of the strong interactions.

However, important applications of path integrals in quantum field theory go far beyond just a convenient derivation of the perturbation theory rules. We consider various non-perturbative approximations for calculations in field theoretical models, variational methods (including the Feynman variational method in the non-relativistic field theory of the polaron), the description of topologically non-trivial field configurations, semiclassical, in particular *instanton*, calculations, the quantization of extended objects (*solitons*) and calculation of quantum anomalies.

The last section of chapter 3 contains some advanced applications of the path-integral technique in the theory of quantum gravity, cosmology, black holes and in string theory, which is believed to be the most plausible candidate (or, at least, a basic ingredient) for a 'theory of everything'.

As we have previously pointed out, the universality of the path-integral approach allows us to apply it without crucial modification to statistical (both classical and quantum) systems. We discuss how to incorporate statistical properties into the path-integral formalism for the study of many-particle systems in chapter 4. At first, we present, for its easier calculation, a convenient path-integral representation of the so-called configuration integral entering the *classical* partition function. In the next section, we pass to quantum systems and, in order to establish a 'bridge' to what we considered in chapter 2, we introduce a path-integral representation for an arbitrary but *fixed* number of *indistinguishable* particles obeying Bose or Fermi statistics. We also discuss the generalization to the case of particles with *parastatistics*.

The next step is the transition to the case of an *arbitrary number* of particles which requires the use of second quantization, and hence, field theoretical methods. Consideration of path-integral methods in quantum field theory in chapter 3 proves to be highly useful in the derivation of the path-integral representation for the partition functions of statistical systems with an arbitrary number of particles. We present some of the most fruitful applications of the path-integral techniques to the study of fundamental problems of quantum statistical physics, such as the analysis of critical phenomena (phase transitions), calculations in field theory at finite (non-zero) temperature or at finite (fixed) energy, as well as the study of non-equilibrium systems and the phenomena of superfluidity and superconductivity. One section is devoted to the presentation of basic elements of the method of *stochastic quantization*, which non-trivially combines ideas borrowed from the theory of stochastic processes (chapter 1), quantum mechanics (chapter 2) and quantum field theory (chapter 3), as well as methods of non-equilibrium statistical mechanics. The last section of this chapter (and the whole book) is devoted to systems defined on lattices. Of course, there are no continuous trajectories on a lattice and, hence, no true path integrals in this case. But since in quantum mechanics as well as in quantum field theory the precise definition of a path integral is heavily based on the discrete approximation, discrete-time or spacetime approximations prove to be the most reliable method of calculations. Then the aim is to pass to the corresponding continuum limit which just leads to what is called a 'path integral'. However, in many cases there are strong reasons for direct investigation of the *discrete approximations* of the path integrals and their calculation, without going to the continuum limit. Such calculations become extremely important and fruitful in situations when there are simply no other suitable exact or approximate ways to reach physical results. This is true, in particular, for the gauge theory of *strong interactions*. We also consider physically discrete systems (in particular, the Ising model) which do not require transition to the continuous limit at all, but which can be analyzed by methods borrowed from the path-integral technique.

For the reader's convenience, each chapter starts with a short review of basic concepts in the corresponding subject. The reader who is familiar with the basic concepts of stochastic processes, quantum mechanics, field theory and statistical physics can skip, without loss, these parts (printed with a specific type in order to distinguish them) and use them in case of necessity, only for clarification of our notation. A few supplements at the end of the book serve basically a similar aim. They contain short information about some mathematical and physical objects necessary for understanding parts of the text, as well as tables of useful ordinary and path integrals. Besides, each section is supplemented by a set of problems (together with more or less detailed hints for their solution), which are integral parts of the presentation of the material. In a few appendixes we have collected mathematical details of the proofs of statements discussed in the main text, which can be skipped for a first reading without essential harm for understanding.

An obvious problem in writing a book devoted to a wide field is that, while trying to describe the diversity of possible ways of calculation, tricks and applications, the book does not become ponderous. For this purpose and for a better orientation of the reader, we have separated the text in the subsections into shorter topics (marked with the sign $\diamond$) and have given each one an appropriate title. We have tried to present the technical methods discussed in the book, whenever possible, accompanied by non-trivial physical applications. Necessarily, these examples, to be tractable in a single book, contain oversimplifications but the reader will find references to the appropriate literature for further details. The present monograph can also be considered as a preparatory course for these original or review articles and specialized books. The diversity of applications of path integrals also explains some non-homogeneity of the text with respect to detailing the presentation and requirements with respect to prior knowledge of the reader. In particular, chapters 1 and 2 include all details, are completely self-contained and require only a very basic knowledge of mathematical analysis and non-relativistic quantum mechanics. For a successful reading of the main part of chapter 3, it is helpful to have some acquaintance with a standard course of quantum field theory, at least at a very elementary level. The last section of this chapter contains advanced and currently developing topics. Correspondingly, the presentation of this part is more fragmentary and without much detail. Therefore, their complete understanding requires rather advanced knowledge in the theory of gravitation and differential geometry and can be achieved only by rather experienced readers. However, even those readers who do not feel fully ready for reading this part are invited to go through it (without trying to absorb all the details), in order to get an idea about this modern and fascinating area of applications of path integrals. Chapter 4, which contains a discussion of path-integral applications for solving various problems in statistical physics, is also necessarily written in a more fragmentary style in comparison with chapters 1 and 2. Nevertheless, all crucial points are covered and though some prior familiarity with the theory of critical phenomena is useful for reading this chapter, we have tried to make the text as self-contained as possible.

Notational conventions

Some general notation:

$\mathbb{Z}$	integers
$\mathbb{Z}_+$	positive integers
$\mathbb{R}$	real numbers
$\mathbb{C}$	complex numbers
$\stackrel{\text{def}}{\equiv}$	definition
$\widehat{A}$	operator
M	matrix
$\mathbb{I}$	identity operator or matrix
x	vector
c^*	complex conjugation of $c \in \mathbb{C}$
$\widehat{A}^\dagger$	Hermitian conjugation of the operator $\widehat{A}$

$\mathsf{M}^\top$	matrix transposition
$\dot{f}(t,x)$	time derivation: $\dot{f}(t,x) \stackrel{\text{def}}{\equiv} \frac{\partial f(t,x)}{\partial t}$
$f'(t,x)$	derivation with respect to a space variable x: $f'(t,x) \stackrel{\text{def}}{\equiv} \frac{\partial f(t,x)}{\partial x}$
$\mathcal{O}(\varepsilon)$	a quantity of the order of ε
$A = \{a \mid F\}$	subset A of elements a (belonging to some larger set) which satisfy the condition F
$\mathbb{P}\{A\}$	probability of the event A
$\mathbb{D}(X)$	dispersion of a random quantity X
$d_W x(\tau)$	Wiener functional integration measure
$d_F x(\tau)$	Feynman functional integration 'measure'
$\mathcal{D}\varphi(x)$	general notation for a functional integration 'measure'
$\mathcal{C}\{\boldsymbol{x}_1,t_1;\boldsymbol{x}_2,t_2\}$	set of trajectories starting at $\boldsymbol{x}(t_1)=\boldsymbol{x}_1$ and having the endpoint $\boldsymbol{x}(t_2)=x_2$
$\mathcal{C}\{x_1,t_1;[AB],t_2\}$	set of trajectories with the starting point $x_1 = x(t_1)$ and ending in the interval $[AB] \in \mathbb{R}$ at the time t_2
$\mathcal{C}\{\boldsymbol{x}_1,t_1;t_2\}$	set of trajectories with an arbitrary endpoint
$\mathcal{C}\{\boldsymbol{x}_1,t_1;\boldsymbol{x}_2,t_2;\boldsymbol{x}_3,t_3\}$	set of trajectories having the starting and endpoint at x_1 and x_3, respectively, and passing through the point x_2 at the time t_2
$W(x,t\mid x_0,t_0)$	transition probability in the theory of stochastic processes
$K(x,t\mid x_0,t_0)$	transition amplitude (propagator) in quantum mechanics
$G(x-y), D(x-y), S(x-y)$	field theoretical Green functions

General comments:

- Some introductory parts of chapters or sections in the book contain preliminaries (basic concepts, facts, etc) on a field where path integrals find applications to be discussed later in the main part of the corresponding chapters or sections. The text of these preliminaries is distinguished by the present specific print.
- The symbol of averaging (mean value) $\langle \cdots \rangle$ acquires quite different physical and even mathematical meaning in different parts of this book (e.g., in the sense of stochastic processes, quantum-mechanical or statistical (classical or quantum) averaging). In many cases we stress its concrete meaning by an appropriate subscript. But essentially, all the averages $\langle A \rangle$ are achieved by path integration of the quantity A with a corresponding functional integral measure.
- We assume the usual summation convention for repeated indices *unless the opposite is indicated explicitly*; in ambiguous cases, we use the explicit sign of summation.
- Operators are denoted by a 'hat': $\widehat{A}, \widehat{B}, \widehat{x}, \widehat{p}, \ldots$ with the only exception that the time-ordering operator (an operator acting on other operators) is denoted by $\mathbf{T}$; for example:

$$\mathbf{T}(\hat{\varphi}(x_1)\hat{\varphi}(x_2)).$$

- Normally, vectors in $\mathbb{R}^n$ and $\mathbb{C}^n$ are marked by bold type: $\boldsymbol{x}$. However, in some cases, when it cannot cause confusion as well as for an easier perception of cumbersome formulas, we use the ordinary print for vectors in spaces of arbitrary dimension. As is customary, four-dimensional vectors of the relativistic spacetime are always denoted by the usual type $x = \{x^0, x^1, x^2, x^3\}$ and the corresponding scalar product reads: $xy \stackrel{\text{def}}{\equiv} g_{\mu\nu} x^\mu y^\nu$, where $g_{\mu\nu} = \text{diag}\{1,-1,-1,-1\}$ is the Minkowski metric. An expression of the type A_μ^2 is the shorthand form for $g_{\mu\nu}A^\mu A^\nu$. If the vector indices $\mu, \nu, \ldots$ take in some expression with only spacelike values 1, 2, 3, we shall denote them by Latin letters $l, k, \ldots$ and use the following shorthand notation: $A_l B_l = \sum_{l=1}^{3} A_l B_l$, where A_l, B_l are the spacelike components of some four-dimensional vectors $A_\mu = \{A_0, A_l\}$, $B_\nu = \{B_0, B_l\}$ in the Minkowski spacetime.

- Throughout the book we use the same notation for probability densities in the case of random variables having continuous values and for probability distributions when random variables have discrete sets of values. We also take the liberty to use the term *probability density* in cases when the type of value (discrete or continuous) is not specified.

List of abbreviations:

BRST	Becchi–Rouet–Stora–Tyutin (symmetry)
CCR	canonical commutation relations
ESKC	Einstein–Smoluchowski–Kolmogorov–Chapman (relation)
OPI	one-particle irreducible (diagram, Green function)
PI	path integral
QFT	quantum field theory
QCD	quantum chromodynamics
QED	quantum electrodynamics
SUSY	supersymmetry
WKB	Wentzel–Kramers–Brillouin (approximation)
YM	Yang–Mills (theory, fields)

Chapter 1

Path integrals in classical theory

The aim of this chapter is to present and to discuss the general concept and mathematical structure of path integrals, introduced for the first time by N Wiener (1921, 1923, 1924, 1930), as a tool for solving problems in the theory of classical systems subject to random influences from the surrounding medium. The most famous and basic example of such a system is a particle performing the so-called *Brownian motion*. This phenomenon was discovered in 1828 by the British botanist R Brown, who investigated the pollen of different plants dispersed in water. Later, scientists realized that small fractions of any kind of substance exhibit the same behaviour, as a result of random fluctuations driven by the medium. The theory of Brownian motion emerged in the beginning of the last century as a result of an interplay between physics and mathematics and at present it has a wide range of applications in different areas, e.g., diffusion in stellar dynamics, colloid chemistry, polymer physics, quantum mechanics.

In section 1.1, we shall discuss Wiener's (path-integral) treatment of Brownian motion which must remain a prototype for us whenever dealing with a path integral. Section 1.2 is devoted to the more general path integral description of various stochastic processes. We shall consider a Brownian particle with inertia, systems of interacting Brownian particles, etc. The central point of this section is the famous and very important Feynman–Kac formula, expressing the transition probability for a wide class of stochastic processes in terms of path integrals. Besides, we shall construct *generating* (also called *characteristic*) *functionals* for probabilities expressed via the path integrals and shortly discuss an application of the path-integral technique in polymer physics. In both sections 1.1 and 1.2, we shall also present calculation methods (including approximate ones) for path integrals.

1.1 Brownian motion: introduction to the concept of path integration

After a short exposition of the main facts from the physics of Brownian motion, we shall introduce in this section the *Wiener measure* and the *Wiener integral*, prove their existence, derive their properties and learn the methods for practical calculations of path integrals.

1.1.1 Brownian motion of a free particle, diffusion equation and Markov chain

The apparently irregular motion that we shall describe, however non-deterministic it may be, still obeys certain rules. The foundations of the strict theory of Brownian motion were developed in the pioneering work by A Einstein (1905, 1906) (these fundamental works on Brownian motion were reprinted in Einstein (1926, 1956)).

◇ **Derivation of the diffusion equation: macroscopic consideration**

The heuristic and simplest way to derive the equation which describes the behaviour of particles in a medium is the following one. Consider a large number of particles which perform Brownian motion along some axis (for simplicity, we consider, at first, *one-dimensional* movement) and which do not interact with each other. Let $\rho(x,t)\,dx$ denote the number of particles in a small interval dx around the position x, at a time t (i.e. the density of particles) and $j(x,t)$ denote the particle current, i.e. the net number of Brownian particles that pass the point x in the direction of increasing values of x per unit of time. It is known as an experimental fact that the particle current is proportional to the gradient of their density:

$$j(x,t) = -D\frac{\partial \rho(x,t)}{\partial x}. \qquad (1.1.1)$$

This relation also serves as the definition of the *diffusion constant* D. If particles are neither created nor destroyed, the density and the current obey the continuity equation

$$\frac{\partial \rho(x,t)}{\partial t} = -\frac{\partial j(x,t)}{\partial x} \qquad (1.1.2)$$

which, due to (1.1.1), can also be written in the form:

$$\frac{\partial \rho(x,t)}{\partial t} = D\frac{\partial^2 \rho(x,t)}{\partial x^2}. \qquad (1.1.3)$$

This is the well-known *diffusion equation*.

◇ **Derivation of the diffusion equation: microscopic approach**

A more profound derivation of the diffusion equation and further insight into the nature of the Brownian motion can be achieved through the microscopic approach. In this approach, we consider a particle which suffers displacements along the x-axis in the form of a series of steps of the same length ℓ, each step being taken in either direction within a certain period of time, say of duration ε. In essence, we may think of both space and time as being replaced by sequences of equidistant sites, i.e. we consider now the *discrete* version of a model for the Brownian motion. Assuming that there is no physical reason to prefer right or left directions, we may postulate that forward and backward steps occur with equal probability $\frac{1}{2}$ (the case of different left and right probabilities is considered in problem 1.1.1, page 49, at the end of this section). Successive steps are assumed to be *statistically independent*. Hence the probability for the transition from $x = j\ell$ to the new position $x = i\ell$ during the time ε is

$$W(i\ell - j\ell, \varepsilon) = \begin{cases} \frac{1}{2} & \text{if } |i-j| = 1 \\ 0 & \text{otherwise} \end{cases} \quad (i, j \in \mathbb{Z}) \qquad (1.1.4)$$

where i and j are integers (the latter fact is expressed in (1.1.4) by the shorthand notation: i, j belong ($\in$) to the set $\mathbb{Z}$ of all positive and negative integers including zero).

The process of discrete random walk considered here represents the basic example of a *Markov chain* (see, e.g., Doob (1953), Gnedenko (1968), Breiman (1968)):

- A sequence of trials forms a Markov chain (more precisely, a *simple* Markov chain) if the conditional probability of the event $A_i^{(s)}$ from the set of K inconsistent events

$A_1^{(s)}, A_2^{(s)}, \ldots, A_K^{(s)}$ at the trial s ($s = 1, 2, 3, \ldots$) depends only on the previous trial and *does not depend* on the results of earlier trials.

This definition can be reformulated in the following way:

- Suppose that some physical system can be in one of the states $A_1, A_2, \ldots, A_K$ and that it can change its state at the moments $t_1, t_2, t_3, \ldots$. In the case of a Markov chain, the probability of transition to a state $A_i(t_s)$, $i = 1, 2, \ldots, K$, at the time t_s, depends on the state $A_i(t_{s-1})$ of the system at t_{s-1} and *does not depend* on states at earlier moments $t_{s-2}, t_{s-3}, \ldots$.

Quite generally, a Markov chain can be characterized by a pair $(W(t_n), w(0))$, where $W = (W_{ij}(t_n))$ stands for what is called a *transition matrix* or a *transition probability* and $w(0) = (w_i(0))$ is the *initial probability distribution*. In other words, $w_i(0)$ is the probability of the event i occurring at the starting time $t = 0$ and $W_{ij}(t_n)$ defines the probability distribution $w_i(t_n)$ at the moment t_n, $n = 1, 2, 3, \ldots$:

$$w_i(t_n) = \sum_j W_{ij}(t_n) w_j(0).$$

Due to the probabilistic nature of w_i and W_{ij}, we always have:

$$0 \leq w_i(0) \leq 1 \qquad \sum_i w_i(0) = 1$$

$$0 \leq W_{ij} \leq 1 \qquad \sum_i W_{ij} = 1.$$

For discrete Brownian motion, the event i is identified with the particle position $x = i\ell$ and the (infinite) matrix $W(\varepsilon)$ has the components:

$$W_{ij}(\varepsilon) = W(i\ell - j\ell, \varepsilon). \tag{1.1.5}$$

After n steps (i.e. after the elapse of time $n\varepsilon$, where n is a non-negative integer, $n \in \mathbb{Z}_+$; $\mathbb{Z}_+$ is the set of all non-negative integers $0, 1, 2, \ldots$) the resulting transition probabilities are defined by the product of n matrices $W(\varepsilon)$:

$$W(i\ell - j\ell, n\varepsilon) = (W^n(\varepsilon))_{ij}. \tag{1.1.6}$$

This is due to the characteristic property of a Markov chain, namely, the statistical independence of successive trials (i.e. transitions to new sites at the moments $t_n = n\varepsilon$, $n = 1, 2, 3\ldots$, in the case of the Brownian motion).

If at the time $t = 0$ the position of the particle is known with certainty, say $x = 0$, we have $w_i(0) = 0$ for $i \neq 0$ and $w_0(0) = 1$, or, using the Kronecker symbol δ_{ij},

$$w_i(0) = \delta_{i0}. \tag{1.1.7}$$

After the time $n\varepsilon \geq 0$, the system has evolved and is described now by the new distribution

$$w_i(n\varepsilon) = \sum_j (W^n(\varepsilon))_{ij} w_j(0)$$

or, in matrix notation,
$$\boldsymbol{w}(n\varepsilon) = \mathsf{W}^n \boldsymbol{w}(0). \tag{1.1.8}$$

Thus W^n, regarded as a function of the relevant time variable n, defines the evolution of the system. In probability theory, this has the evident probabilistic meaning of *conditional probability*: it gives the probability of an event i (in our case, the position of the Brownian particle at the site i of the space lattice) under the condition that the event j (the position of the particle at the site j) has occurred. Together with its property to define the evolution of the Markov chain, this explains the name 'transition probability' for this quantity.

Now we want to derive an explicit expression for this transition probability. To this aim, let us introduce the operators (infinite matrices):

$$\mathsf{R} = \begin{pmatrix} \ddots & 0 & \cdots & \cdots & 0 \\ 1 & 0 & & & \vdots \\ 0 & \ddots & \ddots & & \vdots \\ \vdots & & 1 & 0 & \vdots \\ 0 & \cdots & \cdots & \ddots & \ddots \end{pmatrix}, \quad \mathsf{L} = \begin{pmatrix} \ddots & \ddots & \cdots & \cdots & 0 \\ \vdots & 0 & 1 & & \vdots \\ \vdots & & \ddots & \ddots & 0 \\ \vdots & & & 0 & 1 \\ 0 & \cdots & \cdots & \cdots & \ddots \end{pmatrix} \tag{1.1.9}$$

which shift the particle's position to the right and left respectively, by the amount ℓ. Indeed, these matrices have the elements
$$R_{ij} = \delta_{i(j+1)}, \qquad L_{ij} = \delta_{(i+1)j} \tag{1.1.10}$$
so that, for example, the action of the operator R gives
$$w_i \to w'_i = \sum_j R_{ij} w_j = w_{i-1}$$

which means that the primed distribution is shifted *to the right*. To convince oneself, consider the particular distribution $w_i = \delta_{ik}$ (i.e. the particle is located at the site defined by the number k); after the action of R, we have
$$w'_i = w_{i-1} = \delta_{(i-1)k} = \delta_{i(k+1)}$$
so that now the particle is located at the site $(k+1)$. Analogously, L shifts the distribution to the left. Obviously, $\mathsf{L} = \mathsf{R}^{-1}$ and thus $\mathsf{RL} = \mathsf{LR} = 1$. This commutativity essentially simplifies the calculation of powers of W. First, note that according to (1.1.4) and (1.1.10),
$$\mathsf{W} = \tfrac{1}{2}(\mathsf{R} + \mathsf{L}) \tag{1.1.11}$$
and hence, using the binomial formula,
$$\mathsf{W}^n = \frac{1}{2^n} \sum_{k=0}^{n} \binom{n}{k} \mathsf{R}^k \mathsf{L}^{n-k}$$
$$= \frac{1}{2^n} \sum_{k=0}^{n} \binom{n}{k} \mathsf{R}^{2k-n} = \frac{1}{2^n} \sum_{k=0}^{n} \binom{n}{k} \mathsf{L}^{n-2k}. \tag{1.1.12}$$

From the properties of the matrices L and R, it follows that
$$(R^m)_{ij} = \delta_{i(j+m)}$$
$$(L^m)_{ij} = \delta_{(i+m)j} \qquad m \in \mathbb{Z}$$

so that the transition probabilities after n steps take the form:

$$W(i\ell - j\ell, n\varepsilon) = \begin{cases} 0 & \text{if } |i - j| > n \\ & \text{or } (i - j) + n \text{ is odd} \\ \frac{1}{2^n}\binom{n}{\frac{1}{2}(n+i-j)} & \text{if } |i - j| \leq n \\ & \text{and } (i - j) + n \text{ is even.} \end{cases} \qquad (1.1.13)$$

Note that the process of Brownian motion has three obvious properties, namely, it is:

(i) *homogeneous in space*—the transition probability W is merely a function of the difference $(i - j)$;
(ii) *homogeneous in time*—the transition probability W does not depend on the moment when the particle starts to wander but only on the *difference* between the starting and final time;
(iii) *isotropic*—the transition probability does not depend on the direction in space, i.e. W is left unchanged if (i, j) is replaced by $(-i, -j)$.

If we use the initial distribution as before, i.e. $w_i(0) = \delta_{i0}$, then equations (1.1.13) and (1.1.8) give for the evolution of the distribution

$$w_i(n) = \begin{cases} 0 & \text{if } |i| > n \text{ or } (i + n) \text{ is odd} \\ \frac{1}{2^n}\binom{n}{\frac{1}{2}(n+i)} & \text{if } |i| \leq n \text{ and } (i + n) \text{ is even.} \end{cases} \qquad (1.1.14)$$

Making use of the well-known recursion formula for binomial coefficients,

$$\binom{n+1}{k} = \binom{n}{k} + \binom{n}{k-1}$$

and writing the space index as an argument,

$$w(i\ell, n\varepsilon) \stackrel{\text{def}}{\equiv} w_i(n)$$

we can derive from (1.1.14) the following difference equation:

$$w(x, t + \varepsilon) = \tfrac{1}{2}w(x + \ell, t) + \tfrac{1}{2}w(x - \ell, t) \qquad (1.1.15)$$

with $x \equiv i\ell$ and $t \equiv n\varepsilon$. Equation (1.1.15) can be rewritten as

$$\frac{w(x, t + \varepsilon) - w(x, t)}{\varepsilon} = \frac{\ell^2}{2\varepsilon} \frac{w(x + \ell, t) - 2w(x, t) + w(x - \ell, t)}{\ell^2}. \qquad (1.1.16)$$

Now we can pass to a macroscopic (large scale) description of the random walk by the limiting process $\ell \to 0$, $\varepsilon \to 0$, with the ratio

$$D = \frac{\ell^2}{2\varepsilon} \qquad (1.1.17)$$

held fixed. This process turns x and t into continuous variables: $x \in \mathbb{R}$ (all real numbers), $t \in \mathbb{R}_+$ (non-negative real numbers), which are much closer to our usual view of space and time. As a result, equation (1.1.16) becomes the diffusion equation (1.1.3), with D being the diffusion constant

$$\frac{\partial w(x, t)}{\partial t} = D\frac{\partial^2 w(x, t)}{\partial x^2}. \qquad (1.1.18)$$

The obvious generalization of a finite collection x_i, $i = 1, \ldots, N$ of random variables is a map $t \to x_t$, where t ranges over some interval. Any such map is called a *stochastic process* in continuous time (see, e.g., Doob (1953) and Gnedenko (1968)). More details about stochastic processes and their classification can be found in section 1.2.

The density ρ in (1.1.3) and the distribution $w(x,t)$ are related by a constant factor, namely, by the total number K of Brownian particles which are considered in the macroscopic derivation of the diffusion equation (1.1.3):

$$\rho = Kw.$$

◇ **Multidimensional diffusion equation**

An analogous derivation of the diffusion equation can be carried out for a particle wandering in a space of arbitrary dimension d, with the result:

$$\frac{\partial w(\boldsymbol{x},t)}{\partial t} = D\Delta^{(d)} w(\boldsymbol{x},t) \qquad (1.1.19)$$

where $\boldsymbol{x} = \{x_1, x_2, \ldots, x_d\}$ and $\Delta^{(d)}$ is the d-dimensional Laplacian (in Cartesian coordinates)

$$\Delta^{(d)} = \frac{\partial^2}{\partial x_1^2} + \frac{\partial^2}{\partial x_2^2} + \cdots + \frac{\partial^2}{\partial x_d^2}.$$

We suggest the derivation of the multidimensional equation as an exercise for the reader (see problem 1.1.2, page 49).

Equation (1.1.18) and its multidimensional generalization (1.1.19) (in particular, when a Brownian particle wanders in realistic two- or three-dimensional space) form the basis of Einstein's theory of Brownian motion (Einstein 1905, 1906) (republished in Einstein (1926, 1956)).

The expression (1.1.17) for the diffusion constant shows that in the continuous limit, no meaning can be attributed to the velocity of the Brownian particle, since the condition

$$\frac{\ell^2}{2\varepsilon} \xrightarrow[\ell,\varepsilon \to 0]{} \text{constant}$$

implies that the one-step 'velocity' $\pm \ell/\varepsilon$, in the same limit, becomes infinite

$$\frac{\ell}{\varepsilon} \to \infty.$$

In more mathematical terms, although continuous, a typical Brownian path is nowhere differentiable as a function of time (for more details see the next subsection and problem 1.1.5, page 53).

◇ **Solution of the diffusion equation**

The solution of (1.1.18) with the continuous analog of the initial condition (1.1.7), i.e.

$$w(x,t) \xrightarrow[t \to 0]{} \delta(x) \qquad (1.1.20)$$

($\delta(x)$ is the Dirac δ-function) can be obtained by the Fourier transform

$$w(x,t) = \int_{-\infty}^{\infty} dk\, e^{ikx} \widetilde{w}(k,t). \tag{1.1.21}$$

Using the well-known Fourier representation for the δ-function

$$\delta(x) = \frac{1}{2\pi} \int_{-\infty}^{\infty} dk\, e^{ikx} \tag{1.1.22}$$

we can see that the initial condition (1.1.20), in terms of the Fourier image $\widetilde{w}(k,t)$, has the form

$$\widetilde{w}(k,0) = \frac{1}{2\pi}. \tag{1.1.23}$$

After performing the Fourier transformation, the diffusion equation (1.1.18) becomes

$$\frac{\partial \widetilde{w}(k,t)}{\partial t} = -Dk^2 \widetilde{w}(k,t) \tag{1.1.24}$$

with the obvious solution

$$\widetilde{w}(k,t) = \widetilde{w}(k,0)e^{-Dk^2 t} = \frac{1}{2\pi} e^{-Dk^2 t} \tag{1.1.25}$$

so that the distribution can be represented as follows:

$$w(x,t) = \int_{-\infty}^{\infty} dk\, \frac{1}{2\pi} e^{-Dk^2 t} e^{ikx}. \tag{1.1.26}$$

Shifting the integration variable, $k \to k - ix/(2Dt)$, and using the value of the *Gaussian integral*

$$\int_{-\infty}^{\infty} dx\, e^{-\alpha x^2} = \sqrt{\frac{\pi}{\alpha}} \tag{1.1.27}$$

we obtain

$$w(x,t) = \frac{1}{\sqrt{4\pi Dt}} \exp\left\{-\frac{x^2}{4Dt}\right\}. \tag{1.1.28}$$

By construction, the distribution (1.1.28) is a solution of the diffusion equation (1.1.18) with the initial value (1.1.20) (the reader may also verify this fact directly).

Note that

$$\int_{-\infty}^{\infty} dx\, w(x,t) = \int_{-\infty}^{\infty} dx\, \frac{1}{\sqrt{4\pi Dt}} \exp\left\{-\frac{x^2}{4Dt}\right\} = 1 \tag{1.1.29}$$

which is compatible with the probabilistic interpretation of $w(x,t)$ as being the probability of finding the Brownian particle at the moment t at the place x, if the particle has been at the origin $x = 0$ at the initial time $t = 0$.

The transition probability (1.1.6) in the continuous limit reads

$$W_{ij}^N = W(i\ell - j\ell, N\varepsilon) = W(j\ell, N\varepsilon | i\ell, 0)$$
$$\xrightarrow[\ell,\varepsilon \to 0]{} W(x_t, t | x_0, 0) \tag{1.1.30}$$

or, for an arbitrary initial moment,

$$W(x_t, t|x_0, t_0) \qquad x_t = x(t) \qquad x_0 = x(t_0) \tag{1.1.31}$$

and the evolution of the probability density takes the form

$$w(x_t, t) = \int_{-\infty}^{\infty} dx_0\, W(x_t, t|x_0, t_0) w(x_0, t_0). \tag{1.1.32}$$

Since $w(x_0, t_0)$ is an arbitrary function (satisfying, of course, the normalization condition (1.1.29)), this means, in turn, that the transition probability also satisfies the diffusion equation

$$\frac{\partial W(x_t, t|x_0, t_0)}{\partial t} = D \frac{\partial^2 W(x_t, t|x_0, t_0)}{\partial x_t^2} \qquad t_0 < t \tag{1.1.33}$$

with the initial condition

$$W(x_t, t|x_0, t_0) \xrightarrow[t \to t_0]{} \delta(x_t - x_0) \tag{1.1.34}$$

which follows from (1.1.32). The solution of the diffusion equation (1.1.33) with the initial condition (1.1.34) reads

$$W(x_t, t|x_0, t_0) = \frac{1}{\sqrt{4\pi D(t - t_0)}} \exp\left\{-\frac{(x_t - x_0)^2}{4D(t - t_0)}\right\} \tag{1.1.35}$$

and satisfies the normalization condition

$$\int_{-\infty}^{\infty} dx_t\, W(x_t, t|x_0, t_0) = 1. \tag{1.1.36}$$

This is the *normal (Gaussian)* probability distribution with the mean value (mathematical expectation) x_0 and the dispersion $\mathbb{D} = 2D(t - t_0)$.

The relation (1.1.30) reflects the well-known fact (see, e.g., Gnedenko (1968) and Korn and Korn (1968)) that the *binomial* distribution (1.1.13) converges to the normal distribution in the limit of an infinite number of trials.

In higher-dimensional spaces, the equation, its solution, boundary and normalization conditions have a form which is a straightforward generalization of the one-dimensional case:

$$\frac{\partial W(\boldsymbol{x}_t, t|\boldsymbol{x}_0, t_0)}{\partial t} = D \Delta^{(d)} W(\boldsymbol{x}_t, t|\boldsymbol{x}_0, t_0) \qquad t > t_0 \tag{1.1.37}$$

$$W(\boldsymbol{x}_t, t|\boldsymbol{x}_0, t_0) \xrightarrow[t \to t_0]{} \delta^{(d)}(\boldsymbol{x}_t - \boldsymbol{x}_0) \tag{1.1.38}$$

$$W(\boldsymbol{x}_t, t|\boldsymbol{x}_0, t_0) = \frac{1}{(4\pi D(t - t_0))^{d/2}} \exp\left\{-\frac{(\boldsymbol{x}_t - \boldsymbol{x}_0)^2}{4D(t - t_0)}\right\} \tag{1.1.39}$$

$$\int_{-\infty}^{\infty} d^d x_t\, W(\boldsymbol{x}_t, t|\boldsymbol{x}_0, t_0) = 1 \tag{1.1.40}$$

where $\boldsymbol{x} = \{x_1, \ldots, x_d\}$, $\boldsymbol{x}^2 = x_1^2 + x_2^2 + \cdots + x_d^2$.

Due to the space and time homogeneity of the Brownian motion, the transition probability is only a function of the differences of the variables:

$$W(\boldsymbol{x}_f, t_f|\boldsymbol{x}_0, t_0) = W(\boldsymbol{x}_f - \boldsymbol{x}_0, t_f - t_0) \equiv W(\boldsymbol{x}, t). \tag{1.1.41}$$

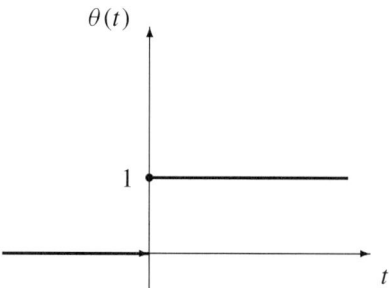

Figure 1.1. Step-function $\theta(t)$.

We shall use this shortened notation along with the transition probabilities with complete indication of all variables.

One more useful remark is that a slight modification of the solution (1.1.35) converts it into the solution of the *inhomogeneous* diffusion equation

$$\frac{\partial W(x,t)}{\partial t} = D\frac{\partial^2 W(x,t)}{\partial x^2} + \delta(t)\delta(x) \qquad (1.1.42)$$

(we return, for simplicity of notation, to the one-dimensional case but all the discussion can be trivially generalized to an arbitrary dimension). Namely, we can extend the function $W(x,t)$ to the complete temporal line $t \in (-\infty, \infty)$, i.e. consider formally the transition probability $W(x_f, t_f | x_0, t_0)$ also for $t_f \leq t_0$. To express the fact that non-vanishing W exists only for positive values of the time variable t, we must multiply the solution (1.1.35) by the *step-function* (see figure 1.1)

$$\theta(t) = \begin{cases} 1 & \text{if } t \geq 0 \\ 0 & \text{if } t < 0. \end{cases} \qquad (1.1.43)$$

The reader may check that the function

$$W(x,t) = \frac{\theta(t)}{\sqrt{4\pi Dt}} \exp\left\{-\frac{x^2}{4Dt}\right\} \qquad -\infty < t < \infty \qquad (1.1.44)$$

satisfies indeed the equation (1.1.42) (see problem 1.1.4, page 51). Thus from the mathematical point of view, the transition probability $W(x,t)$ in (1.1.44) is *the Green function* (or *fundamental solution*) of the diffusion equation (1.1.33) because it satisfies equation (1.1.42) with the δ-functions as an inhomogeneous term.

Knowledge of the transition probability allows us to find the probability density $w(x,t)$ at any time t for any initial density $w(x_0, t_0)$ from the relation (1.1.32) (recall that the density (1.1.28) has been obtained for the δ-functional initial density (1.1.20)).

◇ **Semigroup property of the transition probability: Einstein–Smoluchowski–Kolmogorov–Chapman (ESKC) relation**

Now let us consider the probability densities at three instants of time

$$w(x_0, t_0) \qquad w(x', t') \qquad w(x, t) \qquad t_0 < t' < t.$$

The distribution $w(x', t')$ can be considered as an initial one for $w(x, t)$, while $w(x_0, t_0)$ can serve as an initial one for both distributions $w(x, t)$ and $w(x', t')$. Hence we can write

$$w(x, t) = \int_{-\infty}^{\infty} dx' \, W(x, t | x', t') w(x', t')$$

$$w(x, t) = \int_{-\infty}^{\infty} dx_0 \, W(x, t | x_0, t_0) w(x_0, t_0)$$

$$w(x', t') = \int_{-\infty}^{\infty} dx_0 \, W(x', t' | x_0, t_0) w(x_0, t_0).$$

Combining these equations gives

$$W(x, t | x_0, t_0) = \int_{-\infty}^{\infty} dx' \, W(x, t | x', t') W(x', t' | x_0, t_0) \qquad t_0 < t' < t. \qquad (1.1.45)$$

This is the very important *Einstein–Smoluchowski–Kolmogorov–Chapman (ESKC) relation*. The nature of this relation is quite analogous to the well-known Huygens–Fresnel principle in optics and provides the causal description of the phenomenon under consideration, in our case Brownian motion. It expresses also a general *semigroup property* of the transition probability $W(x, t | x_0, t_0)$, which is automatic for any random motion without memory and with temporal homogeneity (Markov chain or Markov process; see more about random processes in section 1.2).

In general terms, the semigroup property of a set of some objects $F(t)$ depending on a positive parameter t means that there exists a composition law $F(t) * F(t')$, satisfying the rule

$$F(t) * F(t') = F(t + t'). \qquad (1.1.46)$$

In the case of transition probabilities, the composition law is defined by integration over the intermediate coordinates. Using the homogeneity, we can rewrite the ESKC as follows:

$$W(x - x_0, t - t_0) = \int_{-\infty}^{\infty} dx' \, W(x - x', t - t') W(x' - x_0, t' - t_0) \qquad (1.1.47)$$

or, in symbolic form

$$W(t - t_0) \equiv W((t - t') + (t' - t_0)) = W(t - t') * W(t' - t_0)$$

where

$$W(t - t') * W(t' - t_0) \stackrel{\text{def}}{\equiv} \int_{-\infty}^{\infty} dx' \, W(x - x', t - t') W(x' - x_0, t' - t_0)$$

so that the transition probability W does indeed satisfy the semigroup property (1.1.46). We use the term 'semigroup' rather than just 'group' because it is impossible to define any kind of inverse element (this is a necessary condition for a set of objects with a composition law to form a group, see, e.g., Wybourn (1974), Barut and Rączka (1977) and Chaichian and Hagedorn (1998)). In our case, an inverse element would correspond to a movement backward in time, i.e. to the transition probability $W(x_f, t_f | x_0, t_0)$ with $t_f < t_0$. But, as follows from the explicit form (1.1.35), the transition probability $W(x_f - x_0, t_f - t_0)$ for $(t_f - t_0) < 0$ is meaningless (without the step-function factor as in (1.1.44), which makes it just equal to zero for $(t_f - t_0) < 0$): the basic relations (1.1.32) (for the majority of reasonable probability densities) and the composition law (1.1.45) do not exist because of the exponential growing of the integrands.

22 Path integrals in classical theory

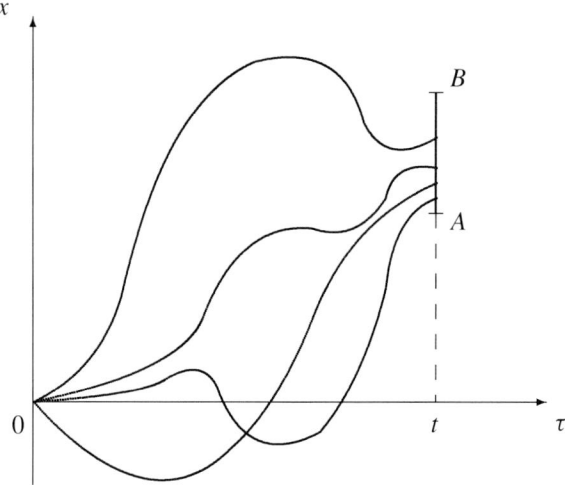

Figure 1.2. Trajectories of a Brownian particle starting from the origin and ending anywhere in the gate AB at the moment t.

1.1.2 Wiener's treatment of Brownian motion: Wiener path integrals

Now we start the discussion of the original approach to the description of Brownian motion by Wiener (1921, 1923, 1924), where the concept of a path integral was first introduced.

◇ Markovian property of Brownian motion, Markov and Wiener stochastic processes

Consider again (for simplicity) one-dimensional Brownian motion. Using the results of the preceding subsection, the probability of a Brownian particle to be at the moment t *anywhere* in the interval $[AB]$ (see figure 1.2) is given by

$$\mathbb{P}\{x(t) \in [AB]\} = \int_A^B dx\, w(x, t). \tag{1.1.48}$$

Complete information about the stochastic process definitely contains more than just knowing the set of probabilities $\mathbb{P}\{x(t) \in [AB]\}$. In particular, the essential characteristic of such a process is a probability of a *compound event*. In the case of Brownian motion, this is the probability that the particle, starting at $x(0) = 0$, successively passes through the gates $A_1 \leq x(t_1) \leq B_1, A_2 \leq x(t_2) \leq B_2, \ldots, A_N \leq x(t_N) \leq B_N$ at the corresponding instants of time $t_1, t_2, \ldots, t_N$, as shown in figure 1.3. The statistical independence of subsequent displacements of the Brownian particle (Markovian property) gives

$$\mathbb{P}\{x(t_1) \in [A_1, B_1], x(t_2) \in [A_2, B_2], \ldots, x(t_N) \in [A_N, B_N]\} \tag{1.1.49}$$

$$= \int_{A_1}^{B_1} dx_1 \frac{\exp\left\{-\frac{x_1^2}{4Dt_1}\right\}}{\sqrt{4\pi Dt_1}} \int_{A_2}^{B_2} dx_2 \frac{\exp\left\{-\frac{(x_2-x_1)^2}{4D(t_2-t_1)}\right\}}{\sqrt{4\pi D(t_2-t_1)}}$$

$$\times \int_{A_3}^{B_3} dx_3 \frac{\exp\left\{-\frac{(x_3-x_2)^2}{4D(t_3-t_2)}\right\}}{\sqrt{4\pi D(t_3-t_2)}} \cdots \int_{A_N}^{B_N} dx_N \frac{\exp\left\{-\frac{(x_N-x_{N-1})^2}{4D(t_N-t_{N-1})}\right\}}{\sqrt{4\pi D(t_N-t_{N-1})}}. \tag{1.1.50}$$

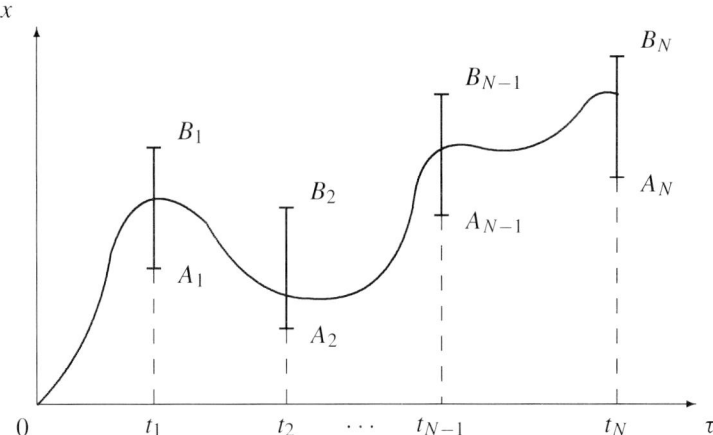

Figure 1.3. A trajectory of a Brownian particle starting from origin and passing through the gates $A_i B_i$ at the times t_i ($i = 1, \ldots, N$).

Recall that the probability $\mathbb{P}\{\mathcal{A}_i, \mathcal{A}_j\}$ that some *independent* events $\mathcal{A}_i$ and $\mathcal{A}_j$ both occur, is given by the product of the probabilities of the simple events: $\mathbb{P}\{\mathcal{A}_i, \mathcal{A}_j\} = \mathbb{P}\{\mathcal{A}_i\}\mathbb{P}\{\mathcal{A}_j\}$. Thus the fact that the right-hand side of (1.1.50) is a product tells us that, given the present position $x(t)$ of the Brownian particle, the distribution of $x(t')$ at some later time t' (in the future) is completely determined and does not depend on the past history of the path taken by the particle. This is the characteristic property of a Markov chain introduced in the preceding subsection. Another way to express this fact is to note that, due to the probability product form of the joint distribution, the characteristic property of the Brownian motion is the independence of the *increments* of particle positions at arbitrary sequence of times $t_1, t_2, \ldots, t_N$.

The Markov property is simply the probabilistic analog of a property familiar from the theory of deterministic dynamical systems: given the initial data, then, by solving the equation of motion, the future state of the system can be obtained without knowing what happened in the past. The present state already contains all the information relevant for the future.

In the limit of continuous time, diminishing the sizes of each gate and infinitely increasing their number, so that

$$(t_i - t_{i-1}) \equiv \Delta t_i \to 0 \qquad 1 \leq i \leq N$$

the position $x(t)$ of a particle depends on the continuous time variable and we obtain what is called a *stochastic process* (for more on stochastic processes, their classification and basic properties, see section 1.2). A stochastic process with independent increments as in (1.1.50) is said to have no memory and is termed a *Markov process*. In general, the definition of a Markov process places no restriction either on the initial distribution $w(x, 0)$ or on the *transition probabilities*:

$$W(x_t, t|x_0, t_0)\, dx_t \stackrel{\text{def}}{\equiv} \mathbb{P}\{x(t) \in [x_t, x_t + dx_t], x(t_0) = x_0\} \qquad t_0 < t$$

except normalization. But in the case of Brownian motion, the initial distribution is of the form (1.1.20) (the initial point can, of course, be arbitrary, i.e. $w(x, t) \xrightarrow[t \to 0]{} \delta(x - x_0)$) and the transition probabilities

24   Path integrals in classical theory

are given by (1.1.35), so that they result in the joint distribution of the form (1.1.50). Such a stochastic process is said to be a *Wiener process*.

◇ **Transition to the limit of an infinite number of 'gates': the Wiener measure and Wiener path integral**

Considering the continuous limit in (1.1.50), we obtain the probability that the Brownian particle moves through an infinite number of infinitesimal 'gates' dx along the trajectory $x(t)$

$$\lim_{\substack{\Delta t_i \to 0 \\ N \to \infty}} \exp\left\{-\sum_{i=1}^{N} \frac{(x_i - x_{i-1})^2}{4D(t_i - t_{i-1})}\right\} \prod_{i=1}^{N} \frac{dx_i}{\sqrt{4\pi D(t_i - t_{i-1})}}$$

$$= \lim_{\substack{\Delta t_i \to 0 \\ N \to \infty}} \exp\left\{-\frac{1}{4D}\sum_{i=1}^{N} \left(\frac{x_i - x_{i-1}}{t_i - t_{i-1}}\right)^2 \Delta t_i\right\} \prod_{i=1}^{N} \frac{dx_i}{\sqrt{4\pi D \Delta t_i}}$$

$$\equiv \exp\left\{-\frac{1}{4D}\int_0^t d\tau\, \dot{x}^2(\tau)\right\} \prod_{\tau=0}^{t} \frac{dx(\tau)}{\sqrt{4\pi D\, d\tau}} \qquad (1.1.51)$$

in other words, we obtain the probability of the particle motion inside the infinitesimally thin tube surrounding the path $x(\tau)$, or simply *moving along the trajectory* $x(\tau)$.

Let us denote by $\mathcal{C}\{\boldsymbol{x}_1, t_1; \mathcal{B}, t_2\}$ the set of trajectories starting at the point $\boldsymbol{x}_1 = \boldsymbol{x}(t_1)$ at the time t_1 and having the endpoint $\boldsymbol{x}(t_2)$ in some domain $\mathcal{B}$ of $\mathbb{R}^d$. In particular:

- $\mathcal{C}\{\boldsymbol{x}_1, t_1; \boldsymbol{x}_2, t_2\}$ denotes the set of trajectories starting at the point $\boldsymbol{x}(t_1) = \boldsymbol{x}_1$ and having the endpoint $\boldsymbol{x}(t_2) = \boldsymbol{x}_2$;
- $\mathcal{C}\{x_1, t_1; [AB], t_2\}$ denotes, in the one-dimensional case, the set of trajectories with the starting point $x_1 = x(t_1)$ and ending in the gate $[AB]$ at the time t_2.

However, in the special case, we shall simplify the notation as follows.

- If a trajectory has an arbitrary endpoint in the interval from $-\infty$ to $+\infty$ for all coordinates, then we shall omit the explicit indication of the whole space $\mathbb{R}^d$: $\mathcal{C}\{\boldsymbol{x}_1, t_1; \mathbb{R}^d, t_2\} \equiv \mathcal{C}\{\boldsymbol{x}_1, t_1; t_2\}$.

Thus for example, $\mathcal{C}\{0, 0; t\}$ denotes the set of trajectories starting at the origin at $t = 0$ and having arbitrary endpoints at t.

This notation is applicable to spaces of arbitrary dimensions, but we continue to consider the one-dimensional space because, being notationally simpler, it contains all the essential points for a path-integral description of the Brownian motion in spaces of higher dimension.

It is clear that to obtain the probability that the particle ends up somewhere in the gate $[AB]$ at the time t, we have to sum the probabilities (1.1.51) over the set $\mathcal{C}\{0, 0; [AB], t\}$ of all the trajectories which end up in the interval $[A, B]$, i.e.

$$\mathbb{P}\{x(t) \in [AB]\} = \int_{\mathcal{C}\{0,0;[AB],t\}} \prod_{\tau=0}^{t} \frac{dx(\tau)}{\sqrt{4\pi D\, d\tau}} \exp\left\{-\frac{1}{4D}\int_0^t d\tau\, \dot{x}^2(\tau)\right\}$$

$$= \int_A^B dx\, \frac{1}{\sqrt{4\pi Dt}} \exp\left\{-\frac{x^2}{4Dt}\right\} \qquad (1.1.52)$$

where the second equality follows from (1.1.48) and (1.1.28). The symbol

$$\int_{\mathcal{C}\{0,0;[AB],t\}}$$

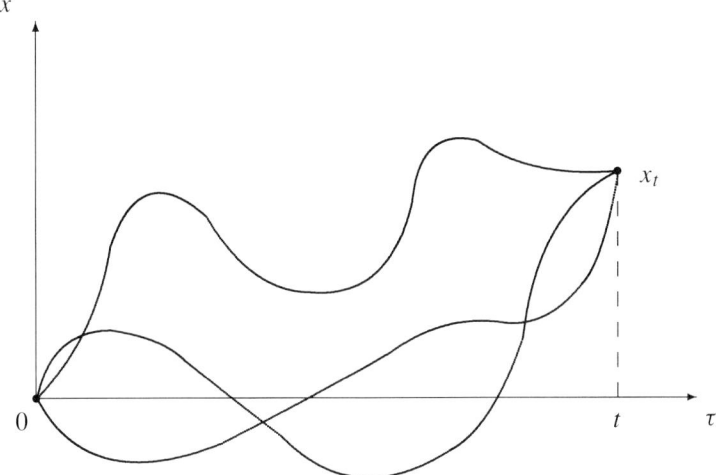

Figure 1.4. Trajectories of the Brownian particle with fixed initial and final points.

formally denotes the summation over the set of trajectories and since this set is continuous, we have used the symbol of an integral. The summation over a set of trajectories of the type (1.1.52) is called the *Wiener path integral*.

In the limiting case $A = B = x_t$, the set $C\{0, 0; x_t, t\}$ consists of paths for which both the initial and final points are fixed (see figure 1.4). The integration over this set obviously gives the transition probability (1.1.28)

$$W(x_t, t|0, 0) = \int_{C\{0,0;x_t,t\}} d_W x(\tau) = \frac{1}{\sqrt{4\pi Dt}} \exp\left\{-\frac{x_t^2}{4Dt}\right\} \qquad (1.1.53)$$

where the integration measure

$$d_W x(\tau) \stackrel{\text{def}}{\equiv} \exp\left\{-\frac{1}{4D}\int_0^t d\tau\, \dot{x}^2(\tau)\right\} \prod_{\tau=0}^{t} \frac{dx(\tau)}{\sqrt{4\pi D\, d\tau}} \qquad (1.1.54)$$

is called the *Wiener measure* (Wiener 1921, 1923, 1924, Paley and Wiener 1934). If we consider a set of trajectories with *arbitrary* endpoints (see figure 1.5), the measure (1.1.54) is called the *unconditional Wiener measure* (or, sometimes, *full Wiener measure*, or *absolute Wiener measure*). From its probabilistic meaning, the normalization condition

$$\int_{C\{x_1,t_1;t_2\}} d_W x(\tau) = 1 \qquad (1.1.55)$$

follows, since the probability that the particle will end up anywhere is equal to unity. Here the class of functions $x(\tau) \in C\{x_1, t_1; t_2\} : [t_1, t_2] \to \mathbb{R}^d$ is still to be defined (i.e. whether the trajectories are smooth or differentiable as functions of the time variable). We shall study this important point later.

The so-called *conditional Wiener measure* corresponds to integrations over sets $C\{0, 0; x_t, t\}$ of paths

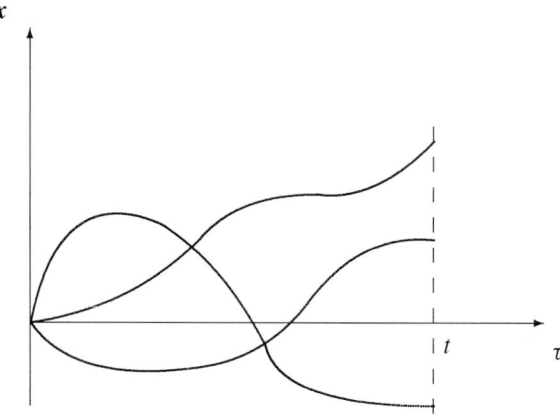

Figure 1.5. Samples of trajectories defining the unconditional Wiener measure.

with a *fixed* endpoint as in figure 1.4. Obviously,

$$\int_{C\{0,0;t\}} d_W x(\tau) = \int_{-\infty}^{\infty} dx_t \int_{C\{0,0;x_t,t\}} d_W x(\tau). \tag{1.1.56}$$

The same is true for Wiener integrals with a functional $F[x(\tau)]$ as the integrand (we shall discuss path integrals with functionals in more detail in the next subsection):

$$\int_{C\{0,0;t\}} d_W x(\tau) F[x(\tau)] = \int_{-\infty}^{\infty} dx_t \int_{C\{0,0;x_t,t\}} d_W x(\tau) F[x(\tau)]. \tag{1.1.57}$$

The conditional measure is directly related to the transition probability (see (1.1.53)).

In terms of the Wiener measure and integral, the Einstein–Smoluchowski–Kolmogorov–Chapman (ESKC) relation (1.1.47) takes the form

$$\int_{C\{x_0,0;x_t,t\}} d_W x(\tau) = \int_{-\infty}^{\infty} dx' \int_{C\{x_0,0;x',t'\}} d_W x(\tau) \int_{C\{x',t';x_t,t\}} d_W x(\tau). \tag{1.1.58}$$

◊ **Similarity between the notions of 'probability' and 'measure'**

Starting from the Brownian transition probability and distribution we have naturally arrived at the measure and integral over the *functional infinite-dimensional space* of all trajectories $x(\tau)$. We would like to emphasize that the appearance of a measure and integration in a probabilistic description of any phenomenon is highly natural and practically unavoidable. The point is that a *measure* (understood as a mathematically rigorous generalization of the intuitive notion of 'volume') and a *probability* satisfy almost the same set of axioms. Without going into details, let us just compare the basic axioms satisfied by a measure and a probability (see table 1.1).

The very idea that probability theory can be formulated on the basis of measure theory appeared for the first time in the classical work by E Borel (1909). The most complete axiomatics of probability theory were developed by Kolmogorov (1938) (also see Kolmogorov (1956)) and e.g., Doob (1953) and Billingsley (1979) for an extensive introduction into the probability theory and its relation to the measure theory). Thus measure theory provides the mathematical background for probability theory. In the case

Table 1.1.

Probability	Measure (axiomatic construction)
The probability $\mathbb{P}\{\mathcal{A}\}$, defined on a class of events $\mathcal{A}_i$, is a function with the properties:	The measure $\mu[S]$, defined on a class of point sets S_i, is a function of the sets with the properties:
(1) $\mathbb{P}\{\mathcal{A}_i\} \geq 0$ (2) for any finite or countable sequence of mutually incompatible events $\mathcal{A}_1, \mathcal{A}_2, \ldots,$ $\mathbb{P}\{\mathcal{A}_1 \cup \mathcal{A}_2 \cup \cdots\} = \mathbb{P}\{\mathcal{A}_1\} + \mathbb{P}\{\mathcal{A}_2\} + \cdots$ (3) $\mathbb{P}\{\mathcal{A}^{(\text{certain})}\} = 1$ for a certain event $\mathcal{A}^{(\text{certain})} = \cup_i \mathcal{A}_i$ (union of all possible events)	(1) $\mu[S_i] \geq 0$ (2) for any finite or countable set of mutually non-intersecting sets $S_1, S_2, \ldots,$ $\mu[S_1 \cup S_2 \cup \cdots] = \mu[S_1] + \mu[S_2] + \cdots$ (3) Probabilistic measures: subclass of measures, such that $\mu[S] = 1$ for the total set $S = \cup_i S_i$ (union of all sets under consideration)

of Brownian motion we have to consider a probability (actually a probability density) of realization of a given trajectory and hence the (Wiener) measure over the set of trajectories (paths) and the corresponding Wiener path integral appears.

From the general point of view of probability theory constructing the (Wiener) path integral merely means generalizing the notion of probability distributions $w(x_1, \ldots, x_n)$ to *functional distributions* $\Omega[f(\tau)]$ describing the probability of finding a function from some appropriate set in the infinitesimal vicinity of (in the infinitesimal tube around) some given function $f(\tau)$ from this set:

$$w(x_1, x_2, \ldots, x_n) \longrightarrow \Omega[f(\tau)] \qquad (1.1.59)$$

$$\langle g(x_1, \ldots, x_n) \rangle_{\text{w}} \stackrel{\text{def}}{\equiv} \int dx_1 \cdots dx_n \, w(x_1, \ldots, x_n) g(x_1, \ldots, x_n)$$

$$\longrightarrow \langle F[f(\tau)] \rangle_{\Omega} \stackrel{\text{def}}{\equiv} \int \mathcal{D}f(\tau) \Omega[f(\tau)] F[f(\tau)]. \qquad (1.1.60)$$

Here and throughout the book the notation $\langle \cdots \rangle$ denotes an expectation (mean) value (in an appropriate sense which varies in different parts of the book) and $\mathcal{D}f(\tau)$ symbolically denotes a functional measure. In the case of Brownian motion, we have

$$\mathcal{D}f(\tau)\Omega[f(\tau)] \equiv d_\text{W} x(\tau).$$

In (1.1.59) and (1.1.60), we have assumed that the probability distributions are normalized. Sometimes it is convenient to use non-normalized functional distributions, writing

$$\langle F[f(\tau)] \rangle_{\Omega} = \frac{\int \mathcal{D}f(\tau) \Omega[f(\tau)] F[f(\tau)]}{\int \mathcal{D}f(\tau) \Omega[f(\tau)]}. \qquad (1.1.61)$$

◇ **Set of trajectories contributing to the Wiener path integral: continuous but non-differentiable**

Of course, the important question concerns the properties of the set of functions which must be averaged over. Wiener (1921, 1923, 1924) has proved that in the case of path integrals (1.1.57) with measure (1.1.54), the set $\mathcal{C}$ of functions which contribute to the integral consists of *continuous* but *non-differentiable* functions. The latter is no longer surprising because consideration of the continuous limit for the Brownian motion in the preceding subsection has shown that the notion of velocity for the Brownian particle is ill defined. Another way to see this is to calculate the mean value $\langle x^2 \rangle$ of the squared shift using the distribution (1.1.28):

$$\langle x^2 \rangle = \int_{-\infty}^{\infty} dx \, x^2 w(x, t) = \frac{t}{2}. \tag{1.1.62}$$

Thus the shift during the period of time t is of the order $\sqrt{\langle x^2 \rangle} \sim \sqrt{t}$ (or, in a more general case, $\sqrt{\langle (x - x_0)^2 \rangle} \sim \sqrt{t - t_0}$) and the speed of the Brownian particle at any moment of time is infinite:

$$\frac{\sqrt{t}}{t} \xrightarrow[t \to 0]{} \infty.$$

On the other hand, we note that the transition probability

$$W(x_t, t | x_0, 0) = \frac{1}{\sqrt{4\pi Dt}} \exp\left\{-\frac{(x(t) - x(0))^2}{4Dt}\right\}$$

satisfies the characteristic property

$$\lim_{t \to 0} W(x_t, t | x_0, 0) = \delta(x_t - x_0). \tag{1.1.63}$$

Since the δ-function $\delta(x_t - x_0)$ is equal to zero for $x_t \neq x_0$, the limit (1.1.63) shows that after the lapse of an infinitesimally small period of time the particle appears to be in the *infinitesimally small* vicinity of the initial point x_0. This means that all the paths are continuous at $t = 0$ and hence at any moment τ ($0 \leq \tau \leq t$) (due to the homogeneity of a Brownian process in time, so that we can start from any moment t_0 which therefore becomes the initial time). That is why we denoted the class of functions under the sign of the path integrals as $\mathcal{C}$ (i.e. *continuous* functions).

Therefore, one of the most important peculiarities of the path integrals is that, contrary to the propensity of our intuition to conceive them as sums over paths which are somehow 'smooth' (as depicted in our simplified figures), they are, in reality, sums over fully 'zigzag-like' trajectories, corresponding to non-differentiable functions (see figure 1.6). Note that this property of the trajectories of a Brownian particle, which can be described more precisely in the framework of *fractal theory* (see, e.g., Mandelbrot (1977, 1982)), has important physical consequences (for example, the chemoreception of living cells and hence their normal functions would be impossible without such a specific property of the Brownian trajectories, see Wiegel (1983)). The fractal corresponding to the Brownian motion possesses the space dimension two, i.e. the trajectory of the Brownian particle is a 'thick' one, having non-zero area.

The formal notation in (1.1.51) and (1.1.52) containing $\dot{x}(\tau)$ is therefore somewhat misleading in the sense that all the important (contributing to the integral) paths are *non-differentiable* in the continuous limit.

Since this point is very important for the correct understanding of path integrals in general and of the Wiener path integral in particular, we summarize once more the essence and possible approaches to definition of the path integral.

In the case of the *Wiener* path integral, there are essentially two approaches for giving a strict definition:

(1) to define the path integral via a finite-dimensional approximation of the form (1.1.50) in the spirit of the general Volterra approach to handling functionals (cf Introduction) and to consider a path integral as the shorthand notation for the appropriate 'continuous' limit (1.1.51) when the number of *time slices* goes to infinity (the quotation marks stand for the peculiarities with the non-differentiability of

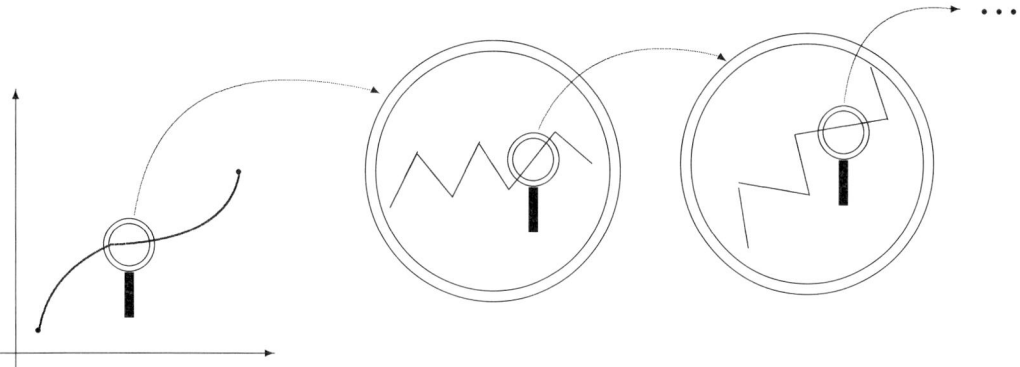

Figure 1.6. Schematic illustration of the fact that the detailed consideration ('under magnifying glasses' with increasing resolution) of Brownian trajectories reveals their fully 'zigzag-like' non-differentiable structure.

essential trajectories); table 1.2 summarizes the symbolic notation used in path-integral calculations from the point of view of finite-dimensional approximations;
(2) to define the Wiener measure in the frame of the axiomatic probabilistic measure theory as a Gaussian-type proper measure on the set of trajectories (paths) of the Brownian particle; in this case, the right-hand side of expression (1.1.54) has a rather symbolic sense, as one more notation for the defined (Wiener) measure $d_W x(t)$ with the property (1.1.55), which allows us to interpret the corresponding integral as a probability.

The second approach is mathematically more refined and we shall discuss it and the related mathematical details in the next subsection. One more reason for the use of differential notation in path integrals will be discussed in section 1.2.7.

1.1.3 Wiener's theorem and the integration of functionals

In view of its fundamental significance, let us formulate the Wiener result in the form of a theorem and present its more rigorous proof (in comparison with the intuitive arguments presented at the end of the preceding subsection), still skipping some minor mathematical details. This proof will give a deeper insight into the peculiarities of path integrals as well as experience in handling them.

Theorem 1.1 (Wiener's theorem). The Wiener path integral is equal to zero over both the set of discontinuous and the set of differentiable trajectories. In more precise mathematical terms, the set of discontinuous as well as the set of differentiable functions have a zero Wiener measure.

Proof. We shall discuss here in detail only the proof of the statement in the theorem about discontinuous functions. The proof for differentiable functions (which is quite analogous) we leave to the reader as an exercise (see problem 1.1.5, page 53).

Consider the set Z^h_{mj} of the functions $x(t)$ on the interval $0 \leq t \leq 1$ which satisfy the inequality

$$\left| x\left(\frac{j+1}{2^m}\right) - x\left(\frac{j}{2^m}\right) \right| > \frac{h}{A^m}. \qquad (1.1.64)$$

Here $h > 0$, $1 < A < \sqrt{2}$, $j = 0, 1, \ldots, 2^m - 1$, $m \in \mathbb{Z}_+$ (non-negative integers). The choice of a unit time interval just simplifies the notation: it is clear that any time interval can be transformed into the

Table 1.2. Path integral notation for the corresponding finite-dimensional approximations; for simplicity we consider all the Δt_i, $i = 1, \ldots, N$, to be equal: $\Delta t_i = \varepsilon$.

$dx(t)$	'infinitesimally small' collection of functions $x(t)$ which obey the relations $x(t_0) = x_0$, $x_1 < x(t_1) < x_1 + dx_1$, $x_2 < x(t_2) < x_2 + dx_2$, $\vdots$ $x_N < x(t_N) < x_N + dx_N$, $x_{N+1} < x(t_{N+1}) < x_{N+1} + dx_{N+1}$, for the *unconditional* Wiener measure; $x_{N+1} = x(t) = x_t$, for the *conditional* Wiener measure. In the finite-dimensional integral, this results in the appearance of the measure $\prod_{j=1}^{N} dx_j$
$\displaystyle\int \prod_{\tau=0}^{t} \frac{dx(\tau)}{\sqrt{4\pi D dt}}$	$\displaystyle\frac{1}{(4\pi D\varepsilon)^{(N+1)/2}} \int_{-\infty}^{\infty} dx_1 \int_{-\infty}^{\infty} dx_2 \ldots \int_{-\infty}^{\infty} dx_{N+1}$, for the *unconditional* Wiener measure $\displaystyle\frac{1}{(4\pi D\varepsilon)^{(N+1)/2}} \int_{-\infty}^{\infty} dx_1 \int_{-\infty}^{\infty} dx_2 \ldots \int_{-\infty}^{\infty} dx_N$, for the *conditional* Wiener measure;
$\displaystyle\int_0^t d\tau\, \dot{x}^2(\tau)$	$\displaystyle\frac{1}{\varepsilon} \sum_{i=0}^{N} (x_{i+1} - x_i)^2$
$\displaystyle\int d_W x(\tau)$	$\displaystyle\frac{1}{(4\pi D\varepsilon)^{(N+1)/2}} \int_{-\infty}^{\infty} dx_1 \int_{-\infty}^{\infty} dx_2 \ldots \int_{-\infty}^{\infty} dx_{N+1}\, e^{-\frac{1}{4D\varepsilon} \sum_{i=0}^{N} (x_{i+1} - x_i)^2}$, for the *unconditional* Wiener measure; $\displaystyle\frac{1}{(4\pi D\varepsilon)^{(N+1)/2}} \int_{-\infty}^{\infty} dx_1 \int_{-\infty}^{\infty} dx_2 \ldots \int_{-\infty}^{\infty} dx_N\, e^{-\frac{1}{4D\varepsilon} \sum_{i=0}^{N} (x_{i+1} - x_i)^2}$, for the *conditional* Wiener measure.

standard one $[0, 1]$ by rescaling. Another convenient simplification which we shall use in this proof is the choice of a unit of length for space measurements (i.e. for the coordinate x), such that the diffusion constant takes the value

$$D = \tfrac{1}{4}. \tag{1.1.65}$$

To prove the statement of the theorem about discontinuous functions, we shall go through the following steps:

- **Step 1**. Estimating the functional measure ('volume') of the functions belonging to the set Z_{mj}^h, i.e. satisfying condition (1.1.64).
- **Step 2**. Estimating the Wiener measure of the union Z^h of all the sets Z_{mj}^h with arbitrary m, j but *fixed* parameter h.
- **Step 3**. Proof that the intersection $Z = \prod_{h=1}^{\infty} Z^h$ of all sets Z^h has vanishing Wiener measure.
- **Step 4**. Proof that any discontinuous function belongs to the intersection Z and hence the set of all discontinuous functions has also vanishing Wiener measure.

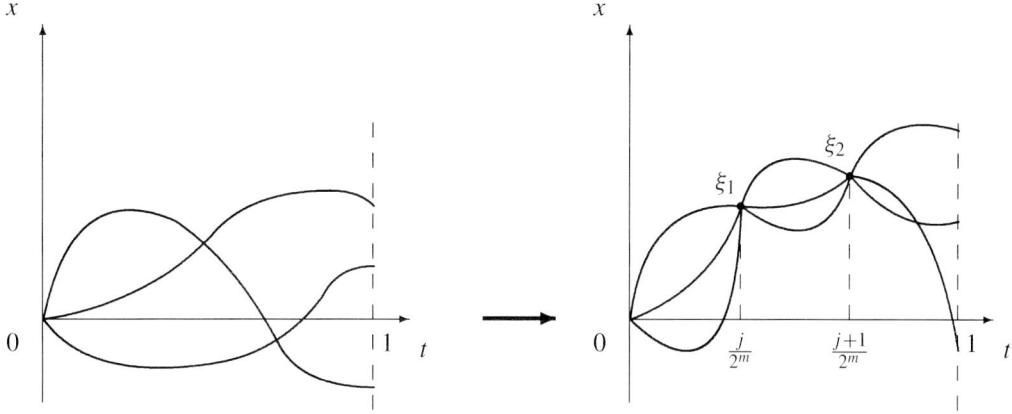

Figure 1.7. Essential trajectories in the path integral (1.1.67) with the characteristic function $\chi_{Z_{mj}^h}$ (1.1.66), in the proof of Wiener's theorem.

Step 1. The Wiener integral over the set Z_{mj}^h of functions with arbitrary endpoints can be written using the *characteristic functional* $\chi_{Z_{mj}^h}$, as

$$\int_{C\{0,0;t\}} d_{\mathrm{W}}x(\tau)\, \chi_{Z_{mj}^h}[x(\tau)] \tag{1.1.66}$$

(do not confuse such functionals with those in the probability theory which have the same name but quite a different meaning, cf section 1.2.8). Recall that the characteristic function of a subset $\mathcal{F}$ of some larger set $\mathcal{F}'$ is defined as follows:

$$\chi_{\mathcal{F}}[f] = \begin{cases} 1 & \text{if } f \in \mathcal{F} \\ 0 & \text{if } f \notin \mathcal{F}. \end{cases}$$

For example, the characteristic function of an interval $[A, B] \subset \mathbb{R}$ has the form

$$\chi_{[A,B]}(x) = \begin{cases} 1 & \text{if } x \in [A, B] \\ 0 & \text{if } x \notin [A, B]. \end{cases}$$

Thus the integration of this function over $\mathbb{R}$ is reduced to the integral only over the interval and is equal to the length (one-dimensional volume or *measure*) of this subset of the whole real line $\mathbb{R}$:

$$\int_{-\infty}^{\infty} dx\, \chi_{[A,B]}(x) = \int_A^B dx = B - A.$$

Analogously, (1.1.66) with the appropriately chosen characteristic functional $\chi_{Z_{mj}^h}[x(\tau)]$ gives the measure ('volume') of functions satisfying (1.1.64).

Using the ESKC relation, the Wiener integral (1.1.66) can be presented in terms of three standard (i.e. without the characteristic functionals) Wiener path integrals of the form (1.1.53), (1.1.55) and can, in this way, be reduced to an ordinary two-dimensional integral (see figure 1.7)

$$\int_{C\{0,0;1\}} d_{\mathrm{W}}x(t)\, \chi_{Z_{mj}^h}[x(t)] = \int_{|\xi_1 - \xi_2| > \frac{h}{A^m}} d\xi_1\, d\xi_2 \int_{C\{0,0;\xi_1, j/2^m\}} d_{\mathrm{W}}x(\tau)$$

$$\times \int_{C\{\xi_1,j/2^m;\xi_2,(j+1)/2^m\}} d_W x(\tau) \int_{C\{\xi_2,(j+1)/2^m;1\}} d_W x(t)$$

$$= \int_{|\xi_1-\xi_2|>\frac{h}{A^m}} d\xi_1\, d\xi_2 \sqrt{\frac{2^m}{\pi j}} e^{-\xi_1^2 2^m/j} \sqrt{\frac{2^m}{\pi}} e^{-(\xi_2-\xi_1)^2 2^m}. \quad (1.1.67)$$

Making the change of variables

$$2^{\frac{m}{2}}(\xi_2 - \xi_1) = \eta \qquad \xi_1 = \zeta$$

and integrating over the variable ζ, expression (1.1.67) can be converted into a one-dimensional integral:

$$\frac{1}{\sqrt{\pi}} \int_{|\eta|>h(\sqrt{2}/A)^m} d\eta\, e^{-\eta^2} = \frac{2}{\sqrt{\pi}} \int_{h(\sqrt{2}/A)^m}^{\infty} d\eta\, e^{-\eta^2} \leq \frac{1}{\sqrt{\pi} h \theta^m} e^{-h^2 \theta^{2m}} \quad (1.1.68)$$

where $\theta = \sqrt{2}/A$, so that $1 < \theta < \sqrt{2}$.

Step 2. Now let us estimate the Wiener measure of the union

$$Z^h = \bigcup_{m=1}^{\infty} \bigcup_{j=0}^{2^m-1} Z_{mj}^h$$

of all the sets Z_{mj}^h. Since these sets with different values of m and j have intersections, the measure of the union Z^h does not exceed the sum of the measures of Z_{mj}^h:

$$\int_{C\{0,0;1\}} d_W x(t)\, \chi_{Z^h} \leq \sum_{m=1}^{\infty} 2^m \int_{C\{0,0;1\}} d_W x(t)\, \chi_{Z_{mj}^h}$$

$$\leq \sum_{m=1}^{\infty} \frac{2^m}{\sqrt{\pi} h \theta^m} e^{-h^2 \theta^{2m}} < \frac{e^{-h^2}}{\sqrt{\pi} h} \sum_{m=1}^{\infty} \left(\frac{2}{e\theta}\right)^m. \quad (1.1.69)$$

The second inequality follows from (1.1.68). To obtain the last inequality, the estimate $h^2 > m/(\theta^{2m} - 1)$, valid for sufficiently large h, has been used.

Step 3. The last sum in (1.1.69) is convergent, so that we have for the measure under consideration

$$\int_{C\{0,0;1\}} d_W x(t)\, \chi_{Z^h} < \frac{e^{-h^2}}{\sqrt{\pi} h} (\text{constant}) \xrightarrow[h\to\infty]{} 0. \quad (1.1.70)$$

This, in turn, means that

$$\int_{C\{0,0;1\}} d_W x(t)\, \chi_Z = 0 \quad (1.1.71)$$

where $Z = \bigcap_{h=1}^{\infty} Z^h$.

Step 4. Now consider any discontinuous function $x(t)$. By the definition of discontinuous functions, for any h there exist two points $t_1 = j/2^m$ and $t_2 = (j+1)/2^m$ for some m and j, such that

$$|x(t_2) - x(t_1)| > h(t_2 - t_1)^{\log_2 A} = h(t_2 - t_1)^{\frac{1}{2}-\varepsilon} \qquad 0 < \varepsilon < \tfrac{1}{2}. \quad (1.1.72)$$

Since $h(t_2 - t_1)^{\log_2 A} = h/(2^{m \log_2 A}) = h/A^m$, any discontinuous function belongs to the set Z_{mj}^h with arbitrary $h > 0$: $x(t) \in Z_{mj}^h\ \forall h$ and hence $x(t) \in Z$. Thus the discontinuous functions have zero measure due to (1.1.71). In other words, the Wiener integral over the set of discontinuous trajectories is equal to zero. On the other hand, the set of all real functions has unit measure (cf (1.1.55)). From this fact we immediately derive the fact that the set of *continuous* functions has unit measure, and, moreover, (1.1.72) shows that functions satisfying the Hölder–Lipschitz condition (see, e.g., Ilyin and Poznyak (1982))

$$|x(t_2) - x(t_1)| < h|t_2 - t_1|^{\frac{1}{2}-\varepsilon} \qquad (1.1.73)$$

also have unit measure (we shall use this fact later, in section 1.2.7).

The proof of Wiener's theorem for the case of differentiable trajectories is considered in problem 1.1.5, page 53.

◇ **Integration of functionals: general approach**

So far, we have discussed mainly the Wiener measure of sets of trajectories. To develop the integration theory further, we begin by considering *simple functionals* $F[x(\tau)]$ for which the path integrals of the form

$$\int_{C\{0,0;t\}} d_W x(\tau) \, F[x(\tau)] \equiv \int_{C\{0,0;t\}} \prod_{\tau=0}^{t} \frac{dx(\tau)}{\sqrt{4\pi D\,d\tau}} e^{-\frac{1}{4D}\int_0^t \dot{x}^2(\tau)\,d\tau} F[x(\tau)] \qquad (1.1.74)$$

can be evaluated immediately. Measurable functionals will be defined as appropriate infinite limits of the simple functionals. The proof of the Wiener theorem prompts the obvious example of a simple functional: this is the characteristic functional of a measurable set. If we take

$$F[x(\tau)] = \chi_Y[x(\tau)]$$

where χ_Y is the characteristic functional of some set Y of the trajectories $x(\tau)$ defined as

$$\chi_Y[x(\tau)] = \begin{cases} 1 & \text{if } x(\tau) \in Y \\ 0 & \text{if } x(\tau) \notin Y \end{cases}$$

then the definition gives

$$\int d_W x(\tau) \, \chi_Y[x(\tau)] = \mu_W(Y)$$

where $\mu_W(Y)$ is the Wiener measure of the set Y (cf explanation in the proof of Wiener's theorem, below equation (1.1.66)).

As an example, let us choose the set $Y(x_1, \ldots, x_N)$ of the trajectories having fixed positions $x_1, \ldots, x_N$ at some sequence $t_1, \ldots, t_N$ of the time variable:

$$\chi_Y[x(\tau); x_1, \ldots, x_N] = \begin{cases} 1 & \text{if } x(t_1) = x_1, \ldots, x(t_N) = x_N \\ 0 & \text{otherwise.} \end{cases}$$

The corresponding path integral is given by the product of the transition probabilities (cf (1.1.50) and (1.1.51)) from the points $x(t_{i-1})$ to the next positions $x(t_i)$ in the sequence:

$$\int_{C\{0,0;t\}} d_W x(\tau) \, \chi_Y[x(\tau); x_1, \ldots, x_N] = \prod_{i=1}^{N} W(x_i, t_i | x_{i-1}, t_{i-1})$$

$$= \prod_{i=1}^{N} \frac{1}{\sqrt{4\pi D(t_i - t_{i-1})}} \exp\left\{-\frac{(x_i - x_{i-1})^2}{4D(t_i - t_{i-1})}\right\}$$

$$x_0 = 0, \ t_0 = 0. \qquad (1.1.75)$$

Now we can consider linear combinations of such characteristic functionals with some coefficients $F_N(x_1, \ldots, x_N)$,

$$F_N[x(\tau)] = \int_{-\infty}^{\infty} \cdots \int_{-\infty}^{\infty} dx_1 \cdots dx_N \, F_N(x_1, \ldots, x_N) \chi_Y[x(\tau); x_1, \ldots, x_N]. \tag{1.1.76}$$

Making use of (1.1.75), the functional integration of these functionals can be easily reduced to the ordinary finite-dimensional integration

$$\int_{C\{0,0;t\}} d_W x(\tau) \, F_N[x(\tau)] = \int_{-\infty}^{\infty} \frac{dx_1}{\sqrt{4\pi D t_1}} \cdots \frac{dx_N}{\sqrt{4\pi D(t_N - t_{N-1})}}$$

$$\times F_N(x_1, \ldots, x_N) \exp\left\{ -\frac{1}{4D} \sum_{i=1}^{N} \frac{(x_i - x_{i-1})^2}{t_i - t_{i-1}} \right\}. \tag{1.1.77}$$

We see that $F_N(x_1, \ldots, x_N)$ should be such a function of $x_1, \ldots, x_N$ that the right-hand side of (1.1.77) exists, e.g., it can be a polynomial.

These *simple functionals* form a vector space $\mathcal{F}$ and the Wiener path integral allows us to define the norm (distance) $\|\cdot\|$ in this vector space:

$$\|F - G\| = \int d_W x(\tau) |F - G| \tag{1.1.78}$$

($|\cdot|$ on the right-hand side is just the absolute value of the difference of the two functionals), where F and G are two simple functionals (i.e. of the type (1.1.76)). Readers with mathematical orientation may easily check that all the axioms for a norm are satisfied by (1.1.78). Having defined the norm (1.1.78), the general problem of extending the path integral to a larger set of functionals can be accomplished by the standard mathematical method. First, we define a sequence $F^{(N)}$ of simple functionals with the property

$$\|F^{(N)} - F^{(M)}\| \xrightarrow[N,M \to \infty]{} 0$$

called the *Cauchy sequence*. The functional $F[x(\tau)]$ is said to be *integrable* if there exists a Cauchy sequence $F^{(N)}$ of simple functionals such that

$$F^{(N)} \xrightarrow[N \to \infty]{} F \tag{1.1.79}$$

with respect to the norm (1.1.78). The path integral is then defined by

$$\int_{C\{0,0;t\}} d_W x(\tau) \, F[x(\tau)] \stackrel{\text{def}}{\equiv} \lim_{N \to \infty} \int_{C\{0,0;t\}} d_W x(\tau) \, F_N[x(\tau)]. \tag{1.1.80}$$

◇ **Practical method of integration of functionals: approximation by piecewise linear functions**

In practice, we can use the fact that the set of functions possessing non-zero Wiener measure can be uniformly approximated by piecewise linear functions $\ell_N(\tau)$ which are linear for $\tau \in (t_{i-1}, t_i)$ and

$$\ell_N(t_i) = x(t_i) \equiv x_i \qquad i = 1, \ldots, N$$

(see figure 1.8). This means that for any ε and any function $x(t)$, there exist points $t_0, t_1, \ldots, t_N$; $N = N(\varepsilon)$ such that

$$|x(\tau) - l_N(\tau)| < \varepsilon \qquad N = N(\varepsilon).$$

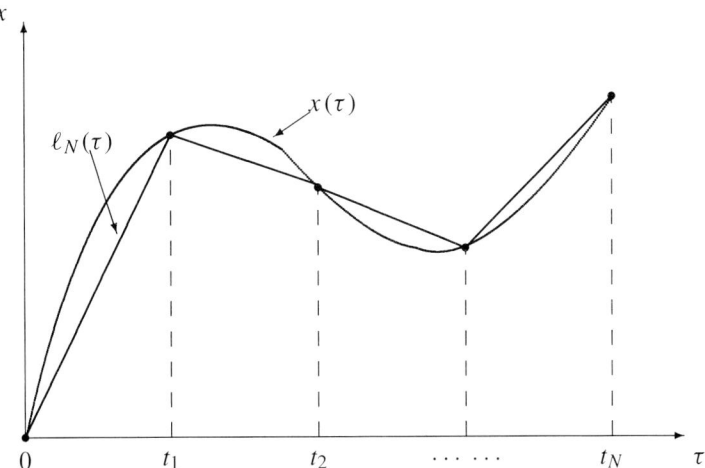

Figure 1.8. Approximation of a trajectory $x(\tau)$ by the piecewise linear function $\ell_N(\tau)$.

Then smooth functionals satisfy the inequality

$$|F[x(\tau)] - F[l_N(\tau)]| < \delta(\varepsilon)$$

where $\delta(\varepsilon) \to 0$ for $\varepsilon \to 0$. However, $F[l_N(\tau)] = F_N(x_1, \ldots, x_N)$, since full information about $l_N(\tau)$ is determined by the set of values $x_1 = x(t_1), \ldots, x_N = x(t_N)$. Therefore,

$$\left| \int_{\mathcal{C}\{0,0;t\}} d_W x(\tau) \, F[x(\tau)] - \int_{\mathcal{C}\{0,0;t\}} d_W x(\tau) \, F_N(x_1, \ldots, x_N) \right|$$

$$\leq \int_{\mathcal{C}\{0,0;t\}} d_W x(\tau) \, |F[x(\tau)] - F_N(x_1, \ldots, x_N)|$$

$$\leq \int_{\mathcal{C}\{0,0;t\}} d_W x(\tau) \, \delta(\varepsilon) = \delta(\varepsilon).$$

The set of simple functionals $F_N(x_1, \ldots, x_N) = F[l_N(\tau)]$ with increasing $N = 1, 2, \ldots$ thus provides a Cauchy sequence approaching the given functional. Consequently, the sequence

$$\int_{\mathcal{C}\{0,0;t\}} d_W x(\tau) \, F_N(x_1, \ldots, x_N) = \int_{-\infty}^{\infty} \frac{dx_1}{\sqrt{4\pi D t_1}} \cdots \frac{dx_N}{\sqrt{4\pi D(t_N - t_{N-1})}} F_N(x_1, \ldots, x_N)$$

$$\times \exp\left\{ -\frac{1}{4D} \sum_{i=1}^{N} \frac{(x_i - x_{i-1})^2}{t_i - t_{i-1}} \right\}$$

($N = 1, 2, 3, \ldots$) can be used for the calculation of the desired Wiener integral:

$$\int_{\mathcal{C}\{0,0;t\}} d_W x(\tau) \, F[x(\tau)] = \lim_{N \to \infty} \int_{-\infty}^{\infty} \frac{dx_1}{\sqrt{4\pi D t_1}} \cdots \frac{dx_N}{\sqrt{4\pi D(t_N - t_{N-1})}} F_N(x_1, \ldots, x_N)$$

$$\times \exp\left\{-\frac{1}{4D}\sum_{i=1}^{N}\frac{(x_i - x_{i-1})^2}{t_i - t_{i-1}}\right\}. \tag{1.1.81}$$

This formula reduces the Wiener path integral to the limit of finite-dimensional integrals.

Since the natural number N is assumed to have arbitrary large (finite but unbounded) values, we might justly speak of the path integral as an extension, to an infinite dimension, of the traditional notion of a finite-dimensional integral. The corresponding finite-dimensional integral on the right-hand side of (1.1.81) is termed a *discrete-time* or *time-sliced approximation* of the path integral on the left-hand side.

◇ **Some mathematical remarks**

We conclude this subsection with a few short mathematical remarks.

- Wiener constructed the functional measure at the beginning of the 1920s (Wiener 1921, 1923, 1924) using an explicit mapping of the space of continuous functions into the interval $(0, 1) \subset \mathbb{R}$ (more precisely, into the interval $(0, 1)$ minus a set of zero measure). Under this mapping, a set of functions (trajectories) passing through fixed gates (such functions are said to belong to cylindrical sets; Wiener also called them 'quasi-integrals') as in figure 1.3 is transformed into the set on the unit interval with an ordinary Lebesgue measure. The latter is numerically equal to the measure of the 'quasi-integrals' defined by (1.1.50). The reader can find this construction in Wiener's original papers and in chapter IX of the book by Paley and Wiener (1934).
- Later, mathematicians comprehensively studied the functional measure using the much more abstract and powerful method of the axiomatic measure theory. The cornerstone of this approach is the important *Kolmogorov theorem* (see Kolmogorov (1956)), stating that for any given set of functions satisfying some self-consistency conditions (in fact, these conditions endow the given set with the properties of the probabilities of compound events) there exists a set Ω (of events) with additive measure μ and a set of measurable functions $X(t, \omega)$, $\omega \in \Omega$, so that the measure μ defines the probability of the corresponding compound event (of the type (1.1.49)). In the case of the Wiener measure, the self-consistency condition mentioned in the Kolmogorov theorem is provided by the ESKC relation. Note that using the abstract approach, we can construct a non-trivial (non-zero) measure on the set of *all* real functions (including discontinuous functions); but in this case, the physically important set of continuous functions proves to be unmeasurable.
- Since the self-consistency of the Wiener measure is based on the ESKC relation (1.1.45), it is interesting to study other solutions of this relation. One class of the solutions has the following form

$$f(x, t|x_0, 0) = \frac{1}{2\pi}\int_{-\infty}^{\infty} dp\, \exp\{ip(x - x_0)\} \exp\{-t|p|^{\alpha}\} \qquad 0 < \alpha \leq 2.$$

The case $\alpha = 2$ corresponds to Brownian motion. If $\alpha < 2$, it turns out that on the set of continuous functions a non-trivial measure, analogous to (1.1.50), cannot be constructed. However, such a measure exists on the set of functions continuous from the left (or from the right).

Further details on the mathematical theory of functional measure can be found in Kac (1959), Kuo (1975), Simon (1979) and Reed and Simon (1975).

1.1.4 Methods and examples for the calculation of path integrals

Basic methods for calculating path integrals are summarized in figure 1.9. In this subsection we shall consider some of them, calculating important concrete examples of path integrals. Other methods are treated in subsequent sections and chapters of the book.

Brownian motion: introduction to the concept of path integration

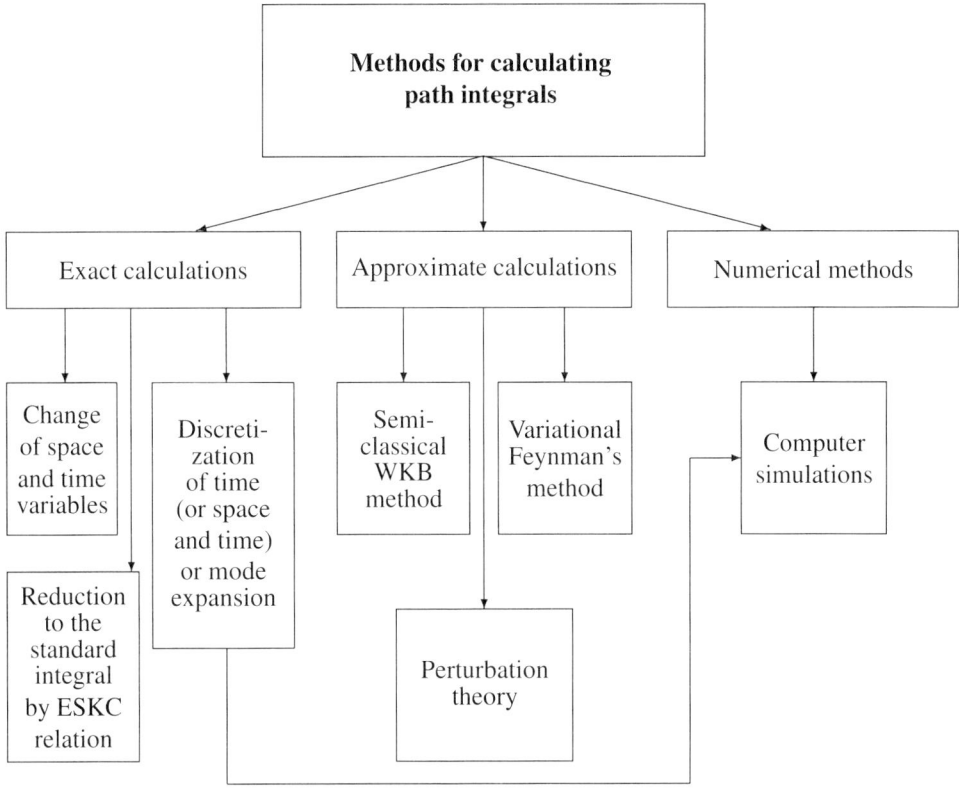

Figure 1.9. Methods for calculations of path integrals.

We start from the discrete-time approximation. This method has special significance for path integrals. First of all, the very definition of the latter is heavily based on such an approximation (especially, in the case of path integrals in quantum mechanics, see the next chapter). Thus, if the multiple integral obtained from a path integral after time-slicing can be calculated exactly, this gives, in the appropriate limit, an exact expression for the initial path integral. Furthermore, if an exact calculation of the multiple integral is impossible, the discrete approximation provides the basis for numerical calculations (as is indicated in figure 1.9).

Example 1.1. To obtain some experience with the calculation of path integrals with the help of the *discrete approximation*, suppose we do not know the right-hand side of (1.1.53) and we would like to calculate it explicitly using (1.1.51). For the sake of simplicity, put $x_0 = x_t = 0$ (i.e. we must integrate over the set of paths $C\{0, 0; 0, t\}$); later we shall see that even more general path integrals can be reduced to this type. We take all the time intervals $t_i - t_{i-1}$ to be equal: $t_i - t_{i-1} = \varepsilon \equiv t/(N+1)$ for any $i = 1, \ldots, N+1$.

The transition probability W in terms of the discrete approximation has the form

$$W(0, t|0, 0) = I_1 \equiv \lim_{\substack{\varepsilon \to 0 \\ N \to \infty}} \frac{1}{(\sqrt{4\pi D\varepsilon})^{N+1}} \int_{-\infty}^{\infty} dx_1 \int_{-\infty}^{\infty} dx_2$$

$$\cdots \int_{-\infty}^{\infty} dx_N \exp\left\{-\frac{1}{4D\varepsilon}\sum_{i=0}^{N}(x_{i+1}-x_i)^2\right\}. \tag{1.1.82}$$

The exponential in (1.1.82) is a bilinear form (recall that $x_0 = x_{N+1} = 0$):

$$\sum_{i=0}^{N}(x_{i+1}-x_i)^2 = \sum_{k,l=1}^{N} x_k A_{kl} x_l \tag{1.1.83}$$

where $\mathbf{A} = (A_{kl})$ is the three-diagonal matrix

$$\mathbf{A} = \begin{pmatrix} 2 & -1 & 0 & 0 & \cdots & \cdots & 0 \\ -1 & 2 & -1 & 0 & \cdots & \cdots & 0 \\ 0 & -1 & 2 & -1 & 0 & \cdots & 0 \\ \vdots & & \ddots & \ddots & \ddots & & \vdots \\ 0 & \cdots & 0 & -1 & 2 & -1 & 0 \\ 0 & \cdots & \cdots & 0 & -1 & 2 & -1 \\ 0 & \cdots & \cdots & \cdots & 0 & -1 & 2 \end{pmatrix}. \tag{1.1.84}$$

The $N \times N$ matrix $\mathbf{A}$ has zero matrix elements apart from those in the main diagonal and in the two neighbouring diagonals.

Now we shall use a formula for the *multidimensional Gaussian integral* (its calculation pertains to problem 1.1.6, page 53):

$$\int_{-\infty}^{\infty} dx_1 \cdots dx_N \exp\left\{-\sum_{i,j}^{N} b_{ij} x_i x_j\right\} = \frac{(\sqrt{\pi})^N}{\sqrt{\det \mathbf{b}}} \tag{1.1.85}$$

which is very important in numerous applications and for what follows in this book. Using this result for the calculation of (1.1.82), we find that

$$W(0,t|0,0) = \lim_{\substack{\varepsilon \to 0 \\ N \to \infty}} \frac{1}{\sqrt{4\pi D\varepsilon \det \mathbf{A}_N}} \tag{1.1.86}$$

where the subscript N indicates that $\mathbf{A}$ is an $N \times N$ matrix. All that is needed now is the determinant of the matrix $\mathbf{A}_N$, which can be found as follows. First, we calculate $\det \mathbf{A}_N$ *by hand* for small N: $\det \mathbf{A}_1 = 2$, $\det \mathbf{A}_2 = 3$. This leads to the guess

$$\det \mathbf{A}_N = N + 1. \tag{1.1.87}$$

If $\det \mathbf{A}_N$ is expanded in the elements of the last column, we find that

$$\det \mathbf{A}_N = 2 \det \mathbf{A}_{N-1} - \det \mathbf{A}_{N-2}.$$

This recursion relation is satisfied by (1.1.87) which proves its general validity. Substituting (1.1.87) into (1.1.86), we obtain the expected result (cf (1.1.53))

$$W(0,t|0,0) = (4\pi Dt)^{-1/2}. \tag{1.1.88}$$

Another method to calculate (1.1.82), which we suggest to the reader as a useful exercise (see problem 1.1.7, page 53), essentially amounts to performing the integrations one by one.

Throughout the remainder of this subsection, we let the diffusion constant be

$$D = \tfrac{1}{4} \qquad (1.1.89)$$

a convention which, if necessary, can always be achieved by an appropriate choice of the time and/or space units.

Example 1.2. Let us calculate (1.1.74) for the simple functional $F[x(\tau)] = F(x(s)) = x(s)$, where s is a fixed time in the interval $[0, t]$ (the path integral with unconditional Wiener measure). To get accustomed to the properties of path integrals we shall consider two ways of calculation.

The *first method* uses the ESKC relation (1.1.58). Identically rewriting the path integral, we find that the functional integration is reduced to the known expressions (1.1.55) and (1.1.53):

$$\begin{aligned} I_2 &= \int_{\mathcal{C}\{x_0,0;t\}} d_W x(\tau)\, x(s) \\ &= \int_{-\infty}^{\infty} dx_s\, x_s \int_{\mathcal{C}\{x_0,0;x_s,s\}} d_W x(\tau) \int_{\mathcal{C}\{x_s,s;t\}} d_W x(\tau) \end{aligned} \qquad (1.1.90)$$

where $x_s = x(s)$. The last path integral in (1.1.90) is equal to unity due to the normalization condition (1.1.55), while (1.1.53) gives

$$I_2 = \int_{-\infty}^{\infty} dx_s\, x_s\, \frac{1}{\sqrt{\pi s}} e^{-(x_s-x_0)^2/s} \cdot 1 = x_0. \qquad (1.1.91)$$

The *second method* to calculate (1.1.90) is based on the discrete approximation formula (1.1.51):

$$\begin{aligned} I_2 \approx I_2^{(n)} &= \int_{-\infty}^{\infty} \prod_{i=1}^{N} \frac{dx_i}{\sqrt{\pi \Delta t_i}} \exp\left\{ -\sum_{i=1}^{N} \frac{(x_i - x_{i-1})^2}{t_i - t_{i-1}} \right\} x_l \\ &= \int_{-\infty}^{\infty} \frac{dx_l}{\sqrt{\pi \Delta t_l}} x_l \int_{-\infty}^{\infty} \prod_{i=1}^{l-1} \frac{dx_i}{\sqrt{\pi \Delta t_i}} \exp\left\{ -\sum_{i=1}^{l} \frac{(x_i - x_{i-1})^2}{t_i - t_{i-1}} \right\} \\ &\quad \times \int_{-\infty}^{\infty} \prod_{i=l+1}^{N} \frac{dx_i}{\sqrt{\pi \Delta t_i}} \exp\left\{ -\sum_{i=l+1}^{N} \frac{(x_i - x_{i-1})^2}{t_i - t_{i-1}} \right\} \\ &= \int_{-\infty}^{\infty} dx_l\, \frac{x_l}{\sqrt{\pi t_l}} \exp\left\{ -\frac{(x_l - x_0)^2}{t_l} \right\} = x_0. \end{aligned} \qquad (1.1.92)$$

In this calculation, we have chosen the time s to lie exactly on the discrete time slice t_l: $s = t_l$ and hence $x(s) = x_l$.

— ○ —

Integration of more general functionals can be worked out by the formula (1.1.80), i.e. using the discrete approximation and limiting procedure.

Example 1.3. Let us calculate the average over the unconditional Wiener measure

$$I_3 = \int_{\mathcal{C}\{0,0;t\}} d_W x(\tau)\, F[x(\tau)]$$

for the functional
$$F[x(\tau)] = F\left(\int_0^t a(\tau)x(\tau)\,d\tau\right).$$

The functions $F(y)$ and $a(\tau)$ must satisfy some conditions which will be revealed later. Formula (1.1.80) for this case reads as follows:
$$I_3 = \lim_{N\to\infty} I_3^{(N)}$$

$$I_3^{(N)} = \int_{-\infty}^{\infty} \frac{dx_1}{\sqrt{\pi\Delta t_1}} \cdots \frac{dx_N}{\sqrt{\pi\Delta t_N}} F\left(\sum_{i=1}^N a_i x_i \Delta t_i\right) \exp\left\{-\sum_{i=1}^N \frac{(x_i - x_{i-1})^2}{\Delta t_i}\right\} \quad (1.1.93)$$

where $a_i = a(t_i)$, $x_i = x(t_i)$. Making the change of variables
$$y_i = x_i - x_{i-1}, \quad i = 1,\ldots,N, \quad (1.1.94)$$

with the unit Jacobian, (1.1.93) can be rewritten as
$$I_3^{(N)} = \int_{-\infty}^{\infty} \frac{dy_1}{\sqrt{\pi\Delta t_1}} \cdots \frac{dy_N}{\sqrt{\pi\Delta t_N}} \exp\left\{-\sum_{i=1}^N \frac{y_i^2}{\Delta t_i}\right\}$$
$$\times F(y_1(a_1\Delta t_1 + \cdots + a_N\Delta t_N) + \cdots + y_N a_N \Delta t_N). \quad (1.1.95)$$

Further, we introduce new variables:
$$z_i = A_i y_i \qquad A_i = \sum_{k=i}^N a_k \Delta t_k \qquad 1 \le i \le N. \quad (1.1.96)$$

In terms of these variables, $I_3^{(N)}$ becomes (taking into account the Jacobian of (1.1.96)):
$$I_3^{(N)} = \int_{-\infty}^{\infty} \frac{dz_1}{\sqrt{\pi A_1^2 \Delta t_1}} \cdots \frac{dz_N}{\sqrt{\pi A_N^2 \Delta t_N}} F(z_1 + \cdots + z_N) \exp\left\{-\sum_{i=1}^N \frac{z_i^2}{A_i^2 \Delta t_i}\right\}. \quad (1.1.97)$$

Thus the problem is reduced to calculating the average of the function whose argument is the sum of variables with the Gaussian distribution. To accomplish the calculations, first introduce $\eta = z_1 + z_2$, $\zeta = z_2$, so that (1.1.97) is transformed into
$$I_3^{(N)} = \int_{-\infty}^{\infty} \frac{d\zeta}{\sqrt{\pi A_2^2 \Delta t_2}} \int_{-\infty}^{\infty} \frac{d\eta}{\sqrt{\pi A_1^2 \Delta t_1}} \int_{-\infty}^{\infty} \frac{dz_3}{\sqrt{\pi A_3^2 \Delta t_3}} \cdots \frac{dz_N}{\sqrt{\pi A_N^2 \Delta t_N}}$$
$$\times F(\eta + z_3 + \cdots + z_N) \exp\left\{-\frac{(\eta-\zeta)^2}{A_1^2 \Delta t_1} - \frac{\zeta^2}{A_2^2 \Delta t_2} - \sum_{i=3}^N \frac{z_i^2}{A_i^2 \Delta t_i}\right\}. \quad (1.1.98)$$

Since the function F does not depend on the variable ζ, we can integrate over this variable in (1.1.98) using Gaussian integration:
$$\int_{-\infty}^{\infty} d\zeta \frac{1}{\sqrt{\pi A_1^2 \Delta t_1}\sqrt{\pi A_2^2 \Delta t_2}} \exp\left\{-\frac{(\eta-\zeta)^2}{A_1^2 \Delta t_1} - \frac{\zeta^2}{A_2^2 \Delta t_2}\right\}$$
$$= \frac{1}{\sqrt{\pi(A_1^2 \Delta t_1 + A_2^2 \Delta t_2)}} \exp\left\{-\frac{\eta^2}{A_1^2 \Delta t_1 + A_2^2 \Delta t_2}\right\}.$$

This allows us to represent (1.1.98) in the form

$$I_3^{(N)} = \int_{-\infty}^{\infty} \frac{d\eta}{\sqrt{\pi A_1^2 \Delta t_1 + \pi A_2^2 \Delta t_2}} \int_{-\infty}^{\infty} \frac{dz_3}{\sqrt{\pi A_3^2 \Delta t_3}} \cdots \frac{dz_N}{\sqrt{\pi A_N^2 \Delta t_N}}$$
$$\times F(\eta + z_3 + \cdots + z_N) \exp\left\{\frac{-\eta^2}{A_1^2 \Delta t_1 + A_2^2 \Delta t_2} - \sum_{i=3}^{N} \frac{z_i^2}{A_i^2 \Delta t_i}\right\}. \quad (1.1.99)$$

The structure of this integral is quite similar to that of (1.1.98) except that the number of the variables of integration is reduced to $N - 1$. After $(N - 1)$ repetitions of this step, we come to the result

$$I_3^{(N)} = \int_{-\infty}^{\infty} \frac{dz}{\sqrt{\pi \sum_{i=1}^{N} A_i^2 \Delta t_i}} F(z) \exp\left\{-\frac{z^2}{\sum_{i=1}^{N} A_i^2 \Delta t_i}\right\}. \quad (1.1.100)$$

In the continuous limit, $\Delta t_i \to 0$, $N \to \infty$, we have

$$\lim_{\Delta t_i \to 0} A_i = \int_{\tau}^{t} ds\, a(s)$$

$$R \equiv \lim_{\Delta t_i \to 0} \sum_{i=1}^{N} A_i^2 \Delta t_i = \int_{0}^{t} d\tau \left[\int_{\tau}^{t} ds\, a(s)\right]^2$$

so that

$$I_3 = \lim_{\Delta t_i \to 0} I_3^{(N)} = \int_{-\infty}^{\infty} dz\, \frac{F(z)}{\sqrt{\pi R}} e^{-z^2/R}. \quad (1.1.101)$$

Now we can formulate the conditions for the functions $a(t)$ and $F(z)$: the former must be an integrable function in the interval $[0, t]$ (cf first equation in (1.1.101)) and $F(z)$ must be continuous and may grow at infinity with the only restriction of the existence of the integral (1.1.101). Explicit restoration of the diffusion constant D reduces to the simple rescaling $R \to 4DR$.

───── ○ ─────

Example 1.4. Consider the Wiener integral I_4 with the functional

$$F[x(\tau)] = \exp\left\{-\int_0^t d\tau\, p(\tau) x^2(\tau)\right\}$$

i.e.

$$I_4 = \int_{C\{0,0;t\}} d_W x(\tau) \exp\left\{-\int_0^t d\tau\, p(\tau) x^2(\tau)\right\}. \quad (1.1.102)$$

Formula (1.1.51) gives (put, for simplicity, $\Delta t_i = \varepsilon = t/N$, $i = 1, \ldots, N$)

$$I_4 = \lim_{N \to \infty} I_4^{(N)}$$

$$I_4^{(N)} = \int_{-\infty}^{\infty} \frac{dx_1}{\sqrt{\pi \varepsilon}} \cdots \frac{dx_N}{\sqrt{\pi \varepsilon}} \exp\left\{-\sum_{i=1}^{N} p_i x_i^2 \varepsilon - \sum_{i=1}^{N} \frac{(x_i - x_{i-1})^2}{\varepsilon}\right\}$$

$$\equiv \int_{-\infty}^{\infty} \frac{dx_1}{\sqrt{\pi \varepsilon}} \cdots \frac{dx_N}{\sqrt{\pi \varepsilon}} \exp\left\{-\sum_{i,j=1}^{N} a_{ij} x_i x_j\right\} \quad (1.1.103)$$

where $\mathbf{a} = (a_{ij})$ is the three-diagonal matrix

$$\mathbf{a} = \begin{pmatrix} a_1 & -\frac{1}{\varepsilon} & 0 & 0 & \cdots & \cdots & 0 \\ -\frac{1}{\varepsilon} & a_2 & -\frac{1}{\varepsilon} & 0 & \cdots & \cdots & 0 \\ 0 & -\frac{1}{\varepsilon} & a_3 & -\frac{1}{\varepsilon} & 0 & \cdots & 0 \\ \vdots & & \ddots & \ddots & \ddots & & \vdots \\ 0 & \cdots & 0 & -\frac{1}{\varepsilon} & a_{N-2} & -\frac{1}{\varepsilon} & 0 \\ 0 & \cdots & \cdots & 0 & -\frac{1}{\varepsilon} & a_{N-1} & -\frac{1}{\varepsilon} \\ 0 & \cdots & \cdots & \cdots & 0 & -\frac{1}{\varepsilon} & a_N \end{pmatrix} \qquad (1.1.104)$$

and

$$a_i = a_{ii} = p_i \varepsilon + \frac{2}{\varepsilon} \qquad i = 1, \ldots, N-1$$

$$a_N = p_N \varepsilon + \frac{1}{\varepsilon}.$$

Application of the result (1.1.85) to our calculation of I_4 gives

$$\begin{aligned} I_4^{(N)} &= \frac{1}{\sqrt{(\varepsilon)^N \det \mathbf{a}}} \\ &= \frac{1}{\sqrt{\det(\varepsilon \mathbf{a})}} \end{aligned} \qquad (1.1.105)$$

and reduces the problem to finding the determinant of the matrix $\varepsilon \mathbf{a}$. This determinant can be calculated by the following trick. Denote by $D_k^{(N)}$ the determinant of the matrix, obtained from $\varepsilon \mathbf{a}$ by removing the first $k-1$ rows and columns, i.e.

$$D_k^{(N)} = \begin{vmatrix} \varepsilon a_k & -1 & 0 & 0 & \cdots & \cdots & 0 \\ -1 & \varepsilon a_{k+1} & -1 & 0 & \cdots & \cdots & 0 \\ 0 & -1 & \varepsilon a_{k+2} & -1 & 0 & \cdots & 0 \\ \vdots & & & \ddots & & & \vdots \\ 0 & \cdots & 0 & -1 & \varepsilon a_{N-2} & -1 & 0 \\ 0 & \cdots & \cdots & 0 & -1 & \varepsilon a_{N-1} & -1 \\ 0 & \cdots & \cdots & \cdots & 0 & -1 & \varepsilon a_N \end{vmatrix}. \qquad (1.1.106)$$

The key observation for the calculation of the determinant is that the expansion of (1.1.106) in the elements of the first row results in the recurrence relation:

$$D_k^{(N)} = [p_k \varepsilon^2 + 2] D_{k+1}^{(N)} - D_{k+2}^{(N)}$$

or

$$\frac{D_k^{(N)} - 2 D_{k+1}^{(N)} + D_{k+2}^{(N)}}{\varepsilon^2} = p_k D_{k+1}^{(N)}. \qquad (1.1.107)$$

Introduce the variable

$$s = \frac{k-1}{N}.$$

In the continuous limit, $N \to \infty$, the determinants become the functions $D_k^{(N)} \xrightarrow[N\to\infty]{} D(s)$ and the finite difference equation (1.1.107) transforms into a differential one:

$$\frac{d^2 D(\tau)}{d\tau^2} = p(\tau) D(\tau). \qquad (1.1.108)$$

The determinant of the matrix $\varepsilon \mathbf{a}$ corresponds to the value of the function $D(s)$ at the origin $D_1^{(N)} \to D(0)$. Thus the value of the path integral is defined by a solution of equation (1.1.108). However, to pick out the particular solution of the equation from the set of all solutions of (1.1.108), we must impose the appropriate boundary conditions. To this aim we note, first, that

$$D(t) = \lim_{N\to\infty} D_N^{(N)} = 1 \qquad (1.1.109)$$

since

$$D_N^{(N)} = p_N \varepsilon^2 + 1.$$

Analogously,

$$\left.\frac{dD(\tau)}{d\tau}\right|_{\tau=t} = \lim_{N\to\infty} \frac{D_N^{(N)} - D_{N-1}^{(N)}}{\varepsilon} = 0 \qquad (1.1.110)$$

since

$$D_{N-1}^{(N)} = \begin{vmatrix} p_{N-1}\varepsilon^2 + 2 & -1 \\ -1 & p_N \varepsilon^2 + 1 \end{vmatrix}$$
$$= p_N p_{N-1} \varepsilon^4 + 2 p_N \varepsilon^2 + p_{N-1} \varepsilon^2 + 1$$

and hence

$$\frac{D_N^{(N)} - D_{N-1}^{(N)}}{\varepsilon} = \mathcal{O}(\varepsilon).$$

Thus the value of the functional integral I_4 from (1.1.102) proves to be

$$I_4 = \frac{1}{\sqrt{D(0)}} \qquad (1.1.111)$$

i.e. defined by the solution of the differential equation (1.1.108) satisfying the boundary conditions (1.1.109) and (1.1.110). Of course, equation (1.1.108) cannot be solved analytically for an arbitrary function $p(s)$.

This method of the determinant calculation has been suggested by Gelfand and Yaglom (1960) and is called the *Gelfand–Yaglom method*.

— ◦ —

In examples 1.2–1.4, we considered the *unconditional Wiener measure*, i.e. integration over paths with arbitrary endpoints. In the following example, we shall integrate the same functional as in example 1.4 but use the *conditional Wiener measure*, i.e. we shall integrate over the set $C\{0, 0; x_t, t\}$ with fixed endpoints x_t of trajectories (the calculation of the conditional Wiener integral of the functional from example 1.2 we suggest as an exercise in problem 1.1.10, page 54).

Example 1.5. Consider the analog of example 1.4 for the *conditional* measure:

$$I_4^{\text{cond}} = \int_{\mathcal{C}\{0,0;x_t,t\}} d_W x(\tau) \exp\left\{-\int_0^t d\tau\, p(\tau)x^2(\tau)\right\}. \tag{1.1.112}$$

In the same manner as example 1.4 was solved, we can show (after rather cumbersome calculations) that the Gelfand–Yaglom method now gives

$$I_4^{\text{cond}} = \frac{1}{\sqrt{\pi \widetilde{D}(0)}} \exp\left\{-x_t^2 \frac{D(0)}{\widetilde{D}(0)}\right\} \tag{1.1.113}$$

where $D(\tau)$ and $\widetilde{D}(\tau)$ satisfy the differential equations

$$\begin{cases} \dfrac{d^2 D(\tau)}{d\tau^2} = p(\tau)D(\tau) \\ D(t) = 1 \\ \left.\dfrac{dD}{d\tau}\right|_{\tau=t} = 0 \end{cases} \tag{1.1.114}$$

and

$$\begin{cases} \dfrac{d^2 \widetilde{D}(\tau)}{d\tau^2} = p(\tau)\widetilde{D}(\tau) \\ \widetilde{D}(t) = 0 \\ \left.\dfrac{d\widetilde{D}}{d\tau}\right|_{\tau=t} = -1. \end{cases} \tag{1.1.115}$$

For the special case $p(s) = k^2$, equations (1.1.114) and (1.1.115) can be straightforwardly solved. We suggest this to the readers as an exercise (problem 1.1.11, page 55). The result is

$$I_4^{\text{cond}}(p(s) = k^2) = \sqrt{\frac{k}{\pi \sinh(kt)}} e^{-kx_t^2 \coth(kt)}. \tag{1.1.116}$$

We shall calculate this path integral by one more method in section 1.2.6 (cf (1.2.130) and (1.2.131)) for an arbitrary time period and an arbitrary initial point (though this is not difficult to do by the Gelfand–Yaglom method as well).

One can check in this particular case that the general relation (1.1.57) is indeed fulfilled:

$$I_4(p(\tau)) = \int_{-\infty}^{\infty} dx_t\, I_4^{\text{cond}}(p(\tau)) = \int_{-\infty}^{\infty} dx_t\, \frac{1}{\sqrt{\pi \widetilde{D}(0)}} \exp\left\{-x_t^2 \frac{D(0)}{\widetilde{D}(0)}\right\}$$

$$= \frac{1}{\sqrt{D(0)}}.$$

Note also that

$$I_4^{\text{cond}}(p(\tau)) \xrightarrow[p(\tau)\to 0]{} \frac{1}{\sqrt{\pi t}} e^{-x_t^2/t}$$

i.e. the path integral in this limit equals the fundamental solution of the diffusion equation or transition probability, as it must because in this limiting case the integral (1.1.112) is reduced just to the conditional Wiener measure.

1.1.5 Change of variables in path integrals

In the case of usual finite-dimensional integrals, changes of integration variables prove to be a powerful method of simplification and sometimes calculation of the integrals. The possibilities of functional changes in path integrals are more restricted. Nevertheless, in some cases, they allow us to reduce certain types of path integral to simpler ones and to calculate them in this way. In the present subsection we shall consider general aspects of functional changes of integration variables and illustrate their possible application by one example. Other applications of this technique pertain to the next subsections.

◇ **General approach to functional change of variables in path integrals**

Recall that in the case of N-dimensional Lebesgue or Riemann integrals, after the substitution of the integration variables $\{x_1, \ldots, x_N\} \longrightarrow \{y_1, \ldots, y_N\}$, such that

$$x_i = x_i(y_1, \ldots, y_N) \qquad i = 1, \ldots, N$$

there appears the Jacobian $J = \frac{\partial(x_1, \ldots, x_N)}{\partial(y_1, \ldots, y_N)}$:

$$\int_{a_1}^{b_1} \cdots \int_{a_N}^{b_N} dx_1 \cdots dx_N \, f(x_1, \ldots, x_N) = \int_{a'_1}^{b'_1} \cdots \int_{a'_N}^{b'_N} dy_1 \cdots dy_N \, J \tilde{f}(y_1, \ldots, y_N)$$

$$\tilde{f}(y_1, \ldots, y_N) = f(x_1(y_1), \ldots, x_N(y_N))$$
$$a_i = x_i(a'_1, \ldots, a'_N)$$
$$b_i = x_i(b'_1, \ldots, b'_N).$$

For example, for the simple substitution $x_i = k_i y_i$ with the constant coefficients k_i, the Jacobian is

$$J = \prod_{i=1}^{N} k_i.$$

It is obvious that in the limit $N \to \infty$ the Jacobian becomes zero (if all $k_i < 1$) or infinite (if all $k_i > 1$) and thus it is ill defined even for such a simple substitution.

However, there exist functional substitutions which lead to a finite Jacobian in the Wiener integral. One possibility is provided by the *Fredholm integral equation* of the second kind:

$$y(t) = x(t) + \lambda \int_a^b ds \, K(t, s) x(s) \qquad (1.1.117)$$

where $K(t, s)$ is a given function of t and s and is called the *kernel* of the integral equation. Let us remember some facts from the theory of integral equations (see, e.g., Tricomi (1957) and Korn and Korn (1968)).

The solution $x(t)$ of (1.1.117) can be represented in the form

$$x(t) = y(t) - \lambda \int_a^b ds \, R(t, s; \lambda) y(s). \qquad (1.1.118)$$

The so-called *resolvent kernel* $R(t, s; \lambda)$ satisfies the equation

$$R(t, s; \lambda) + \lambda \int_a^b du \, K(t, u) R(u, s; \lambda) = K(t, s).$$

The resolvent kernel $R(t, s; \lambda)$ can be expressed in terms of two series in the parameter λ:

$$R(t, s; \lambda) = \frac{D(t, s; \lambda)}{D(\lambda)} \qquad (1.1.119)$$

where

$$D(\lambda) = \sum_{k=0}^{\infty} \frac{\lambda^k}{k!} C_k \qquad (1.1.120)$$

$$D(t, s; \lambda) = \sum_{k=0}^{\infty} \frac{\lambda^k}{k!} D_k(t, s)$$

with C_k, $D_k(t, s)$ defined by the recursion relations

$$C_k = \int_a^b ds\, D_{k-1}(s, s) \qquad k = 1, 2, \ldots$$

$$D_k(t, s) = C_k K(t, s) - k \int_a^b dr\, K(t, r) D_{k-1}(r, s) \qquad k = 1, 2, \ldots$$

with the initial values

$$C_0 = 1$$
$$D_0(t, s) = K(t, s).$$

We shall interpret the integral equation (1.1.117) as a functional substitution $y(t) \to x(t)$.

The particular case of the Fredholm equation with $b \to t$, called the *Volterra integral equation*, proves to be especially important for functional changes of variables in the Wiener path integral. The Volterra equation reads

$$y(t) = x(t) + \lambda \int_a^t ds\, K(t, s) x(s). \qquad (1.1.121)$$

This equation is obtained from (1.1.117) for the integral kernel $K(t, s)$ satisfying the condition (see figure 1.10)

$$K(t, s) = 0 \qquad \text{for} \qquad s > t.$$

If the kernel $K(t, s)$ does not depend on t in equation (1.1.121), the resulting equation,

$$y(t) = x(t) + \lambda \int_a^t ds\, K(s) x(s) \qquad (1.1.122)$$

is equivalent to the differential one

$$\dot{y}(t) = \dot{x}(t) + \lambda K(t) x(t). \qquad (1.1.123)$$

This special case is the most important for the applications in this subsection. Note that, in the case of equation (1.1.122), the kernel $K(t, s)$ must be discontinuous. Otherwise, it proves to be zero on the diagonal $K(t, t) = 0$ and hence equation (1.1.122) becomes trivial. Since variation of values of a function at a single point (that is on a set of measure zero) does not change the value of an integral, we can define the value of K at $s = t$ in different ways. The most natural way is to define it through averaging:

$$K(t) = \frac{\lim_{s \to t-0} K(t, s) + \lim_{s \to t+0} K(t, s)}{2} = \frac{\lim_{s \to t-0} K(t, s)}{2}. \qquad (1.1.124)$$

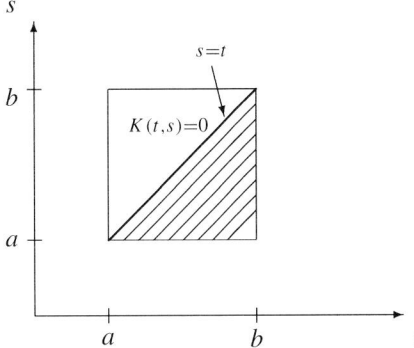

Figure 1.10. Domain of non-zero values of the kernel $K(t,s)$ for the Volterra equation (lower right hatched triangle) and of its zero values (upper left empty triangle).

The Jacobian $J(\lambda)$ of the transformation (1.1.122) can be calculated easily with the help of the discrete approximation of this relation:

$$y_i = x_i + \lambda(K_1 x_1 + \cdots + K_{i-1} x_{i-1} + \tfrac{1}{2} K_i x_i)\varepsilon. \tag{1.1.125}$$

Here $i = 1, \ldots, N$ and $\varepsilon = (b-a)/N$. The Jacobian of (1.1.125)

$$J_N(\lambda) = \begin{vmatrix} 1 + \lambda K_1 \varepsilon/2 & 0 & \cdots & \cdots & 0 \\ \lambda K_1 \varepsilon & 1 + \lambda K_2 \varepsilon/2 & 0 & \cdots & \vdots \\ \vdots & \vdots & \ddots & \cdots & 0 \\ \lambda K_1 \varepsilon & \lambda K_2 \varepsilon & \cdots & \cdots & 1 + \lambda K_N \varepsilon/2 \end{vmatrix}$$

is equal to the product of the diagonal elements due to its triangular structure:

$$J_N(\lambda) = \prod_{i=1}^{N}(1 + \lambda K_i \varepsilon/2) = \prod_{i=1}^{N} \exp\{\lambda K_i \varepsilon/2 + \mathcal{O}(\varepsilon^2)\} \tag{1.1.126}$$

so that

$$J(\lambda) = \lim_{N \to \infty} J_N(\lambda) = \exp\left\{\frac{\lambda}{2} \int_a^b ds\, K(s)\right\}. \tag{1.1.127}$$

It is seen that the value of the Jacobian and, hence, the value of the path integral obtained after the change of variables depends on a choice of the prescription for $K(s)$ at $s = t$. In the discrete-time approximation, the chosen definition (1.1.124) through averaging is equivalent to the so-called *midpoint prescription*. We shall discuss this prescription in more detail in sections 2.2.5 and 2.5.1 in the framework of quantum-mechanical path integrals. At the moment, we only note that we can *a posteriori* justify this choice by the fact that transition probabilities obtained by a change of variables with this prescription satisfy the normalization condition (see, e.g., section 1.2.2 and example 1.8, page 84).

◇ An example of application of the functional changes of variables

Now we shall apply the transformation (1.1.125) for the calculation of the path integral I_3 from example 1.4, page 41, by another method, namely, by making use of the change of variables.

To this aim, we start from the (unconditional) Wiener integral (on the standard time interval $[0, t]$)

$$\int_{C\{0,0;t\}} d_W y(\tau) = 1 \qquad (1.1.128)$$

and use the substitution corresponding to the Volterra equation (1.1.125) of the special form

$$y(\tau) = x(\tau) - \int_0^\tau ds \, \frac{\dot{D}(s)}{D(s)} x(s) \qquad (1.1.129)$$

where $D(s)$ is the solution of the Cauchy problem:

$$\frac{d^2 D(s)}{ds^2} + p(s) D(s) = 0$$
$$D(t) = 1 \qquad (1.1.130)$$
$$\left.\frac{dD}{ds}\right|_{s=t} = 0.$$

The result of the substitution is

$$\int_{C\{0,0;t\}} \prod_{\tau=0}^t \frac{dx(\tau)}{\sqrt{\pi d\tau}} \exp\left\{-\int_0^t d\tau \left(\dot{x} - \frac{\dot{D}}{D} x\right)^2\right\} \exp\left\{-\frac{1}{2}\int_0^t d\tau \, \frac{\dot{D}}{D}\right\} = 1 \qquad (1.1.131)$$

(the dot means time derivative; do not confuse $D(s)$ here with the diffusion constant which we have chosen in this section to be $1/4$). The second exponential in the integrand is the Jacobian. Let us calculate it explicitly:

$$\exp\left\{-\frac{1}{2}\int_0^t d\tau \, \frac{\dot{D}}{D}\right\} = \exp\left\{-\frac{1}{2}\int_0^t d\ln D(\tau)\right\}$$
$$= \exp\left\{-\frac{1}{2}(\ln D(t) - \ln D(0))\right\} = \sqrt{D(0)}$$

having taken into account the boundary condition (1.1.130). The first exponent in (1.1.131) can be transformed as follows:

$$\int_0^t d\tau \left(\dot{x} - \frac{\dot{D}}{D} x\right)^2 = \int_0^t d\tau \, \dot{x}^2 - \int_0^t dx^2 \, \frac{\dot{D}}{D} + \int_0^t d\tau \left(\frac{\dot{D}}{D}\right)^2 x^2$$
$$= \int_0^t d\tau \, \dot{x}^2 - \left.\frac{\dot{D}}{D} x^2\right|_{\tau=0}^{\tau=t} + \int_0^t d\tau \left[\frac{d}{d\tau}\left(\frac{\dot{D}}{D}\right)\right] x^2 + \int_0^t d\tau \left(\frac{\dot{D}}{D}\right)^2 x^2$$
$$= \int_0^t d\tau \, \dot{x}^2 + \int_0^t \frac{\ddot{D} D - (\dot{D})^2}{D^2} x^2 + \int_0^t d\tau \left(\frac{\dot{D}}{D}\right)^2 x^2$$
$$= \int_0^t d\tau \, \dot{x}^2 + \int_0^t d\tau \, \frac{\ddot{D}}{D} x^2.$$

Recalling that $D(s)$ satisfies equation (1.1.130), we have

$$\int_0^t d\tau \left(\dot{x} - \frac{\dot{D}}{D} x\right)^2 = \int_0^t d\tau \, \dot{x}^2 + \int_0^t d\tau \, p(\tau) x^2(\tau)$$

so that the equality (1.1.131) takes the form

$$\sqrt{D(0)} \int_{\mathcal{C}\{0,0;t\}} d_W x(\tau) \exp\left\{\int_0^t d\tau\, p(\tau) x^2(\tau)\right\} = \sqrt{D(0)} I_4 = 1$$

which exactly corresponds to the result of example 1.4, i.e. $I_4 = 1/\sqrt{D(0)}$.

1.1.6 Problems

Problem 1.1.1. Derive the diffusion equation for the so-called *Bernoullian random walk* for which the probabilities p and q of left- and right-hand moves of the Brownian particle on a line are different ($p \neq q$, $p + q = 1$). Compare the solution of this equation for the initial condition $w(x, 0) = \delta(x)$ with that of equation (1.1.18).

Hint. The matrices R and L of the right- and left-hand shifts (cf (1.1.9)) must be substituted now by qR and pL, respectively. Calculations analogous to those in equations (1.1.11)–(1.1.15) now give

$$w(x, t + \varepsilon) = pw(x + \ell, t) + qw(x - \ell, t).$$

The continuous limit may be applied as before, i.e. $\ell \to 0$, $\varepsilon \to 0$, $\ell^2(2\varepsilon)^{-1} \to D$ (the diffusion constant). However, to avoid misbehaviour of the distribution function $w(x, t)$, we require that the following limit

$$v = \lim_{\substack{\ell \to 0 \\ \varepsilon \to 0}} \frac{\ell}{\varepsilon}(p - q)$$

be a finite quantity. As a result, the ordinary diffusion equation is replaced by

$$\frac{\partial w_v(x, t)}{\partial t} = \left(D\frac{\partial^2}{\partial x^2} + v\frac{\partial}{\partial x}\right) w_v(x, t).$$

The parameter v plays the role of a mean *drift velocity*. It is easy to check that by the Galilean transformation

$$x' = x - vt$$

this equation is reduced to the diffusion equation (1.1.18). This means that the solution $w_v(x, t)$ of this equation is related to that of (1.1.28) as follows:

$$w_v(x, t) = w(x - vt, t).$$

Problem 1.1.2. Derive the evolution of the probability density for a Brownian particle wandering in a multidimensional space and the corresponding diffusion equation (1.1.19) using the generalization of the discrete transition operator W used in section 1.1.1 for the one-dimensional case.

Hint. Consider the transition operator $\widehat{W}$ in the d-dimensional space (cf (1.1.8) and (1.1.11))

$$\widehat{W}w(\boldsymbol{x}) \stackrel{\text{def}}{=} \frac{1}{2d}\sum_{k=1}^{d}\{w(\boldsymbol{x} + \ell\boldsymbol{e}_k) + w(\boldsymbol{x} - \ell\boldsymbol{e}_k)\} \qquad (1.1.132)$$

where $\boldsymbol{e}_k$ ($k = 1, \ldots, d$) are vectors of an orthonormal basis in $\mathbb{R}^d$.

The ordinary space is now represented by the discrete variables $\boldsymbol{x} = \{n_1\ell, n_2\ell, \ldots, n_d\ell\}$, $n_I \in \mathbb{Z}$. Making use of the Fourier transformation

$$w(\boldsymbol{x}) = \int_{-\pi/\ell}^{\pi/\ell} dp_1\, dp_2 \cdots dp_d\, e^{i\boldsymbol{px}}\, \tilde{w}(\boldsymbol{p})$$

$$\tilde{w}(\boldsymbol{p}) = \frac{\ell}{2\pi} \sum_{n_I \in \mathbb{Z}} e^{-i\boldsymbol{px}} w(\boldsymbol{x})$$

$$\boldsymbol{px} \stackrel{\text{def}}{\equiv} p_1 x_1 + \cdots + p_d x_d$$

we can diagonalize the operator $\widehat{W}$ in (1.1.132):

$$\widehat{W}\tilde{w}(\boldsymbol{p}) = \lambda(\boldsymbol{p})\tilde{w}(\boldsymbol{p})$$

$$\lambda(\boldsymbol{p}) = \frac{1}{d}\sum_{k=1}^{d} \cos(p_k \ell)$$

so that after N steps we obtain

$$\widehat{W}(\boldsymbol{x}, N\varepsilon) \stackrel{\text{def}}{\equiv} \widehat{W}^N(\boldsymbol{x}) = c \int_{-\pi/\ell}^{\pi/\ell} d^d p\, e^{i\boldsymbol{px}} \lambda^N(\boldsymbol{p})$$

where c is the normalization constant. The power of the eigenvalues of $\widehat{W}$ can be estimated as follows:

$$\lambda^N(\boldsymbol{p}) = \exp\left\{N \ln\left(\frac{1}{d}\sum_{k=1}^{d}\cos(p_k\ell)\right)\right\}$$

$$= \exp\left\{\frac{t}{\varepsilon} \ln\left(1 - \frac{\ell^2}{2d}\boldsymbol{p}^2 + \mathcal{O}(\ell^4)\right)\right\}$$

$$= \exp\{-Dt\boldsymbol{p}^2 + \mathcal{O}(\ell^2)\}$$

where ε is the time discretization scale, $\boldsymbol{p}^2 \stackrel{\text{def}}{\equiv} \sum_k p_k^2$ and D is the generalization of (1.1.17) given now by

$$D = \lim_{\substack{\ell \to 0 \\ \varepsilon \to 0}} \frac{\ell^2}{2\varepsilon d}.$$

In the continuous limit, $\ell, \varepsilon \to 0$, using the inverse Fourier transformation and the normalization condition, we find that the multidimensional probability density at a time t has the form

$$w(\boldsymbol{x}, t) = (2\pi)^{-d} \int_{-\infty}^{\infty} d^d p\, e^{i\boldsymbol{px}} e^{-Dt\boldsymbol{p}^2}$$

$$= (4\pi Dt)^{-d/2} \exp\left\{-\frac{\boldsymbol{x}^2}{4Dt}\right\}.$$

and satisfies the d-dimensional diffusion equation (1.1.19).

Problem 1.1.3. Show that *the Poisson distribution*

$$P_n(t) = e^{-\lambda t} \frac{(\lambda t)^n}{n!}$$

satisfies the semigroup property (1.1.46) (i.e. the ESKC relation (1.1.45)), where the composition law (cf (1.1.46)) is defined as

$$\sum_{k=0}^{n} P_k(\tau) P_{n-k}(t-\tau) = \sum_{k=0}^{n} e^{-\lambda \tau} \frac{(\lambda \tau)^k}{k!} e^{-\lambda(t-\tau)} \frac{(\lambda(t-\tau))^{n-k}}{(n-k)!}.$$

Thus the Poisson distribution defines a Markovian stochastic process with the transition probability $W(n,t|n_0,t_0) \equiv P_{n-n_0}(t-t_0)$. Of course, this is not a Gaussian and hence not a Wiener process.

Solution. We have to prove the equality:

$$W(k,\tau|0,0) * W(n,t|k,\tau) = W(n,t|0,0)$$

that is

$$\sum_{k=0}^{n} P_k(\tau) P_{n-k}(t-\tau) = P_n(t).$$

To this aim, we write:

$$\begin{aligned}\sum_{k=0}^{n} P_k(\tau) P_{n-k}(t-\tau) &= \sum_{k=0}^{n} e^{-\lambda \tau} \frac{(\lambda \tau)^k}{k!} e^{-\lambda(t-\tau)} \frac{(\lambda(t-\tau))^{n-k}}{(n-k)!} \\ &= e^{-\lambda t} \lambda^n \sum_{k=0}^{n} \frac{\tau^k (t-\tau)^{n-k}}{k!(n-k)!} \\ &= e^{-\lambda t} \frac{\lambda^n}{n!} \sum_{k=0}^{n} \frac{n!}{k!(n-k)!} \tau^k (t-\tau)^{n-k}. \end{aligned} \quad (1.1.133)$$

The last sum can be recognized as the right-hand side of the binomial formula:

$$[\tau + (t-\tau)]^n = \sum_{k=0}^{n} \frac{n!}{k!(n-k)!} \tau^k (t-\tau)^{n-k}$$

so that this sum is equal to t^n. Using this fact in (1.1.133), we obtain the required result:

$$\sum_{k=0}^{n} P_k(\tau) P_{n-k}(t-\tau) = e^{-\lambda t} \lambda^n \frac{t^n}{n!} = P_n(t).$$

Problem 1.1.4. Show that the transition probability (1.1.44) is indeed the *fundamental solution* (*Green function*) of the diffusion equation, i.e. it satisfies equation (1.1.42).

Solution. If $t > 0$, both the function (1.1.44) and the equation (1.1.42) coincide with the solution (1.1.28) and the homogeneous diffusion equation (1.1.18), respectively. Hence in this domain the function (1.1.44) satisfies (1.1.42).

Consider some test-function $f \in \mathcal{D}(\mathbb{R}^2)$, where $\mathcal{D}(\mathbb{R}^2)$ is the set of compactly supported (i.e. vanishing outside a compact domain) functions of t and x. Using the definition of the derivative of distributions (generalized functions), the condition

$$W(x,t) = 0 \qquad \text{if } t < 0$$

and integrating by parts, we obtain

$$\int_{-\infty}^{\infty} dt \int_{-\infty}^{\infty} dx\, f(x,t) \left(\frac{\partial W(x,t)}{\partial t} - D\frac{\partial^2 W(x,t)}{\partial x^2} \right)$$

$$= -\int_{0}^{\infty} dt \int_{-\infty}^{\infty} dx\, W(x,t) \left(\frac{\partial f(x,t)}{\partial t} + D\frac{\partial^2 f(x,t)}{\partial x^2} \right)$$

$$= -\lim_{\varepsilon \to +0} \int_{\varepsilon}^{\infty} dt \int_{-\infty}^{\infty} dx\, W(x,t) \left(\frac{\partial f(x,t)}{\partial t} + D\frac{\partial^2 f(x,t)}{\partial x^2} \right)$$

$$= \lim_{\varepsilon \to +0} \left[\int_{-\infty}^{\infty} dx\, W(x,\varepsilon) f(x,\varepsilon) + \int_{\varepsilon}^{\infty} dt \int_{-\infty}^{\infty} dx\, f(x,t) \left(\frac{\partial W(x,t)}{\partial t} - D\frac{\partial^2 W(x,t)}{\partial x^2} \right) \right]$$

$$= \lim_{\varepsilon \to +0} \int_{-\infty}^{\infty} dx\, W(x,\varepsilon) f(x,0) + \lim_{\varepsilon \to +0} \int_{-\infty}^{\infty} dx\, W(x,\varepsilon)[f(x,\varepsilon) - f(x,0)]$$

$$= \lim_{\varepsilon \to +0} \int_{-\infty}^{\infty} dx\, W(x,\varepsilon) f(x,0).$$

In this derivation, we have used the fact that $W(x,t)$ at $t > 0$ is the solution of the homogeneous diffusion equation (1.1.33), while the last equality is a consequence of the relation

$$\left| \lim_{\varepsilon \to +0} \int_{-\infty}^{\infty} dx\, W(x,\varepsilon)[f(x,\varepsilon) - f(x,0)] \right| \leq \text{constant} \cdot \varepsilon \int_{-\infty}^{\infty} dx\, W(x,\varepsilon) = \text{constant} \cdot \varepsilon$$

(in the latter equation we have used the normalization condition (1.1.29)). Since

$$W(x,t) \xrightarrow[t \to 0]{} \delta(x)$$

according to the characteristic property (1.1.63), we obtain the equality

$$\int_{-\infty}^{\infty} dt \int_{-\infty}^{\infty} dx\, f(x,t) \left(\frac{\partial W(x,t)}{\partial t} - D\frac{\partial^2 W(x,t)}{\partial x^2} \right) = f(0,0)$$

which is equivalent to equation (1.1.44) by the definition of the δ-function.

Another way to reach the same result is simply to use the *known* result (which, in fact, we have proven earlier!) from the theory of distributions (distributions here mean generalized functions)

$$\frac{d}{dt}\theta(t) = \delta(t).$$

Then it is easy to see that if $\widetilde{w}(\lambda,t)$ satisfies (1.1.24), the function $\theta(t)\widetilde{w}(\lambda,t)$ is the solution of

$$\frac{\partial}{\partial t}(\theta(t)\widetilde{w}(\lambda,t)) = -D\lambda^2 \theta(t)\widetilde{w}(\lambda,t) - \delta(t)\widetilde{w}(\lambda,t). \qquad (1.1.134)$$

Taking into account that the solution (1.1.25) of the initial equation at the origin has the value $\widetilde{w}(\lambda,0) = 1/2\pi$, equation (1.1.134) can be rewritten in the form

$$\dot{\widetilde{w}}(\lambda,t) = -D\lambda^2 \widetilde{w}(\lambda,t) - \frac{1}{2\pi}\delta(t)$$

which is equivalent to (1.1.42) (after the Fourier transform).

Problem 1.1.5. Prove the Wiener theorem for differentiable functions: the set of differentiable functions has zero measure.

Hint. Formulate the condition of differentiability as follows:
$$(x(t_2) - x(t_1))|_{t_2 \to t_1} \leq h(t_2 - t_1)$$
and, using as a guide the proof of the Wiener theorem for discontinuous functions, show that the path integral over the corresponding characteristic function of the form (1.1.66) is reduced to the one-dimensional integral
$$\frac{2}{\sqrt{\pi}} \int_0^{h(\sqrt{2}/A)^m} d\eta \, e^{-\eta^2}$$
(similar to that in (1.1.68)), where $A > \sqrt{2}$ and m is an arbitrary positive integer. Hence this integral vanishes in the limit $m \to \infty$.

Problem 1.1.6. Calculate the multidimensional Gaussian integral (1.1.85)
$$\int_{-\infty}^{\infty} dx_1 \cdots dx_N \exp\left\{ -\sum_{i,j}^{N} b_{ij} x_i x_j \right\}$$
where $\mathbf{b} = (b_{ij})$ is an arbitrary positive definite, real symmetric matrix.

Solution. Since $\mathbf{b}$ is a real symmetric matrix, it can be diagonalized by an orthogonal transformation C_{ij}:
$$x_i = \sum_{j=1}^{N} C_{ij} y_j \qquad \det \mathbf{C} = 1 \qquad (1.1.135)$$
so that
$$\mathbf{b} \longrightarrow \mathbf{B} = \mathbf{C}^\top \mathbf{b} \mathbf{C} \qquad B_{ij} = \lambda_i \delta_{ij}.$$
In these new variables y_i, the integral is reduced to the product of the standard Gaussian integral (1.1.27)
$$\int_{-\infty}^{\infty} dy_1 \cdots dy_N \exp\left\{ -\sum_{i}^{N} \lambda_i y_i^2 \right\} = \prod_{i=1}^{N} \int_{-\infty}^{\infty} dy_i \exp\{-\lambda_i y_i^2\} = \frac{(\sqrt{\pi})^N}{\sqrt{\lambda_1 \cdots \lambda_N}}. \qquad (1.1.136)$$

The product $\lambda_1 \cdots \lambda_N$ of the diagonal elements of the matrix $\mathbf{B}$ is obviously equal to its determinant, which, due to the relations (1.1.135), is equal to the determinant of the matrix $\mathbf{b}$. Thus for the n-dimensional Gaussian integral we obtain
$$\int_{-\infty}^{\infty} dx_1 \cdots dx_N \exp\left\{ -\sum_{i,j}^{N} b_{ij} x_i x_j \right\} = \frac{(\sqrt{\pi})^N}{\sqrt{\det \mathbf{b}}}. \qquad (1.1.137)$$

Problem 1.1.7. Calculate (1.1.82) performing the integrations over $x_1, x_2, \ldots, x_N$ one by one.

Hint. Use the formula
$$\int_{-\infty}^{\infty} dx \, \exp\{-a(x-x')^2 - b(x-x'')^2\} = \left(\frac{\pi}{a+b}\right)^{1/2} \exp\left\{-\frac{ab}{a+b}(x'-x'')^2\right\}$$

54   *Path integrals in classical theory*

which reflects the semigroup property (ESKC relation) of such Gaussian distributions.

Problem 1.1.8. Calculate the Wiener integral over trajectories from the set $C\{0, 0; t\}$ (i.e. with the unconditional measure) with the functional

$$F[x(\tau)] = x(s)x(\rho)$$

where s and ρ are two fixed moments of time with $0 < s < \rho < t$.

Solution. Manipulations analogous to those in example 1.2, page 39 give

$$\begin{aligned}
I_2 &= \int_{C\{0,0;t\}} d_W x(\tau)\, x(s)x(\rho) \\
&= \int_{-\infty}^{\infty} dx_s\, dx_\rho\, x_s x_\rho \int_{C\{0,0;x_s,s\}} d_W x(\tau) \int_{C\{x_s,s;x_\rho,\rho\}} d_W x(\tau) \int_{C\{x_\rho,\rho;t\}} d_W x(\tau) \\
&= \int_{-\infty}^{\infty} dx_s\, dx_\rho\, x_s x_\rho \frac{1}{\sqrt{\pi s}} e^{-\frac{x_s^2}{s}} \frac{1}{\sqrt{\pi(\rho-s)}} e^{-\frac{(x_\rho - x_s)^2}{\rho - s}} = \frac{s}{2}.
\end{aligned}$$

(Another possibility is to use the discrete approximation as in example 1.2.)

The path integral I_2 is in fact the correlation function

$$\langle x(s) x(\rho) \rangle_W \stackrel{\text{def}}{\equiv} \int_{C\{0,0;t\}} d_W x(\tau)\, x(s) x(\rho) \tag{1.1.138}$$

with the average taken over the Wiener measure $d_W x(\tau)$, so that we could write in general, for both $s < \rho$ and $\rho < s$,

$$\langle x(s)x(\rho) \rangle_W = \tfrac{1}{2} \min(s, \rho). \tag{1.1.139}$$

Problem 1.1.9. We have stressed that the Brownian motion is a homogeneous stochastic process, that is, its characteristics do not depend on the overall shifting of time variables. However, the correlation function (1.1.138) found in the preceding problem looks, at first sight, as if it would be *not* invariant with respect to such a time shifting (see (1.1.139)). Explain this apparent contradiction.

Hint. Calculate the correlation function for an *arbitrary* initial time t_0, that is, calculate the following path integral:

$$\int_{C\{0,t_0;t\}} d_W x(\tau)\, x(s)x(\rho) \qquad t_0 < s < \rho < t.$$

Problem 1.1.10. Calculate the conditional integral (1.1.57) for the functional $F[x(\tau)] = F(x(s)) = x(s)$, where s is a fixed point in the interval $[0, t]$.

Solution. Making use of the ESKC relation (1.1.58) in the same manner as in example 1.2, we obtain

$$\begin{aligned}
I_2^{\text{cond}} &= \int_{C\{0,0;x_t,t\}} d_W x(\tau)\, x(s) \\
&= \int_{-\infty}^{\infty} dx_s\, x_s \int_{C\{0,0;x_s,s\}} d_W x(\tau) \int_{C\{x_s,s\ x_t,t\}} d_W x(\tau) \tag{1.1.140}
\end{aligned}$$

so that (1.1.53) gives

$$I_2^{\text{cond}} = \int_{-\infty}^{\infty} dx_s \, x_s \frac{e^{-\frac{x_s^2}{s}}}{\sqrt{\pi s}} \frac{e^{-\frac{(x_t-x_s)^2}{t-s}}}{\sqrt{\pi(t-s)}}$$

$$= \frac{x_t s}{\sqrt{\pi} t^{3/2}} e^{-\frac{x_t^2}{t}}. \qquad (1.1.141)$$

Recall that the corresponding unconditional path integral is equal to zero (cf example 1.2).

Problem 1.1.11. Calculate the integrals I_4 and I_4^{cond}, defined in (1.1.102) and (1.1.112), respectively, by the Gelfand–Yaglom method for the particular case of the function $p(\tau): p(\tau) = k^2 = \text{constant}$.

Hint. The general solution of the equation

$$\frac{d^2 f}{d\tau^2} = k^2 f$$

is

$$f = C_1 e^{k\tau} + C_2 e^{-k\tau}.$$

The boundary conditions (1.1.114) and (1.1.115) give

$$D(0) = \cosh(kt)$$
$$\tilde{D}(0) = \frac{1}{k} \sinh(kt).$$

Substitution of these expressions into (1.1.113) results in (1.1.116). For the unconditional integral, the formula (1.1.111) gives $I_4 = 1/\sqrt{\cosh(kt)}$.

Problem 1.1.12. Solve the Fredholm equation (1.1.117) for $K(t,s) \equiv 1, a = 0, b = 1$.

Solution. Equation (1.1.117) now reads

$$y(t) = x(t) + \lambda \int_0^1 ds \, x(s).$$

Using (1.1.120) and (1.1.121) or directly solving the equation for the resolvent kernel in this simple case, we obtain

$$R = \frac{1}{1+\lambda}$$

and the solution (1.1.118) has the form

$$x(t) = y(t) - \frac{\lambda}{1+\lambda} \int_0^1 ds \, y(s).$$

1.2 Wiener path integrals and stochastic processes

In this section, after a short excursion into the theory of stochastic processes, we shall consider the path-integral method of analysis of basic stochastic equations and some applications to physical processes (a Brownian particle with inertia and a system of interacting Brownian particles). The central point of this section is the *Feynman–Kac formula*, expressing transition probabilities for a wide class of stochastic processes in terms of path integrals. Also, we shall discuss the Wiener path integral in the space of velocities and the Wiener path integral with constraints which, besides its direct meaning, serves as the prototype for the extremely fruitful application of path-integral techniques in gauge-field theories (see chapter 3). We shall continue the study of methods of calculations for path integrals, in particular, the semiclassical approximation and the Fourier mode expansion. One more practically very useful object which will be introduced in this section is the characteristic (generating) functional for Wiener processes. Its analog plays an essential role in the path-integral formulation of quantum field theories (see chapter 3). In conclusion of this section, we shortly consider an instructive and physically important application of path integrals in polymer physics, which illustrates the fact that the path-integral methods developed for the description of Brownian motion can be successfully applied to quite different physical problems.

1.2.1 A short excursion into the theory of stochastic processes

Recall that for the description of a random event in probability theory we introduce (see, e.g., Doob (1953), Feller (1951, 1961) and Gnedenko (1968)) a suitable 'probability space' (space of elementary events) Ω, i.e. the set of all possible realizations of a given phenomenon and to any subset $A \subset \Omega$ we ascribe a non-negative number $\mathbb{P}\{A\}$, the probability of the event A. In particular, the probability space which describes the Brownian motion of a particle consists of all possible trajectories and ascribing probabilities to their subsets leads to the construction of the Wiener measure as we discussed in the preceding section. Mathematically this situation is described via the introduction of a *random variable* ξ, defined as a function $\xi = \xi(v)$, $v \in \Omega$, on a probability space Ω. In fact, random variables serve as coordinates on spaces of elementary events Ω and, in the same way as coordinates of any space, put into correspondence with a point of a space a real number (if this is a one-dimensional space) or a set of numbers (for higher-dimensional spaces). Analogously to the freedom in the choice of coordinate frames for an arbitrary space, a set of (basic) random variables on Ω can be defined in many different ways.

◇ **Distribution, partition function and probability density**

If ξ takes a discrete set of values (finite or countable) $x_1, x_2, \ldots, x_N, \ldots$, then we can introduce the set of probabilities

$$p_k = \mathbb{P}\{v|\xi(v) = x_k\} \qquad k = 1, 2, \ldots, N \qquad (1.2.1)$$

called a *probability distribution* for the values of a random variable ξ. As we have already noted in the preceding section, probability is closely related to (or in fact defines) the measure on a space of elementary events. Practically, it is constructed using *a partition function*

$$F_\xi(x) = \mathbb{P}\{v|\xi(v) \leq x\} \qquad -\infty < x < \infty \qquad (1.2.2)$$

so that the probability of a random variable ξ to be in an interval $[A, B]$ is expressed in terms of the Lebesgue–Stiltjes integral:

$$\mathbb{P}\{v|A \leq \xi(v) \leq B\} = \int_A^B dF_\xi(x) \qquad (1.2.3)$$

(such a function, of course, can be introduced for discrete sets of values of the random variable, and in this case $F_\xi = \sum_{k; x_k \leq x} p_k$). If it is desirable to consider a few random variables simultaneously (for example, three coordinates $x(t), y(t), z(t)$ of a Brownian particle at a moment t), we introduce their joint distribution function

$$F_{\xi_1,\ldots,\xi_m}(x_1,\ldots,x_m) = \mathbb{P}\{v|\xi_1(v) \leq x_1, \ldots, \xi_m(v) \leq x_m\}. \tag{1.2.4}$$

In the case of a continuous distribution, the integral (1.2.3) turns into the Riemann integral

$$\mathbb{P}\{v|A < \xi(v) < B\} = \int_A^B dx\, w(x) \tag{1.2.5}$$

where the positive function $w(x) \geq 0$ is the *probability density*, which is related to the partition function as

$$F_\xi(x) = \int_{-\infty}^x dy\, w(y). \tag{1.2.6}$$

If the partition function is differentiable, the inverse formula is

$$w(x) = \frac{\partial F_\xi(x)}{\partial x}. \tag{1.2.7}$$

Partition functions and probability densities should obviously satisfy the normalization condition

$$\int_{-\infty}^\infty dF_\xi(x) = \int_{-\infty}^\infty dx\, w(x) = 1. \tag{1.2.8}$$

A probability density of a few random variables is defined analogously.

Thus we can say that from the mathematical point of view, a space of elementary events Ω together with a set of random variables (coordinates) and a set of attributed probabilities defines the space with the probabilistic (i.e. subject to the normalization condition (1.2.8)) measure.

◇ **Definition and classification of stochastic processes**

A collection of random variables ξ_t defined on the same probability space Ω and labelled by the elements of some set $\mathcal{T}$ is called a *random* or *stochastic function*. For example, the coordinates of the Brownian particle are stochastic functions defined on the half line $[0, \infty)$ of a time variable. For the special sets $\mathcal{T}$, we use the following terminology:

- if $\mathcal{T}$ is the set of integers $\mathbb{Z}$, the stochastic function is called a *stochastic chain* or a *stochastic sequence*;
- if $\mathcal{T}$ is a multidimensional space $\mathbb{R}^d$, the stochastic function is called a *stochastic field*;
- if $\mathcal{T}$ is the line $\mathbb{R}$ or an interval, the stochastic function is called a *stochastic process*.

An example of a stochastic chain is the discrete-time approximation of the Brownian motion discussed in the preceding section, while its continuous limit corresponds to a stochastic process. In this section, we shall consider the latter object.

All the information about a stochastic process is contained in the joint probability distributions

$$w_N(\xi_1, t_1; \xi_2, t_2; \ldots; \xi_N, t_N) \tag{1.2.9}$$

for collections of random variables $\xi(t_1), \xi(t_2), \ldots, \xi(t_N)$ with arbitrary N. They define the *mathematical expectation* (or *mean value*) $\langle f(\xi_1, \ldots, \xi_N) \rangle$ of a function $f(\xi_1, \ldots, \xi_N)$ of random variables:

$$\langle f(\xi_1, \ldots, \xi_N) \rangle \stackrel{\text{def}}{\equiv} \int d\xi_1 \, d\xi_2 \cdots d\xi_N \, w_N(\xi_1, t_1; \ldots; \xi_N, t_N) f(\xi_1, \ldots, \xi_N) \qquad (1.2.10)$$

(with the obvious substitution of the integral by the sum in the case of a discrete distribution). For a deterministic process $\xi = f(t)$, the probability densities are expressed by δ-functions, e.g., $w_1(\xi, t) = \delta(\xi - f(t))$.

It is worth noting that an explicit form of the joint distribution w_N depends on the way of averaging under consideration, for example:

- in the case of the correlation function $\langle x(s)x(\rho) \rangle_W$ for Brownian motion with *arbitrary* final points (averaging over the unconditional Wiener measure), the joint distribution has the form (cf problem 1.1.8, page 54)

$$w_2 = W(x_s, s|x_0, 0) W(x_\rho, \rho|x_s, s)$$

- in the case of trajectories with a fixed final point (averaging over the conditional Wiener measure), the joint probability reads as

$$w_2 = W(x_s, s|x_0, 0) W(x_\rho, \rho|x_s, s) W(x_t, t|x_\rho, \rho).$$

However, in contrast to the discrete sets of random variables as in (1.2.9), in many cases it is more convenient to use the probability functional of $\omega[\xi(\tau)]$ of a random variable ξ, i.e. the continuous analog of the joint probability density (1.2.9). We proceeded along this way in the preceding section for discussing Brownian motion (by having considered the probability that the Brownian particle moves through infinitesimal gates dx along a given trajectory $x(t)$, cf (1.1.51)) and reached the notion of the Wiener path integral.

All these definitions are easily generalized to multicomponent stochastic processes $\boldsymbol{\xi}(t) = \{\xi_1(t), \ldots, \xi_m(t)\}$.

As we have already mentioned, there are several classes of stochastic processes which are of special importance in physics:

- *Stationary processes*: processes with joint (multipoint) probability characteristics which do not depend on the choice of the starting moment or, in other words, are invariant with respect to time translations. This means, in turn, that all characteristics depend only on time differences, in particular, $\langle \xi(t) \rangle = $ constant and $\langle \xi(t_2)\xi(t_1) \rangle = f(t_2 - t_1)$.
- *Markov processes*: processes for which *all* joint (multipoint) probability characteristics are expressed only in terms of a one-point probability density $w(\xi_t, t)$ and a two-point *transition probability* $W(\xi_{t_2}, t_2|\xi_{t_1}, t_1)$. The notion of transition probability implies that it satisfies the semigroup property (1.1.46) or, in the terminology of the theory of stochastic process, the ESKC relation (1.1.45). In more general words, this means that in a Markov process, the future depends on the past *only* through the present moment state.
- *Gaussian (normal) processes*: processes for which *all* joint (multipoint) probability characteristics are expressed only in terms of a first $w_1(\xi)$ and second (transition) $w_2(\xi_2, \xi_1) = W(\xi_2, t_2|\xi_1, t_1)$ probability density (and hence only in terms of the mean value $\langle \xi(t) \rangle$ and *correlation function* $\langle \xi(t_2)\xi(t_1) \rangle$), both joint probabilities having necessarily

Gaussian (called also *normal*) form,

$$w_1(\xi) \equiv w(\xi) = \frac{1}{\sqrt{2\pi\Lambda}} \exp\left\{-\frac{(\xi-a)^2}{2\Lambda}\right\} \tag{1.2.11}$$

$$w_2(\xi_2, \xi_1) = \frac{1}{\sqrt{2\pi \det(\lambda_{ij}^{-1})}} \exp\left\{-\frac{1}{2}\sum_{i,j=1}^{2} \lambda_{ij}(\xi_i - a_i)(\xi_j - a_j)\right\} \tag{1.2.12}$$

where $a, \Lambda, a_i, \lambda_{ij}$ ($i, j = 1, 2$) are the parameters of the distributions.

- *Wiener process*: a stationary *Markov* and *Gaussian* process with independent increments $\xi_i - \xi_{i-1}$ ($i = 1, \ldots, n$) for any instant of time $t_0 < t_1 < \cdots < t_n$, is called a *Wiener* stochastic process.

The reader may verify that the increments in the positions of the Brownian particle are indeed independent (cf the remark after (1.1.50) and problem 1.2.1). So, as follows from the discussion in the preceding section, Brownian motion is a Wiener process.

In what follows, we shall mention also the process called *white noise*:

- *White noise process*: a stationary stochastic process $\xi(t)$ with completely independent random variables at different moments, so that the correlation function is given by the δ-function

$$\langle \xi(t_2)\xi(t_1)\rangle = \Phi_0 \delta(t_2 - t_1) \tag{1.2.13}$$

with a constant *spectral density*

$$\Phi(\omega) = \int d(t_2 - t_1)\, \langle \xi(t_2)\xi(t_1)\rangle e^{i\omega(t_2-t_1)} = \Phi_0. \tag{1.2.14}$$

The latter property explains the adjective 'white': white light also has constant spectral density.

◇ **The Fokker–Planck equation for stochastic processes**

In the preceding section we derived the diffusion equations (1.1.3), (1.1.19) and (1.1.33) using natural physical assumptions and then convinced ourselves that the corresponding transition probability (1.1.35), i.e. the fundamental solution of the diffusion equation, satisfies the ESKC (semigroup) relation (1.1.45). From the point of view of the general theory and a classification of stochastic processes given earlier, it is natural to reverse the steps and to start from the fundamental properties of stochastic processes and then to derive the corresponding equations. For Markov processes, and their particular subset of Wiener processes, such a fundamental property is just the ESKC relation. It is instructive to find additional assumptions which allow us to derive the corresponding differential (in particular, diffusion) equations from the ESKC relation. Such an equation can be derived from the very definition of a Wiener process, i.e. from the combination of the Markovian property (ESKC relation) with the Gaussian form of distributions and independence of increments. However, we prefer to use, together with the ESKC relation, the equivalent but physically more transparent additional assumptions.

Let a transition probability obey the conditions

$$\lim_{t\to\tau}\left\langle \frac{x_t - x_\tau}{t - \tau}\right\rangle = A(x_\tau, \tau) \tag{1.2.15}$$

$$\lim_{t\to\tau}\left\langle\frac{(x_t-x_\tau)^2}{t-\tau}\right\rangle = 2B(x_\tau,\tau) \tag{1.2.16}$$

$$\lim_{t\to\tau}\left\langle\frac{(x_t-x_\tau)^3}{t-\tau}\right\rangle = 0 \tag{1.2.17}$$

where $\langle\cdots\rangle$ means the statistical averaging

$$\langle f(x_t-x_\tau)\rangle = \int_{-\infty}^{\infty} dx_t\, f(x_t-x_\tau)W(x_t,t|x_\tau,\tau).$$

Note that we consider here the mean values of the powers of increments, therefore the initial point x_τ should not be used for integration (in contrast with the case of a correlation function, e.g., $\langle f(x_t)h(x_\tau)\rangle = \int dx_t\, dx_\tau\, W(x_t,t|x_\tau,\tau)W(x_\tau,\tau|x_0,0))$, cf (1.2.10)).

It is obvious that the first condition (1.2.15) states that the mean velocity of the particle is equal to $A(x_t,t)$ and that the condition (1.2.16) contains the information about the influence of random hits from the medium (since if there were no hits at all, the problem would be fully deterministic and the width of the distribution, $B(x_t,t)$, would be zero). The reader may easily check that for a Gaussian stochastic process with appropriate dependence of the exponent on the time variable, the conditions (1.2.15)–(1.2.17) are indeed satisfied (see problem 1.2.4, page 111).

To derive an equation for the transition probability $W(x_t,t|x_0,0)$ or, equivalently, for the corresponding probability density $w(x,t)$ from the ESKC relation (1.1.47), we multiply the latter by an arbitrary function $g(x_t)$ vanishing at infinity together with its derivative,

$$g(x_t)\xrightarrow[x_t\to\pm\infty]{}0 \qquad g'(x_t)\xrightarrow[x_t\to\pm\infty]{}0 \tag{1.2.18}$$

and integrate the result over x_t:

$$\int_{-\infty}^{\infty} dx_t\, g(x_t)W(x_t,t|x_0,0) = \int_{-\infty}^{\infty} dx_t\, dx_\tau\, g(x_t)W(x_t,t|x_\tau,\tau)W(x_\tau,\tau|x_0,0). \tag{1.2.19}$$

Using for $g(x_t)$ the Taylor series with a remainder term

$$g(x_t) = g(x_\tau) + (x_t-x_\tau)g'(x_\tau) + \tfrac{1}{2}(x_t-x_\tau)^2 g''(x_\tau) + \tfrac{1}{6}(x_t-x_\tau)^3 g'''(\xi)$$

where ξ lies somewhere between x_τ and x_t, the right-hand side of (1.2.19) can be written as

$$\int_{-\infty}^{\infty} dx_\tau\, g(x_\tau)W(x_\tau,\tau|x_0,0) + \int_{-\infty}^{\infty} dx_\tau\, g'(x_\tau)W(x_\tau,\tau|x_0,0)\langle(x_t-x_\tau)\rangle$$

$$+ \int_{-\infty}^{\infty} dx_\tau\, g''(x_\tau)W(x_\tau,\tau|x_0,0)\tfrac{1}{2}\langle(x_t-x_\tau)^2\rangle$$

$$+ \int_{-\infty}^{\infty} dx_\tau\, g'''(\xi)W(x_\tau,\tau|x_0,0)\tfrac{1}{6}\langle(x_t-x_\tau)^3\rangle$$

(we have used here the normalization condition (1.1.36)).

Moving the first term of the latter expression to the left-hand side of (1.2.19), dividing the obtained equality by $(t-\tau)$ and taking the limit $t\to\tau$, we arrive at the equation

$$\int_{-\infty}^{\infty} dx_t\, g(x_t)\frac{\partial W(x_t,t|x_0,0)}{\partial t} = \int_{-\infty}^{\infty} dx_t\, g'(x_t)A(x_t,t)W(x_t,t|x_0,0)$$

$$+ \int_{-\infty}^{\infty} dx_t\, g''(x_t)B(x_t,t)W(x_t,t|x_0,0). \tag{1.2.20}$$

Table 1.3. Expectation values specific for a Wiener stochastic process (D is the diffusion constant).

Mean value	$\langle x(t) \rangle_W = 0$
Correlation function	$\langle x(t')x(t) \rangle_W = 2D \min(t', t)$
Dispersion $\mathbb{D}(x(t))$	$\langle [x(t) - \langle x(t) \rangle_W]^2 \rangle_W = \langle x^2(t) \rangle_W = 2Dt$
Dispersion of increments $\mathbb{D}(x(t') - x(t)), t' > t$	$\langle [(x(t') - x(t)) - \langle x(t') - x(t) \rangle_W]^2 \rangle_W$
	$= \langle [x(t') - x(t)]^2 \rangle_W$
	$= 2D(t' - t)$

After integration by parts with the help of conditions (1.2.18), the equation reads

$$\int_{-\infty}^{\infty} dx_t \, g(x_t) \left[\frac{\partial W}{\partial t} + \frac{\partial (AW)}{\partial x_t} - \frac{\partial^2 (BW)}{\partial x_t^2} \right] = 0$$

which, due to the arbitrariness of $g(x_t)$, gives the *Fokker–Planck equation*

$$\frac{\partial W}{\partial t} + \frac{\partial (AW)}{\partial x_t} - \frac{\partial^2 (BW)}{\partial x_t^2} = 0 \qquad (1.2.21)$$

(mathematicians usually call equation (1.2.21) 'the second Kolmogorov equation').

In the particular case

$$A(x_t, t) = 0 \qquad B(x_t, t) = D = \text{constant} \qquad (1.2.22)$$

the Fokker–Planck equation (1.2.21) just becomes the diffusion equation (1.1.18).

In addition to conditions (1.2.15)–(1.2.17) and (1.2.22), the Wiener process has the specific expectation values depicted in table 1.3 (cf problems 1.1.8, page 54 and 1.2.1, page 111, as well as equation (1.1.62)).

◇ **Microscopic approach to the description of stochastic processes: Langevin equation**

In section 1.1, we presented both the macroscopic (based on the density ρ, cf (1.1.3)) and the microscopic (probabilistic, cf (1.1.18)) derivations of the diffusion equation. We did this from the point of view of the physics of the Brownian motion. In the framework of the general theory of stochastic processes, it is the ESKC and Fokker–Planck equations which provide the *macroscopic* derivation of the corresponding differential equations. The microscopic approach starts from the stochastic *Langevin equation*.

Equations describing the behaviour of a system under the influence of random forces are called *stochastic equations*. The analysis of stochastic equations consists in determining the statistical characteristics of their solutions, e.g., expectation (mean) values, correlation functions, probability density. In particular, such equations provide another approach to the derivation of basic relations describing the Brownian motion.

The cornerstone of deterministic classical mechanics is the fundamental Newton equation which, in the case of one particle moving in some medium, has the form

$$m\ddot{x} + \eta\dot{x} = F \qquad (1.2.23)$$

where m is the mass of the particle, $\eta > 0$ is the friction coefficient for the motion in the medium and F is an external force. In the theory of Brownian motion, we assume the existence of a *random force* which describes the hits by molecules of the medium (in general, in addition to the deterministic external force F). The modification of the Newton equation due to the inclusion of a random force Φ leads to the so-called *Langevin equation*

$$m\ddot{x} + \eta\dot{x} = F + \Phi. \tag{1.2.24}$$

For sufficiently large time intervals ($t \gg m/\eta$), we can neglect the term $m\ddot{x}$ (i.e. consider a particle without inertia), so that the Langevin equation becomes

$$\dot{x} = f + \phi \tag{1.2.25}$$
$$f = F/\eta \qquad \phi = \Phi/\eta.$$

In simple cases, the time derivative of a stochastic process, as in (1.2.25),

$$\dot{x}(t) \equiv \frac{dx(t)}{dt}$$

produces, by definition, a stochastic process $\dot{x}(t)$ with its possible realizations being the derivatives of possible realizations of the initial stochastic process $x(t)$ itself. The expectation value and correlation functions for the derivatives are then defined as follows:

$$\langle \dot{x}(t) \rangle = \frac{d}{dt}\langle x(t) \rangle \tag{1.2.26}$$

$$\langle \dot{x}(t')\dot{x}(t) \rangle = \frac{d}{dt'}\frac{d}{dt}\langle x(t')x(t) \rangle \tag{1.2.27}$$

$$\langle \dot{x}(t')x(t) \rangle = \frac{d}{dt'}\langle x(t')x(t) \rangle. \tag{1.2.28}$$

However, an attentive reader may worry about the apparent logical inconsistency: we claimed that the Langevin equation, containing the *time derivative*, is the (microscopic) basis for the Fokker–Planck equation and, in particular, for the diffusion equation. On the other hand, as we proved in section 1.1.3, the essential (having non-zero measure) trajectories are *non-differentiable*.

To resolve this inconsistency and to bring rigorous mathematical sense to the Langevin equation, we must use the well-developed *theory of stochastic equations*, in particular, the so-called stochastic integral and differential calculi. It goes without saying that the 'derivative' in (1.2.25) is merely a formal stochastic object with certain correlation functions (according to (1.2.27)), e.g., for a Wiener process (see table 1.3):

$$\langle \dot{x}(t')\dot{x}(t) \rangle = \frac{d}{dt'}\frac{d}{dt}2D\min(t',t)$$
$$= 2D\delta(t'-t) \tag{1.2.29}$$

(see problem 1.2.5, page 113). According to the definition (see (1.2.13)), the process $\dot{x}(t)$ is referred to as white noise (emphasizing that the Fourier transformed correlation function (1.2.14) is constant). To give a precise meaning to $\dot{x}(t)$, we introduce the concept of a *generalized stochastic process* in analogy with generalized functions (distributions); see Hida (1980) for details. In fact, a process like $\dot{x}(t)$ is defined as the functional on an appropriate space of test functions $f(t)$:

$$\dot{x}[f] = \int_0^\infty dt\, f(t)\dot{x}(t) = \int dx(t)\, f(t). \tag{1.2.30}$$

The last expression is referred to as a *stochastic integral*. Roughly speaking, this integral is defined as the limit of the sum

$$\sum_j f(\tilde{t}_j)(x(t_{j+1}) - x(t_j)) \qquad \tilde{t}_j \in [t_j, t_{j+1}]$$

when $\Delta t_j \equiv t_{j+1} - t_j \to 0$ for all j. The reader can become better acquainted with some of the properties of such integrals, including the important *Ito integral*, by considering problem 1.2.17, page 120. We will not go into details of this rather involved mathematical theory and refer interested readers to the literature (Rosenblatt 1962, Gihman and Skorohod 1972, Kuo 1975, Hida 1980, Ikeda and Watanabe 1981, Chung and Williams 1983).

For us, it is enough to consider equations of the Langevin type as symbolic relations (bearing in mind that, in principle, they all can be well defined). Such a point of view is enough to relate different stochastic processes taking place under the influence of the same random force from the side of a medium (i.e. occurring in media with identical physical characteristics, such as temperature). If two processes $x(t)$ and $y(t)$ are described by stochastic equations with the same random force and hence with identical right-hand sides of the Langevin equations, we can equate the left-hand sides of the equations as well and establish in this way a relation between different processes. In the following subsection we shall use this method to define path-integral representations for the transition probabilities of more general, than ordinary Brownian motion, stochastic processes.

1.2.2 Brownian particles in the field of an external force: treatment by functional change of variables in the path integral

Consider the Langevin equation (1.2.24) with an external force of the special (harmonic) kind

$$F = -\eta k x(t)$$

so that (1.2.25) takes the form

$$\dot{x}(t) + k x(t) = \phi(t). \tag{1.2.31}$$

Brownian motion of a particle subject to an external harmonic force is called an *Ornstein–Uhlenbeck process* (Uhlenbeck and Ornstein 1930) (see also Chandrasekhar (1943)).

Comparing this equation with that for the case of a zero deterministic external force,

$$\dot{y} = \phi(t) \tag{1.2.32}$$

and *the same random force* $\phi(t)$ (i.e. both particles, described by $x(t)$ and $y(t)$, are in the same medium under the same conditions such as temperature), we obtain

$$\dot{y}(t) = \dot{x}(t) + k x(t). \tag{1.2.33}$$

The integral form of this relation, assuming $y(0) = x(0)$, reads as

$$y(t) = x(t) + k \int_0^t ds \, x(s) \tag{1.2.34}$$

and the inverse we read as

$$x(t) = \int_0^t ds \, e^{-k(t-s)} \dot{y}(s) + y(0) e^{-kt}$$

$$= y(s)e^{-k(t-s)}|_{s=0}^{s=t} - k\int_0^t ds\, e^{-k(t-s)} y(s) + y(0)e^{-kt}$$

$$= y(t) - k\int_0^t ds\, e^{-k(t-s)} y(s) \qquad (1.2.35)$$

so that there is a one-to-one continuous correspondence between the trajectories $x(t)$ and $y(t)$. The integration in (1.2.35) has to be understood in the sense of stochastic calculus (cf the discussion in the preceding subsection). Fortunately, these integrals possess many of the properties of usual integrals; in particular, the integration by parts which we used in (1.2.35) is correct for integrands *linear* in a stochastic variable (see problem 1.2.17, page 120). Therefore, we are treating them loosely as if they were the usual integrals of ordinary functions.

Let us, for a moment, consider the case of a single random variable ξ. The probability of finding its value in an infinitesimal vicinity of some value x is given by $dF_\xi(x) = \omega_\xi(x)\, dx$, where $dF_\xi(x)$ is the corresponding partition function and $\omega_\xi(x)$ is the probability density (see (1.2.2)–(1.2.7)). If we are interested in some new random variable ζ which is in one-to-one correspondence with ξ defined by a function $\zeta = f(\xi)$, the probability to find ζ in an infinitesimal vicinity of $y = f(x)$ is related to that for ξ by the obvious relation:

$$dF_\zeta(y) \equiv \omega_\zeta(y)\, dy = \omega_\zeta(f(x))\, df(x)$$

$$= \omega_\zeta(f(x)) \frac{\partial f}{\partial x} dx = \omega_\xi(x)\, dx = dF_\xi(x).$$

Note that $\partial f/\partial x$ plays the role of a Jacobian in the one-dimensional case.

The straightforward generalization of this consideration to the infinite-dimensional (functional) case, using the discrete-time approximation at the intermediate step, allows us to derive (Beilinson 1959) the probability density for paths $x(\tau)$ from the probability density for $y(\tau)$ ($0 \leq \tau \leq t$), the latter being described by the Wiener measure (1.1.54), i.e.

$$d_W y(\tau) = \exp\left\{-\int_0^t d\tau\, \dot{y}^2(\tau)\right\} \prod_{\tau=0}^t \frac{dy(\tau)}{\sqrt{\pi\, d\tau}} \qquad (1.2.36)$$

(we use the agreement $D = 1/4$ as in section 1.1.4). Now, to obtain the probability density for the stochastic process $x(t)$, we can use the functional change of variable (1.2.34), which gives

$$d_W y(\tau) = e^{kt/2} \exp\left\{-\int_0^t d\tau\, (\dot{x} + kx)^2\right\} \prod_{\tau=0}^t \frac{dx(\tau)}{\sqrt{\pi\, d\tau}} \qquad (1.2.37)$$

where the first factor in the right-hand side of (1.2.37) corresponds to the Jacobian of the transformation (1.2.34) (cf (1.1.127)). The initial points of the trajectories $x(\tau)$ and $y(\tau)$ are fixed and coincide by construction. The endpoints x_t and y_t of the trajectories should also be kept fixed and we put $y_t = x_t$, so that relations (1.2.33) and (1.2.34) are valid for all points on the trajectories except the endpoints (they are fixed and not random variables). Thus the transition probability for the process $x(t)$ is expressed via the path integral

$$W(x_t, t|x_0, 0) = e^{kt/2} \int_{C\{x_0,0;x_t,t\}} \prod_{\tau=0}^t \frac{dx(\tau)}{\sqrt{\pi\, d\tau}} \exp\left\{-\int_0^t d\tau\, (\dot{x} + kx)^2\right\}$$

$$= \exp\{kt/2 - k(x_t^2 - x_0^2)\} \int_{C\{x_0,0;x_t,t\}} d_W x(\tau) \exp\left\{-\int_0^t d\tau\, k^2 x^2\right\}. \qquad (1.2.38)$$

The latter equality follows from

$$\int_0^t dt\,(\dot{x}+kx)^2 = \int_0^t dt\,\dot{x}^2 + k\int_0^t d(x^2) + k^2\int_0^t dt\,x^2$$
$$= \int_0^t dt\,\dot{x}^2 + k(x_t^2 - x_0^2) + k^2\int_0^t dt\,x^2$$

and from the fact that we integrate over the set of paths with fixed endpoints, i.e. we have a conditional Wiener path integral. Path integrals of the type (1.2.38) have been considered in example 1.5, page 43 (for $x_0 = 0$) and the reader may integrate (1.2.38) by this method. We shall calculate the path integral (1.2.38) explicitly in section 1.2.6 by the variational method (cf (1.2.132)).

As a result, we have obtained that the transition probability of a Brownian particle moving in the field of a harmonic force can be expressed through the path integral (1.2.38) with the functional

$$\exp\left\{-k^2\int_0^t d\tau\,x^2\right\}.$$

Note that the exponent can be written via the harmonic potential

$$V(x) = \tfrac{1}{2}\eta k^2 x^2.$$

This is the first example which shows that the characteristics of more complicated (than just diffusion) processes are expressed by Wiener path integrals of the corresponding functionals. An especially important expression of this kind is the *Feynman–Kac formula* for a particular equation, called the *Bloch equation*, which is of the Fokker–Planck type. This equation, besides its direct meaning for a specific type of stochastic process, is important because of its close relation with the Schrödinger equation in quantum mechanics. We shall consider this equation and the Feynman–Kac formula in detail in section 1.2.5. To avoid confusion, note that in quantum mechanics, the Bloch equation directly contains the potential of an external force acting on a particle. Brownian motion in the field of an external force, which we discuss at the moment, is described by the Langevin equation of the type (1.2.31) (or, more generally, by (1.2.39)) and has non-zero drift velocity (1.2.15) and hence the Fokker–Planck equation (1.2.21) with a first-order space derivative term (compare this with the Bloch equation in section 1.2.5). To relate the Fokker–Planck equation to the Bloch equation we must make an additional transformation (cf (1.2.90) in section 1.2.5). This, in turn, makes a distinction between the expression (1.2.38) and the Feynman–Kac formula.

◇ **Brownian particles in a field of general non-stationary and nonlinear external forces**

For a more general non-stationary and nonlinear external force $F = -\eta f(x(t), t)$ entering the Langevin equation, (1.2.25) takes the form

$$\dot{x}(t) + f(x(t), t) = \phi(t) \tag{1.2.39}$$

which can be analyzed by the change of variable

$$y(t) = x(t) + \int_0^t ds\,f(x(s), s). \tag{1.2.40}$$

The Jacobian J of this transformation can be calculated by the discrete-time approximation analogously to the one for (1.1.122). The result is

$$J(\varepsilon) = \begin{vmatrix} 1 + f'(x_1, \varepsilon)\varepsilon/2 & 0 & \cdots & \cdots & 0 \\ f'(x_1, \varepsilon) & 1 + f'(x_2, 2\varepsilon)\varepsilon/2 & 0 & \cdots & 0 \\ \vdots & & \ddots & & \vdots \\ \vdots & & & \ddots & \vdots \\ f'(x_1, \varepsilon) & f'(x_2, 2\varepsilon) & \cdots & \cdots & 1 + f'(x_N, N\varepsilon)\varepsilon/2 \end{vmatrix}$$

where $f' \equiv \frac{\partial f}{\partial x}$ and $\varepsilon = t/N$. Due to triangularity, the determinant is equal to the product of the diagonal elements (cf (1.1.126))

$$J(\varepsilon) = \prod_{j=1}^{N}\left(1 + \frac{\varepsilon f'(x_j, j\varepsilon)}{2N}\right). \qquad (1.2.41)$$

In the continuous limit, the determinant reads as

$$J = \lim_{\varepsilon \to 0} \exp\left\{\frac{1}{2}\sum_{j=1}^{N} \varepsilon f'(x_j, j\varepsilon)\right\}$$

$$= \exp\left\{\frac{1}{2}\int_0^t ds\, f'(x(s), s)\right\}. \qquad (1.2.42)$$

Thus the transition probability for the stochastic process defined by the Langevin equation (1.2.39) is expressed via the path integral

$$W(x_t, t | x_0, 0) = \int_{C\{x_0, 0; x_t, t\}} \prod_{\tau=0}^{t} \frac{dx(\tau)}{\sqrt{\pi\, d\tau}} \exp\left\{-\int_0^t d\tau\, [\dot{x} + f(x(\tau), \tau)]^2\right\}$$

$$\times \exp\left\{\frac{1}{2}\int_0^t d\tau\, f'(x(\tau), \tau)\right\}. \qquad (1.2.43)$$

This transition probability describes the behaviour of a single inertialess (the mass m being neglected, cf (1.2.25)) Brownian particle moving in the field of an arbitrary non-stationary, nonlinear external force.

1.2.3 Brownian particles with interactions

Let us consider two Brownian particles. If they have no interactions and hence their motion corresponds to two independent stochastic processes

$$\dot{y}_1 = \phi_1 \qquad \dot{y}_2 = \phi_2 \qquad (1.2.44)$$

the transition probability is given by the product of the Wiener path integrals

$$W(y_{1t}, t; y_{2t}, t | y_{10}, 0; y_{20}, 0) = \int_{C\{y_{10}, 0; y_{1t}, t\}} d_W y_1(\tau) \int_{C\{y_{20}, 0; y_{2t}, t\}} d_W y_2(\tau)$$

$$= W(y_{1t}, t | y_{10}, 0) W(y_{2t}, t | y_{20}, 0). \qquad (1.2.45)$$

If two Brownian particles interact with each other and move in the presence of a linear external force, the stochastic equations take the form

$$\dot{x}_1 + k_{11}x_1 + k_{12}x_2 = \phi_1$$
$$\dot{x}_2 + k_{21}x_1 + k_{22}x_2 = \phi_2. \qquad (1.2.46)$$

By introducing the matrices

$$\mathsf{k} = \begin{pmatrix} k_{11} & k_{12} \\ k_{21} & k_{22} \end{pmatrix} \tag{1.2.47}$$

$$\boldsymbol{x} = \begin{pmatrix} x_1 \\ x_2 \end{pmatrix} \qquad \boldsymbol{\phi} = \begin{pmatrix} \phi_1 \\ \phi_2 \end{pmatrix} \qquad \boldsymbol{y} = \begin{pmatrix} y_1 \\ y_2 \end{pmatrix} \tag{1.2.48}$$

equations (1.2.44) and (1.2.46) can be rewritten in matrix form:

$$\frac{\partial \boldsymbol{y}}{\partial t} = \boldsymbol{\phi} \tag{1.2.49}$$

$$\frac{\partial \boldsymbol{x}}{\partial t} + \mathsf{k}\boldsymbol{x} = \boldsymbol{\phi} \tag{1.2.50}$$

which is quite similar to equations (1.2.32) and (1.2.31). Therefore, to derive the transition probability for two Brownian particles with Langevin equations (1.2.46), we can use the same method of functional substitution $\boldsymbol{y}(t) \to \boldsymbol{x}(t)$ as that used in the preceding section.

Since the external stochastic process $\boldsymbol{\phi}$ is the same for both variables $\boldsymbol{x}(t)$ and $\boldsymbol{y}(t)$, we write

$$\dot{\boldsymbol{y}}(t) = \dot{\boldsymbol{x}}(t) + \mathsf{k}\boldsymbol{x}(t) \tag{1.2.51}$$

or, in integral form,

$$\boldsymbol{y}(t) = \boldsymbol{x}(t) + \mathsf{k} \int_0^t d\tau\, \boldsymbol{x}(\tau) \tag{1.2.52}$$

assuming $\boldsymbol{y}(0) = \boldsymbol{x}(0)$. The Jacobian of this transformation can be calculated by the same method of discrete-time approximation $t_{i+1} - t_i \equiv \varepsilon$ for equation (1.2.52) (cf the preceding subsection and (1.1.125)) with the result

$$J_N = \begin{vmatrix} \mathbb{I} + \mathsf{k}\varepsilon/2 & 0 & \cdots & \cdots & 0 \\ \mathsf{k}\varepsilon & \mathbb{I} + \mathsf{k}\varepsilon/2 & 0 & \cdots & 0 \\ \vdots & \cdots & \ddots & \cdots & \vdots \\ \mathsf{k}\varepsilon & \mathsf{k}\varepsilon & \cdots & \cdots & \mathbb{I} + \mathsf{k}\varepsilon/2 \end{vmatrix}$$

$$= \prod_{i=1}^N \begin{vmatrix} 1 + k_{11}\varepsilon/2 & k_{12}\varepsilon/2 \\ k_{21}\varepsilon/2 & 1 + k_{22}\varepsilon/2 \end{vmatrix} \tag{1.2.53}$$

where $\mathbb{I}$ is the two-dimensional unit matrix. Rewriting the factors in (1.2.53) up to the second order in ε,

$$\begin{vmatrix} 1 + k_{11}\varepsilon/2 & k_{12}\varepsilon/2 \\ k_{21}\varepsilon/2 & 1 + k_{22}\varepsilon/2 \end{vmatrix} = 1 + \frac{1}{2}(k_{11} + k_{22})\varepsilon + \mathcal{O}(\varepsilon^2)$$

$$= \exp\{\tfrac{1}{2}\varepsilon \operatorname{Tr}\mathsf{k} + \mathcal{O}(\varepsilon^2)\}$$

we obtain, in the limit $N \to \infty$, the Jacobian

$$J = \lim_{N \to \infty} \prod_{i=1}^N \exp\{\tfrac{1}{2}\varepsilon \operatorname{Tr}\mathsf{k} + \mathcal{O}(\varepsilon^2)\} = \exp\{\tfrac{1}{2}t \operatorname{Tr}\mathsf{k}\}. \tag{1.2.54}$$

Thus, performing the substitution in the path integral (1.2.45) and taking into account the Jacobian (1.2.54), we finally obtain

$$W(\boldsymbol{x}_t, t|\boldsymbol{x}_0, 0) = \exp\left\{\tfrac{1}{2}t \operatorname{Tr}\mathsf{k}\right\} \int_{C\{x_0,0;x_t,t\}} \prod_{\tau=0}^t \frac{dx_1(\tau)}{\sqrt{\pi\, d\tau}} \prod_{\tau=0}^t \frac{dx_2(\tau)}{\sqrt{\pi\, d\tau}}$$

$$\times \exp\left\{-\int_0^t d\tau \sum_{i=1}^2 \left(\dot{x}_i + \sum_{j=1}^2 k_{ij} x_j\right)^2\right\}. \tag{1.2.55}$$

Generalization to the case of K Brownian particles is quite straightforward: we just introduce the K-dimensional vectors $\boldsymbol{x}$, $\boldsymbol{y}$ and $\boldsymbol{\phi}$ and the result has the same general form (1.2.55), except that the sums in the exponent are extended to K terms.

◇ Interacting Brownian particles subject to nonlinear and non-stationary external forces

A treatment of K Brownian particles subject to nonlinear and non-stationary external forces also can be carried out using the matrix method and results for a single particle which we have discussed in the preceding subsection. The corresponding Langevin equations read as

$$\dot{x}_i(t) + f_i(x(t), t) = \phi_i(t) \qquad i = 1, \ldots, K$$

or, in the matrix form,

$$\dot{\boldsymbol{x}}(t) + \boldsymbol{f}(\boldsymbol{x}(t), t) = \boldsymbol{\phi}(t). \tag{1.2.56}$$

The functional change of variables $\boldsymbol{y} \to \boldsymbol{x}$ (1.2.52) in the time-discrete approximation has a block-triangular Jacobian of the form

$$J(\varepsilon) = \begin{vmatrix} \mathsf{A}(\varepsilon) & 0 & \cdots & \cdots & 0 \\ * & \mathsf{A}(2\varepsilon) & 0 & \cdots & 0 \\ \vdots & & \ddots & & \vdots \\ \vdots & & & \ddots & \vdots \\ * & * & \cdots & * & \mathsf{A}(N\varepsilon) \end{vmatrix}$$

where

$$\mathsf{A}(\varepsilon) = \begin{pmatrix} 1 + \frac{1}{2}\frac{\partial f_1(\boldsymbol{x},\varepsilon)}{\partial x_1}\varepsilon & \frac{1}{2}\frac{\partial f_1(\boldsymbol{x},\varepsilon)}{\partial x_2}\varepsilon & \cdots & \cdots & \frac{1}{2}\frac{\partial f_1(\boldsymbol{x},\varepsilon)}{\partial x_K}\varepsilon \\ \frac{1}{2}\frac{\partial f_2(\boldsymbol{x},\varepsilon)}{\partial x_1}\varepsilon & 1 + \frac{1}{2}\frac{\partial f_2(\boldsymbol{x},\varepsilon)}{\partial x_2}\varepsilon & \cdots & \cdots & \cdots \\ \vdots & & \ddots & & \vdots \\ \vdots & & & \ddots & \vdots \\ \frac{1}{2}\frac{\partial f_K(\boldsymbol{x},\varepsilon)}{\partial x_1}\varepsilon & \frac{1}{2}\frac{\partial f_K(\boldsymbol{x},\varepsilon)}{\partial x_2}\varepsilon & \cdots & \cdots & 1 + \frac{1}{2}\frac{\partial f_K(\boldsymbol{x},\varepsilon)}{\partial x_K}\varepsilon \end{pmatrix} \tag{1.2.57}$$

and the asterisks $*$ stand for matrix blocks which are of no importance because, due to the triangular form, the determinant is equal to the product of the determinants of the matrices A:

$$\det \mathsf{A}(n\varepsilon) = \lim_{\varepsilon \to 0} \prod_{j=1}^K \left(1 + \frac{1}{2}\frac{\partial f_j}{\partial x_j} n\varepsilon\right) \qquad n = 1, \ldots, N$$

$$J = \lim_{\varepsilon \to 0} J(\varepsilon) = \exp\left\{\frac{1}{2} \sum_{i=1}^K \int_0^t d\tau \, \frac{\partial f_i(\boldsymbol{x}(\tau), \tau)}{\partial x_i}\right\}.$$

Thus the transition probability in this case reads as

$$W(\boldsymbol{x}_t, t | \boldsymbol{x}_0, 0) = \int_{\mathcal{C}\{\boldsymbol{x}_0, 0; \boldsymbol{x}_t, t\}} \prod_{\tau=0}^t \frac{dx_1(\tau)}{\sqrt{\pi \, d\tau}} \cdots \prod_{\tau=0}^t \frac{dx_K(\tau)}{\sqrt{\pi \, d\tau}}$$

$$\times \exp\left\{-\sum_{i=1}^{K}\int_{0}^{t}d\tau\,(\dot{x}_i + f_i(x(\tau),\tau))^2\right\}$$

$$\times \exp\left\{\frac{1}{2}\sum_{i=1}^{K}\int_{0}^{t}d\tau\,\frac{\partial f_i(x(\tau),\tau)}{\partial x_i}\right\}. \quad (1.2.58)$$

1.2.4 Brownian particles with inertia: a Wiener path integral with constraint and in the space of velocities

We are now ready to consider the complete Langevin equation (1.2.24), i.e. with an account of the mass term. The main difference from the stochastic equations considered earlier is that this is now a *second-order* time differential equation. Thus the first natural step to solving it (i.e. to find the corresponding transition probability) is to represent it as a couple of first-order differential equations by introducing the new variable $v(t)$:

$$\dot{v} + \frac{\eta}{m}v - \frac{1}{m}F = \frac{1}{m}\Phi$$

$$\dot{x} - v = 0. \quad (1.2.59)$$

If the right-hand side of the second equation contained a non-trivial stochastic process (i.e. a random force), this system would be of the type considered in the preceding subsection and we could apply the method of functional change of variables developed there. To achieve this, we can consider (1.2.59) as the limiting case $T_2 \to 0$ of the system

$$\dot{v} + \frac{\eta}{m}v - \frac{1}{m}F = T_1\frac{1}{m}\Phi_1$$

$$\dot{x} - v = T_2\Phi_2. \quad (1.2.60)$$

For similarity, we have also introduced an additional coefficient T_1 into the first equation (we can put $T_1\Phi_1 = \Phi$ or consider (1.2.59) as the limit at $T \to 1$, $\Phi_1 = \Phi$). It should be mentioned that the introduction of the coefficients T_1, T_2 is not just a formal trick: in fact, T_1 and T_2 represent the temperatures of the two thermal baths in which each of the 'two' Brownian particles with the 'trajectories' $x(\tau)$ and $v(\tau)$ move, respectively. One of the main results of Einstein's theory of Brownian motion (Einstein 1905) is the expression for the diffusion constant: for a spherical particle of radius R, this is given by the expression

$$D = \frac{k_B T}{6\pi\eta R} \quad (1.2.61)$$

(k_B is the Boltzmann constant, η is the friction coefficient). Therefore, the diffusion constant is proportional to the temperature and, for our present aim, it is convenient to choose such a unit of time that

$$D = \frac{T}{4}. \quad (1.2.62)$$

The dependence of the diffusion constant on temperature can be physically understood. Indeed, it is intuitively clear that the intensity of an external stochastic force is proportional to the temperature of the medium, $\phi \sim T$, since the very source of this force is the thermal motion of molecules of the medium. This shows that the appearance of the coefficients T_1, T_2 in (1.2.60) has an obvious physical meaning. More precisely, this means that the mean value (1.2.16), the last term in the Fokker–Planck equation (1.2.21) and hence the diffusion constant are proportional to the temperature.

70   Path integrals in classical theory

An explicit account of the temperature dependence of the diffusion constant according to (1.2.62) slightly modifies the expression for the statistical characteristics of Brownian processes. For example, the transition probability for several Brownian particles moving under the influence of stochastic and external deterministic forces, considered in the preceding subsection, has the form

$$W(\bm{x}_t, t|\bm{x}_0, 0) = \int_{\mathcal{C}\{\bm{x}_0, 0; \bm{x}_t, t\}} \prod_{\tau=0}^{t} \frac{dx_1(\tau)}{\sqrt{\pi\, d\tau\, T_1}} \cdots \prod_{\tau=0}^{t} \frac{dx_K(\tau)}{\sqrt{\pi\, d\tau\, T_K}}$$

$$\times \exp\left\{ -\sum_{i=1}^{K} \int_0^t d\tau\, \frac{(\dot{x}_i + f_i(\bm{x}(\tau),\tau))^2}{T_i} \right\}$$

$$\times \exp\left\{ \frac{1}{2}\sum_{i=1}^{K} \int_0^t d\tau\, \frac{\partial f_i(\bm{x}(\tau),\tau)}{\partial x_i} \right\}. \quad (1.2.63)$$

Here T_i is the temperature for the ith Brownian particle as in (1.2.60).

Since the method of treatment of the Brownian particles moving in the presence of an external deterministic force has been discussed in detail in the preceding subsections, we consider here the simplified case of a particle with inertia, subject only to a stochastic force, i.e. the particular case of the Langevin equation (1.2.24):

$$\ddot{x} = \phi$$

which, after a transition to a couple of first-order equations and the explicit introduction of temperatures (including the final limit of $T_2 \to 0$), becomes

$$\dot{v} = T_1 \phi_1$$
$$\dot{x} - v = T_2 \phi_2. \quad (1.2.64)$$

For this particular case, (1.2.63) reads

$$W^{T_1, T_2}(x_t, v_t, t|x_0, v_0, 0) = \int_{\mathcal{C}\{x_0, v_0, t_0; x_t, v_t, t\}} \prod_{\tau=0}^{t} \frac{dv(\tau)}{\sqrt{\pi\, d\tau\, T_1}} \prod_{\tau=0}^{t} \frac{dx(\tau)}{\sqrt{\pi\, d\tau\, T_2}}$$

$$\times \exp\left\{ -\frac{1}{T_1}\int_0^t d\tau\, \dot{v}^2 - \frac{1}{T_2}\int_0^t d\tau\, (\dot{x}-v)^2 \right\}. \quad (1.2.65)$$

Consider a part of the integrand in (1.2.65) in the discrete-time approximation:

$$\prod_{\tau=0}^{t} \frac{1}{\sqrt{\pi\, d\tau\, T_2}} \exp\left\{ -\frac{1}{T_2}\int_0^t d\tau\, (\dot{x}-v)^2 \right\}$$

$$\to \left(\prod_{i=0}^{n} \frac{1}{\sqrt{\pi\, \Delta\tau\, T_2}} \right) \exp\left\{ -\frac{1}{T_2}\sum_{i=0}^{N}\left(\frac{x_{i+1}-x_i}{\Delta\tau}-v_i\right)^2 \Delta\tau \right\}$$

$$\xrightarrow[T_2 \to 0]{} \prod_{i=0}^{N} \frac{1}{\Delta\tau}\delta\left(\frac{x_{i+1}-x_i}{\Delta\tau}-v_i\right) \quad (1.2.66)$$

$$\xrightarrow[\Delta\tau \to 0]{} \prod_{\tau=0}^{t} \frac{1}{d\tau}\delta(\dot{x}(\tau)-v(\tau)) \quad (1.2.67)$$

$$\equiv \delta[\dot{x}(\tau)-v(\tau)]\prod_{\tau=0}^{t}\frac{1}{d\tau}. \quad (1.2.68)$$

Here $\delta[f(s)]$ denotes the δ-*functional*, i.e. the infinite products of Dirac δ-functions at each slice of time τ as expressed in (1.2.67) and (1.2.68).

Inserting this result into (1.2.65), we obtain

$$W(x_t, v_t, t | x_0, v_0, 0) = \lim_{T_1 \to 1, T_2 \to 0} W^{T_1, T_2}(x_t, v_t, t | x_0, v_0, 0)$$

$$= \int_{C\{x_0, v_0, 0; x_t, v_t, t\}} \prod_{\tau=0}^{t} \frac{dv(\tau)}{\sqrt{\pi \, d\tau}} \prod_{\tau=0}^{t} \frac{dx(\tau)}{d\tau} \exp\left\{-\int_0^t d\tau \, \dot{v}^2\right\} \delta[\dot{x} - v] \quad (1.2.69)$$

$$= \int_{C\{v_0, 0; v_t, t\}} \prod_{\tau=0}^{t} \frac{dv(\tau)}{\sqrt{\pi \, d\tau}} \exp\left\{-\int_0^t d\tau \, \dot{v}^2\right\} \delta\left(x(t) - x(0) - \int_0^t d\tau \, v(\tau)\right) \quad (1.2.70)$$

i.e. a path integral with the constraint

$$x(t) - x(0) = \int_0^t d\tau \, v(\tau). \quad (1.2.71)$$

(The value $T_1 = 1$ is provided by the appropriate choice of time unit.) To convince ourselves that the integration of the delta-functional in the double path integral (1.2.69) indeed gives the single path integral (1.2.70), we can use the discrete-time approximation (see problem 1.2.6, page 113).

The variable v obviously has the meaning of velocity for the Brownian particle and the condition (1.2.71) states that the integration of the velocity over a time interval gives the resulting shift of the particle. Thus we have reduced the problem of calculating the transition probability for a Brownian particle with inertia to the integration over its velocities with the constraint (1.2.71). From the formal point of view, the Wiener integral obtained corresponds to integration with the conditional Wiener measure and over the subspace of functions subject to the constraint (1.2.71).

This means that taking into account the non-zero mass (inertia) of the Brownian particle allows us to introduce the notion of velocity for Brownian motion. Of course, the trajectories contributing to the Wiener integral (1.2.69) (having non-zero measure) are still non-differentiable, but now these are trajectories in the *combined* space of random coordinates and random velocities.

In practice, this integral can be calculated by the variational method which we shall discuss in section 1.2.6. We postpone the calculation until then and until problem 1.2.7, page 113, while here we present only the result:

$$W(x_t, v_t, t | x_0, v_0, 0) = \frac{\sqrt{12}}{\pi t^2} \exp\left\{-\frac{12}{t^3}\left((x_t - x_0) - \frac{v_t + v_0}{2}t\right)^2 - \frac{(v_t - v_0)^2}{t}\right\}. \quad (1.2.72)$$

It is clear that after the integration of this probability over x_t, we must obtain the fundamental solution of the diffusion equation in the space of velocities, i.e.

$$\frac{1}{\sqrt{\pi t}} \exp\left\{-\frac{(v_t - v_0)^2}{t}\right\}$$

which can be checked by explicit calculations. Note also that (1.2.72) has the normalization

$$\int_{-\infty}^{\infty} dx_t \int_{-\infty}^{\infty} dv_t \, W(x_t, v_t, t | x_0, v_0, 0) = 1. \quad (1.2.73)$$

72   Path integrals in classical theory

The complete transition probability (1.2.72) is the solution of the Fokker–Planck equation

$$\frac{\partial W}{\partial t} + v\frac{\partial W}{\partial x} = \frac{1}{4}\frac{\partial^2 W}{\partial v^2} \qquad (1.2.74)$$

(problem 1.2.8, page 114).

Another property of the transition probability (1.2.72) is that in the limit $t \to 0$, we have

$$W(x_t, v_t, t|x_0, v_0, 0) \xrightarrow[t\to 0]{} \delta\left(x_t - x_0 - \frac{v_t + v_0}{2}t\right)\delta(v_t - v_0)$$

since

$$\frac{\sqrt{12}}{\sqrt{\pi t^3}}\exp\left\{-\frac{12}{t^3}\left((x_t - x_0) - \frac{v_t + v_0}{2}t\right)^2\right\} \xrightarrow[t\to 0]{} \delta\left(x_t - x_0 - \frac{v_t + v_0}{2}t\right)$$

$$\frac{1}{\sqrt{\pi t}}\exp\left\{-\frac{(v_t - v_0)^2}{t}\right\} \xrightarrow[t\to 0]{} \delta(v_t - v_0).$$

Thus there is a correlation between coordinate shifts and the average velocity $(v_t + v_0)/2$.

1.2.5 Brownian motion with absorption and in the field of an external deterministic force: the Bloch equation and Feynman–Kac formula

In sections 1.2.2–1.2.4 we obtained the path-integral representations for the transition probabilities of more general processes than the motion of just a free Brownian particle using relations between different Langevin equations. Thus we used the *microscopic* approach. Another possible, and even more powerful, way is to use the 'macroscopic' Fokker–Planck equations for more general processes and find their solutions in terms of path integrals. This way leads to the well-known *Feynman–Kac formula*.

◇ The Bloch equation and its solution in terms of path integrals

Let us consider Brownian motion in a medium where Brownian particles can be annihilated with a probability density $V(t, x)$ per unit of time. To derive the corresponding differential equation, we may proceed as in section 1.1.1. The particle current $j(x, t)$ is still given by (1.1.1) but instead of the continuity equation (1.1.2), the balance equation for the number of Brownian particles per unit of volume, i.e. for the density $\rho(t, x)$, becomes

$$\frac{\partial \rho}{\partial t} = -\frac{\partial j}{\partial x} - V\rho. \qquad (1.2.75)$$

Hence the density of particles is governed by the partial differential equation

$$\frac{\partial \rho}{\partial t} = D\frac{\partial^2 \rho}{\partial x^2} - V\rho \qquad (1.2.76)$$

which is called the *Bloch equation*, and the corresponding transition probability $W_B(x_t, t|x_{t_0}, t_0)$ is defined for $t > t_0$ as the solution of this equation under the initial condition

$$W_B(x_t, t|x_0, t_0)|_{t=t_0} = \delta(x_t - x_0). \qquad (1.2.77)$$

This transition probability can also be represented by a Wiener path integral which heuristically can be estimated in the following way. Consider an arbitrary particle path from the set $\mathcal{C}\{x_0, 0; x_t, t\}$. The probability that the Brownian particle will survive this path without being absorbed equals

$$\mathbb{P}[x(\tau)] = \exp\left\{-\int_0^t d\tau\, V(x(\tau), \tau)\right\} \tag{1.2.78}$$

where the integration is carried out along the chosen path. It is natural to assume that the transition probability is the sum of these expressions over all paths from x_0 to x_t:

$$W_B(x_t, t|x_0, 0) = \int_{\mathcal{C}\{x_0, 0; x_t, t\}} d_W x(\tau) \exp\left\{-\int_0^t d\tau\, V(x(\tau))\right\}. \tag{1.2.79}$$

Later, we shall rigorously prove this very important statement, called the *Feynman–Kac formula*, but before then let us discuss the relation between $W_B(x_t, t|x_0, t_0)$ and the transition probability for the diffusion equation (i.e. (1.2.76) with $V = 0$) which we shall denote now by $W_D(x_t, t|x_0, t_0)$ (the subscript D refers to 'diffusion').

◇ **Iterative representation for the solution of the Bloch equation in terms of the solution for the diffusion equation**

It is convenient to remove formally the condition $t > t_0$ by adding instead the condition

$$W_B(x_t, t|x_0, t_0) = W_D(x_t, t|x_0, t_0) = 0 \qquad \text{for } t < t_0$$

or multiplying W_B and W_D by the step-function (1.1.43). In this case, as we have discussed in section 1.1.1, the transition probabilities become the Green functions of the corresponding differential equations in the strict mathematical sense, i.e. they satisfy the *inhomogeneous* equations

$$\left[\frac{\partial}{\partial t} - D\frac{\partial^2}{\partial x^2}\right] W_D(x_t, t|x_0, t_0) = \delta(t - t_0)\delta(x - x_0) \tag{1.2.80}$$

and

$$\left[\frac{\partial}{\partial t} - D\frac{\partial^2}{\partial x^2} + V(x, t)\right] W_B(x_t, t|x_0, t_0) = \delta(t - t_0)\delta(x - x_0). \tag{1.2.81}$$

This implies that W_B is the solution of the integral equation

$$W_B(x_t, t|x_0, t_0) = W_D(x_t, t|x_0, t_0) - \int_{-\infty}^{\infty} dx' \int_{-\infty}^{\infty} dt'\, W_D(x_t, t|x', t') V(x', t') W_B(x', t'|x_0, t_0) \tag{1.2.82}$$

as can be verified by applying the differential operator

$$\left[\frac{\partial}{\partial t} - D\frac{\partial^2}{\partial x^2}\right]$$

to both sides of the equality and using (1.2.80) to perform the integration over x' and t'.

Consider the discrete approximation for the path integral (1.2.79)

$$W_B(x_t, t|x_0, 0) = \lim_{\substack{N \to \infty \\ \varepsilon \to 0}} (4\pi D\varepsilon)^{-(N+1)/2} \int_{-\infty}^{\infty} dx_1 \int_{-\infty}^{\infty} dx_2 \ldots \int_{-\infty}^{\infty} dx_N$$

$$\times \exp\left\{-\frac{1}{4D\varepsilon}\sum_{j=0}^{N}(x_{j+1} - x_j)^2 - \varepsilon\sum_{j=1}^{N} V(x_j, t_j)\right\} \qquad \varepsilon = \frac{t - t_0}{N + 1}. \tag{1.2.83}$$

74   Path integrals in classical theory

The second exponential factor can be expanded into a power series:

$$\exp\left\{-\varepsilon\sum_{j=1}^{N}V(x_j,t_j)\right\} = 1 - \varepsilon\sum_{j=1}^{N}V(x_j,t_j) + \frac{1}{2}\varepsilon^2\sum_{j=1}^{N}\sum_{k=1}^{N}V(x_j,t_j)V(x_k,t_k) - \cdots. \qquad (1.2.84)$$

After the integrations over the intermediate x-coordinates are performed, we find that the perturbation expansion

$$W_B(x_t,t|x_0,t_0) = W_D(x_t,t|x_0,t_0) - \varepsilon\sum_{j=1}^{N}\int_{-\infty}^{\infty}dx_j\,W_D(x_t,t|x_j,t_j)V(x_j,t_j)W_D(x_j,t_j|x_0,t_0)$$

$$+ \frac{1}{2!}\varepsilon^2\sum_{j=1}^{N}\sum_{k=1}^{N}\int_{-\infty}^{\infty}dx_j\int_{-\infty}^{\infty}dx_k\,W_D(x_t,t|x_j,t_j)$$

$$\times V(x_j,t_j)W_D(x_j,t_j|x_k,t_k)V(x_k,t_k)W_D(x_k,t_k|x_0,t_0) - \cdots. \qquad (1.2.85)$$

In the limit $\varepsilon \to 0$ the sum $\varepsilon\sum_j$ can be replaced by the integral $\int_{t_0}^{t}dt_j$ and the factor $1/k!$ can be omitted if the time variables are ordered

$$t_0 < t_j < t,$$
$$t_0 < t_k < t_j < t,$$
$$\vdots$$

This *chronological ordering* of the time variables is automatically realized due to the condition

$$W_D(x,t|x_0,t_0) = 0 \quad \text{if } t < t_0.$$

Hence the expansion (1.2.85) can be rewritten in the form

$$W_B(x_t,t|x_0,t_0) = W_D(x_t,t|x_0,t_0) - \int_{-\infty}^{\infty}dx'\int_{-\infty}^{\infty}dt'\,W_D(x_t,t|x',t')V(x',t')W_D(x',t'|x_0,t_0)$$

$$+ \int_{-\infty}^{\infty}dx'\int_{-\infty}^{\infty}dt'\int_{-\infty}^{\infty}dx''\int_{-\infty}^{\infty}dt''\,W_D(x_t,t|x',t')$$

$$\times V(x',t')W_D(x',t'|x'',t'')V(x'',t'')W_D(x'',t''|x_0,t_0) - \cdots \qquad (1.2.86)$$

which is nothing but the iterative solution of the integral equation (1.2.82) for the transition probability W_B which satisfies also the differential equation (1.2.81). This shows that the continuous limit (1.2.79) of the discrete approximation (1.2.83) is indeed the path-integral representation for the transition probability satisfying the Bloch equation (1.2.76) (or (1.2.81)). In our consideration we skipped some mathematical details which the reader may find in Kac (1959).

It is namely this connection among Wiener integrals, differential equations and integral equations which can be frequently used to show that physical systems that bear no resemblance to each other are nevertheless mathematically highly similar. In particular, the Bloch equation is closely related to the Schrödinger equation, which is a cornerstone in quantum mechanics (the latter turns into the former after the transition to the imaginary time $t \to it$). We shall heavily use this important fact in the next chapter. In the case of quantum mechanics, the function $V(x)$ has the meaning of potential energy.

◇ Description of a Brownian particle moving in the field of an external force by the Bloch equation

Within classical physics, the direct physical motivation, apart from the absorption, for investigating the Bloch equation comes from the consideration of a Brownian particle moving in the field of an external force.

If the friction coefficient η (cf (1.2.24)) is sufficiently large, the drift velocity is directly defined by an external force F, so that in (1.2.15) we have

$$A(x_t, t) = \frac{F}{\eta}.$$

This results in the Fokker–Planck equation (1.2.21) with a *non-vanishing* first-order space derivative term. For the sake of generality, let us write the three-dimensional generalization of the Fokker–Planck equation (1.2.21) with an external force $F(r)$:

$$\frac{\partial W}{\partial t} = D \triangle W - \eta^{-1} \operatorname{div}(W F) \tag{1.2.87}$$

where $r = (x, y, z)$ denotes the three Cartesian coordinates, $F(r)$ denotes the external force, $\triangle$ and div denote the Laplacian operator and the divergence, respectively:

$$\triangle W \stackrel{\text{def}}{\equiv} \frac{\partial^2 W}{\partial x^2} + \frac{\partial^2 W}{\partial y^2} + \frac{\partial^2 W}{\partial z^2} \tag{1.2.88}$$

$$\operatorname{div} K \stackrel{\text{def}}{\equiv} \frac{\partial K_x}{\partial x} + \frac{\partial K_y}{\partial y} + \frac{\partial K_z}{\partial z} \tag{1.2.89}$$

(the reader may derive the three-dimensional analog of the Fokker–Planck equation (1.2.21) as a simple exercise).

The most straightforward way to derive a path-integral representation for the solution of (1.2.87) is to write

$$W(r_t, t | r_0, t_0) = \exp\left\{\frac{3\pi R}{k_B T} \int_{r_0}^{r_t} dr \cdot F\right\} \widetilde{W}(r_t, t | r_0, t_0) \tag{1.2.90}$$

where k_B is the Boltzmann constant, T is the temperature and R is the radius of the particle. In this formula, the line integral follows any continuous path which starts at r_0 and ends at r_t. The value of the line integral is independent of the form of this contour, provided the external force F is conservative, i.e. F can be written as the gradient of a scalar function:

$$F = -\nabla \Phi$$

where

$$\nabla \stackrel{\text{def}}{\equiv} \left\{\frac{\partial}{\partial x}, \frac{\partial}{\partial y}, \frac{\partial}{\partial z}\right\}.$$

When (1.2.90) is substituted into (1.2.87), we find that $\widetilde{W}$ must be a solution of the three-dimensional generalization of the Bloch equation (1.2.76) (or (1.2.81), if the time variable is extended to the whole real line)

$$\frac{\partial \widetilde{W}}{\partial t} = D \triangle \widetilde{W} - V(r) \widetilde{W} \tag{1.2.91}$$

where

$$V \stackrel{\text{def}}{\equiv} \frac{3\pi R}{2\eta k_B T} F^2 + \frac{\operatorname{div} F}{2\eta}$$

$$= \frac{1}{4\eta^2 D} F^2 + \frac{\operatorname{div} F}{2\eta} \tag{1.2.92}$$

and where we have used the Einstein relation (1.2.61) which connects the diffusion constant, the friction coefficient η and the Boltzmann constant. Note that V in (1.2.75), (1.2.91) and (1.2.92) *is not* the potential energy of a particle in the field of the external force $\boldsymbol{F}$.

In the limit $t \to 0$, the function $\widetilde{W}$ has to approach $\delta(\boldsymbol{r}_t - \boldsymbol{r}_0)$. Hence $\widetilde{W}$ is given by the three-dimensional generalization of the Feynman–Kac formula (1.2.79) and using (1.2.90), we find for $t > t_0$ the transition probability of a Brownian particle subject to an external force $\boldsymbol{F}$ as follows:

$$W(\boldsymbol{r}_t, t | \boldsymbol{r}_0, t_0) = \exp\left\{\frac{1}{2\eta D}\int_{\boldsymbol{r}_0}^{\boldsymbol{r}_t} d\boldsymbol{r} \cdot \boldsymbol{F}\right\} \int_{\mathcal{C}\{\boldsymbol{r}_0, 0; \boldsymbol{r}_t, t\}} d_W x(\tau)\, d_W y(\tau)\, d_W z(\tau)\, \exp\left\{-\int_0^t d\tau\, V(\boldsymbol{r}(\tau))\right\}. \tag{1.2.93}$$

The reader may verify that the one-dimensional version of (1.2.93) gives, in the case of a *harmonic* external force, the same path-integral representation for the transition probability as the one derived in section 1.2.2 using the microscopic approach (cf (1.2.38)) by the functional change of random variables (problem 1.2.10, page 115).

Note that the same formula (1.2.93) is valid for non-conservative forces (which cannot be represented as a gradient of a potential). We shall not prove the formula for this case, referring the reader to the original work (Hunt and Ross 1981). However, to become acquainted with the situation involving non-conservative forces, we suggest solving problem 1.2.11, page 115.

◇ **Proof of the Feynman–Kac theorem**

Now we shall present a more rigorous proof of the Feynman–Kac formula (Feynman 1942, 1948, Kac 1949) following Dynkin (1955) (further mathematical refinements can be found in Evgrafov (1970)). In other words, we shall show that the Wiener path integral

$$W_B(x_t, t | x_0, 0) = \int_{\mathcal{C}\{x_0, 0; x_t, t\}} d_W x(s)\, \exp\left\{-\int_0^t ds\, V(x(s))\right\} \tag{1.2.94}$$

is the fundamental solution of the *Bloch equation*

$$\frac{\partial W}{\partial t} = \frac{1}{4}\frac{\partial^2 W}{\partial x_t^2} - V(x_t) W. \tag{1.2.95}$$

For simplicity, we have again put

$$D = \tfrac{1}{4}.$$

First of all, we note that the transition probability (1.2.94) satisfies the ESKC relation (1.1.47). Indeed,

$$\int_{-\infty}^{\infty} dx_s\, W_B(x_t, t | x_s, s) W_B(x_s, s | x_0, 0) = \int_{-\infty}^{\infty} dx_s \int_{\mathcal{C}\{x_0, 0; x_s, s\}} d_W x(\tau)\, \exp\left\{-\int_0^s d\tau\, V(x(\tau))\right\}$$

$$\times \int_{\mathcal{C}\{x_s, s; x_t, t\}} d_W x(s')\, \exp\left\{-\int_s^t d\tau'\, V(x(\tau'))\right\}$$

$$= \int_{-\infty}^{\infty} dx_s \int_{\mathcal{C}\{x_0, 0; x_s, s; x_t, t\}} d_W x(\tau)\, \exp\left\{-\int_0^t d\tau\, V(x(\tau))\right\}$$

$$= \int_{\mathcal{C}\{x_0, 0; x_t, t\}} d_W x(\tau)\, \exp\left\{-\int_0^t d\tau\, V(x(\tau))\right\}$$

$$= W_B(x_t, t | x_0, 0).$$

Here $\mathcal{C}\{x_0, 0; x_\tau, \tau; x_t, t\}$ denotes the set of trajectories having the start and endpoint at x_0 and x_t, respectively, and passing through the point x_τ at the time τ.

Taking into account that at $t \to 0$ the integrand of (1.2.94) is reduced to the purely Wiener measure, we can easily derive the initial condition which W_B satisfies:

$$W_B(x_t, t|x_0, 0) \xrightarrow[t \to 0]{} \delta(x_t - x_0). \tag{1.2.96}$$

The functional entering (1.2.94) satisfies the identity

$$\exp\left\{-\int_0^t d\tau\, V(x(\tau))\right\} = 1 - \int_0^t d\tau \left(V(x(\tau)) \exp\left\{-\int_0^\tau ds\, V(x(s))\right\}\right) \tag{1.2.97}$$

(which can be proved using differentiation by t; the value of the constant is found from the condition of coincidence of both sides of the relation at $t = 0$). This identity is correct for all continuous functions $x(s)$ and can be integrated with Wiener measure:

$$\int_{\mathcal{C}\{x_0,0;x_t,t\}} d_W x(\tau) \exp\left\{-\int_0^t d\tau\, V(x(\tau))\right\} = \int_{\mathcal{C}\{x_0,0;x_t,t\}} d_W x(\tau)$$
$$- \int_{\mathcal{C}\{x_0,0;x_t,t\}} d_W x(s) \left[\int_0^t d\tau \left(V(x(\tau)) \exp\left\{-\int_0^\tau ds\, V(x(s))\right\}\right)\right]. \tag{1.2.98}$$

The first term in the right-hand side of (1.2.98) is the fundamental solution of the diffusion equation and the second one is a well-defined Wiener integral (because the integrand is an exponentially decreasing continuous functional), so that we can change the order of the integrations:

$$\int_{\mathcal{C}\{x_0,0;x_t,t\}} d_W x(\tau) \left[\int_0^t ds \left(V(x(s)) \exp\left\{-\int_0^s d\tau\, V(x(\tau))\right\}\right)\right]$$
$$= \int_0^t ds \int_{\mathcal{C}\{x_0,0;x_t,t\}} d_W x(\tau) \left[V(x(s)) \exp\left\{-\int_0^s d\tau\, V(x(\tau))\right\}\right]$$
$$= \int_0^t ds \int_{-\infty}^\infty dx_s \int_{\mathcal{C}\{x_0,0;x_s,s;x_t,t\}} d_W x(\tau) \left[V(x_s) \exp\left\{-\int_0^s d\tau\, V(x(\tau))\right\}\right]$$
$$= \int_0^t ds \int_{-\infty}^\infty dx_s\, V(x_s) \int_{\mathcal{C}\{x_0,0;x_s,s\}} d_W x(\tau) \exp\left\{-\int_0^s d\tau\, V(x(\tau))\right\} \int_{\mathcal{C}\{x_s,s;x_t,t\}} d_W x(s')$$
$$= \int_0^t ds \int_{-\infty}^\infty dx_s\, V(x_s) W_B(x_s, s|x_0, 0) W_D(x_t, t|x_s, s) \tag{1.2.99}$$

where $W_D(x_t, t|x_\tau, \tau)$ is the transition probability (fundamental solution) (1.1.35) for the diffusion equation. Thus equation (1.2.98) takes the form

$$W_B(x_t, t|x_0, 0) = W_D(x_t, t|x_0, 0) - \int_0^t d\tau \int_{-\infty}^\infty dx_\tau\, W_D(x_t, t|x_\tau, \tau) V(x_\tau) W_B(x_\tau, \tau|x_0, 0). \tag{1.2.100}$$

Since W_B (as well as W_D) satisfies the ESKC equation (hence it corresponds to a Markovian process), there exists an operator of finite shift in time corresponding to some first order (in time derivative) differential equation (analogously to W_D which defines a shift in time for the diffusion equation), i.e.

$$\frac{\partial W_B}{\partial t} = \widehat{L}_B W_D. \tag{1.2.101}$$

Here $\widehat{L}_B$ is the infinitesimal operator (generator) of time shifts. It is clear that this can be calculated by the formula

$$\widehat{L}_B = \lim_{t\to 0}\frac{W_B - \mathbb{I}}{t} \tag{1.2.102}$$

where $\mathbb{I}$ is the unit (identity) operator. In this relation, we consider $W_B(x_t, t|x_0, 0)$ as an integral kernel of the operator $\widehat{W}_B$, acting on an arbitrary function $g(x)$, as follows:

$$[\widehat{W}_B g](x_t) = \int_{-\infty}^{\infty} dx_0\, W_B(x_t, t|x_0, 0)g(x_0).$$

Taking into account (1.2.100), we have

$$\widehat{L}_B = \widehat{L}_D - \lim_{t\to 0}\frac{1}{t}\int_0^t d\tau \int_{-\infty}^{\infty} dx_\tau\, V(x_\tau) W_B(x_\tau, \tau|x_0, 0) W_D(x_t, t|x_\tau, \tau) \tag{1.2.103}$$

where

$$\widehat{L}_D = \frac{1}{4}\frac{\partial^2}{\partial x^2}$$

is the infinitesimal operator for the diffusion equation. The initial conditions for W_B and W_D (i.e. $W_{B,D}(x_t, t|x_0, 0) \xrightarrow[t\to 0]{} \delta(x_t - x_0)$) give

$$\lim_{t\to 0}\frac{1}{t}\int_0^t d\tau \int_{-\infty}^{\infty} d\xi\, dx_\tau\, W_D(x_t, t|x_\tau, \tau) V(x_\tau) W_B(x_\tau, \tau|\xi, 0) g(\xi) = V(x_0)g(x_0)$$

(note that $0 \le \tau \le t$ and hence $\tau \xrightarrow[t\to 0]{} 0$), from which we immediately derive

$$\widehat{L}_B = \frac{1}{4}\frac{\partial^2}{\partial x^2} - V(x) \tag{1.2.104}$$

i.e. the infinitesimal operator corresponding to W_B coincides with the right-hand side of the Bloch equation (1.2.95).

1.2.6 Variational methods of path-integral calculations: semiclassical and quadratic approximations and the method of hopping paths

Let us return to the general case of a Brownian particle moving in the field of an external force with the transition probability defined by (1.2.93). The corresponding probability density functional for the particle to follow a specific trajectory $r(t)$ has the form (cf (1.1.51) in the case of a zero external force and one-dimensional space)

$$\Omega[r(\tau)] = \exp\left\{\frac{1}{2\eta D}\int_{r_0}^{r_t} dr \cdot F - \frac{1}{4D}\int_{t_0}^{t} d\tau \left(\frac{dr}{d\tau}\right)^2 - \int_{t_0}^{t} d\tau\, V(r(\tau))\right\}. \tag{1.2.105}$$

Writing

$$\int_{r_0}^{r} dr \cdot F = \int_{t_0}^{t} d\tau\, F \cdot \frac{dr}{d\tau}$$

and using the explicit form (1.2.92) of the function $V(r)$, we find the alternative and very suggestive expression

$$\Omega = \exp\left\{-\frac{1}{4D}\int_{t_0}^{t} d\tau\, L(r(\tau))\right\} \tag{1.2.106}$$

where
$$L(\boldsymbol{r}(\tau)) = \left(\frac{d\boldsymbol{r}}{d\tau} - \frac{\boldsymbol{F}}{\eta}\right)^2 + 2\frac{D}{\eta}\,\mathrm{div}\,\boldsymbol{F}. \qquad (1.2.107)$$

In order to extract some physical interpretation of the theory, it is natural to consider first the limit $D \to 0$, in which the fluctuations due to Brownian motion are switched off. In this limit we can neglect the second term on the right-hand side of (1.2.107) and, as the coefficient $(4D)^{-1}$ becomes infinite, (1.2.106) shows that the only contributing trajectory satisfies the condition

$$\int_{t_0}^{t} d\tau \left(\frac{d\boldsymbol{r}}{d\tau} - \frac{\boldsymbol{F}}{\eta}\right)^2 = 0 \qquad (1.2.108)$$

that is, the solution of
$$\frac{d\boldsymbol{r}}{d\tau} = \frac{\boldsymbol{F}}{\eta}. \qquad (1.2.109)$$

But this is exactly the equation of motion of a particle in an external field of force $\boldsymbol{F}$ in the presence of *very large* friction in the medium (i.e. neglecting the term with second-order time derivative). Hence, in the limit $D \to 0$, the path integral leads back to the deterministic dynamical equation of ordinary classical mechanics.

For a finite value of the diffusion constant D, all trajectories $\boldsymbol{r}(\tau)$ will contribute to the path integral (1.2.93). The largest contribution will come from the trajectory for which the integral in the exponential in (1.2.106) is as small as possible. According to the rules of the variational calculus (see, e.g., Courant and Hilbert (1953) and Sagan (1992)), a trajectory with fixed endpoints, which provides the minimum of the functional

$$S \stackrel{\mathrm{def}}{\equiv} \int_{t_0}^{t} d\tau\, L(\boldsymbol{r}(\tau))$$

in the exponent, satisfies the *extremality* condition

$$\delta S \equiv \delta \int_{t_0}^{t} d\tau\, L = 0 \qquad (1.2.110)$$

with both ends $\boldsymbol{r}(t_0) = \boldsymbol{r}_0$ and $\boldsymbol{r}(t) = \boldsymbol{r}$ fixed and where δ is the functional variation. This condition is equivalent to the *Euler–Lagrange equations*

$$\begin{aligned}
\frac{\partial L}{\partial x} - \frac{\partial}{\partial \tau}\frac{\partial L}{\partial \dot{x}} &= 0 \\
\frac{\partial L}{\partial y} - \frac{\partial}{\partial \tau}\frac{\partial L}{\partial \dot{y}} &= 0 \\
\frac{\partial L}{\partial z} - \frac{\partial}{\partial \tau}\frac{\partial L}{\partial \dot{z}} &= 0.
\end{aligned} \qquad (1.2.111)$$

Hence, for small values of the diffusion constant, an asymptotic expression for the transition probability is simply given by

$$W(\boldsymbol{r}_t, t|\boldsymbol{r}_0, t_0) \simeq \phi(t - t_0)\exp\left\{-\frac{1}{4D}\int_{t_0}^{t} d\tau\, L[\boldsymbol{r}_\mathrm{c}(\tau)]\right\} \qquad (1.2.112)$$

where $\boldsymbol{r}_\mathrm{c}(\tau)$ is the solution (classical trajectory) of (1.2.111). Note that the diffusion constant is a dimensionfull quantity, therefore the smallness is understood in the sense that D is small in comparison with the functional S: $D/S \ll 1$. Calculation of the pre-exponential factor $\phi(t - t_0)$ (also called the *fluctuation factor*) is outside the scope of the semiclassical approximation evaluation of the path integrals. Note, however, that this factor does not depend on the initial and final points of the Brownian motion (strictly speaking, this is correct only for quadratic functions $L(x(\tau))$; see explanations later). Hence, if we are looking for a path-integral solution of an equation for some *probability*, the factor $\phi(t - t_0)$ can be determined from the requirement that the integral of W over the whole three-dimensional space should be equal to unity for any $t - t_0$ (the normalization condition).

In classical mechanics, the Euler–Lagrange equations (1.2.111) determine the actual trajectory of a classical particle. That is why the approximation (1.2.112) for the transition probability and the corresponding path integral, which takes into account only the solution of (1.2.111), is called the *semiclassical approximation*. Note, however, that the function L in (1.2.111) does not have the direct meaning of a Lagrangian of the Brownian particle under consideration (cf (1.2.92) and (1.2.107)). In the case of quantum mechanics, where the name of this approximation, semiclassical, has come from, the function L has indeed the direct meaning of the Lagrangian and $r_c(\tau)$ is the corresponding true classical trajectory (see the next chapter).

The general approach to the calculations of path integrals by the variational method has been suggested by R Feynman (1942, 1948, 1972a) and Feynman and Hibbs (1965).

We shall discuss different versions of variational calculations throughout the book. In this subsection, we shall deal with examples of *semiclassical*, *quadratic* and *hopping path* approximations. Further development of the variational methods for path integration the reader will find in sections 2.1.4, 2.2.3, 3.3.2, 3.3.3 and 3.3.5.

◇ **Examples of calculations by the semiclassical method**

To simplify formulae we confine ourselves to the one-dimensional case; generalization to higher-dimensional spaces is straightforward. In the one-dimensional case the functionals which we are going to integrate have the form

$$F[x(\tau)] = \exp\left\{-\int_0^t d\tau\, f(\dot{x}(\tau), x(\tau), \tau)\right\} \tag{1.2.113}$$

where $f(\dot{x}, x, \tau)$ is some function. The classical trajectory $x_c(\tau)$ satisfies the extremal condition

$$\delta \int_0^t d\tau\, f(\dot{x}(\tau), x(\tau), \tau)\Big|_{x(\tau) = x_c(\tau)} = 0 \quad \text{with } x_c(0) = x_0 \text{ and } x_c(t) = x_t \text{ fixed} \tag{1.2.114}$$

and it is a solution of the *Euler–Lagrange equation*

$$\frac{\partial f(\dot{x}_c, x_c, \tau)}{\partial x_c} - \frac{d}{d\tau}\frac{\partial f(\dot{x}_c, x_c, \tau)}{\partial \dot{x}_c} = 0 \tag{1.2.115}$$

$$x_c(0) = x_0 \quad x_c(t) = x_t.$$

Now let us turn to some examples of functionals and corresponding integrals.

Example 1.6. Calculate the basic path integral (i.e. transition probability; again put $D = 1/4$)

$$W(x_t, t | x_0, 0) = \int_{C\{x_0, 0; x_t, t\}} \prod_{\tau=0}^{t} \frac{dx(\tau)}{\sqrt{\pi dt}} \exp\left\{-\int_0^t dt\, \dot{x}^2\right\} \tag{1.2.116}$$

by the semiclassical method.

Wiener path integrals and stochastic processes 81

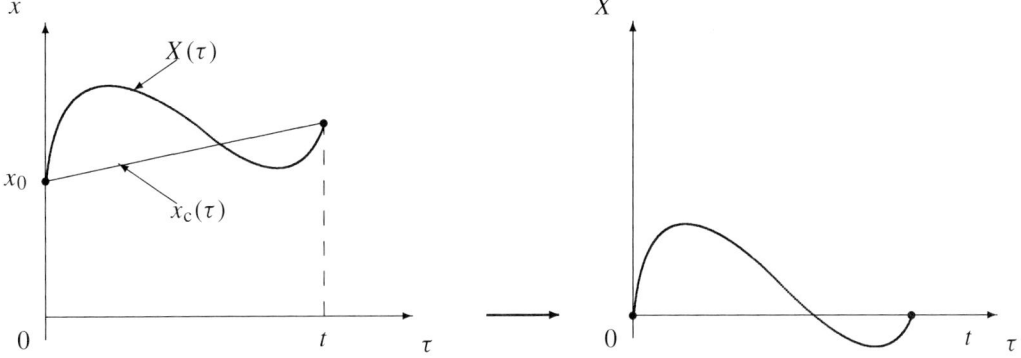

Figure 1.11. Detachment of the classical and fluctuation parts of free Brownian motion.

The Euler–Lagrange equation (1.2.115) in this case has the simple form

$$\ddot{x}_c(\tau) = 0 \qquad x_c(0) = x_0 \qquad x_c(t) = x_t \qquad (1.2.117)$$

and the solution is

$$x_c(\tau) = x_0 + \frac{\tau}{t}(x_t - x_0). \qquad (1.2.118)$$

Let us shift the trajectory $x(\tau)$ by the extremal (classical) trajectory $x_c(\tau)$ through the change of variables $x(\tau) \to X(\tau)$:

$$x(\tau) = x_c(\tau) + X(\tau) \qquad (1.2.119)$$

with $x_c(0) = x_0$, $x_c(t) = x_t$, so that

$$X(0) = X(t) = 0 \qquad (1.2.120)$$

(see figure 1.11).

Since $x_c(\tau)$ is a fixed trajectory, the difference in trajectories, $X(\tau)$, is now a new variable to be integrated over in the path integral. The Jacobian of the transformation $x(\tau) \to X(\tau)$ is obviously equal to unity. Inserting (1.2.119) into (1.2.116), we obtain

$$W(x_t, t|x_0, 0) = \exp\left\{-\int_0^t d\tau\, \dot{x}_c^2\right\} \int_{C\{0,0;0,t\}} \prod_{\tau=0}^t \frac{dX(\tau)}{\sqrt{\pi d\tau}} \exp\left\{-\int_0^t d\tau\, (\dot{X}^2 + 2\dot{x}_c\dot{X})\right\}$$

$$= \exp\left\{-\int_0^t dt\, \dot{x}_c^2\right\} \int_{C\{0,0;0,t\}} \prod_{\tau=0}^t \frac{dX(\tau)}{\sqrt{\pi d\tau}} \exp\left\{-\int_0^t d\tau\, \dot{X}^2\right\} \qquad (1.2.121)$$

since

$$\int_0^t d\tau\, \dot{x}_c\dot{X} = 0.$$

The last equation is the consequence of (1.2.120) and of the fact that for the extremal trajectory $x_c(\tau)$ given by (1.2.118), its time derivative is independent of τ:

$$\dot{x}_c(\tau) = \frac{x_t - x_0}{t}.$$

We note that the path integral over the new variable $X(\tau)$ in (1.2.121) is now only a function of time, i.e.

$$\int_{C\{0,0;0,t\}} \prod_{\tau=0}^{t} \frac{dX(\tau)}{\sqrt{\pi d\tau}} \exp\left\{-\int_0^t d\tau\, \dot{X}^2\right\} = \phi(t). \qquad (1.2.122)$$

The reason is that the path integral in (1.2.122) is performed over the trajectories with fixed endpoints (conditional integral) and thus it depends only on $X(0)$, $X(t)$ and on the time t. However, because the endpoints have been chosen by construction to be zero (cf (1.2.119) and (1.2.120)), the path integral depends only on the final time t.

Sometimes, the Gaussian random process $X(\tau)$, describing fluctuations of the Brownian particle around the classical path $x_c(\tau)$, is called a *Brownian bridge*.

The explicit form of the extremal path (1.2.118) gives

$$\int_0^t d\tau\, \dot{x}_c^2 = \frac{(x_t - x_0)^2}{t}$$

and the path integral (1.2.116) proves to be the following:

$$W(x_t, t|x_0, 0) = \phi(t) \exp\left\{-\frac{(x_t - x_0)^2}{t}\right\}. \qquad (1.2.123)$$

Since we know that the path integral (1.2.116) defines the transition probability, the function $\phi(t)$ is found from the normalization

$$\int_{-\infty}^{\infty} dx_t\, W(x_t, t|x_0, 0) = 1 \qquad (1.2.124)$$

which gives

$$\phi(t) = \frac{1}{\sqrt{\pi t}}.$$

Thus, in this example, the semiclassical method gives the *exact result* ((1.1.53) with $D = 1/4$).

In the limit $t \to 0$, we can directly check that the path integral (1.2.116) indeed satisfies the normalization condition (1.2.124), because for $t \to 0$, (1.2.116) can be written as

$$\lim_{t\to 0} W(x_t, t|x_0, 0) = \lim_{t\to 0} \frac{1}{\sqrt{\pi t}} e^{-x^2/t} = \delta(x_t - x_0). \qquad (1.2.125)$$

A rigorous calculation of the *fluctuation factor* $\phi(t - t_0)$ requires more complicated methods, e.g., the time-slicing (discretization) method which we used in example 1.1, page 37.

— o —

Example 1.7. Calculate the Wiener path integral

$$I_7 = \int_{C\{x_0,0;x_t,t\}} d_W x(\tau) \exp\left\{-k^2 \int_0^t d\tau\, x^2\right\}. \qquad (1.2.126)$$

Acting exactly as in example 1.6, we shall find first the extremal trajectory. The Euler equation (1.2.115) now reads

$$\ddot{x}_c - k^2 x_c = 0 \qquad (1.2.127)$$

with the solution
$$x_c(\tau) = C_1 e^{k\tau} + C_2 e^{-k\tau}. \tag{1.2.128}$$

The integration constants are defined from the boundary conditions:
$$x_0 = C_1 + C_2 \qquad x_t = C_1 e^{kt} + C_2 e^{-kt}$$

which give
$$C_1 = \frac{x_t - x_0 e^{-kt}}{e^{kt} - e^{-kt}} = 2\frac{x_t - x_0 e^{-kt}}{\sinh(kt)}$$

$$C_2 = -\frac{x_t - x_0 e^{kt}}{e^{kt} - e^{-kt}} = -2\frac{x_t - x_0 e^{kt}}{\sinh(kt)}.$$

Then, in analogy with the preceding example 1.6, we obtain
$$\int_0^t d\tau \, (\dot{x}^2 + k^2 x^2) = \int_0^t d\tau \, (\dot{x}_c^2 + k^2 x_c^2) + \int_0^t d\tau \, (\dot{X}^2 + k^2 X^2) \tag{1.2.129}$$

where the classical part proves to be the following:
$$\int_0^t d\tau \, (\dot{x}_c^2 + k^2 x_c^2) = k \frac{(x_t^2 + x_0^2)(e^{kt} + e^{-kt}) - 4x_0 x_t}{e^{kt} - e^{-kt}}$$
$$= k \frac{(x_t^2 + x_0^2) \cosh(kt) - 2x_0 x_t}{\sinh(kt)}.$$

The cross-terms in (1.2.129) (containing both x_c and X) have again disappeared due to the classical equation of motion (1.2.127) (which is not an accidental fact as we shall explain soon) and the path integral takes the form
$$I_7 = \phi(t) \exp\left\{-k \frac{(x_t^2 + x_0^2) \cosh(kt) - 2x_0 x_t}{\sinh(kt)}\right\}. \tag{1.2.130}$$

The function $\phi(t)$ is again found from the normalization condition analogous to (1.2.124) and reads as
$$\phi(t) = \sqrt{\frac{k}{\pi \sinh(kt)}}. \tag{1.2.131}$$

For the particular values $x_0 = 0$ the result (1.2.130), (1.2.131) coincides with that obtained in section 1.1.4 (cf (1.1.116)) which was computed in that section by using the discrete-time approximation and the Gelfand–Yaglom method.

This path integral gives the transition probability $W_k^{(\text{harm})}$ for the Bloch equation with the harmonic function $V(x) = k^2 x^2$. Since this result is important practically, we rewrite it restoring explicitly the diffusion constant
$$W_k^{(\text{harm})}(x_t, t|x_0, t_0) = \int_{C\{x_0, t_0; x_t, t\}} d_W x(\tau) \exp\left\{-k^2 \int_{t_0}^t d\tau \, x^2\right\}$$
$$= \left[\frac{k}{2\pi \sqrt{D} \sinh(2k\sqrt{D}(t - t_0))}\right]^{\frac{1}{2}}$$
$$\times \exp\left\{-k \frac{(x_t^2 + x_0^2) \cosh(2k\sqrt{D}(t - t_0)) - 2x_0 x_t}{2\sqrt{D} \sinh(2k\sqrt{D}(t - t_0))}\right\}. \tag{1.2.132}$$

One can also check that $W_k^{(\text{harm})}$ (1.2.132) indeed satisfies the characteristic property of a fundamental solution:
$$W_k^{(\text{harm})}(x_t, t|x_0, 0) \xrightarrow[t \to 0]{} \delta(x_t - x_0).$$

Also, we see that in the limit $k \to 0$, (1.2.132) becomes the fundamental solution (1.1.53) of the diffusion equation. The semiclassical method again produces an *exact* result.

— ○ —

Example 1.8. Apply the variational method of calculation to the following Wiener integral
$$I_8 = \int_{C\{x_0,0;x_t,t\}} \prod_{\tau=0}^{t} \frac{dx(\tau)}{\sqrt{\pi dt}} \exp\left\{kt/2 - \int_0^t dt\,(\dot{x} + kx)^2\right\} \tag{1.2.133}$$

which defines the transition probability of a Brownian particle in the field of an external harmonic force (cf section 1.2.2, (1.2.38)).

For the calculation of this integral we can use the results of the preceding example. Indeed, rewrite I_8 as follows

$$I_8 = \exp\{kt/2\} \int_{C\{x_0,0;x_t,t\}} d_W x(\tau) \exp\left\{-k^2 \int_0^t d\tau\, x^2\right\} \exp\left\{-2k \int_0^t d\tau\, \dot{x}x\right\}$$
$$= \exp\{kt/2\} \exp\{-k(x_t^2 - x_0^2)\} \int_{C\{x_0,0;x_t,t\}} d_W x(\tau) \exp\left\{-k^2 \int_0^t d\tau\, x^2\right\}$$
$$= I_7 \exp\{kt/2\} \exp\{-k(x_t^2 - x_0^2)\}.$$

The result of the preceding example and simple algebraic calculations give the expression for the transition probability $W_k^{(\text{h.f.})}$ of a Brownian particle subject to a harmonic force:

$$W_k^{(\text{h.f.})}(x_t, t|x_0, t_0) = \left[\frac{k}{2\pi D(1 - e^{-2k(t-t_0)})}\right]^{\frac{1}{2}} \exp\left\{-\frac{k}{2D} \frac{(x_t - x_0 e^{-k(t-t_0)})^2}{(1 - e^{-2k(t-t_0)})}\right\}. \tag{1.2.134}$$

— ○ —

Example 1.9. Calculate the Wiener integral (1.2.70)
$$W(x_t, v_t, t|x_0, v_0, t_0) = \int_{C\{v_0,0;v_t,t\}} \prod_{\tau=0}^{t} \frac{dv(\tau)}{\sqrt{\pi d\tau}} \exp\left\{-\int_0^t d\tau\, \dot{v}^2\right\} \delta\left[x(t) - x(0) - \int_0^t d\tau\, v(\tau)\right] \tag{1.2.135}$$

i.e. the path integral with the constraint (1.2.71):
$$x(t) - x(0) = \int_0^t d\tau\, v(\tau) \tag{1.2.136}$$

by the variational method with Lagrange multiplier.

In order to use the semiclassical method of calculation, we have to find an extremum of the exponential in (1.2.135) under the condition (1.2.136). The standard way to do this is the method of the Lagrange multiplier (Courant and Hilbert 1953). According to this method, we have to add to the functional $\int_0^t d\tau\, \dot{v}^2(\tau)$ the constraint (1.2.136) with an initially indefinite multiplier λ:

$$\int_0^t d\tau\, \dot{v}^2(\tau) + \lambda \left[\int_0^t d\tau\, v(\tau) - (x - x_0) \right]$$

($x_0 = x(0)$, $x = x(t)$) and to find the extremum of this new λ-dependent functional. This is given by the Euler equation

$$2\ddot{v} + \lambda = 0$$

with the obvious general solution

$$v(\tau) = -\frac{\lambda}{4}\tau^2 + C_1 \tau + C_2. \qquad (1.2.137)$$

The integration constants C_1, C_2 and the Lagrange multiplier λ are defined by the boundary conditions

$$\begin{cases} v(0) = v_0 \\ v(t) = v_t \end{cases} \Rightarrow \begin{cases} C_2 = v_0 \\ -\lambda t^2/4 + C_1 t + C_2 = v \end{cases}$$

and by the constraint

$$\int_0^t d\tau \left(-\frac{\lambda}{4}\tau^2 + C_1 \tau + C_2 \right) = x - x_0$$

or

$$-\frac{\lambda}{12} t^3 + \frac{1}{2} C_1 t^2 + C_2 t = x - x_0.$$

These equations give

$$\begin{aligned} C_1 &= \frac{v - v_0}{t} + \frac{6}{t^2}\left[x - x_0 - \frac{1}{2}(v + v_0)t \right] \\ C_2 &= v_0 \\ \lambda &= \frac{24}{t^3}\left[x - x_0 - \frac{1}{2}(v + v_0)t \right]. \end{aligned} \qquad (1.2.138)$$

Substitution of the solution (1.2.137) with the constants (1.2.138) into the exponential of the path integral (1.2.135) results in the expression

$$W(x_t, v_t, t | x_0, v_0, 0) = \phi(t) \exp\left\{ -\frac{12}{t^3}\left(x_t - x_0 - \frac{v_t + v_0}{2} t \right)^2 - \frac{(v_t - v_0)^2}{t} \right\}. \qquad (1.2.139)$$

The normalization condition (which can be checked in a way analogous to (1.2.125)) for the transition probability (1.2.135),

$$\int_{-\infty}^{\infty} dx_t\, dv_t\, W(x_t, v_t, t | x_0, v_0, 0) = 1 \qquad (1.2.140)$$

gives the normalization constant $\phi(t)$ and, finally, the transition probability for a Brownian particle with inertia proves to be

$$W(x_t, v_t, t | x_0, v_0, 0) = \frac{\sqrt{12}}{\pi t^2} \exp\left\{ -\frac{12}{t^3}\left((x_t - x_0) - \frac{v_t + v_0}{2} t \right)^2 - \frac{(v_t - v_0)^2}{t} \right\}. \qquad (1.2.141)$$

86   Path integrals in classical theory

◇ Calculations of path integrals in the quadratic approximation

The next approximation of the transition probability which is suggested by the path-integral representation (1.2.79), (1.2.94) (Feynman–Kac formula) is generally known as the *quadratic approximation*. This method forms an extension of the semiclassical approximation discussed earlier.

Let us consider the transition probability

$$W(x,t|x_0,t_0) = \int_{\mathcal{C}\{x_0,t_0;x,t\}} \prod_{\tau=0}^{t} \frac{dx(\tau)}{\sqrt{\pi d\tau}} \exp\left\{-\frac{1}{4D}\int_0^t d\tau\,[\dot{x}^2(\tau) + 4DV(x(\tau))]\right\} \quad (1.2.142)$$

specified by the Gaussian functional with a function $V(x)$. Write, as usual,

$$x(\tau) = x_c(\tau) + X(\tau) \quad (1.2.143)$$

where $x_c(\tau)$ stands for the most probable classical trajectory solved from (1.2.111) and (1.2.115), and $X(\tau)$ denotes the deviation from the most probable path (so that $X(0) = X(t) = 0$). The next step is to expand the exponential in (1.2.142) in powers of X. For an arbitrary functional $F[x(\tau)]$, $\tau \in [0, t]$, the expansion reads

$$F[x(\tau)] = F[x_c(\tau) + X(\tau)] = F[x_c(\tau)] + \int_0^t ds\,\frac{\delta F[x(\tau)]}{\delta x(s)}\bigg|_{x(s)=x_c(s)} X(s)$$

$$+ \frac{1}{2!}\int_0^t ds\,du\,\frac{\delta^2 F[x(\tau)]}{\delta x(s)^2}x(u)\bigg|_{x(s)=x_c(s)} X(s)X(u) + \cdots \quad (1.2.144)$$

where

$$\frac{\delta F[x(\tau)]}{\delta x(s)} \qquad \frac{\delta^2 F[x(\tau)]}{\delta x(s)^2}x(u)$$

are *functional derivatives*. In fact, this expansion defines the functional derivatives as the factors at the corresponding powers of X. Recall, that in analogy with ordinary derivatives, a functional derivative $\delta F[x(\tau)]/\delta x(\tau)$ defines the linear part of the difference

$$F[x(\tau) + \eta(\tau)] - F[x(\tau)] = \int_0^t ds\,\frac{\delta F[x(\tau)]}{\delta x(s)}\eta(s) + \cdots \quad (1.2.145)$$

if the variation $\eta(\tau)$ is small enough. Another way to illustrate this analogy is to use differentials of functions

$$df(x_1,\ldots,x_n) = \sum_{i=1}^n \frac{\partial f}{\partial x_i}dx_i \quad \leftrightarrow \quad \delta F[x(\tau)] = \int_0^t ds\,\frac{\delta F[x(\tau)]}{\delta x(s)}\delta x(s). \quad (1.2.146)$$

In the discrete approximation $\tau \in [t_0, t] \longrightarrow \{t_0, t_1, \ldots, t_N = t\}$, where $t_i - t_{i-1} = \varepsilon$ for all $i = 1, 2, \ldots, N$, a functional becomes the function:

$$F[x(\tau)] \longrightarrow F(x_0,\ldots,x_N) \qquad x_j = x(t_j).$$

and the functional derivative turns into partial derivatives:

$$\frac{\delta F}{\delta x(s)} \longrightarrow \frac{1}{\varepsilon}\frac{\partial F}{\partial x_k} \qquad s \approx t_k. \quad (1.2.147)$$

The latter correspondence follows from the continuous limit of the equality

$$F(\ldots, x_i + \eta_i, x_{i+1} + \eta_{i+1}, \ldots) - F(\ldots, x_i, x_{i+1}, \ldots) = \sum_i \frac{\partial F}{\partial x_i} \eta_i.$$

In the concrete case of the integrand (1.2.142), the zero-order term gives rise to the factor

$$\exp\left\{-\frac{1}{4D}\int_0^t d\tau\, [\dot{x}_c^2(\tau) + 4DV(x_c(\tau))]\right\} \qquad (1.2.148)$$

in front of the path integral. The linear term gives no contribution because of the definition of $x_c(\tau)$: the first derivative of a functional

$$F[x(\tau)] = \int d\tau\, f(\dot{x}(\tau), x(\tau), \tau)$$

is equivalent to the left-hand side of the Euler–Lagrange equation (1.2.115) and hence

$$\left.\frac{\delta F[x(\tau)]}{\delta x(s)}\right|_{x=x_c} = 0.$$

The second-order term is of the form

$$-\frac{1}{4D}\int_0^t d\tau\, [\dot{X}^2(\tau) + 4DV_c''(x_c(\tau))X^2(\tau)]. \qquad (1.2.149)$$

The *quadratic approximation* consists in neglecting all terms of order *higher than the second* in the expansion of the exponentials. It will be a good approximation if the path probability decreases rapidly with an increase in the deviation of a path from the extremal (classical) one, i.e. if D is small compared with the integral in (1.2.149).

Collecting the obtained results and using the discrete-time approximation for the path integral, we find the expression

$$W(x,t|x_0,t_0) \simeq F(x,t;x_0,t_0)\exp\left\{-\frac{1}{4D}\int_0^t d\tau\,[\dot{x}_c^2(\tau) + 4DV(x_c)]\right\} \qquad (1.2.150)$$

$$F(x,t;x_0,t_0) = \lim_{\substack{\varepsilon \to 0 \\ N \to \infty}} (4\pi\varepsilon D)^{-(N+1)/2}\int_{-\infty}^{\infty} dX_1 \int_{-\infty}^{\infty} dX_2$$

$$\ldots \int_{-\infty}^{\infty} dX_N \exp\left\{-\frac{1}{4D\varepsilon}\sum_{j=0}^N (X_{j+1} - X_j)^2 - \frac{1}{2}\varepsilon\sum_{j=0}^N V_j'' X_j^2\right\} \qquad (1.2.151)$$

where $V_j'' = V''(x_c(\tau_j))$, $X_0 = X_{N+1} = 0$. The multiple integral can be evaluated by generalization of the *Gelfand–Yaglom method*. Write the exponential in (1.2.151) in the form

$$-\frac{1}{4D\varepsilon}\sum_{k,l=1}^N X_k B_{kl} X_l.$$

The $N \times N$ matrix **B** has zero matrix elements apart from those in the main diagonal and the two neighboring diagonals:

$$B_{kk} = 2 + 2D\varepsilon^2 V_k''$$
$$B_{k,k+1} = B_{k,k-1} - 1.$$

88   Path integrals in classical theory

Denoting its eigenvalues by λ_j, we find

$$F(x, t; x_0, t_0) = \lim_{\substack{\varepsilon \to 0 \\ N \to \infty}} (4\pi\varepsilon D)^{-(N+1)/2} \prod_{j=1}^{N} \left(\frac{4\pi\varepsilon D}{\lambda_j}\right)^{1/2}$$

$$= \lim_{\substack{\varepsilon \to 0 \\ N \to \infty}} (4\pi\varepsilon D \det \mathbf{B}_N)^{-1/2}. \tag{1.2.152}$$

If the determinant $\det \mathbf{B}_N$ of the matrix $\mathbf{B}$ is expanded in the elements of its last column, the following relation is obtained

$$\det \mathbf{B}_N = (2 + 2\varepsilon^2 DV_N'') \det \mathbf{B}_{N-1} - \det \mathbf{B}_{N-2}. \tag{1.2.153}$$

In the absence of an external field we found in section 1.1.4 (cf (1.1.87)) that the determinant of $\mathbf{B}_N$ equals $N+1$. Therefore, let us introduce the quantity

$$C_N = \frac{\det \mathbf{B}_N}{N+1} \tag{1.2.154}$$

which obeys the recurrent relation

$$C_N - 2C_{N-1} + C_{N-2} = -\frac{2}{N+1}(C_N - C_{N-2}) + 2\varepsilon^2 D \frac{NV_N''}{N+1} C_{N-1}. \tag{1.2.155}$$

In the limit $N \to \infty$, $\varepsilon \to 0$, $(N+1)\varepsilon = t - t_0$, the discrete variable C_N tends to a continuous function $C(x, t)$ which obeys the ordinary differential equation:

$$\frac{\partial^2 C}{\partial t^2} + \frac{2}{(t - t_0)} \frac{\partial C}{\partial t} = 2DV''(x_c(t))C. \tag{1.2.156}$$

The initial conditions are

$$C(x, t_0) = 1 \qquad \left.\frac{\partial C}{\partial t}\right|_{t=t_0} = 0. \tag{1.2.157}$$

Substitution of these results into (1.2.152) gives

$$F(x, t) = [4\pi D(t - t_0)C(x, t)]^{-1/2}. \tag{1.2.158}$$

The last step in our evaluation of the quadratic approximation of the path integral consists in the explicit calculation of the function $C(x, t)$. Introduce

$$H(x, t) = (t - t_0)C(x, t) \tag{1.2.159}$$

as the unknown function, which is found to obey

$$\frac{\partial^2 H}{\partial t^2} = 2DV''(x_c(t))H \tag{1.2.160}$$

with the boundary conditions

$$H(x, t_0) = 0 \qquad \left.\frac{dH}{dt}\right|_{t=t_0} = 1. \tag{1.2.161}$$

As usual, the Gelfand–Yaglom method reduces the calculation of the determinant to the solution of the second-order differential equation. If instead of the factor $2DV''(x_c(t))$ on the right-hand side of (1.2.160) stood an arbitrary function of time $p(t)$, as e.g., in (1.1.108), the equation could not be solved. Fortunately, the factor in (1.2.160) is quite specific: it depends on time only through $x_c(t)$ and the second derivative of V,

$$V''(x_c(t)) = \left.\frac{d^2V(x)}{dx^2}\right|_{x=x_c(t)}.$$

Thus it is natural to choose x_c as the independent variable, instead of t. To relate these two variables, let us consider the integrand in (1.2.148) as a Lagrangian:

$$L_f \stackrel{\text{def}}{\equiv} \dot{x}_c^2(\tau) + 4DV(x_c(\tau)) \tag{1.2.162}$$

for some *fictitious* particle (remember that $V(x)$ does not have the direct meaning of potential of an external force acting on a real Brownian particle). Then, as is well known (see, e.g., ter Haar (1971)), there is a conserved quantity, namely, the total energy E of this fictitious particle, which reads

$$E = \dot{x}_c^2 - 4DV.$$

Thus,

$$\frac{dx_c}{dt} = (E + 4DV)^{1/2} \tag{1.2.163}$$

and equations (1.2.160) and (1.2.161) turn into the following one

$$(E + 4DV)H'' + 2DV'H' = 2DV''H \tag{1.2.164}$$

with the initial conditions

$$H|_{x_c=x_0} = 0 \qquad \left.\frac{dH}{dx_c}\right|_{x_c=x_0} = (E + 4DV)^{-1/2}. \tag{1.2.165}$$

Since (1.2.164) can actually be written as

$$[(E + 4DV)H]'' = [6DV'H]' \tag{1.2.166}$$

the solution of the equation is

$$H(x_c) = (E + 4DV(x_0))^{1/2}(E + 4DV(x_c))^{1/2}\int_{x_0}^{x_c} dy\,(E + 4DV(y))^{-3/2} \tag{1.2.167}$$

in the case $x_c > x_0$. Substitution of the last formula and (1.2.158) and (1.2.159) into (1.2.150) gives, for the quadratic approximation of the path integral (1.2.142):

$$W(x, t|x_0, t_0) = \left[4\pi D(E + 4DV(x_0))^{1/2}(E + 4DV(x))^{1/2}\int_{x_0}^{x} dv\,(E + 4DV(v))^{-3/2}\right]^{-1/2}$$
$$\times \exp\left\{-\frac{1}{4D}\int_{t_0}^{t} d\tau\,[\dot{x}_c^2 + 4DV(x_c)]\right\}. \tag{1.2.168}$$

Note that this formula is exact in the case when $V(x)$ is a polynomial of at most second order in x (and coincides with the semiclassical approximation), since then no term of higher than second order occurs and the path integral is a Gaussian one.

90   Path integrals in classical theory

Example 1.10. Using the quadratic approximation, consider *the Ornstein–Uhlenbeck process* (cf section 1.2.2, equation (1.2.31)), i.e. a Brownian motion in the field of the harmonic force

$$F(x) = -\eta k x \qquad (1.2.169)$$

so that the function $V(x)$ takes the form

$$V(x) = \frac{k^2}{4D}x^2 - \frac{k}{2}. \qquad (1.2.170)$$

In this simple case the extremal (classical) trajectory $x_c(\tau)$ can be solved from the Euler–Lagrange equation in a straightforward way (cf examples 1.7 and 1.8) and has the form

$$x_c(\tau) = x_0 e^{-k\tau} + A(e^{k\tau} - e^{-k\tau}) \qquad (1.2.171)$$
$$A = (e^{kt} - e^{-kt})^{-1}(x - x_0 e^{-kt}). \qquad (1.2.172)$$

The function in the first exponent in (1.2.93) reads

$$\frac{1}{2\eta D}\int_{x_0}^{x} dv\, F(v) = \frac{k}{4D}(x_0^2 - x^2) \qquad (1.2.173)$$

while the function in the exponent in (1.2.150) can be written as

$$\int_0^t d\tau\, [\dot{x}_c^2 + 4DV(x_c)] = -2kDt + k(1 - e^{-kt})^{-1}(x - x_0 e^{-kt})^2 - k(1 - e^{kt})^{-1}(x - x_0 e^{kt})^2. \qquad (1.2.174)$$

The factor F in front of the exponent in (1.2.150) becomes

$$F(x, t) = (4\pi D H(x, t))^{-1/2}$$

and can be calculated from (1.2.160) and (1.2.161) which give the differential equation

$$\frac{d^2 H}{dt^2} = k^2 H \qquad (1.2.175)$$

with the solution

$$H(t) = \frac{1}{2k}(e^{kt} - e^{-kt}). \qquad (1.2.176)$$

The combination of (1.2.173), (1.2.174), (1.2.176) and (1.2.167) gives, for the full transition probability of the Ornstein–Uhlenbeck process, again the expression (1.2.134) obtained in example 1.8 by the semiclassical method. Both methods give for this simple path integral an exact result. The reason for this fact is that $V(x)$ is quadratic and the second derivative $V''(x_c)$ in this case does not actually depend on x_c, so that equation (1.2.149) and hence the forefactor F in (1.2.150) are independent of x_t and x_0. Thus, if $V(x)$ is a quadratic polynomial, the factor $F = F(t - t_0)$ can be actually found from the normalization condition, as in the semiclassical calculations.

— ○ —

More complicated examples of calculations in the framework of the quadratic approximation will be considered in subsequent sections of the book.

◇ **Hopping path approximation for Wiener path integrals ('instantons' in the theory of stochastic processes)**

In this subsection we have presented the heuristic and qualitative arguments that for a small diffusion constant D (compared with the integral in the exponent of the Wiener integral), the semiclassical and quadratic approximations give main contributions to path integrals for a wide class of functions $V(x)$. It is possible to develop a complete perturbation expansion in powers of the small parameter D with quantitative estimation of the relative contributions of different terms in the series. However, before doing this, we have to learn more technical methods for handling path integrals. That is why we postpone this more rigorous discussion of the variational expansion of path integrals until chapters 2 and 3. Note, also, that the exponent of integrands in classical physics (in contrast with the quantum-mechanical case) contains, in general, terms with different powers of the small parameter D (see, e.g., (1.2.92)). This makes the rigorous analysis a bit more complicated. Nevertheless, it should not be too difficult for readers to carry out such a consideration after reading chapters 2 and 3. In this subsection, we continue the qualitative non-rigorous discussion of variational methods and now we shall discuss one more important generalization of the semiclassical approximation.

In general, the Euler–Lagrange equations (1.2.111) and (1.2.115) may have several extremal solutions and all of them may give essential contributions to the semiclassical and quadratic approximations for transition probabilities. Besides, it would be erroneous to assume that the semiclassical or quadratic approximations are good for all time intervals $t - t_0$, even if D is small. For large times we have to use an extension of the quadratic approximation which is called *the method of hopping paths* (Wiegel 1975, 1986). In quantum mechanics, this method of path-integral calculations has been developed in connection with the theory of *instantons* (Polyakov 1975, Belavin *et al* 1975); we shall discuss this topic in chapter 3.

The method of hopping paths takes into account, in addition to classical trajectories, a certain set of *large fluctuations* by writing the transition probability in the form

$$W(x, t|x_0, t_0) \simeq C \sum_k \exp\left\{-\frac{1}{4D} \int_{t_0}^{t} d\tau \, L_f(x_k(\tau))\right\} \qquad (1.2.177)$$

instead of the expressions (1.2.150) and (1.2.151). In this formula, C is a constant (which usually can not be evaluated explicitly), and the summation extends over all those functions $x_k(\tau)$ which are *approximate* solutions of the extremal path condition, i.e. Euler–Lagrange equations (1.2.111) and (1.2.115). The function L_f is defined as in (1.2.162) but for the approximate solution x_k. We stress that, in general, x_k is not the exact solution of these equations, which in the case of one dimension and an integrand of the form (1.2.142), has the explicit form

$$\ddot{x} = 2DV'. \qquad (1.2.178)$$

Example 1.11. Let us consider an application of this method for the important case of a bistable *double-well 'potential'* for which the force F_f (i.e. the right-hand side of (1.2.178); cf also (1.2.162)) acting on the *fictitious* particle is given by

$$F_f(x) \stackrel{\text{def}}{\equiv} 2DV' = -\alpha(x-a)(x-b)(x-c) \qquad a < b < c \qquad (1.2.179)$$

where $\alpha > 0, a, b, c$ are some real constants. The general shape of the 'potential' V and the corresponding force F_f are plotted in figure 1.12. The actual external force F acting on the Brownian particle under consideration can be found from the one-dimensional version of (1.2.92), that is, from

$$V = \frac{1}{4\eta^2 D} F^2 + \frac{F'(x)}{2\eta}.$$

92   Path integrals in classical theory

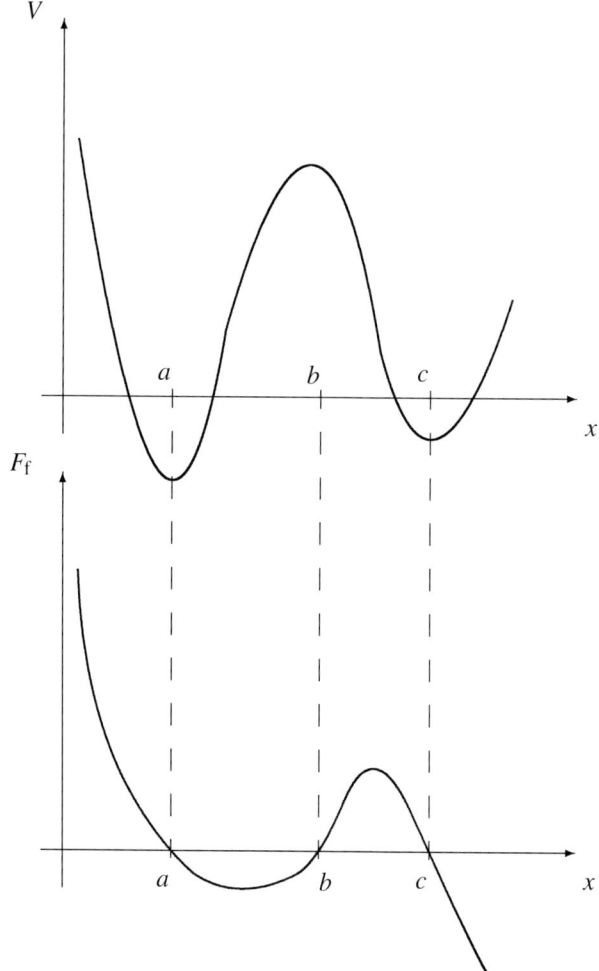

Figure 1.12. General shape of a bistable potential $V(x)$ and the corresponding fictitious force $F_f(x)$.

The constant functions
$$x_1(\tau) = a \quad \text{and} \quad x_2(\tau) = c$$
are the solutions of (1.2.178). For such a solution, the time integral of the 'Lagrangian' which appears in the exponential of (1.2.177) is equal to

$$\int_{t_0}^{t} d\tau\, L_f(x_k) = 4DV(x_k)(t - t_0) \qquad k = 1, 2. \tag{1.2.180}$$

A typical 'hopping-path' solution $x_k(\tau)$ has the form shown in figure 1.13.

The value of x_k hops from a to c and back, at the times $t_1, t_2, t_3, \ldots$. The contribution of a hopping path is determined by the expression (1.2.177).

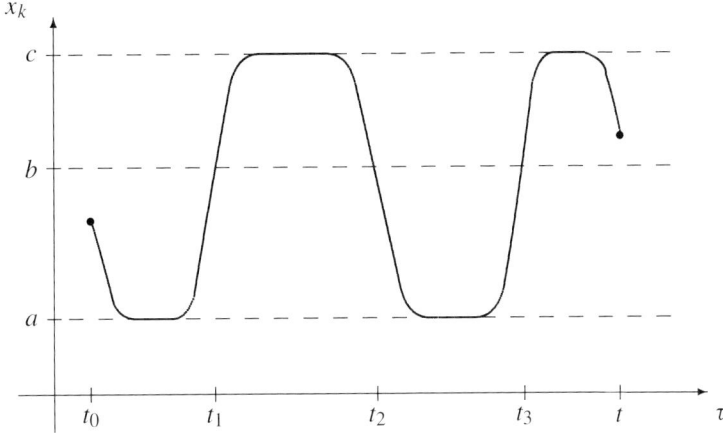

Figure 1.13. A typical 'hopping-path' fluctuation $x_k(\tau)$ which dominates the path integral.

However, the exact calculation of the time integral for the complicated hopping trajectory can hardly be carried out exactly. Instead, we use the fact that a particle, moving along the hopping trajectory as shown in figure 1.13, spends a considerable time at rest in the minima $x = a$, $x = b$ of the potential $V(x)$. Thus even for the hopping path we can start from expression (1.2.180) and look for an appropriate correction reflecting the fact that $x_k(\tau)$ is not constant during the transition from one minimum to another. Skipping the details of the calculations (see Wiegel (1975)), we just state the result: to every transition between the two wells in the potential V, there appears a correction factor R:

$$R \simeq K \exp\left\{-\frac{\gamma}{D}\right\} \qquad (1.2.181)$$

where K and γ are constants if D (the diffusion constant) is small enough. Note that we can easily understand the general character of the dependence of R on the diffusion constant. Indeed, a hop from one well to another is possible only by medium constituents hitting the particle. So, the larger the influence the side of the medium has (i.e. the larger the value of D), the larger the contribution of hopping paths is to the transition probability. In contrast to this case, when the 'activity' of the medium is switched off, i.e. $D \to 0$, the system becomes a deterministic one and due to the conservation of energy, the particle stays all the time in one of the wells. Hence, in this case, the correction factor R turns out to be zero. To accept the exponential form of the correction factor, the reader may invoke his/her acquaintance with the quantum-mechanical tunneling effect, where probability also has a typical exponential form (cf chapter 3).

Thus for small values of the diffusion constant, the hopping-path approximation (1.2.177) for the transition probability is given by

$$W(x,t|x_0,t_0) \simeq C \sum_{n=0}^{\infty} R^{2n} \int dt_1 \, dt_2 \ldots dt_{2n} \, \exp\{-V(a)[(t_1 - t_0) + (t_3 - t_2) + \cdots + (t - t_{2n})]$$
$$- V(c)[(t_2 - t_1) + (t_4 - t_3) + \cdots + (t_{2n} - t_{2n-1})]\}. \qquad (1.2.182)$$

In this formula, we have assumed that x_0 and x are both close to a. The integrations are taken over the times at which the particle passes through b and are subject to the ordering:

$$t_0 < t_1 < t_2 < \cdots < t_{2n} < t. \qquad (1.2.183)$$

The multiple integral in (1.2.182) is evaluated by taking the *Laplace transform* of all functions with respect to the time differences in their arguments:

$$\int_{t_0}^{\infty} dt\, W(x - x_0, t - t_0) \exp\{-s(t - t_0)\} = \widetilde{W}(x - x_0, s) \qquad (1.2.184)$$

$$\int_{t_0}^{\infty} dt_1\, \exp\{-V(a)(t_1 - t_0) - s(t_1 - t_0)\} = \frac{1}{V(a) + s} \qquad (1.2.185)$$

$$\int_{t_1}^{\infty} dt_2\, \exp\{-V(c)(t_2 - t_1) - s(t_2 - t_1)\} = \frac{1}{V(c) + s}. \qquad (1.2.186)$$

Using the fact that the Laplace transform of a convolution of two functions equals the product of the Laplace transforms of the factors, we find that

$$\widetilde{W}(x - x_0, s)|_{x \approx x_0 \approx a} \simeq C \sum_{N=0}^{\infty} R^{2n} [V(c) + s]^{-n} [V(a) + s]^{-n-1}$$

$$= C[V(c) + s]([V(a) + s][V(c) + s] - R^2)^{-1}. \qquad (1.2.187)$$

The transition probability itself follows from the inverse Laplace transform:

$$W(x, t|x_0, t_0)|_{x \approx x_0 \approx a} \simeq \frac{1}{2\pi i} \int_{s-i\infty}^{s+i\infty} ds\, e^{s(t-t_0)} \frac{C[V(c) + s]}{[V(a) + s][V(c) + s] - R^2}. \qquad (1.2.188)$$

The integrand has two poles at

$$s_{\pm} = -\tfrac{1}{2}[V(a) + V(c)] \pm \tfrac{1}{2}\sqrt{[V(a) - V(c)]^2 + 4R^2}.$$

When the contour of integration is deformed into two small circles around these poles, we find that

$$W(x, t|x_0, t_0)|_{x \approx x_0 \approx a} \simeq C_+ \exp\{s_+(t - t_0)\} + C_- \exp\{s_-(t - t_0)\} \qquad (1.2.189)$$

where C_+ and C_- are two constants whose values can be calculated with the aid of the Cauchy theorem. The long-time behaviour of the transition probability is determined by the larger of the two frequencies s_+ and s_-. The larger one is s_+, for the special case when the two minima in the potential have equal depths ($V(a) = V(c) \equiv V_0$), and gives the asymptotic form

$$W(a, t|a, t_0)|_{t \to \infty} \simeq C_+ \exp\{[-V_0 + R](t - t_0)\}. \qquad (1.2.190)$$

This can be interpreted as the product of a factor $\exp\{-V_0(t - t_0)\}$, which would arise if a Brownian particle simply stayed in the vicinity of $x = a$, multiplied by a factor $\exp\{R(t - t_0)\}$, due to penetration of the barrier around $x = b$. Since R has a singularity at $D = 0$ (cf (1.2.181)), this provides us with the first example of an approximate evaluation of a path integral which leads to a non-analytic (singular) dependence of the result on the characteristic parameter of the problem.

We shall return to this method of calculating path integrals in chapter 3 by considering the so-called instantons in quantum mechanics and discuss there more details about this technique.

1.2.7 More technicalities for path-integral calculations: finite-difference calculus and Fourier decomposition

In this subsection, we intend to show that a path integral can be transformed into an infinite-dimensional integral over a (countable) set of variables, namely over the Fourier expansion coefficients of a stochastic process $X(\tau)$ (the Brownian bridge) representing fluctuations of a Brownian particle (deviations from the classical trajectory). The background of this method is the possibility to expand in a Fourier series any function from the set which gives a non-zero contribution to the Wiener path integral. This follows from the fact that these functions satisfy the Hölder–Lipschitz condition (cf the proof of the Wiener theorem, inequality (1.1.73)). It is known (see, e.g., Ilyin and Poznyak (1982)) that such functions have a convergent Fourier expansion. Therefore, after standard decomposition of the path into classical (extremal trajectory) $x_c(\tau)$ and fluctuation $X(\tau)$ parts,

$$x(\tau) = x_c(\tau) + X(\tau) \tag{1.2.191}$$
$$x_c(0) = x_0 \qquad x_c(t) = x_t \tag{1.2.192}$$
$$X(0) = X(t) = 0 \tag{1.2.193}$$

the periodic function $X(\tau)$ can be represented in the form of a *Fourier series*

$$X(\tau) = \sqrt{\frac{2}{t}} \sum_{n=1}^{\infty} a_n \sin\left(\pi n \frac{\tau}{t}\right). \tag{1.2.194}$$

Now we can consider the coefficients a_n, $n = 1, 2, 3, \ldots$, as the (countable set of) coordinates of the trajectories.

This is another set of independent Gaussian random variables (see problem 1.2.12, page 117) describing a Brownian motion and we would like to use them as integration variables in the path-integral expressions for probabilities. To this aim, we have to find a relation between the Wiener measures in terms of the functions $x(\tau)$ and their Fourier coefficients, loosely speaking, between $\prod_{\tau=0}^{t} dx(\tau)$ and $\prod_{n=1}^{\infty} da_m$. But $x(\tau)$ depends on the uncountable set of 'parameters' or 'coordinates' (i.e. values of $x(\tau)$ at all $\tau \in [0, t]$) in the set of all essential functions to be integrated over. In contrast with this, the set of Fourier coefficients is countable. Thus, it is clear that the correct transformation of the path integral into an infinite-dimensional integral over Fourier coefficients requires some 'regularization' of the former, its reduction to an integral over at least countable set of variables. Perhaps, the reader has already guessed that we are going to use again the time slicing which is, in fact, at the heart of the very definition of path integrals.

◇ **The preliminary technical step: finite difference calculus**

Before doing the actual calculations, let us introduce some useful notation. A useful technique for manipulating sums on a sliced time axis is based on the *finite-difference operators* $\partial^{(\varepsilon)}$ and $\bar{\partial}^{(\varepsilon)}$ (also called *lattice derivatives*):

$$\partial^{(\varepsilon)} x(t) \stackrel{\text{def}}{\equiv} \frac{1}{\varepsilon}[x(t+\varepsilon) - x(t)]$$
$$\bar{\partial}^{(\varepsilon)} x(t) \stackrel{\text{def}}{\equiv} \frac{1}{\varepsilon}[x(t) - x(t-\varepsilon)]. \tag{1.2.195}$$

In the continuum limit, they reduce to the usual time derivative ∂_t

$$\partial^{(\varepsilon)}, \bar{\partial}^{(\varepsilon)} \xrightarrow[\varepsilon \to 0]{} \partial_t$$

if these difference operators act on differentiable functions. For the coordinates $x_j \stackrel{\text{def}}{\equiv} x(t_j)$ we shall use the short notation

$$\partial^{(\varepsilon)} x_j \stackrel{\text{def}}{\equiv} \frac{1}{\varepsilon}(x_{j+1} - x_j) \qquad 0 \leq j \leq N$$

$$\bar{\partial}^{(\varepsilon)} x_j \stackrel{\text{def}}{\equiv} \frac{1}{\varepsilon}(x_j - x_{j-1}) \qquad 1 \leq j \leq N+1.$$

(1.2.196)

The lattice derivatives have properties quite similar to ordinary derivatives. In particular, they satisfy the rule of *summation by parts* which is analogous to integration by parts:

$$\varepsilon \sum_{j=1}^{N+1} p_j \bar{\partial}^{(\varepsilon)} x_j = p_j x_j \Big|_0^{N+1} - \varepsilon \sum_{j=0}^{N} (\partial^{(\varepsilon)} p_j) x_j \qquad (1.2.197)$$

(see problem 1.2.13). The particularly useful cases of this formula are:

- if $x_{N+1} = x_0 = 0$,

$$\sum_{j=1}^{N+1} p_j \bar{\partial}^{(\varepsilon)} x_j = -\sum_{j=0}^{N} (\partial^{(\varepsilon)} p_j) x_j \qquad (1.2.198)$$

- if both $p_0 = p_{N+1}$ and $x_0 = x_{N+1}$,

$$\sum_{j=1}^{N+1} p_j \bar{\partial}^{(\varepsilon)} x_j = -\sum_{j=1}^{N+1} (\partial^{(\varepsilon)} p_j) x_j. \qquad (1.2.199)$$

For functions which have vanishing endpoints, (1.2.198) gives the relation

$$\sum_{j=1}^{N+1} (\bar{\partial}^{(\varepsilon)} x_j)^2 = \sum_{j=1}^{N+1} (\partial^{(\varepsilon)} x_j)^2 = -\sum_{j=0}^{N} x_j \partial^{(\varepsilon)} \bar{\partial}^{(\varepsilon)} x_j \qquad (1.2.200)$$

which can be presented in standard matrix notation as

$$-\sum_{j=0}^{N} x_j \partial^{(\varepsilon)} \bar{\partial}^{(\varepsilon)} x_j = -\sum_{j,k=0}^{N} x_j (\partial^{(\varepsilon)} \bar{\partial}^{(\varepsilon)}) x_k \qquad (1.2.201)$$

with the $(N+1) \times (N+1)$ matrix

$$\partial^{(\varepsilon)} \bar{\partial}^{(\varepsilon)} = \bar{\partial}^{(\varepsilon)} \partial^{(\varepsilon)} = \frac{1}{\varepsilon^2} \begin{pmatrix} 2 & -1 & 0 & 0 & \cdots & \cdots & 0 \\ -1 & 2 & -1 & 0 & \cdots & \cdots & 0 \\ 0 & -1 & 2 & -1 & 0 & \cdots & 0 \\ \vdots & & \ddots & \ddots & \ddots & & \vdots \\ 0 & \cdots & 0 & -1 & 2 & -1 & 0 \\ 0 & \cdots & \cdots & 0 & -1 & 2 & -1 \\ 0 & \cdots & \cdots & \cdots & 0 & -1 & 2 \end{pmatrix}. \qquad (1.2.202)$$

We have already dealt with this matrix in section 1.1.4, equation (1.1.84), where it is denoted by A. Of course, this is not accidental: the right-hand side of (1.2.200) is nothing but the compact form (up to the obvious factor $-\varepsilon/4D$) of the exponential in the Wiener measure (cf (1.1.82)). This is obviously the lattice version of the double-time derivative ∂_t^2, to which it reduces in the continuum limit.

A further common property of lattice and ordinary derivatives is that they both can be diagonalized by going to Fourier components. Performing the decomposition

$$x(t) = \int_{-\infty}^{\infty} d\omega \, e^{-i\omega t} x(\omega) \tag{1.2.203}$$

we can easily find that the Fourier components are eigenfunctions of the discrete derivatives (the imaginary unit 'i' is introduced here in the left-hand sides for later convenience):

$$i\partial^{(\varepsilon)} e^{-i\omega t} = \Omega e^{-i\omega t} \tag{1.2.204}$$
$$i\bar{\partial}^{(\varepsilon)} e^{-i\omega t} = \bar{\Omega} e^{-i\omega t} \tag{1.2.205}$$

with the eigenvalues

$$\Omega = \frac{i}{\varepsilon}(e^{-i\omega\varepsilon} - 1) \tag{1.2.206}$$

$$\bar{\Omega} = -\frac{i}{\varepsilon}(e^{i\omega\varepsilon} - 1). \tag{1.2.207}$$

It is clear that in the continuum limit

$$\Omega, \bar{\Omega} \xrightarrow[\varepsilon \to 0]{} \omega$$

i.e. both Ω and $\bar{\Omega}$ become eigenvalues of $i\partial_t$. From (1.2.206) and (1.2.207) it follows that $\bar{\Omega} = \Omega^*$, so that the operator $-\partial^{(\varepsilon)}\bar{\partial}^{(\varepsilon)} = -\bar{\partial}^{(\varepsilon)}\partial^{(\varepsilon)}$ is real and has non-negative eigenvalues,

$$-\partial^{(\varepsilon)}\bar{\partial}^{(\varepsilon)} e^{-i\omega t} = \frac{1}{\varepsilon^2}[2 - 2\cos(\omega\varepsilon)]e^{-i\omega t} \tag{1.2.208}$$

$$[2 - 2\cos(\omega\varepsilon)] \geq 0.$$

◇ **The Wiener measure in terms of the Fourier coefficients**

If we consider the decomposition of the fluctuations $X(\tau)$ around the classical solution, the boundary conditions (1.2.193) impose certain restrictions on the coefficients of (1.2.203), reducing it to the sine-series (1.2.194) with discrete values for the frequencies ω.

For the time-sliced version, a further restriction arises from the fact that the series has to represent $X(\tau)$ only at the discrete points $X(\tau_j)$, $j = 0, \ldots, N+1$. It is therefore sufficient to carry the sum only up to $m = N$ and to expand $X(\tau_j)$ as

$$X(\tau_j) = \sum_{m=1}^{N} \sqrt{\frac{2}{(N+1)\varepsilon}} \sin(\nu_m \tau_j) a_m \tag{1.2.209}$$

where we have introduced for convenience

$$\nu_m \stackrel{\text{def}}{\equiv} \frac{\pi m}{(N+1)\varepsilon}.$$

The expansion functions are orthogonal,

$$\frac{2}{N+1} \sum_{j=1}^{N} \sin(\nu_m(\tau_j - t_0)) \sin(\nu_k(\tau_j - t_0)) = \delta_{mk} \qquad 0 < m, k < N+1 \tag{1.2.210}$$

and complete,
$$\frac{2}{N+1}\sum_{m=1}^{N}\sin(\nu_m(\tau_j - t_0))\sin(\nu_m(\tau_l - t_0)) = \delta_{jl} \qquad (1.2.211)$$

(see problem 1.2.14, page 117).

Due to the orthogonality relation (1.2.210), the Jacobian of the transformation $\{X_j\} \to \{a_m\}$ is equal to $\varepsilon^{-N/2}$, implying that

$$\prod_{j=1}^{N} dX_j = \prod_{m=1}^{N} \frac{da_m}{\sqrt{\varepsilon}}. \qquad (1.2.212)$$

Indeed, the substitution (1.2.209) has the form

$$X(\tau_j) = \frac{1}{\sqrt{\varepsilon}}\sum_{m} M_{jm} a_m$$

where M_{jm} is an orthogonal matrix (due to (1.2.210)) and hence has a unit determinant.

Thus the time-sliced Wiener measure $d_W^{(\varepsilon)} X(\tau)$ can be written in terms of the Fourier coefficients:

$$d_W^{(\varepsilon)} X(\tau) \equiv \frac{1}{(4\pi D\varepsilon)^{-(N+1)/2}}\left[\prod_{j=1}^{N} dX_j\right]\exp\left\{-\frac{\varepsilon}{4D}\sum_{k=0}^{N}(\bar{\partial}^{(\varepsilon)} X_k)^2\right\}$$

$$= \frac{1}{\sqrt{4\pi D\varepsilon}}\prod_{m=1}^{N}\left[\frac{da_m}{\sqrt{4\pi D\varepsilon^2}}\exp\left\{-\frac{1}{4D}\Omega_m \bar{\Omega}_m a_m^2\right\}\right]. \qquad (1.2.213)$$

◇ Calculation of path integrals via integration over Fourier coefficients

The calculation of the fluctuation factor $\phi(t)$ in (1.2.122) (example 1.6, section 1.2.6) now goes as follows:

$$\phi^{(N)}(t - t_0) = \frac{1}{\sqrt{4\pi D\varepsilon}}\left[\prod_{m=1}^{N}\int_{-\infty}^{\infty}\frac{da_m}{\sqrt{4\pi D\varepsilon^2}}\exp\left\{-\frac{1}{4D}\Omega_m \bar{\Omega}_m a_m^2\right\}\right]$$

$$= \frac{1}{\sqrt{4\pi D\varepsilon/M}}\prod_{m=1}^{N}\frac{1}{\sqrt{\varepsilon^2\Omega_m \bar{\Omega}_m}}. \qquad (1.2.214)$$

To calculate the product, we use the relation (see Gradshteyn and Ryzhik (1980), formula 1.396.2)

$$\prod_{m=1}^{N}\left(1 + b^2 - 2b\cos\frac{m\pi}{N+1}\right) = \frac{b^{2(N+1)} - 1}{b^2 - 1} \qquad (1.2.215)$$

which, in the limit $b \to 1$, gives

$$\prod_{m=1}^{N}\varepsilon^2\Omega_m\bar{\Omega}_m = \prod_{m=1}^{N} 2\left(1 - \cos\frac{m\pi}{N+1}\right) = N + 1. \qquad (1.2.216)$$

The fluctuation factor is, therefore,

$$\phi^{(N)}(t - t_0) = \frac{1}{\sqrt{4\pi D(N+1)\varepsilon}}$$

$$= \frac{1}{\sqrt{4\pi D(t - t_0)}} = \phi(t - t_0). \qquad (1.2.217)$$

The latter equation is due to the fact that the time-sliced expression has proved to be independent of the number of slices.

This result, of course, coincides with the correct pre-exponential factor of the transition probability calculated earlier in section 1.2.6.

Note that the Wiener measure (1.2.213), in terms of Fourier coefficients, still keeps a trace of time-slicing because of the factors $(4\pi D\varepsilon)^{-1/2}$ and $(4\pi D\varepsilon^2)^{-1/2}$. However, if we consider a ratio of the form

$$\frac{\int d_W x(\tau) \, F[x(\tau)]}{\int d_W x(\tau)} \quad (1.2.218)$$

the ε-dependent factors in the numerator and denominator cancel each other. Such ratios of path integrals naturally appear both in classical (see next subsection) and especially in quantum mechanics (we shall discuss the quantum-mechanical case in the next chapter). When dealing with the ratios (1.2.218), we can consider the path integrals purely in terms of Fourier coefficients and their calculation reduces to the evaluation of infinite-dimensional integrals,

$$\int_{-\infty}^{\infty} \prod_{m=1}^{\infty} da_m \, \exp\left\{-\frac{1}{4D} \int_0^t d\tau \, \dot{x}^2\right\} F[x(\tau)] \quad (1.2.219)$$

where both the exponential and the functional $F[x(\tau)]$ should be expressed in terms of a_m, $m = 1, 2, 3, \ldots, \infty$. In particular,

$$-\frac{1}{4D}\int_0^t d\tau \, \dot{x}^2 = -\frac{(x_t - x_0)^2}{4Dt} + \frac{1}{4D}\int_0^t d\tau \left(X \frac{d^2}{d\tau^2} X\right)$$

$$= -\frac{(x_t - x_0)^2}{4Dt} - \frac{1}{4D}\sum_{N=1}^{\infty}\left(\frac{\pi n}{t}\right)^2 a_n^2. \quad (1.2.220)$$

Here $(\pi n/t)^2$, $n = 1, 2, 3, \ldots$, are the eigenvalues of the non-negative operator $(-\frac{d^2}{d\tau^2})$:

$$\left(-\frac{d^2}{d\tau^2}\right)\psi_n(\tau) = \left(\frac{\pi n}{t}\right)^2 \psi_n(\tau) \quad n = 1, 2, 3, \ldots \quad (1.2.221)$$

where $\psi_n(\tau)$ are the orthonormal functions

$$\psi_n(\tau) = \sqrt{\frac{2}{t}} \sin\left(\pi n \frac{\tau}{t}\right)$$

entering the expansion (1.2.194). In these formulae we have treated $X(\tau)$ as if it were a differentiable function (by acting on it with a differential operator, integrating by parts, etc) in contrast to the Wiener theorem. The resolution of this apparent contradiction is the following: $X(\tau)$ is indeed differentiable for a *finitely approximated* function $X(\tau)$,

$$X^{(N)}(\tau) = \sum_{n=1}^{N} a_n \sqrt{\frac{2}{t}} \sin\left(\pi n \frac{\tau}{t}\right) \quad (1.2.222)$$

so that we must write (1.2.220) more correctly as

$$-\frac{1}{4D}\int_0^t d\tau \, (\dot{x}^{(N)})^2 = -\frac{(x_t - x_0)^2}{4Dt} - \frac{1}{4D}\sum_{n=1}^{N}\left(\frac{\pi n}{t}\right)^2 a_n^2. \quad (1.2.223)$$

This fact justifies the use of the differential notation in path integrals. The interested reader can find more refined mathematical discussion of the relation between differential operators and non-differentiability of functions with non-zero Wiener measure in Reed and Simon (1975) (on the basis of the theory of self-adjoint operators).

If $F[x(\tau)] \equiv 1$, the integration (1.2.219) obviously results in a determinant of the operator $(-\frac{d^2}{d\tau^2})$ on the space of functions with vanishing boundary values:

$$\lim_{N\to\infty} \prod_{n=1}^{N} \left(\frac{\pi n}{t}\right)^2 = \det\left(-\frac{d^2}{d\tau^2}\right). \tag{1.2.224}$$

It is clear that a straightforward computation of the determinant (1.2.224) would give a meaningless infinite result. Sometimes, the reader may even meet in the literature statements such that this way of computation is completely wrong and inappropriate. However, this is not the case. The point is that to calculate the determinant (1.2.224), we must use some *regularization*, in which case no contradiction arises.

A particular gain from the use of the *Fourier decomposition* (also called *mode expansion*), is that now we can use not only the time-slicing (discrete) approximation to regularize path integrals but also other methods of regularization. For instance, such a practically convenient method uses *the Riemann ζ-function* $\zeta(z)$ (Jahnke and Emde 1965),

$$\zeta(z) \stackrel{\text{def}}{\equiv} \sum_{N=1}^{\infty} \frac{1}{n^z} \qquad z \in \mathbb{C}, \ \operatorname{Re} z > 1. \tag{1.2.225}$$

The ζ-regularization is based on the following observation: if the series

$$\sum_{n=1}^{\infty} \log a_n$$

is convergent (in the usual sense), then the following identity holds:

$$\sum_{n=1}^{\infty} \log a_n = \frac{d}{ds} \sum_{n=1}^{\infty} a_n^{-s} \bigg|_{s=0}.$$

However, the right-hand side can have meaning even in situations when the left-hand side is not convergent in the usual sense. Then the right-hand side can be used as a definition for the sum of a divergent (in the usual sense) series. In our particular case, we have $a_n = \pi^2 n^2/t^2$, so that the sum in the right-hand side is given as

$$\left(\frac{t^2}{\pi^2}\right)^s \sum_{n=1}^{\infty} \frac{1}{n^s} = \left(\frac{t^2}{\pi^2}\right)^s \zeta(s)$$

where $\zeta(z)$ is the Riemann ζ-function (1.2.225). The latter can be analytically continued to the whole complex plane, except for the point $z=1$. This means that the right-hand side is well defined and this provides the background for the following evaluation of the determinant (see problem 1.2.16):

$$\ln \det\left(-\frac{d^2}{d\tau^2}\right) = -\lim_{N\to\infty} \frac{d}{ds} \sum_{n=1}^{N} \left(\frac{\pi^2 n^2}{t^2}\right)^{-s} \bigg|_{s=0}$$

$$= -\zeta(0) \ln \frac{t^2}{\pi^2} - 2\zeta'(0)$$

$$= \ln t + \text{(constant)} \tag{1.2.226}$$

and hence
$$\det\left(-\frac{d^2}{d\tau^2}\right) = Ct \qquad (1.2.227)$$

where C is an inessential (independent of t) constant. In (1.2.226) we have used the value of the Riemann ζ-function at the origin (Jahnke and Emde 1965), $\zeta(0) = -1/2$, obtained by analytical continuation from the domain (Re $z > 1$) of convergence of the series.

The representation and approximation of path integrals by *mode expansion* is especially important for rather complicated gauge theories, e.g., quantum gravity, and, in particular, for path-integral calculations of gauge anomalies. We shall discuss these applications in chapter 3.

1.2.8 Generating (or characteristic) functionals for Wiener integrals

In many probabilistic problems it proves to be more convenient to deal with Fourier or Laplace transforms of distributions (Doob 1953, Feller 1951, 1961, Gnedenko 1968, Korn and Korn 1968). In probability theory these transforms have special names:

- For a given probability distribution $w(x)$, the *characteristic function* $Z_c(q)$ is defined as the Fourier transform,
$$Z_c(q) \stackrel{\text{def}}{\equiv} \langle e^{iqx} \rangle = \int_{-\infty}^{\infty} dx\, e^{iqx} w(x) \qquad (1.2.228)$$
where q is a real variable, $-\infty < q < \infty$ (for discrete distributions, the integral is substituted by a sum).

- For a given probability distribution $w(x)$, the *generating function* $Z_g(s)$ is defined as the Laplace transform,
$$Z_g(s) \stackrel{\text{def}}{\equiv} \langle e^{sx} \rangle = \int_{-\infty}^{\infty} dx\, e^{sx} w(x) \qquad (1.2.229)$$
where s is a complex variable with values providing absolute convergence for the integral (1.2.229) (or the sum, if the distribution $w(x)$ is discrete).

From their very definition (as Fourier or Laplace transforms), it is clear that characteristic or generating functions uniquely define probability distributions. The main technical advantage of these objects is that they offer a simple calculation of mean values:

$$\langle x^k \rangle = \frac{1}{i^k} \frac{d^k}{dq^k} Z(q) \bigg|_{q=0} \qquad (1.2.230)$$
$$= \frac{d^k}{ds^k} \Xi(s) \bigg|_{s=0}. \qquad (1.2.231)$$

The definitions of Z_c and Z_g are readily generalized to the case of stochastic processes $x(\tau)$, where they are called *characteristic*, $\mathcal{Z}_c[\eta(\tau)]$, and *generating*, $\mathcal{Z}_g[j(\tau)]$, *functionals*:

$$\mathcal{Z}_c[\eta(\tau)] = \left\langle \exp\left\{ i \int_0^t d\tau\, \eta(\tau) x(\tau) \right\} \right\rangle \qquad (1.2.232)$$
$$\mathcal{Z}_g[j(\tau)] = \left\langle \exp\left\{ \int_0^t d\tau\, j(\tau) x(\tau) \right\} \right\rangle \qquad (1.2.233)$$

where $\eta(\tau)$ and $j(\tau)$ are called *source functions*. These functionals are convenient for calculations of various correlation functions by use of variational derivatives,

$$\langle x(t_1)x(t_2)\cdots x(t_n)\rangle = \frac{\delta}{\delta j(t_1)}\frac{\delta}{\delta j(t_2)}\cdots\frac{\delta}{\delta j(t_n)}\mathcal{Z}_c[\eta(\tau)]\bigg|_{j=0}. \qquad (1.2.234)$$

Of course, the averaging in (1.2.232)–(1.2.234) is carried out with the functional probability distribution $\Omega[f(\tau)]$ (cf (1.1.60)), e.g., of the purely Wiener (1.1.51) or of a more general (1.2.105) form.

As seen from (1.2.228)–(1.2.231), the relation between $\mathcal{Z}_c$ and $\mathcal{Z}_g$ is quite close and simple (in the domain where both functions exist). Therefore, following the custom in quantum mechanics and quantum field theory, we shall use the name '*generating functional*' for both $\mathcal{Z}_c[\eta(\tau)]$ and $\mathcal{Z}_g[\eta(\tau)]$. Moreover, we shall use, in this book, the same notation, $\mathcal{Z}$, for them; an explicit form of the functionals (with or without the imaginary unit in the exponent) will be clear from the context.

◇ **Example: generating functional and probability distribution for Gaussian random fields**

Generally speaking, the variables τ, t in (1.2.232) and (1.2.233) must not necessarily have the meaning of time only and, moreover, the random variable x may depend on a few coordinates. In this case, it becomes a *stochastic field* (the terminology introduced in section 1.2.1). In this general case, the generating functional reads as

$$\mathcal{Z}[\eta(\boldsymbol{y})] = \left\langle \exp\left\{i\int d^d y\, \eta(\boldsymbol{y})\varphi(\boldsymbol{y})\right\}\right\rangle \qquad (1.2.235)$$

where $\varphi(\boldsymbol{y})$ is some random field, $\boldsymbol{y} = \{y_1,\ldots,y_d\}$ are coordinates in $\mathbb{R}^d$. The correlation functions are obtained by functional derivation

$$\langle \varphi(\boldsymbol{y}_1)\varphi(\boldsymbol{y}_2)\cdots\varphi(\boldsymbol{y}_n)\rangle = \frac{1}{i^n}\frac{\delta^n \mathcal{Z}}{\delta\eta(\boldsymbol{y}_1)\cdots\delta\eta(\boldsymbol{y}_n)}\bigg|_{\eta=0}. \qquad (1.2.236)$$

Conversely, if all correlation functions are known, the generating functional can be found from the summation of its Taylor expansion:

$$\mathcal{Z}[\eta(\boldsymbol{y})] = 1 + \sum_{n=1}^{\infty}\frac{i^n}{n!}\int d^d y_1 d^d y_2\cdots d^d y_n\langle\varphi(\boldsymbol{y}_1)\varphi(\boldsymbol{y}_2)\cdots\varphi(\boldsymbol{y}_n)\rangle\eta(\boldsymbol{y}_1)\cdots\eta(\boldsymbol{y}_n). \qquad (1.2.237)$$

Straightforward *extension* of the definition of a Gaussian stochastic *process* (section 1.2.1) to a *field*, implies that the *Gaussian* random field satisfies the decomposition property

$$\langle\varphi(\boldsymbol{y}_1)\varphi(\boldsymbol{y}_2)\cdots\varphi(\boldsymbol{y}_{2l+1})\rangle = 0 \qquad l = 0, 1, 2, \ldots, \infty$$

$$\langle\varphi(\boldsymbol{y}_1)\varphi(\boldsymbol{y}_2)\cdots\varphi(\boldsymbol{y}_{2l})\rangle = \sum \prod_{\alpha=1}^{l}\langle\varphi(\boldsymbol{y}_{i_\alpha})\varphi(\boldsymbol{y}_{j_\alpha})\rangle \qquad (1.2.238)$$

where the sum runs over all the different ways in which the $2l$ indices $1, 2, \ldots, 2l$ can be subdivided into l non-ordered pairs $(i_1, j_1), (i_2, j_2), \ldots, (i_n, j_n)$. Thus (1.2.237) can be rewritten as

$$\mathcal{Z}[\eta(\boldsymbol{y})] = 1 + \sum_{l=1}^{\infty}\frac{(-1)^l}{(2l)!}\frac{(2l)!}{2^l l!}\left[\int d^d y\, d^d y'\, \eta(\boldsymbol{y})\langle\varphi(\boldsymbol{y})\varphi(\boldsymbol{y}')\rangle\eta(\boldsymbol{y}')\right]^l$$

$$= \exp\left\{-\frac{1}{2}\left[\int d^d y\, d^d y'\, \eta(\boldsymbol{y})\langle\varphi(\boldsymbol{y})\varphi(\boldsymbol{y}')\rangle\eta(\boldsymbol{y}')\right]\right\}. \qquad (1.2.239)$$

Therefore, for Gaussian stochastic fields or processes the generating functional is uniquely defined by the two-point correlation functions (called also *covariances*). One possible application of this result is the representation of the *classical* canonical partition function in terms of functional integrals (we shall consider this application in more detail in chapter 4, section 4.2).

Let us express the probability distribution $\Omega[\varphi(x)]$, corresponding to the generating functional (1.2.239), in terms of the two-point correlation function $\langle\varphi(y)\varphi(y')\rangle$. To this aim, consider $\langle\varphi(y)\varphi(y')\rangle$ as the integral kernel of an operator acting on the appropriate Hilbert space of functions $u(y)$ in $\mathbb{R}^d$,

$$u(y) \longrightarrow \int_{-\infty}^{\infty} d^d y' \, \langle\varphi(y)\varphi(y')\rangle u(y') \tag{1.2.240}$$

and assume that this operator is positive definite, i.e. all the eigenvalues λ_k of the equation

$$\int_{-\infty}^{\infty} d^d y' \, \langle\varphi(y)\varphi(y')\rangle u_k(y') = \lambda_k u_k(y)$$

are positive. The eigenfunctions $u_k(y)$ are assumed to form an orthogonal and complete set, so that

$$\varphi(y) = \sum_k a_k u_k(y) \qquad \eta(y) = \sum_k b_k u_k(y)$$
$$a_k = \int_{-\infty}^{\infty} dy \, \varphi(y) u_k(y) \qquad b_k = \int_{-\infty}^{\infty} dy \, \eta(y) u_k(y). \tag{1.2.241}$$

The substitution of (1.2.241) into (1.2.239) gives

$$\mathcal{Z}[b_k] = \exp\left\{-\frac{1}{2}\sum_k \lambda_k b_k^2\right\} \tag{1.2.242}$$

while the substitution of (1.2.241) into (1.2.235) results in

$$\mathcal{Z}[b_k] = \prod_k \int_{-\infty}^{\infty} da_k \, \Omega[a_k] \exp\left\{i\sum_k a_k b_k\right\}. \tag{1.2.243}$$

Comparing (1.2.242) and (1.2.243) and having in mind the value of the Gaussian integral, we obtain

$$\Omega[a_k] = \prod_k \frac{1}{\sqrt{2\pi\lambda_k}} \exp\left\{-\frac{a_k^2}{2\lambda_k}\right\} \tag{1.2.244}$$

and hence

$$\Omega[\varphi(y)] = C^{-1} \exp\left\{-\frac{1}{2}\left[\int d^d y \, d^d y' \, \varphi(y)\langle\varphi(y)\varphi(y')\rangle^{-1}\varphi(y')\right]\right\} \tag{1.2.245}$$

where C is the normalization constant.

Thus we see that the probability distribution functional $\Omega[\varphi(y)]$ of a Gaussian random field $\varphi(y)$ has in the exponent of (1.2.245) the *inverse* of the correlation function, while the generating functional $\mathcal{Z}[\eta(y)]$, given by (1.2.239), has in the corresponding exponent the correlation function itself.

◇ Explicit calculation of the generating functional for quadratic exponent (path integral for a driven harmonic oscillator)

In the last part of this subsection we shall show that generating functionals are very useful for the calculation of a wide class of path integrals. This method of calculation is crucial for the development of perturbation theory in statistical and quantum physics, especially in quantum field theory (see chapter 3).

Path integrals in classical theory

In fact, the path integrals which we are going to calculate now are various correlation functions. Therefore, in the first step, we must calculate a generating functional and then, by subsequent functional differentiation, we can obtain any correlation functions (values of the corresponding path integrals). However, before that we have to decide which functional probability distribution $\Omega[x(\tau)]$ to choose: it should be as general as possible but still allow an *explicit* calculation of the corresponding generating functional. From our experience with finite-dimensional integrals and the examples of the path integrals considered in the preceding sections, we may conclude that the exponential in $\Omega[x(\tau)]$ must be maximally *quadratic* in the random variable.

Thus we consider the following generating functional:

$$\mathcal{Z}[\eta(\tau)] = \int_{C\{x_0,0;x,t\}} d_W x(\tau) \exp\left\{-\int_0^t d\tau\, V_\eta(x,\tau)\right\} \quad (1.2.246)$$

where

$$V_\eta(x,\tau) = k^2 x^2(\tau) - \eta(\tau) x(\tau) \qquad k \in \mathbb{R}. \quad (1.2.247)$$

Recall that at $\eta(\tau) = 0$ this path integral is closely related to the transition probability of a Brownian particle moving in the field of an external harmonic force (cf (1.2.93) and (1.2.92)) and the path integral with non-zero $\eta(\tau)$ is called sometimes the generating functional for a *driven harmonic oscillator*.

There are different ways to calculate the integral (1.2.246). One way of calculating it would be to use the semiclassical approximation which gives an exact result in the case of Gaussian integrals, as discussed in detail in section 1.2.6. However, to provide the reader with a variety of methods and to illustrate the general discussion on the Fourier mode expansion method presented in the preceding subsection, we shall calculate (1.2.246) by means of Fourier decomposition.

First, we rewrite the integral (1.2.246) in the form

$$\int_{C\{x_0,0;x_t,t\}} d_W x(\tau) \exp\left\{-\int_0^t d\tau\, V_\eta(x,\tau)\right\}$$

$$= \left[\int_{C\{x_0,0;x_t,t\}} d_W x(\tau)\right] \frac{\int_{C\{x_0,0;x_t,t\}} d_W x(\tau) \exp\{-\int_0^t d\tau\, V_\eta(x,\tau)\}}{\int_{C\{x_0,0;x_t,t\}} d_W x(\tau)}. \quad (1.2.248)$$

After the scaling of the time variable τ as

$$s \stackrel{\text{def}}{\equiv} \frac{\tau}{t} \qquad 0 \leq s \leq 1 \quad (1.2.249)$$

the ratio in (1.2.248) becomes

$$\frac{\int_{C\{x_0,0;x_t,t\}} \prod_{s=0}^1 dx(s) \exp\left\{-\left[\frac{1}{4Dt}\int_0^1 ds\left(\frac{dx}{ds}\right)^2 + t\int_0^1 ds\, V_\eta(x(s),ts)\right]\right\}}{\int_{C\{x_0,0;x_t,t\}} \prod_{s=0}^1 dx(s) \exp\left\{-\frac{1}{4Dt}\int_0^1 ds\left(\frac{dx}{ds}\right)^2\right\}}. \quad (1.2.250)$$

(All the other common factors in the Wiener measures in the numerator and denominator drop out.) Since we are going to calculate the integrals *exactly*, we can decompose paths into two parts, $x_c(\tau)$ and $X(\tau)$, in an arbitrary way, i.e. $x_c(\tau)$ need not necessarily be the most probable (extremal, classical) trajectory. We choose it to be the classical trajectory for a *free* particle, so that

$$x_c(s) = x_0 + (x_t - x_0)s \quad (1.2.251)$$

$$x_c(0) = x_0$$

$$x_c(s)|_{s=1} = x_c(\tau)|_{\tau=t} = x_t.$$

We still use the notation $x_c(s)$, though it is really the classical trajectory only for the exponential in the denominator of (1.2.250). Let us rewrite the latter in terms of Fourier coefficients

$$\exp\left\{-\frac{(x-x_0)^2}{4Dt}\right\}\prod_{m=1}^{N}\int da_m \exp\left\{-\frac{1}{4Dt}\Omega_m\bar{\Omega}_m a_m^2\right\} \qquad (1.2.252)$$

where the a_m, $m = 1, 2, 3, \ldots, N$ are defined by the Fourier expansion of $X(s)$:

$$X(s) = \sum_{n=1}^{N}\sqrt{2}a_n \sin(\pi ns).$$

To avoid possible errors, we have used the time-slicing regularization with $N+1$ sites on the interval $s \in [0, 1]$; Ω_m, $\bar{\Omega}_m$ are eigenvalues of the finite differences (cf the preceding subsection).

Now we make the change of Fourier variables

$$a_m \to b_m \equiv m\pi a_m \qquad m = 1, 2, 3, \ldots, N \qquad (1.2.253)$$

with the Jacobian

$$J = \prod_{m=1}^{N}\frac{1}{\pi m}. \qquad (1.2.254)$$

We perform in the numerator of (1.2.250) the same transition to the Fourier variables a_m and the substitution (1.2.253), so that the Jacobians (1.2.254) in the numerator and denominator of the ratio cancel out. Now we can pass to the continuous limit $\varepsilon \to 0$, $N \to \infty$ (ε is the value of the time slice). As a result, the product of the integrals in (1.2.252) becomes

$$\prod_{m=1}^{\infty}\int db_m \exp\left\{-\frac{b_m^2}{4Dt}\right\} = \prod_{m=1}^{\infty}\sqrt{4\pi Dt}. \qquad (1.2.255)$$

Let us turn to calculating the numerator. After decomposing the paths into the background $x_c(s)$ and fluctuation $X(s)$ parts, the function V_η takes the form

$$V_\eta = k^2 x^2(s) - \eta(s)x(s)$$
$$= k^2[f(s) + X(s)]^2 - \frac{1}{4k^2}\tilde{\eta}^2(s) \qquad (1.2.256)$$

where

$$\tilde{\eta}(s) \stackrel{\text{def}}{\equiv} \eta(\tau(s))$$
$$f(s) \stackrel{\text{def}}{\equiv} x_c(s) - \frac{1}{2k^2}\eta(s).$$

The Fourier decomposition of the functional in the exponent reads

$$\int_0^1 ds\, V_\eta(x, s) = -\frac{1}{4k^2}I + k^2\sum_{n=1}^{\infty}\left(f_n + \frac{b_n}{\pi n}\right)^2. \qquad (1.2.257)$$

Here

$$f(s) = \sum_{n=1}^{\infty} f_n \sqrt{2} \sin(n\pi s)$$

$$f_n = \frac{\sqrt{2}}{n\pi}(x_0 - (-1)^n x) - \frac{1}{2k^2}\eta_n$$

$$\eta_n = \sqrt{2} \int_0^1 ds\, \tilde{\eta}(s) \sin(n\pi s)$$

$$I = \sum_{n=1}^{\infty} \eta_n^2 = \int_0^1 ds\, \tilde{\eta}^2(s)$$

(we assume that $I < \infty$). After the substitution

$$b_m \to c_m = \frac{b_m}{\sqrt{2Dt}} \qquad m = 1, 2, 3, \ldots, N$$

its Jacobian, $\prod_{m=1}^{\infty} \sqrt{4Dt}$, cancels out (1.2.255), up to the factor $\prod_{m=1}^{\infty}(1/\sqrt{2\pi})$, and the whole ratio (1.2.250) becomes

$$\exp\left\{-\frac{t}{4k^2}I\right\} \prod_{m=1}^{\infty} \frac{1}{2\pi} \int_{-\infty}^{\infty} dc_m \exp\left\{-\frac{c_m^2}{2} - k^2 t \left(f_m + \frac{\sqrt{2Dt}}{\pi n} c_m\right)^2\right\}.$$

The integrals under the product sign in the latter equation are given by

$$I_n \stackrel{\text{def}}{\equiv} \frac{1}{2\pi} \int_{-\infty}^{\infty} dc_m \exp\left\{-\frac{c_m^2}{2} - k^2 t \left(f_m + \frac{\sqrt{2Dt}}{\pi n} c_m\right)^2\right\}$$

$$= \left(1 + \frac{4Dk^2 t^2}{\pi^2 n^2}\right)^{-\frac{1}{2}} \exp\left\{-k^2 t f_n^2 \left(1 + \frac{4Dk^2 t^2}{\pi^2 n^2}\right)^{-1}\right\}. \tag{1.2.258}$$

Note that the first factor in the right-hand side of (1.2.248) is nothing but the transition probability W_D for a free Brownian particle. Combining this factor with the ratio in that equation, we arrive at the following expression for the generating functional:

$$\mathcal{Z}[\eta(\tau); (x_t, t | x_0, 0)] = W_D(x_t, t | x_0, 0) \exp\left\{-\frac{t}{4k^2} I\right\} \prod_{n=1}^{\infty} I_n. \tag{1.2.259}$$

However, this formula is still inconvenient for practical use. To convert (1.2.259) into an explicit expression, we first note that (Gradshteyn and Ryzhik 1980)

$$\prod_{n=1}^{\infty} \left(1 + \frac{y^2}{n^2 \pi^2}\right) = \frac{\sinh y}{y}.$$

Insertion of this into (1.2.258) and (1.2.259) yields

$$\mathcal{Z}[\eta(\tau); (x, t | x_0, 0)] = \left[\frac{k}{2\pi \sqrt{D} \sinh(2k\sqrt{Dt})}\right]^{\frac{1}{2}}$$

$$\times \exp\left\{-\frac{(x-x_0)^2}{4Dt}\right\} \exp\left\{k^2 t \sum_{n=1}^{\infty}\left[\eta_n^2 - f_n^2\left(1 - \frac{4Dk^2t^2}{n^2\pi^2}\right)^{-1}\right]\right\}.$$
(1.2.260)

Finally, the formulae

$$\sum_{n=1}^{\infty}(-1)^n \frac{2y}{y^2 + n^2\pi^2} = \frac{1}{y} - \frac{1}{\sinh y}$$

$$\sum_{n=1}^{\infty}\frac{2y}{y^2 + n^2\pi^2} = \frac{1}{y} - \frac{\cosh y}{\sinh y}$$

$$2\sum_{n=1}^{\infty}\frac{\sin(n\pi s)\sin(n\pi s')}{y^2 + n^2\pi^2} = \frac{\sinh(y\min(s,s'))\sinh(y(1-\max(s,s')))}{y\sinh y} \quad (1.2.261)$$

together with the relations

$$2\sum_{n=1}^{\infty}\frac{n\pi \sin(n\pi s)}{y^2 + n^2\pi^2} = \frac{\sinh(y(1-s))}{\sinh y}$$

$$2\sum_{n=1}^{\infty}(-1)^{n+1}\frac{n\pi \sin(n\pi s)}{y^2 + n^2\pi^2} = \frac{\sinh(ys)}{\sinh y}$$

which follow from (1.2.261), allow us to present (1.2.260) in the form (for completeness, we restore an arbitrary initial time t_0)

$$\mathcal{Z}[\eta(\tau);(x_t,t|x_0,t_0)]$$

$$= \left[\frac{k}{2\pi\sqrt{D}\sinh(2k\sqrt{D}(t-t_0))}\right]^{\frac{1}{2}} \exp\left\{-k\frac{(x_t^2+x_0^2)\cosh(2k\sqrt{D}(t-t_0)) - 2x_0 x_t}{2\sqrt{D}\sinh(2k\sqrt{D}(t-t_0))}\right\}$$

$$\times \exp\left\{\sqrt{D}\int_{t_0}^{t}d\tau \int_{t_0}^{\tau}d\tau'\, \eta(\tau)\eta(\tau')\frac{\sinh(2k\sqrt{D}(t-\tau))\sinh(2k\sqrt{D}(\tau'-t_0))}{k\sinh(2k\sqrt{D}(t-t_0))}\right\}$$

$$\times \exp\left\{\int_{t_0}^{t}d\tau\, \eta(\tau)\frac{x_0 \sinh(2k\sqrt{D}(t-\tau)) + x\sinh(2k\sqrt{D}(\tau-t_0))}{\sinh(2k\sqrt{D}(t-t_0))}\right\} \quad (1.2.262)$$

$$= W_k^{(\text{harm})}(x_t,t|x_0,0)\exp\left\{\int_{t_0}^{t}d\tau \int d\tau'\, \eta(\tau)G_{t,t_0}(\tau,\tau')\eta(\tau')\right\}$$

$$\times \exp\left\{\int_{t_0}^{t}d\tau\, \eta(\tau)x_c^{(\text{harm})}(\tau)\right\} \quad (1.2.263)$$

where $W_k^{(\text{harm})}(x_t,t|x_0,0)$ is the transition probability (1.2.132) for the harmonic force, $G_{t,t_0}(\tau,\tau')$ defines the exponent bilinear in external source $\eta(\tau)$, $x_c^{(\text{harm})}$ defines the exponent linear in $\eta(\tau)$. The latter is nothing but the solution of the classical equation with harmonic force:

$$\ddot{x}_c^{(\text{harm})} = 4Dk^2 x_c^{(\text{harm})}.$$

With the explicit expression (1.2.263) for the generating functionals in hand, we can easily calculate a variety of different path integrals (correlation functions), differentiating (1.2.263) by $\eta(\tau)$ and then

108   *Path integrals in classical theory*

putting this external source, if necessary, to zero (also, if we deal with a *free* Brownian particle, putting $k = 0$). For example,

$$\begin{aligned}\langle x(t_2)x(t_1)\rangle_k &\stackrel{\text{def}}{\equiv} \int_{\mathcal{C}\{x_0,t_0;x_t,t\}} d_W x(\tau)\, x(t_1)x(t_2) \exp\left\{-k^2 \int_{t_0}^{t} d\tau\, x^2(\tau)\right\} \\ &= \frac{\delta}{\delta\eta(t_2)} \frac{\delta}{\delta\eta(t_1)} \mathcal{Z}[\eta(\tau);(x,t|x_0,t_0)]\bigg|_{\eta=0} \\ &= [G_{t,t_0}(t_2,t_1) + x_c^{(\text{harm})}(t_2)x_c^{(\text{harm})}(t_1)] W_k^{(\text{harm})}(x,t|x_0,t_0). \end{aligned} \quad (1.2.264)$$

Here we have assumed that $t_0 < t_1 < t_2 < t$.

The explicit form of this expression, together with a number of other path integrals obtained from the generating functional $Z[\eta(\tau);(x,t|x_0,0)]$, is presented in supplement II (volume II).

1.2.9 Physics of macromolecules: an application of path integration

Macromolecules are very long, chainlike molecules. Both artificially synthesized macromolecules forming a variety of technologically important polymers and natural biological ones exist. The latter are the basic material for living beings. Essentially, they are strings of small groups of atoms which are called monomers or repeating units. The successive repeating units of a macromolecule are strongly bound by chemical bonds, hence their mutual distance is almost constant. However, in many cases, the repeating units can easily be rotated with respect to each other, so that the molecule as a whole—although of a fixed length—is extremely flexible. These two features of real macromolecules form the basis of the random walk model of macromolecule statistics and, in turn, of the application of path-integral techniques for solving some problems in the statistical physics of macromolecules. In fact, different forms of these long chains (with fixed or free ends or with the ends subject to some constraints) serve as the physically most illuminating illustration of the concept of path integration.

The aim of this subsection is to give the reader an idea about this branch of physical chemistry and make him/her familiar with path-integral methods of solving some problems. The reader may find a more extensive discussion in, e.g., Wiegel (1983, 1986), Kleinert (1995), Barber and Ninham (1970) and de Gennes (1969).

◇ The random walk model

In the *random walk model*, a configuration of a macromolecule consisting of N identical repeating units is represented by a random walk of a fictitious particle with N steps, each of a fixed length ℓ. This length ℓ equals the average distance between two successive repeating units in the molecule. Depending on the interactions between the repeating units, we should impose certain constraints on the random walk configurations. In the first approximation, we take into account only interactions responsible for a fixed distance between the units of macromolecules and all other monomer–monomer interaction effects are neglected altogether. This corresponds to free *discrete* random walks.

Denote the probability density that a free random walk which starts at the origin of coordinates will reach the point r after N steps by $w(r, N)$. The first of these probability distributions reads as

$$w(\boldsymbol{r}, 1) = (4\pi \ell^2)^{-1} \delta(|\boldsymbol{r}| - \ell) \quad (1.2.265)$$

because the wandering fictitious particle can appear with equal probability at any point of the sphere of radius ℓ with its centre at the origin. For $N > 1$, the distributions can be calculated in a recursive way from the relation

$$w(\boldsymbol{r}, N+1) = (4\pi \ell^2)^{-1} \int d^3 r'\, w(\boldsymbol{r}', N) \delta(|\boldsymbol{r} - \boldsymbol{r}'| - \ell) \quad (1.2.266)$$

which expresses the fact that the probability $w(\boldsymbol{r}, N+1)d^3r$ of the event that the step $N+1$ ends in the vicinity d^3r of the point $\boldsymbol{r}$ equals the probability $w(\boldsymbol{r}', N)\,d^3r$ multiplied by the transition probability from $\boldsymbol{r}'$ to $\boldsymbol{r}$, integrated over the whole volume. The actual calculation of functions $w(\boldsymbol{r}, N)$ can be carried out by a Fourier transformation with the result

$$w(\boldsymbol{r}, N) = \frac{1}{2\pi}\int d^3k\, e^{-i\boldsymbol{k}\boldsymbol{r}}\left(\frac{\sin k\ell}{k\ell}\right)^N \qquad (1.2.267)$$

as an exact integral representation for the probability distribution.

We shall derive explicitly only the asymptotic behaviour of $w(\boldsymbol{r}, N)$. For $N \gg 1$, the probability density $w(\boldsymbol{r}', N)$ in (1.2.266) is a smooth function of $\boldsymbol{r}'$ and can be expanded in a Taylor series around $\boldsymbol{r}' = \boldsymbol{r}$:

$$w(\boldsymbol{r}', N) = w(\boldsymbol{r}, N) + \sum_{i=1}^{3}(r'_i - r_i)\frac{\partial w}{\partial r_i} + \frac{1}{2}\sum_{i,j=1}^{3}(r'_i - r_i)(r'_j - r_j)\frac{\partial^2 w}{\partial r_i \partial r_j} + \cdots \qquad (1.2.268)$$

$$w(\boldsymbol{r}, N+1) = w(\boldsymbol{r}, N) + \frac{\partial w(\boldsymbol{r}, N)}{\partial N} + \cdots \qquad (1.2.269)$$

where r_i denotes the ith Cartesian coordinate of the vector $\boldsymbol{r}$ and where we treated N as a *continuous* variable. Substitution of the expansion into (1.2.266) gives

$$\frac{\partial w}{\partial N} \simeq \frac{\ell^2}{6}\triangle w + \mathcal{O}(\ell^4 \triangle^2 w) \qquad (1.2.270)$$

where $\triangle$ denotes the Laplacian operator

$$\triangle = \frac{\partial^2}{\partial x^2} + \frac{\partial^2}{\partial y^2} + \frac{\partial^2}{\partial z^2}.$$

Comparing the left-hand side of (1.2.270) with the first term on the right-hand side, we conclude that in the limit $N \to \infty$, we should keep $N\ell^2$ fixed. Since $\partial/\partial N \sim 1/N$ and $\ell^2 \sim 1/N$, we obtain that $\partial w/\partial N$ and $\ell^2 \triangle w$ are of order N^{-1}, but $\ell^4 \triangle^2 w$ is of order N^{-2}. Hence in the limit $N \to \infty$, $\ell \to 0$, and $N\ell^2 = $ constant, we obtain the equation

$$\frac{\partial w}{\partial N} = \frac{\ell^2}{6}\triangle w \qquad (1.2.271)$$

which is nothing but the diffusion equation with the 'diffusion constant' $\ell^2/6$.

Thus, as it has been shown in this chapter, the probability density of the endpoint of the chain molecule in the form of a Wiener integral reads as

$$w(\boldsymbol{r}, N) = W(\boldsymbol{r}, N|0, 0)$$

$$= \int_{C\{0,0;\boldsymbol{r},N\}}\prod_{v=0}^{N}\frac{dx(v)}{\sqrt{2\ell^2\pi\,dv/3}}\frac{dy(v)}{\sqrt{2\ell^2\pi\,dv/3}}\frac{dz(v)}{\sqrt{2\ell^2\pi\,dv/3}}e^{-\int_0^N dv\, L_0} \qquad (1.2.272)$$

where

$$L_0 = \frac{3}{2\ell^2}\left(\frac{d\boldsymbol{r}}{dv}\right)^2. \qquad (1.2.273)$$

The integrand in the Wiener integral (1.2.272) is a measure for the density of polymer configurations which are close to some continuous curve $\boldsymbol{r}(v)$. Of course, if a finer description of the chain configurations is necessary, we should use neither the differential equation (1.2.271) nor the path integral (1.2.272) but instead use the recursive relation (1.2.266).

From this short consideration of the path-integral technique in polymer physics, we can derive two lessons:

- Different shapes of polymer chains play the role of paths over which path integrals are integrated, so that the paths acquire direct physical meaning, as distinct from the original notion of particle trajectories; these paths can really be observed and thus they are illuminating manifestations of the concept of path integration ('incarnation' of path integrals).
- The 'time' variable on paths may have different physical meanings: for a Brownian particle and in quantum mechanics (as we shall learn in the next chapter) it does indeed denote the time variable; in the statistical physics of macromolecules, it is a number of monomers; in quantum statistical physics (see chapter 4), it takes the meaning of inverse temperature.

◇ **The Feynman–Kac formula in the theory of polymers**

The Bloch equation and the Feynman–Kac formula also find important applications in polymer physics. For example, consider now the case in which space is filled with 'dust', i.e. with small impenetrable particles of any size which act as an excluded volume for the macromolecules. Let $V(r)$ denote the fraction of space from which the endpoints of the successive repeating units of the molecule are excluded. In this case, the recursive relation (1.2.266) for free polymers has to be replaced by

$$w(r, N+1) = (4\pi\ell^2)^{-1}[1 - V(r)] \int d^3r' \, w(r', N)\delta(|r - r'| - \ell) \tag{1.2.274}$$

for polymers in such a 'dust-filled' space. Following the procedure for the derivation of (1.2.270), we obtain

$$\frac{\partial w}{\partial N} \simeq -Vw + \frac{\ell^2}{6}\Delta w + (1 - V)\mathcal{O}(\ell^4 \Delta^2 w). \tag{1.2.275}$$

Hence in the limit

$$N \to \infty \qquad \ell \to 0 \qquad V \to 0$$

$$N\ell^2 = \text{constant} \tag{1.2.276}$$

$$\frac{V}{\ell^2} = \text{constant}$$

the differential equation becomes

$$\frac{\partial w}{\partial N} = \frac{\ell^2}{6}\Delta w - Vw \tag{1.2.277}$$

i.e. the Bloch equation. The corresponding Feynman–Kac formula reads as

$$w(r, N) = \int_{C\{0,0;r,N\}} \prod_{v=0}^{N} \frac{dx(v)}{\sqrt{2\ell^2\pi \, dv/3}} \frac{dy(v)}{\sqrt{2\ell^2\pi \, dv/3}} \frac{dz(v)}{\sqrt{2\ell^2\pi \, dv/3}} \exp\left\{-\int_0^N dv \, L_V\right\} \tag{1.2.278}$$

where

$$L_V = \frac{3}{2\ell^2}\left(\frac{dr}{dv}\right)^2 + V(r). \tag{1.2.279}$$

When the forces between the repeating units of a macromolecule are taken into account (e.g., forces restricting the angles of their mutual positions), the calculation of the number of chain configurations becomes much more difficult and requires more involved consideration. But path-integral methods still prove to be very powerful and fruitful. For details, we refer the reader to de Gennes (1969) and Wiegel (1986).

1.2.10 Problems

Problem 1.2.1. Prove that the increments of a Brownian path are independent.

Hint. As follows from the very definition of transition probabilities (cf (1.1.50)), we have

$$\mathbb{P}\{x(t_1) - x(t_0) \in [a_1, b_1], x(t_2) - x(t_1) \in [a_2, b_2], \ldots, x(t_n) - x(t_{n-1}) \in [a_n, b_n]\}$$
$$= \prod_{i=1}^{n} \int_{a_i}^{b_i} dy_i \, W(y_i; t_i - t_{i-1})$$

where

$$W(y_i; t_i - t_{i-1}) \stackrel{\text{def}}{=} W(x_i - x_{i-1}; t_i - t_{i-1})$$
$$= W(x_i, t_i | x_{i-1}, t_{i-1}).$$

On the other hand,

$$\int_{a_i}^{b_i} dy_i \, W(y_i; t_i - t_{i-1}) = \mathbb{P}\{(x_i - x_{i-1}) \in [a_i, b_i]\}.$$

Thus

$$\mathbb{P}\{x(t_1) - x(t_0) \in [a_1, b_1], x(t_2) - x(t_1) \in [a_2, b_2], \ldots, x(t_n) - x(t_{n-1}) \in [a_n, b_n]\}$$
$$= \prod_{i=1}^{n} \mathbb{P}\{x_i - x_{i-1}) \in [a_i, b_i]\}$$

and this proves the independence of increments.

Problem 1.2.2. Derive the Fokker–Planck equation for higher-dimensional (e.g., a three-dimensional) space.

Hint. In a three-dimensional space, the equation has the form

$$\frac{\partial W}{\partial t} + \text{div}(WA) = \triangle WB.$$

Problem 1.2.3. Derive the Fokker–Planck equation corresponding to the Langevin equation for a Brownian particle with inertia.

Hint. The general form of the equation is

$$\frac{\partial W}{\partial t} + v \frac{\partial W}{\partial x} + \frac{\partial}{\partial v}[AW] = \frac{\partial^2 BW}{\partial v^2}$$

where $W = W(x, v, t)$, while $A = A(x, v, t)$ and $B = B(x, v, t)$ are statistical characteristics of the process, analogous to those in (1.2.15) and (1.2.16).

Problem 1.2.4. Show that a space- and time-homogeneous Gaussian stochastic process, with an appropriate dependence on time of the exponent of the probability distribution, satisfies the conditions (1.2.15)–(1.2.17). Write down explicitly the form of the exponent for such a Gaussian distribution.

Solution. Space and time homogeneity of a Gaussian process implies that the transition probability takes the form

$$W(x_t, t | x_0, t_0) = \sqrt{\frac{f(t-t_0)}{\pi}} e^{-f(t-t_0)(x_t - x_0)^2}$$

where $f(t)$ is some function of time (the first factor, $\sqrt{f(t-t_0)/\pi}$, is defined by the normalization condition). Thus we have to calculate the integrals (for simplicity, put $t_0 = 0$)

$$\left\langle \frac{(x_t - x_0)^n}{t} \right\rangle = \frac{1}{t}\sqrt{\frac{f(t)}{\pi}} \int_{-\infty}^{\infty} dx_t \, (x_t - x_0)^n e^{-f(t)(x_t - x_0)^2}$$

$$= \frac{1}{t}\sqrt{\frac{f(t)}{\pi}} \int_{-\infty}^{\infty} dx \, x^n e^{-f(t)x^2}.$$

Such an integral can be calculated using the standard trick: we first calculate the integral

$$I(\alpha) \stackrel{\text{def}}{\equiv} \int_{-\infty}^{\infty} dx \, e^{-bx^2 + \alpha x} = e^{\frac{\alpha^2}{4b}} \int_{-\infty}^{\infty} dx \, e^{-(\sqrt{b}x - \frac{\alpha}{2\sqrt{b}})^2}$$

$$= \frac{1}{\sqrt{b}} e^{\frac{\alpha^2}{4b}} \int_{-\infty}^{\infty} dy \, e^{-y^2} = \frac{\sqrt{\pi}}{\sqrt{b}} e^{\frac{\alpha^2}{4b}}$$

by the completion of the square in the exponent.

Then the integral of interest is obtained by differentiation:

$$\int_{-\infty}^{\infty} dx \, x^n e^{-bx^2} = \frac{d^n}{d\alpha^n} I(\alpha) \bigg|_{\alpha=0}.$$

All the integrals with odd powers of x vanish just because of the antisymmetry of the integrand:

$$\int_{-\infty}^{\infty} dx \, x^{2n+1} e^{-bx^2} = 0 \qquad n = 0, 1, 2, \ldots$$

so that the conditions (1.2.15) and (1.2.17) are satisfied automatically with A in (1.2.15) equal to zero. The averaging of x^2 gives

$$\lim_{t \to 0} \left\langle \frac{(x_t - x_0)^2}{t} \right\rangle = \lim_{t \to 0} \frac{1}{t}\sqrt{\frac{f(t)}{\pi}} \int_{-\infty}^{\infty} dx \, x^2 e^{-fx^2}$$

$$= \lim_{t \to 0} \frac{1}{2tf(t)} = 2B = \text{constant}$$

and hence

$$f(t) \xrightarrow[t \to 0]{} \frac{1}{4Bt}. \qquad (1.2.280)$$

Averaging the higher powers gives

$$\lim_{t \to 0} \left\langle \frac{(x_t - x_0)^n}{t} \right\rangle = \lim_{t \to 0} \frac{1}{2t f^{n/2}(t)} \to 0 \qquad n = 4, 6, \ldots.$$

Note that the transition probability (1.1.35) obviously satisfies condition (1.2.280), so that the corresponding Fokker–Planck equation coincides with the diffusion equation.

Problem 1.2.5. Prove the relation (1.2.29) which means that the *derivative* of a Wiener process is the *white noise*.

Hint. We must show that
$$\frac{d}{dt'}\frac{d}{dt}\min(t',t) = \delta(t'-t).$$
This can be achieved by the representation of the minimum function in the form
$$\min(t',t) = t'\widetilde{\theta}(t-t') + t\widetilde{\theta}(t'-t)$$
where $\widetilde{\theta}(t)$ is the so-called *symmetric* step-function:
$$\widetilde{\theta}(t) \stackrel{\text{def}}{\equiv} \begin{cases} 0 & t<0 \\ \frac{1}{2} & t=0 \\ 1 & t>0. \end{cases} \qquad (1.2.281)$$

Then the equality
$$\frac{d}{dt}\widetilde{\theta}(t) = \delta(t)$$
gives the required result.

Problem 1.2.6. Prove that the δ-functional $\delta[\dot{x}(\tau) - v(\tau)]$, when inserted into the path integral (1.2.69), leads indeed to the constraint (1.2.70).

Hint. If, instead of taking the $\varepsilon \to 0$ limit, we insert (1.2.66)
$$\prod_{i=1}^{N} \frac{1}{\varepsilon}\delta\left(\frac{x^{i+1}-x^i}{\varepsilon} - v^i\right) = \prod_{i=1}^{N} \delta(x^{i+1} - x^i - v^i\varepsilon)$$
into the corresponding discrete approximation of the path integral (1.2.69) and then integrate gradually over $x_2^1, x_2^2, \ldots, x_2^N$, we obtain the δ-function in (1.2.70).

Problem 1.2.7. Calculate the integral (1.2.69) for a particle with inertia using the variational method and integral representation for the δ-function, providing the constraint (1.2.71).

Solution. We start from the Fourier transform of the δ-function in (1.2.69):
$$\delta(a) = \frac{1}{2\pi}\int_{-\infty}^{\infty} dw\, e^{-iwa} \qquad (1.2.282)$$
so that (1.2.70) becomes
$$W(x_t, v_t, t|x_0, v_0, 0) = \frac{1}{2\pi}\int_{-\infty}^{\infty} dw\, \exp\{-iw(x_t - x_0)\}$$
$$\times \int_{C\{v_0,0;v_t,t\}} \prod_{\tau=0}^{t} \frac{dv(\tau)}{\sqrt{\pi\, d\tau}} \exp\left\{-\int_0^t d\tau\,(\dot{v}^2(\tau) - iwv(\tau))\right\}$$
$$= B(t)\int_{-\infty}^{\infty} dw\, \exp\{-iw(x_t - x_0)\}\exp\left\{-\int_0^t d\tau\,(\dot{v}_c^2(\tau) - iwv_c(\tau))\right\} \quad (1.2.283)$$

where $B(t)$ is the factor which does not depend either on the values of path endpoints (cf section 1.2.6), or on the Fourier variable w (because the latter enters only the linear part of the exponent). As usual, v_c is the solution of the Euler equation (extremal 'trajectory'; actually, in this case, extremal velocity):

$$2\ddot{v}_c + iw = 0 \qquad (1.2.284)$$
$$v_c(0) = v_0 \qquad v_c(t) = v_t.$$

Simple calculations give the solution

$$v_c = -\frac{iw}{4}\tau^2 + a\tau + b \qquad (1.2.285)$$

$$a = \frac{v_t - v_0}{t} + \frac{iw}{4} \qquad b = v_0. \qquad (1.2.286)$$

The value of the functional in the exponent of the integrand in (1.2.283) is

$$\int_0^t d\tau\, (\dot{v}_c^2 - iwv_c) = \frac{w^2 t^3}{48} - iwt\frac{v_t + v_0}{2} + \frac{(v_t - v_0)^2}{t}.$$

Thus the transition probability proves to be

$$W(x_t, v_t, t|x_0, v_0, 0) = B(t)\exp\left\{-\frac{(v_t - v_0)^2}{t}\right\}$$

$$\times \int_{-\infty}^{\infty} dw\, \exp\left\{-\frac{w^2 t^3}{48} + iwt\frac{v_t + v_0}{2} - iw(x_t - x_0)\right\}. \qquad (1.2.287)$$

The integral (1.2.287) is Gaussian and can be calculated:

$$W(x_t, v_t, t|x_0, v_0, 0) = B(t)\sqrt{\frac{48\pi}{t^3}}\exp\left\{-\frac{12}{t^3}\left(x_t - x_0 - \frac{v_t + v_0}{2}t\right)^2 - \frac{(v_t - v_0)^2}{t}\right\}. \qquad (1.2.288)$$

The normalization condition for the transition probability,

$$\int_{-\infty}^{\infty} dx_t\, dv_t\, W(x_t, v_t, t|x_0, v_0, 0) = 1 \qquad (1.2.289)$$

gives the normalization constant $B(t)$, and finally the transition probability proves to be

$$W(x_t, v_t, t|x_0, v_0, 0) = \frac{\sqrt{12}}{\pi t^2}\exp\left\{-\frac{12}{t^3}\left((x_t - x_0) - \frac{v_t + v_0}{2}t\right)^2 - \frac{(v_t - v_0)^2}{t}\right\}. \qquad (1.2.290)$$

Problem 1.2.8. Check that the transition probability (1.2.72) (or (1.2.290)) is the fundamental solution of the Fokker–Planck equation

$$\frac{\partial W}{\partial t} + v\frac{\partial W}{\partial x} = \frac{1}{4}\frac{\partial^2 W}{\partial v^2}.$$

Try to give a physical interpretation for this equation.

Hint. To simplify the calculations, we can put $x_0 = v_0 = 0$. For a particle with inertia, it is the velocity that is directly subject to the action of a random force (cf (1.2.59)). Physically, this follows from the fact that the position of a particle with inertia cannot be changed abruptly. Thus the transition probability (or the corresponding density, cf (1.1.1)–(1.1.3)) must satisfy the diffusion equation with respect to the velocity, but the time derivative must now be the total derivative:

$$\frac{d}{dt}W = \frac{\partial W}{\partial t} + \frac{\partial x}{\partial t}\frac{\partial W}{\partial x}$$
$$= \frac{\partial W}{\partial t} + v\frac{\partial W}{\partial x}.$$

Problem 1.2.9. Derive the transition probability for the stochastic equation

$$\ddot{x} + kx = \phi \qquad (1.2.291)$$

and check that it is the fundamental solution of the Fokker–Planck equation

$$\frac{\partial W}{\partial t} + v\frac{\partial W}{\partial x} - \frac{\partial}{\partial v}(kvW) = \frac{1}{4}\frac{\partial^2 W}{\partial v^2}.$$

Hint. Represent the stochastic equation as the system

$$\dot{v} + kx = T_1\phi_1$$
$$\dot{x} - v = T_2\phi_2$$

which is equivalent to (1.2.291) in the limit $T_2 \to 0$ and combine the methods considered in the sections 1.2.2–1.2.4.

Problem 1.2.10. Show that the formula (1.2.38), derived in the microscopic approach based on the Langevin equations and the change of variables, and (1.2.93), derived from the Fokker–Planck equation (1.2.87), are consistent.

Hint. Write down the one-dimensional version of (1.2.93), calculate the integral in the exponent of the forefactor and the function V from (1.2.92).

Problem 1.2.11. Consider the Brownian motion on a plane subject to an external force with the components

$$F_x = -\omega\eta y \qquad F_y = \omega\eta x. \qquad (1.2.292)$$

Check that this force is non-conservative and find the transition probability.

Solution (Wiegel 1986). The inequality of the derivatives,

$$\frac{\partial F_x}{\partial y} \neq \frac{\partial F_y}{\partial x}$$

shows that the components F_x, F_y cannot be represented as derivatives of a scalar function (potential) and hence this is a non-conservative force field. The classical equation of motion (1.2.109) reads

$$\dot{x}_c = -\omega y_c$$
$$\dot{y}_c = \omega x_c. \qquad (1.2.293)$$

In the absence of fluctuations ($D = 0$), the trajectory of the system is the circle

$$x_c^2 + y_c^2 = x_c^2(0) + y_c^2(0)$$

116   *Path integrals in classical theory*

through the initial position $x_c(0)$, $y_c(0)$ and the particle moves with constant angular velocity ω. In the presence of Brownian motion, the potential (1.2.92) has the form

$$V = \frac{\omega^2}{4D}(x^2 + y^2)$$

and the equations of motion (1.2.111) read as

$$\frac{d^2 x_c}{d\tau^2} = \omega^2 x_c - 2\omega \frac{dy_c}{d\tau}$$

$$\frac{d^2 y_c}{d\tau^2} = \omega^2 y_c - 2\omega \frac{dx_c}{d\tau}.$$

The general solution of this equation is a spiral:

$$x_c(\tau) = r_0 \cos(\omega\tau + \phi_0) + s_0\tau \cos(\omega\tau + \psi_0)$$
$$y_c(\tau) = r_0 \sin(\omega\tau + \phi_0) + s_0\tau \sin(\omega\tau + \psi_0)$$

in which the constants r_0, s_0, ϕ_0, ψ are determined by the initial position and velocity at $t_0 = 0$, in the following way. For $\tau = 0$, we have

$$x_c(0) = r_0 \cos\phi_0 \qquad y_c(0) = r_0 \sin\phi_0$$

so r_0 and ϕ_0 are the polar coordinates of the initial position in the (x, y)-plane. Also, for the velocity, we find that

$$\dot{x}_c(0) = -\omega y_c(0) + s_0 \cos\psi_0$$
$$\dot{y}_c(0) = \omega x_c(0) + s_0 \sin\psi_0.$$

The extremal value of the Lagrangian is found upon substitution of the solution $x_c(\tau)$, $y_c(\tau)$ into (1.2.107) and integration over τ; this gives

$$\int_0^t d\tau\, L[r_c(\tau)] = \frac{s_0^2 t}{4D}. \tag{1.2.294}$$

The parameter s_0 is determined by the requirement that the trajectory passes through r at the final time t. This condition gives the equations

$$x_c(t) = r_0 \cos(\omega t + \phi_0) + s_0 t \cos(\omega t + \psi_0)$$
$$y_c(t) = r_0 \sin(\omega t + \phi_0) + s_0 t \sin(\omega t + \psi_0)$$

from which we find that

$$s_0^2 t^2 = [x_c - r_0 \cos(\omega t + \phi_0)]^2 + [y_c - r_0 \sin(\omega t + \phi_0)]^2.$$

Substitution of (1.2.294) and the last formula into (1.2.112) gives

$$W(x, y, t | x_0, y_0, 0) = \Phi(t) \exp\left\{\frac{1}{4Dt}[[x_c - r_0 \cos(\omega t + \phi_0)]^2 + [y_c - r_0 \sin(\omega t + \phi_0)]^2]\right\}. \tag{1.2.295}$$

The function $\Phi(t)$ is found from the normalization of (1.2.295) to be

$$\Phi(t) = \frac{1}{4\pi Dt}.$$

This is an *exact* result for the transition probability because for the force (1.2.292) the exponential is quadratic and, as we have studied in section 1.2.6, the semiclassical method used here gives exact results.

Problem 1.2.12. Show that a set of Fourier coefficients of a process $X(\tau)$ representing Brownian stochastic fluctuations, i.e. when $X(\tau) = x(\tau) - x_c(\tau)$, are *independent* random variables.

Hint. Using the fact that

$$\langle X_{\tau'} X_\tau \rangle = \frac{1}{2} \min(\tau, \tau') - \left(\frac{x_t}{t}\right)^2 \tau\tau'$$

(we have put $x_0 = 0$ for simplicity) and that

$$a_n = \sqrt{\frac{2}{t}} \int_0^t d\tau \, \sin\left(n\pi \frac{\tau}{t}\right) X(\tau)$$

prove the equality

$$\langle a_{n'} a_n \rangle = \delta_{nn'}$$

which is equivalent to the required statement.

Problem 1.2.13. Prove the formula (1.2.197) for summation by parts.

Hint. Write the left-hand side sum of the relation (1.2.197) explicitly and rearrange terms so as to obtain the right-hand side of the relation.

Problem 1.2.14. Prove the orthogonality and completeness relations (1.2.210) and (1.2.211).

Hint. The orthogonality relation follows by rewriting the left-hand side of (1.2.210) in the form

$$\frac{2}{N+1} \frac{1}{2} \operatorname{Re} \sum_{j=0}^{N+1} \left[\exp\left\{ \frac{i\pi(m-k)}{(N+1)} j \right\} - \exp\left\{ \frac{i\pi(m+k)}{(N+1)} j \right\} \right]$$

where the sum has been extended by two trivial ($j = 0$ and $j = N + 1$) terms. The sum in this expression being a geometric series, it can be easily summed. For $m = k$ it is equal to unity, while for $m \neq k$ it yields

$$\frac{2}{N+1} \frac{1}{2} \operatorname{Re} \sum_{j=0}^{N+1} \left[\frac{1 - \exp\{i\pi(m-k)\} \exp\{i\pi(m-k)/(N+1)\}}{1 - \exp\{i\pi(m-k)\}} \right.$$

$$\left. - \frac{1 - \exp\{i\pi(m+k)\} \exp\{i\pi(m+k)/(N+1)\}}{1 - \exp\{i\pi(m+k)\}} \right]. \quad (1.2.296)$$

It is easy to check that

$$\frac{1 - \exp\{i\pi(m-k)\} \exp\{i\pi(m-k)/(N+1)\}}{1 - \exp\{i\pi(m-k)\}} = \begin{cases} 1 & \text{for even } m-k \neq 0 \\ \text{purely imaginary} & \text{for odd } m-k. \end{cases}$$

For the second term in the square brackets in (1.2.296), this also holds true for even and odd $m + k \neq 0$, respectively. Since $m \pm k$ are either both even or both odd, the right-hand side of (1.2.296) vanishes for $m \neq k$. The proof of the completeness relation (1.2.211) can be carried out similarly.

Problem 1.2.15. Using the Feynman–Kac formula, find the probability that a trajectory $x(\tau)$ of the Brownian particle (moving in a one-dimensional space) satisfies the condition $x(\tau) \leq a$ for all τ:

$0 \leq \tau \leq t$. In other words, find the time which is necessary for the Brownian particle to reach the point $x = a$ with a given probability.

Solution (Schulman 1981). The probability

$$p(t) \stackrel{\text{def}}{\equiv} \mathbb{P}\{x(t) = a, x(\tau) < a, \tau < t\}$$

is given by the Wiener integral with unconditional measure and the characteristic functional $\chi_a[x(\tau)]$ of the trajectories subject to the condition

$$x(\tau) \leq a$$

as an integrand, i.e. if

$$\chi_a[x(\tau)] = \begin{cases} 1 & \text{if } x(\tau) \leq a, 0 \leq \tau \leq t \\ 0 & \text{otherwise} \end{cases}$$

then

$$p(t) = \int_{C\{0,0;t\}} d_W x(\tau) \, \chi_a[x(\tau)]. \tag{1.2.297}$$

We can write this characteristic functional in terms of the fictitious potential $V_a(x)$:

$$\chi_a[x(\tau)] = \exp\left\{-\int_0^t d\tau \, V_a(x(\tau))\right\}$$

with

$$V_a(x) = \begin{cases} 0 & \text{if } x < a \\ +\infty & \text{if } x \geq a. \end{cases}$$

This form of the characteristic functional converts the path integral (1.2.297) into the standard Feynman–Kac form and we can deduce immediately the differential equation for the corresponding transition probability:

$$\frac{\partial W(x_t, t|0, 0)}{\partial t} = D\frac{\partial^2 W(x_t, t|0, 0)}{\partial x_t^2} - V_a(x_t) W(x_t, t|0, 0)$$

which we should complete by the obvious boundary condition

$$W(a, t|0, 0) = 0. \tag{1.2.298}$$

The solution of this equation with the correct boundary behaviour can be found by the standard method of images (also called the reflection method). Taking into account that in the domain $x \leq a$ the potential equals zero, we conclude that the transition probability must be a combination of the standard transition probabilities for the ordinary diffusion equation which corresponds to two types of path: (i) going directly from the initial point to x; and (ii) bouncing off the infinite 'wall' at $x_t = a$ (see figure 1.14) before reaching the position x_t at the time t.

Thus the solution takes the form

$$W_a(x_t, t|0, 0) = \frac{1}{\sqrt{4\pi D t}} \left[\exp\left\{-\frac{x_t^2}{4Dt}\right\} - \exp\left\{-\frac{(2a - x_t)^2}{4Dt}\right\}\right]$$

and the probability under consideration is

$$p(t) = \int_{-\infty}^{a} dx_t \, W_a(x_t, t|0, 0)$$

Wiener path integrals and stochastic processes 119

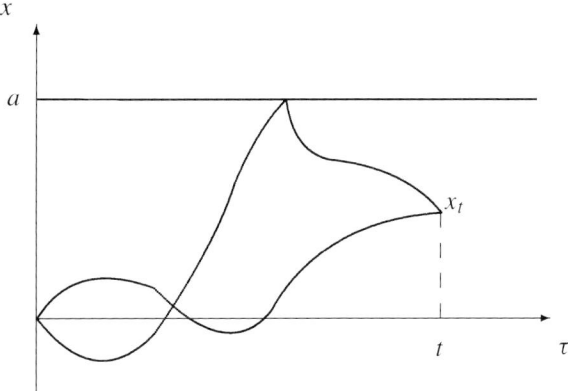

Figure 1.14. Two typical paths in the constrained region considered in problem 1.2.15: one going directly to x, the other bouncing off the 'wall'.

$$= \frac{1}{\sqrt{4\pi Dt}} \int_{-\infty}^{a} dx_t \left[\exp\left\{-\frac{x_t^2}{4Dt}\right\} - \exp\left\{-\frac{(2a-x_t)^2}{4Dt}\right\} \right]$$

$$= \frac{2}{\sqrt{4\pi Dt}} \int_{0}^{a} dx_t \, \exp\left\{-\frac{x_t^2}{4Dt}\right\}.$$

Obviously, the method used here could be generalized to deal with all sorts of questions about confinement in various regions. We shall consider these types of problem in more detail in the next chapter on quantum mechanics.

Problem 1.2.16. Calculate the determinant of the operator

$$\left(-\frac{d^2}{d\tau^2}\right)$$

in the space of functions $X(\tau)$ with zero boundary conditions, $X(0) = X(t) = 0$, with the help of ζ-regularization (in other words, derive the formula (1.2.227)).

Solution.

$$\ln \det \left(-\frac{d^2}{d\tau^2}\right) = \ln \prod_{N=1}^{\infty} \left(\frac{\pi^2 n^2}{t^2}\right)$$

$$= \sum_{N=1}^{\infty} \ln \left(\frac{\pi^2 n^2}{t^2}\right)$$

$$= -\frac{d}{ds} \sum_{N=1}^{\infty} \left(\frac{\pi^2 n^2}{t^2}\right)^{-s} \bigg|_{s=0}$$

$$= -\frac{d}{ds} \left(\frac{t^2}{\pi^2}\right)^s \sum_{n=1}^{\infty} \frac{1}{n^{2s}} \bigg|_{s=0}$$

$$= -\frac{d}{ds}\left(\frac{t^2}{\pi^2}\right)^s \zeta(2s)\bigg|_{s=0}.$$

The latter equality follows from the definition of the Riemann ζ-function (1.2.225). Thus we obtain

$$\ln\det\left(-\frac{d^2}{d\tau^2}\right) = -\zeta(0)\ln\frac{t^2}{\pi^2} - 2\zeta'(0)$$

and, using the value of the ζ-function at the origin (see, e.g., Jahnke and Emde (1965)):

$$\zeta(0) = -\tfrac{1}{2}$$

we have finally

$$\ln\det\left(-\frac{d^2}{d\tau^2}\right) = (\text{constant})t.$$

Problem 1.2.17 (Stochastic integrals). Let $x(\tau)$, $t_0 \leq \tau \leq t$, be a trajectory of a Brownian particle and $f(x(\tau))$ some smooth function. The so-called *Ito stochastic integral* (Ito 1951) (see also, e.g., Kuo (1975) and Simon (1979)) is defined as follows:

$$\int_{x_0}^{x_t} d_1 x\, f(x(\tau)) \stackrel{\text{def}}{\equiv} \lim_{N\to\infty} \sum_{j=0}^{N-1} (x_{j+1} - x_j) f(x_j)$$

$$x_j = x(\tau_j) \qquad \tau_j = t_0 + \frac{j}{N}(t - t_0)$$

$$x_N = x(t) = x_t \qquad x_0 = x(t_0).$$

Prove the following important formula (the analog of *Stokes formula* for the Ito integral):

$$\int_{x_0}^{x_t} d_1 x\, \frac{\partial f(x(\tau))}{\partial x} = f(x_t) - f(x_0) - \frac{1}{2}\int_{t_0}^{t} d\tau\, \frac{\partial^2 f(x(\tau))}{\partial x^2}. \qquad (1.2.299)$$

Note that for a linear function $f(x)$, the Ito integral satisfies the ordinary Stokes formula (first two terms in (1.2.299)). This shows, in particular, the correctness of the transformation (1.2.35) for the Ito stochastic integral.

Consider also the more general definition of stochastic integrals:

$$\int_{x_0}^{x_t} d_S x\, f(x(\tau)) \stackrel{\text{def}}{\equiv} \lim_{N\to\infty} \sum_{j=0}^{N-1} (x_{j+1} - x_j) f(\widetilde{x}_j) \qquad (1.2.300)$$

where

$$\widetilde{x}_j = x_j + \lambda(x_{j+1} - x_j) \qquad 0 \leq \lambda \leq 1 \qquad (1.2.301)$$

and find the analog of the Stokes formula for them.

Solution. Consider the identity

$$f(x_t) - f(x_0) = \sum_{j=0}^{N-1} [f(x_{j+1}) - f(x_j)]$$

and, using the assumed smoothness of $f(x)$, expand each term around $\widetilde{x}_j$ (cf (1.2.301)). This gives

$$f(x_t) - f(x_0) = \sum_{j=0}^{N-1}[f'(\widetilde{x}_j)(x_{j+1}-x_j) + \tfrac{1}{2}f''(\widetilde{x}_j)(1-2\lambda)(x_{j+1}-x_j)^2 + \mathcal{O}((x_{j+1}-x_j)^3)]. \quad (1.2.302)$$

If we choose $\lambda = 0$, the first term within the square brackets on the right-hand side, in the limit $N \to \infty$, becomes the Ito integral. The second term in the sum is of the order $(\Delta x)^2$ and for an ordinary Riemann integral would disappear. But Brownian paths are non-differentiable and

$$\Delta x \sim \sqrt{\Delta \tau}.$$

Therefore,

$$\lim_{N \to \infty} \sum_j f''(x_j)\Delta x_j = \lim_{N \to \infty} \sum_j f''(x(\tau_j))\Delta \tau$$

$$= \int_{t_0}^{t} d\tau \, \frac{\partial^2 f(x(\tau))}{\partial x^2}.$$

This proves the *Ito formula* (1.2.299). In fact, equation (1.2.302) immediately gives analogs of the Stokes formula for arbitrary λ and, in particular, proves transformation (1.2.35) (ordinary form of the Stokes formula in the case of *linear* functions) for stochastic integrals (1.2.300).

Further details and a rigorous mathematical treatment of the Ito integral can be found in Kuo (1975) and Simon (1979).

Chapter 2

Path integrals in quantum mechanics

The aim of this chapter is to expand the concept of the path integral to quantum mechanics. The *Wiener path integral* which we considered in chapter 1 and the *Feynman path integral* in quantum mechanics have many common features. We essentially use this similarity in the first section of the present chapter reducing, in fact, some quantum-mechanical problems to consideration of the Bloch equation and the corresponding Wiener path integral.

On the other hand, there exists an essential difference between the Wiener path integral in the theory of stochastic processes and quantum-mechanical path integrals. The origin of this distinction is the appearance of a new fundamental object in quantum mechanics, namely, the probability *amplitude*. A quantum-mechanical path integral does not express a probability itself directly, but rather the probability amplitude. As a result, it cannot be constructed as an integral defined by some functional *probability measure* in contrast with the case of the Wiener integral (recall the close interrelation between the notions of probability and measure which we discussed in chapter 1). Moreover, path integrals derived from the basic principles of quantum mechanics prove to be over *discontinuous* paths in the *phase space* of the system and only in relatively simple cases can they be reduced to the (Feynman) path integrals over *continuous* trajectories in the *configuration space*. We shall discuss this topic in sections 2.2 and 2.3.

The natural application of path integrals in quantum mechanics is to solve problems with *topological constraints*, e.g., when a particle moves in some restricted domain of the entire space $\mathbb{R}^d$ or with non-trivial, say periodic, boundary conditions (e.g., a particle on a circle or torus). Though this type of problem can be considered by operator methods, the path-integral approach makes the solution simpler and much more transparent. We shall consider these topics in section 2.4.

We are strongly confined to work in the framework of phase-space path integrals (or, at least, to start from them) in the case of systems with *curved* phase spaces. The actuality of such a problem is confirmed, e.g., by the fact that the celebrated *Coulomb problem* (quantum mechanics of atoms) can be solved via the path-integral approach only within a formalism developed on a phase space with curvilinear coordinates (section 2.5).

The last section of this chapter is devoted to generalizations of the path-integral construction to the case of particles (or, more generally, quantum-mechanical excitations) described by operators with anticommutative (fermionic) or even more general defining relations (instead of the canonical Heisenberg commutation relations). Path integrals over anticommuting (Grassmann) variables are especially important for gauge field and supersymmetric theories of the fundamental interactions in quantum field theory, as well as in the theory of superconductivity, which we shall discuss in subsequent chapters.

The very important topic of *path integrals with constraints* in quantum mechanics will be postponed until chapter 3, since its immediate use will be made in the treatment of quantum gauge-field theories in the same chapter.

2.1 Feynman path integrals

In this section, after a short exposition of the postulates and main facts from the conventional operator formulation of quantum mechanics, we shall prove the relevance of path integrals for quantum-mechanical problems (i.e. prove the *Feynman–Kac formula* in quantum mechanics), derive the important properties of the Hamiltonian operator and describe the semiclassical *Bohr–Sommerfeld quantization condition*.

The reader familiar with the basic concepts of quantum mechanics can skip, with no loss, section 2.1.1, using it, in case of necessity, only for clarification of our notation.

2.1.1 Some basic facts about quantum mechanics and the Schrödinger equation

Quantum mechanics (see, e.g., the books by Dirac (1947), von Neumann (1955) or any textbook on the subject, e.g., Davydov (1976) or Landau and Lifshitz (1981)) describes the behaviour of microscopic objects (elementary particles, atoms or molecules) and processes with physical characteristics (with the dimension of an action) comparable in values with the Planck constant, $\hbar = 1.05 \times 10^{-27}$ erg s. For macroscopic objects, the Planck constant can be considered to be of negligible value and thus the classical mechanics describing such objects corresponds to the limit $\hbar \to 0$ of quantum mechanics.

To measure some physical quantity related to a micro-object, we have to use a macroscopic device which changes its macroscopic and quasistable state as a result of the interaction with the microscopic object. A result of this measurement cannot be predicted definitively even if all possible information about the conditions, under which the measurement takes place, is available. Thus the results of quantum-mechanical measurements are *random quantities* and quantum mechanics studies their statistical characteristics. This fact brings quantum mechanics close to the theory of stochastic processes and explains the generality of some mathematical methods (in particular, path integrals) applicable to the analysis of stochastic processes both in classical physics and in quantum mechanics. Physical quantities, whose possible values can be measured by some physical experiment, are called *observables*.

◇ **Postulates of quantum mechanics**

We recall here some basic postulates of quantum mechanics.

Postulate 2.1. Quantum-mechanical states are described by non-zero vectors of a complex separable Hilbert space $\mathcal{H}$, two vectors describing the same state if they differ from each other only by a non-zero complex factor. To any observable, there corresponds a linear self-adjoint operator on $\mathcal{H}$.

The space $\mathcal{H}$ is called the *space of states* of the physical system and the elements of $\mathcal{H}$ are called *state vectors*. We will always imply that state vectors have unit length (i.e. unit norm in $\mathcal{H}$). Any operator will be denoted by a 'hat': $\widehat{A}$.

The observables $A_1, \ldots, A_n$ are called simultaneously measurable if their values can be determined with *arbitrary* precision *simultaneously*, so that in any state $\psi \in \mathcal{H}$, the random variables $A_1, \ldots, A_n$ have a joint probability density (or probability distribution, if the random variables take a discrete set of values), $w_\psi(\lambda_1, \ldots, \lambda_n)$.

Postulate 2.2. Observables are simultaneously measurable if the corresponding self-adjoint operators commute with each other. The joint probability density (probability distribution) of simultaneously measurable observables in a state $\psi \in \mathcal{H}$ has the form

$$w_\psi(\lambda_1, \ldots, \lambda_n) = \langle \psi_{\lambda_1,\ldots,\lambda_n} | \psi \rangle^* \langle \psi_{\lambda_1,\ldots,\lambda_n} | \psi \rangle$$
$$= |\langle \psi_{\lambda_1,\ldots,\lambda_n} | \psi \rangle|^2 \tag{2.1.1}$$

where $\langle \cdots | \cdots \rangle$ denotes the scalar product in $\mathcal{H}$ and $\psi_{\lambda_1,\ldots,\lambda_n}$ are the common eigenfunctions of the operators $\widehat{A}_1, \ldots, \widehat{A}_n$, i.e.

$$\widehat{A}_i \psi_{\lambda_1,\ldots,\lambda_n} = \lambda_i \psi_{\lambda_1,\ldots,\lambda_n} \qquad i = 1, \ldots, n.$$

Recall that for an operator $\widehat{A}$, with a dense domain of definition $\mathcal{D} \subset \mathcal{H}$, the *conjugate operator* $\widehat{A}^\dagger$ is defined by the equality

$$\langle \widehat{A}^\dagger \psi | \phi \rangle = \langle \psi | A \phi \rangle.$$

The operator $\widehat{A}$ is *symmetric* on the linear set $\mathcal{D} \subset \mathcal{H}$ if

$$\langle A\varphi | \psi \rangle = \langle \varphi | A\psi \rangle$$

holds for any $\varphi, \psi \in \mathcal{D}$. If the domains of definition of $\widehat{A}$ and $\widehat{A}^\dagger$ coincide, a symmetric operator is called *self-adjoint* or *Hermitian* and we write $\widehat{A} = \widehat{A}^\dagger$.

Thus to any physical quantity there corresponds a Hermitian (self-adjoint) operator $\widehat{A}$, which has a complete set of eigenfunctions $\psi_a \equiv \psi_{\lambda_a}$, satisfying

$$\widehat{A} \psi_a(x) = \lambda_a \psi_a(x) \tag{2.1.2}$$

with real eigenvalues λ_a (we shall use the simplified notation ψ_a when speaking about the eigenvectors of a single operator; the index a labels different eigenvalues of $\widehat{A}$). The set of all eigenvalues is called the *spectrum of an operator* $\widehat{A}$, which could be a discrete and/or a continuous subset of $\mathbb{R}$ (the reality of the eigenvalues follows from the self-adjointness of the operators corresponding to observables).

Since the set $\{\psi_a\}$ of all eigenvectors of a self-adjoint operator $\widehat{A}$ is complete in $\mathcal{H}$ and can be orthonormalized, any vector $\psi \in \mathcal{H}$ can be represented by the series:

$$\psi = \sum_a c_a \psi_a \qquad c_a \in \mathbb{C} \tag{2.1.3}$$

where a runs over all the eigenvalues of $\widehat{A}$ (if the operator $\widehat{A}$ has a continuous spectrum, the sum is substituted by an integral). The coefficients c_a of this expansion are expressed through the scalar product

$$c_a = \langle \psi_a | \psi \rangle \tag{2.1.4}$$

which, according to postulate 2.2, gives the *probability amplitude* that a measurement of the observable A gives the value λ_a if the system is in the state ψ. The corresponding probability is equal to

$$w_\psi^a = |c_a|^2 = |\langle \psi_a | \psi \rangle|^2. \tag{2.1.5}$$

Thus, the mean value of the quantity A in the state ψ reads as

$$\langle \widehat{A} \rangle_\psi \equiv \langle \psi | \widehat{A} | \psi \rangle = \sum_a \lambda_a w_\psi^a = \sum_a \lambda_a |c_a|^2 \tag{2.1.6}$$

and the dispersion reads
$$(\mathbb{D}_\psi A)^2 = \langle (\widehat{A} - \langle \widehat{A} \rangle_\psi)^2 \rangle_\psi. \tag{2.1.7}$$

Note that a mathematically rigorous formulation (von Neumann 1955) of these postulates uses the so-called spectral theory of operators (Reed and Simon 1975) and, in particular, the unity resolution by projection operators.

The most important physical characteristic of any quantum-mechanical system is its energy. The corresponding operator is denoted by $\widehat{H}$ and is called the *Hamiltonian* operator of the system. The following postulate states that $\widehat{H}$ defines the time evolution of a quantum system.

Postulate 2.3. Let a state of a system, at some time t_0, be described by a vector $\psi(t_0)$. Then, at any moment t, the state of the system is described by the vector
$$\psi(t) = \widehat{U}(t, t_0)\psi(t_0) \tag{2.1.8}$$

where $\widehat{U}(t, t_0)$ is a unitary operator: $\widehat{U}^\dagger \widehat{U} = \widehat{U}\widehat{U}^\dagger = 1$, called the *evolution operator*. The wavefunction $\psi(t)$ is differentiable if it belongs to the domain of definition $\mathcal{D}_H$ of the Hamiltonian $\widehat{H}$ at least at $t = t_0$ and, in this case, the following relation holds:
$$i\hbar \frac{\partial \psi(t)}{\partial t} = \widehat{H}\psi(t) \tag{2.1.9}$$

where $\hbar$ is the Planck constant.

Relation (2.1.9) is the basic equation of quantum mechanics and is called the *Schrödinger equation*.

If two time translations are performed successively, the corresponding operators $\widehat{U}$ are related by the following composition law:
$$\widehat{U}(t, t_0) = \widehat{U}(t, t')\widehat{U}(t', t_0) \qquad t' \in (t_0, t). \tag{2.1.10}$$

This composition law is the analog of the Markovian property for transition probabilities in the theory of stochastic processes (cf chapter 1). Note, however, that, strictly speaking, the Markovian property concerns *probabilities* while the operator $\widehat{U}$ governs an evolution of probability *amplitudes*. Another distinction from the case of stochastic processes is that now the operators $\widehat{U}$ form a representation of the Abelian *group* of time translations (recall that the transition probabilities of chapter 1 define only a *semi*group). The reason for this distinction is evident for the particular case of a conservative system. For such systems, Hamiltonians do not depend on time and the postulate 2.3 can be reformulated as follows:

Postulate 2.3'. Evolution operators $\widehat{U}(t, t_0)$ form the continuous one-parameter group generated by the operator $-i\widehat{H}/\hbar$, i.e.
$$\psi(t) = \widehat{U}(t, t_0)\psi(t_0) = e^{-\frac{i}{\hbar}\widehat{H}(t-t_0)}\psi(t_0). \tag{2.1.11}$$

It is clear that due to the purely imaginary (oscillating) exponent, the sign of the time difference in (2.1.11) is not essential for the general properties of the final state (e.g., for its normalizability). Thus we might consider an evolution in the backward direction which is defined just by the inverse operator
$$\widehat{U}(t_0, t) = \widehat{U}^{-1}(t, t_0).$$

In fact, the existence of the reverse operator for any $\widehat{U}$ is stated in postulate 2.3 where this is declared to be a *unitary* operator.

From the group property of the evolution operator, we can derive the following statement: a state of a system does not change in time (does not evolve) if it is represented by an eigenfunction of the Hamiltonian operator.

Such a state is called *stationary* and the eigenfunction equation

$$\widehat{H}\psi_n = E_n\psi_n \qquad (2.1.12)$$

is called the *stationary Schrödinger equation*.

In the case of time-dependent Hamiltonians (non-conservative systems) the relation between the evolution operator and the Hamiltonian is a little bit more involved and requires the notion of *a time-ordering operator* **T** which, when applied to an arbitrary product of operators depending on the time variable,

$$\widehat{O}_{i_n}(t_{i_n}) \cdots \widehat{O}_{i_1}(t_{i_1})$$

reorders the time successively. More explicitly, we define

$$\mathbf{T}(\widehat{O}_{i_n}(t_{i_n}) \cdots \widehat{O}_{i_1}(t_{i_1})) = \widehat{O}_n(t_n)\widehat{O}_{n-1}(t_{n-1}) \cdots \widehat{O}_1(t_1) \qquad (2.1.13)$$

where the times $t_n, \ldots, t_1$ satisfy the causal condition

$$t_1 < t_2 < \cdots < t_{n-1} < t_n.$$

We use for the time-ordering operator **T** a different notation from that for any other operator (boldface print instead of 'hat'), because **T** acts on other *operators*, not on the Hilbert space vectors.

Using this time-ordering operator, the evolution operator for a non-conservative quantum-mechanical system can be expressed through the corresponding time-dependent Hamiltonian as follows (see problem 2.1.1, page 149):

$$\widehat{U}(t, t_0) = \mathbf{T}\exp\left\{-\frac{i}{\hbar}\int_{t_0}^{t} dt\, \widehat{H}(t)\right\}. \qquad (2.1.14)$$

In any case (for time-dependent or time-independent Hamiltonians), the evolution operator satisfies the equation

$$\left(i\hbar\frac{\partial}{\partial t} - \widehat{H}(t)\right)\widehat{U}(t, t_0) = 0 \qquad (2.1.15)$$

$$\widehat{U}(t_0, t_0) = \mathbb{I}. \qquad (2.1.16)$$

The group (Markovian-like) property (2.1.10) of the operators $\widehat{U}$ is the fundamental property of the evolution operators. It is the most substantial property and provides the basis for the existence of the path-integral formalism in quantum mechanics (in analogy with the derivation of path integrals for Markovian stochastic processes in chapter 1).

The next postulate expresses the important *superposition principle*: if a system can be in a state ψ or in a state φ, it can also be in states corresponding to a superposition of these states, i.e. to their arbitrary linear combinations

$$\Phi = \alpha_1\psi + \alpha_2\varphi \qquad \alpha_1, \alpha_2 \in \mathbb{C}.$$

However, there are exceptions from this situation, the so-called *superselection rules*, separating the Hilbert space of a system in the direct sum of subspaces which correspond to different eigenvalues of some specific observables, e.g., the electric charge.

Postulate 2.4. To any non-zero vector of a Hilbert space $\mathcal{H}$ belonging to some superselection sector, there is a corresponding state of the system and any self-adjoint operator corresponds to a certain observable.

A sum of vectors from different superselection subspaces (*superselection sectors*) do not correspond to any physical state. We shall not concentrate on the mathematically rigorous definition of superselection rules (see, e.g., Jauch (1968)), since this is not important for our aims, and restrict ourselves to the previous intuitive remark.

◇ **Dirac's notation in quantum mechanics**

Dirac proposed an effective and convenient notation in quantum mechanics. In his notation, a state vector in the Hilbert space $\mathcal{H}$ is denoted by $|\psi\rangle$ and is called a 'ket'. The scalar product of two vectors, $|\varphi\rangle$ and $|\psi\rangle$ is $\langle\varphi|\psi\rangle$, where $\langle\varphi|$, called 'bra', denotes the vector conjugated to $|\varphi\rangle$. The mean value of an operator $\widehat{A}$ in the state $|\psi\rangle$, in Dirac notation, is

$$\langle\widehat{A}\rangle_\psi = \bar{A} = \langle\psi|\widehat{A}|\psi\rangle.$$

The *wavefunction* in the x-representation, $\psi(x)$, is interpreted as

$$\psi(x) = \langle x|\psi\rangle \qquad (2.1.17)$$

so that it is the projection of $|\psi\rangle$ onto the state $|x\rangle$ with the sharp value of the position, i.e. $|x\rangle$ is the eigenvector of the position operator $\widehat{x}$ with the eigenvalue x:

$$\widehat{x}|x\rangle = x|x\rangle.$$

Similarly, a wavefunction $\tilde{\psi}(p)$ in the p-representation of states reads as

$$\tilde{\psi}(p) = \langle p|\psi\rangle$$

with $|p\rangle$ being the eigenvector of the momentum operator $\widehat{p}$:

$$\widehat{p}|p\rangle = p|p\rangle.$$

We note that

$$\langle x|p\rangle = \frac{1}{\sqrt{2\pi\hbar}}e^{ipx/\hbar}.$$

The orthogonality relations for the eigenstates $|x\rangle$ and $|p\rangle$ can be written as

$$\langle x|x'\rangle = \delta(x - x') \qquad (2.1.18)$$
$$\langle p|p'\rangle = \delta(p - p'). \qquad (2.1.19)$$

Note that in this chapter, as well as in the next one, we consider only *pure* states, which can be represented by a single vector of the Hilbert space. Consideration of *mixed* states we postpone until chapter 4.

128   Path integrals in quantum mechanics

◇ **Correspondence principle**

It is natural to require that in the case of macroscopic objects the laws of quantum mechanics become close to the laws of classical mechanics. More generally and more precisely, this condition is formulated as follows: if the de Broglie wavelength of a particle is much smaller than the characteristic sizes of the problem under consideration (e.g., characteristic length of variation of the potential energy), the movement of the particle must be close to the one defined by the classical trajectory. Recall that the de Broglie wavelength λ_{Br} of a particle with momentum p is defined as

$$\lambda_{Br} \stackrel{\text{def}}{\equiv} \frac{\hbar}{|p|}.$$

This requirement expresses the important *principle of correspondence*.

An analogous limit is well known in wave optics: if the wavelength λ is close to zero, the optical phenomena can be described by the so-called *geometric optics* where ray propagation is defined by the *Fermat principle*, which is very similar to the *principle of least action* in classical mechanics. Using this analogy, we can expect that the wavefunction of a particle will behave like

$$\psi(x,t)|_{\lambda_{Br} \to 0} \approx C \exp\left\{\frac{i}{\hbar}S(x,t)\right\} \qquad C = \text{constant} \qquad (2.1.20)$$

where S is the action. If

$$S \gg \hbar \qquad (2.1.21)$$

the exponent in (2.1.20) becomes rapidly oscillating due to variation of the particle position x, and then the trajectory of the particle is defined by the minimum of the action S. This important fact proves to be especially illuminating in the path-integral approach to quantum mechanics. The reader, being already familiar with the semiclassical approximation of Wiener integrals, may guess the reasons for this property. We shall discuss the peculiarities of the quantum-mechanical case (mainly due to the purely imaginary exponent in (2.1.20)) in this chapter (section 2.2.3).

◇ **Canonical quantization**

The central part of the quantum-mechanical formalism is the *canonical quantization*, based on Hamiltonian (or canonical) mechanics. Recall, that in *classical* Hamiltonian mechanics, the basic variables are the (generalized) coordinates x_k ($k = 1, \ldots, d$; with d being the number of degrees of freedom of the system) and the canonically conjugate momenta p_k which enter the Hamiltonian $H = H(x_k, p_k)$ and are obtained from the Lagrangian $L(x_k, \dot{x}_k)$ as

$$p_k = \frac{\partial L}{\partial \dot{x}_k}. \qquad (2.1.22)$$

In classical mechanics, the equation of motion for a dynamical variable $F(x_k, p_k)$, without explicit dependence on time, can be written in the form

$$\frac{\partial F}{\partial t} = \{H, F\} \qquad (2.1.23)$$

where

$$\{A, B\} = \sum_{k=1}^{d} \left(\frac{\partial A}{\partial p_k}\frac{\partial B}{\partial x_k} - \frac{\partial A}{\partial x_k}\frac{\partial B}{\partial p_k}\right) \qquad (2.1.24)$$

is the so-called classical *Poisson bracket*. In particular, (2.1.24) gives

$$\{p_k, x_l\} = \delta_{kl} \qquad k, l = 1, \ldots, d$$

$$\{x_k, x_l\} = 0 \qquad \{p_k, p_l\} = 0$$

$$\frac{\partial x_k}{\partial t} = \{H, x_k\} = \frac{\partial H}{\partial p_k} \qquad (2.1.25)$$

$$\frac{\partial p_k}{\partial t} = \{H, p_k\} = -\frac{\partial H}{\partial x_k}.$$

The canonical quantization consists in the substitution of the canonical variables x_k, p_k by the operators $\hat{x}_k, \hat{p}_k$, acting on the Hilbert space of states of the system, and in the substitution of the Poisson brackets for them by commutation relations, so that the quantum analog of (2.1.25) becomes

$$[\hat{x}_k, \hat{p}_l] = i\hbar \delta_{kl} \qquad k, l = 1, \ldots, d$$
$$[\hat{x}_k, \hat{x}_l] = [\hat{p}_k, \hat{p}_l] = 0 \qquad (2.1.26)$$

and the quantum equations of motion read as

$$-i\hbar \frac{\partial}{\partial t} \widehat{F}(\hat{x}_k, \hat{p}_k) = [\widehat{H}(\hat{x}_k, \hat{p}_k), \widehat{F}(\hat{x}_k, \hat{p}_k)] \qquad (2.1.27)$$

where $\widehat{H}$ is the quantum Hamiltonian. Here the commutator of operators is

$$[\widehat{A}, \widehat{B}] \stackrel{\text{def}}{\equiv} \widehat{A}\widehat{B} - \widehat{B}\widehat{A}. \qquad (2.1.28)$$

The number of degrees of freedom is not important for our discussion and, for simplicity, we shall consider the case $d = 1$ in what follows.

For this simple system, all observables have the form $\widehat{A} = A(\hat{x}, \hat{p})$, where $A(x, p)$ is some function. It is clear that, in general, substitution of the variables x, p in the functions $A(x, p)$ by the operators $\hat{x}, \hat{p}$ is an ambiguous operation. In particular, for the monomial $x^l p^k$, we can choose *different* orders of non-commuting operators, e.g., $\hat{x}^l \hat{p}^k$ or $\hat{p}^k \hat{x}^l$ or some other. Different prescriptions lead to different quantum systems (with the same classical limit). Thus an ordering rule is an essential part of the quantization procedure. We shall discuss this point in detail, as well as its influence on path-integral construction, in section 2.3. Note, however, that for functions of the form

$$A(x, p) = A_1(p) + A_2(x)$$

choice of ordering is not essential. Fortunately, the Hamiltonians of many important systems have just this form. In this and the next sections, we shall consider this type of system.

◇ **Elementary properties of position and momentum operators**

In the coordinate representation (2.1.17), the operators $\hat{x}$ and $\hat{p}$ corresponding to the position and momentum of a particle and satisfying (2.1.26) have the form

$$\hat{x}\psi(x) = x\psi(x) \qquad (2.1.29)$$
$$\hat{p}\psi(x) = -i\hbar\partial_x \psi(x). \qquad (2.1.30)$$

Here $\psi(x)$ are some functions from the Hilbert space $\mathcal{H} = \mathcal{L}^2(\mathbb{R})$ of *square-integrable functions*, with the scalar product

$$\langle\varphi|\psi\rangle = \int dx\, \varphi^*(x)\psi(x). \qquad (2.1.31)$$

The operator $\widehat{p}$ is unbounded and has a complete system of generalized eigenfunctions

$$\widehat{p}\psi_p(x) = p\psi_p(x) \qquad p \in \mathbb{R}. \qquad (2.1.32)$$

In the representation (2.1.30), the eigenfunctions are defined by the differential equation

$$-i\hbar\partial_x \psi_p(x) = p\psi_p(x)$$

which gives

$$\psi_p(x) = \frac{1}{\sqrt{2\pi\hbar}} e^{ipx/\hbar} \qquad p \in \mathbb{R} \qquad (2.1.33)$$

(with a suitable normalization). This is a *generalized* eigenfunction, because $\psi_p(x) \notin \mathcal{L}^2(\mathbb{R})$, i.e. it is not a square-integrable function. Instead, it is a distribution: if $\varphi \in \mathcal{S}$ (*the set of Schwarz test functions*: all derivatives $\varphi^{(k)}(x) \stackrel{\text{def}}{\equiv} \frac{d^k\varphi}{dx^k}$ exist and $\lim_{x\to\infty} x^n \varphi^{(k)}(x) = 0$ for any n and k), then there exist the integrals

$$\widetilde{\varphi}(p) \stackrel{\text{def}}{\equiv} \int dx\, \psi_p^*(x)\varphi(x) = \frac{1}{\sqrt{2\pi\hbar}} \int e^{-ipx/\hbar}\varphi(x)\, dx \qquad (2.1.34)$$

where $\widetilde{\varphi}(p)$ is the Fourier transform of $\varphi(x)$. The last equation can be inverted and

$$\varphi(x) = \int dp\, \widetilde{\varphi}(p)\psi_p(x) = \frac{1}{\sqrt{2\pi\hbar}} \int dp\, e^{ipx/\hbar}\widetilde{\varphi}(p) \qquad (2.1.35)$$

as is well known from the theory of Fourier transforms. This tells us that any function $\varphi(x)$ from $\mathcal{S}$ (and after suitable generalization, even from $\mathcal{H}$) can be expanded in the set of eigenfunctions $\psi_p(x)$, $p \in \mathbb{R}$, i.e. the set of eigenfunctions of $\widehat{p}$ is complete.

Similarly, the operator $\widehat{x}$ has a complete set of generalized eigenfunctions

$$\widehat{x}\psi_y(x) = y\psi_y(x) \qquad (2.1.36)$$

which are given as the Dirac δ-function:

$$\psi_y(x) = \delta(x - y). \qquad (2.1.37)$$

This can be rewritten as

$$\varphi(x) = \int \varphi(y)\psi_y(x)\, dy \qquad (2.1.38)$$

and means the completeness of the set $\psi_y(x)$, $y \in \mathbb{R}$.

The mean value of an operator $\widehat{A}$, made of $\widehat{x}$ and $\widehat{p}$ in a state described by a wavefunction $\psi(x)$, is given by

$$\langle A\rangle_\psi = \int dx\, \psi^*(x)\widehat{A}\psi(x)$$

with the dispersion $\mathbb{D}_\psi A$,

$$(\mathbb{D}_\psi A)^2 \equiv \langle(\widehat{A} - \langle\widehat{A}\rangle_\psi)^2\rangle_\psi = \int_{-\infty}^{\infty} dx\, \psi^*(x)(\widehat{A} - \langle\widehat{A}\rangle_\psi)^2 \psi(x) \qquad (2.1.39)$$

which is in general non-zero.

Taking into account the fact that the eigenfunctions of the position operator $\hat{x}$ are δ-functions, we can derive the physical interpretation of the integral:

$$\mathbb{P} = \int_a^b dx \, |\psi(x)|^2 \qquad [a,b] \subset \mathbb{R} \tag{2.1.40}$$

is the probability that the position of the particle described by $\psi(x)$ is in the interval $[a,b]$. Analogously, the integral

$$\tilde{\psi}(p) = \frac{1}{\sqrt{2\pi\hbar}} \int_{-\infty}^{\infty} dx \, e^{-ipx/\hbar} \psi(x) \tag{2.1.41}$$

has the following interpretation: $|\tilde{\psi}(p)|^2$ is the probability density for the distribution of the particle momentum in the state $\psi(x)$.

◇ **The basic quantum-mechanical problem: the harmonic oscillator**

The first substantial problem in quantum mechanics is that of the *quantum harmonic oscillator*. One can say that a huge number of quantum-mechanical problems are somehow related to this fundamental system. For example, describing movement of an atom in molecules or in a solid body is reduced to this problem. Another important example is the quantum electromagnetic field which is nothing but an infinite collection of harmonic oscillators (each field frequency corresponds to an oscillator).

In classical mechanics, the term *harmonic oscillator* is used to denote a particle moving in the quadratic potential

$$V(x) = \frac{m\omega^2}{2} x^2$$

where x is the coordinate of the particle. The total energy, expressed in terms of the phase-space variables x and p, i.e. the Hamiltonian, reads as

$$H(p,x) = \frac{p^2}{2m} + \frac{m\omega^2}{2} x^2. \tag{2.1.42}$$

In the one-dimensional case and in the coordinate representation, the Schrödinger equation for a harmonic oscillator has the form

$$i\hbar \frac{\partial \psi}{\partial t} = -\frac{\hbar^2}{2m} \frac{\partial^2 \psi}{\partial x^2} + \frac{m\omega^2}{2} x^2 \psi_p(x). \tag{2.1.43}$$

The essential peculiarity of the Hamiltonian (2.1.42) is the factorization property: after introducing the complex coordinates in the phase space:

$$a = \frac{1}{\sqrt{2}} \left(\sqrt{\frac{m\omega}{\hbar}} x + \frac{i}{\sqrt{\hbar m\omega}} p \right) \tag{2.1.44}$$

$$a^* = \frac{1}{\sqrt{2}} \left(\sqrt{\frac{m\omega}{\hbar}} x - \frac{i}{\sqrt{\hbar m\omega}} p \right) \tag{2.1.45}$$

the Hamiltonian function (2.1.42) takes the form

$$H(p,x) = \hbar\omega a^* a. \tag{2.1.46}$$

In quantum mechanics, the coordinates (2.1.44) and (2.1.45) become the Hermitian conjugate operators $\widehat{a}, \widehat{a}^\dagger$:

$$\widehat{a} = \frac{1}{\sqrt{2}}\left(\sqrt{\frac{m\omega}{\hbar}}\widehat{x} + \frac{i}{\sqrt{\hbar m\omega}}\widehat{p}\right)$$
$$\widehat{a}^\dagger = \frac{1}{\sqrt{2}}\left(\sqrt{\frac{m\omega}{\hbar}}\widehat{x} - \frac{i}{\sqrt{\hbar m\omega}}\widehat{p}\right).$$
(2.1.47)

The canonical commutation relation (2.1.26) implies the basic relation for these new operators:

$$[\widehat{a}, \widehat{a}^\dagger] = 1.$$
(2.1.48)

In terms of these operators, the quantum Hamiltonian is written as

$$\widehat{H}(p, x) = \hbar\omega(\widehat{a}^\dagger \widehat{a} + \tfrac{1}{2}).$$
(2.1.49)

and satisfies the commutation relations

$$[\widehat{H}, \widehat{a}] = -\hbar\omega\widehat{a}$$
(2.1.50)
$$[\widehat{H}, \widehat{a}^\dagger] = \hbar\omega\widehat{a}^\dagger.$$
(2.1.51)

The essence of the solvability of the harmonic oscillator problem consists in two facts. First, if $\psi \in \mathcal{L}^2(\mathbb{R})$ is an eigenvector of $\widehat{H}$ with an eigenvalue λ and $\widehat{a}^\dagger\psi \neq 0$, then $\widehat{a}^\dagger\psi$ is the eigenvector with the eigenvalue $\lambda + 1$. Second, the function

$$\psi_0(x) = (\kappa\sqrt{\pi})^{-1/2} \exp\left\{-\frac{x^2}{2\kappa^2}\right\} \qquad \kappa = \sqrt{\frac{\hbar}{m\omega}}$$

is the eigenfunction with the lowest eigenvalue (*ground state*):

$$\widehat{H}\psi_0 = \frac{\hbar\omega}{2}\psi_0$$
(2.1.52)
$$\widehat{a}\psi_0 = 0.$$
(2.1.53)

All other eigenfunctions ψ_k can be found by the repeated action of the operator $\widehat{a}^\dagger$:

$$\psi_{k+1} = \frac{1}{\sqrt{k+1}}\widehat{a}^\dagger\psi_k.$$
(2.1.54)

Also, it is easy to check that the operator $\widehat{a}$ lowers the eigenvalues of the Hamiltonian by one unit:

$$\psi_{k-1} = \frac{1}{\sqrt{k}}\widehat{a}\psi_k.$$
(2.1.55)

Thus the operators $\widehat{a}, \widehat{a}^\dagger$ are called *lowering* and *raising* operators, respectively. Another name for these operators is *annihilation* and *creation* operators. Correspondingly, the state ψ_0 is also named the *vacuum state* (absence of particles).

It is known that the functions $\psi_k(x)$ have the form $\psi_k(x) = \text{constant} \times \psi_0 H_k(x)$, where $H_k(x)$ is some polynomial. The functions ψ_k are called *Hermite functions* and the polynomials $H_k(x)$ are called *Hermite polynomials*. Being eigenfunctions of the symmetric operator $\widehat{H}$, the functions $\psi_k(x)$ form an orthonormal system and a complete basis of $\mathcal{L}^2(\mathbb{R})$.

◇ A particle moving in an arbitrary potential

For a particle moving in some potential $V(x)$, the Hamiltonian has the form

$$\widehat{H} = \frac{1}{2m}\widehat{p}^2 + V(\widehat{x})$$
$$= -\frac{\hbar^2}{2m}\partial_x^2 + V(x) \qquad (2.1.56)$$

(in the second line we have used the coordinate representation). In what follows, we shall assume that this Hamiltonian is a Hermitian operator. The Schrödinger differential equation should be supplemented by the initial condition

$$\psi(x,t)|_{t=t_0} = \psi(x,t_0) \equiv \psi_0(x) \qquad (2.1.57)$$

where $\psi_0(x) \equiv \psi(x,t_0)$ is the wavefunction of the particle at the time $t = t_0$.

In the case of unbounded motion in the space (all $x \in \mathbb{R}$ are allowed), there are no boundary conditions, and some restrictions at infinity follow just from the normalization condition

$$\|\psi\|^2 = \int_{-\infty}^{\infty} dx\, |\psi(x,t)|^2 = 1 \qquad (2.1.58)$$

(in other words, the functions $\psi(x)$ are square integrable: $\psi \in \mathcal{L}^2(\mathbb{R})$). If the evolution operator $\widehat{U}(t,t_0)$ (2.1.8) is unitary (this is a necessary self-consistency condition in quantum mechanics), the normalization (2.1.58) is satisfied at any time t.

The solution of the Schrödinger equation becomes simple, if we know the eigenfunctions of the Hamiltonian, i.e. the functions $\varphi_n(x)$, $n = 0, 1, 2, \ldots$, satisfying the equation

$$\widehat{H}\varphi_n(x) = E_n\varphi_n(x). \qquad (2.1.59)$$

Here we have assumed, for simplicity, that $\widehat{H}$ has only a discrete spectrum, generalizations to continuous or combined (discrete and continuous) spectra being straightforward. The solution of the Schrödinger equation (2.1.9) is then given as

$$\psi(x,t) = \sum_n c_n e^{-iE_n t/\hbar}\varphi_n(x) \qquad (2.1.60)$$

where c_n are defined by the initial condition

$$\psi_0(x) = \sum_n c_n e^{-iE_n t_0/\hbar}\varphi_n(x) \qquad (2.1.61)$$

and read as

$$c_n = \langle \varphi_n|\psi_0\rangle e^{iE_n t_0/\hbar}. \qquad (2.1.62)$$

Here the orthonormality of the systems of eigenfunctions has been used.

Thus, the solution of the eigenvalue problem (2.1.59) (the stationary Schrödinger equation) is equivalent to the full solution of the quantum-mechanical problem in question. But, in general, this is a complicated problem, which can be solved exactly only in few cases. Therefore, powerful approximate methods are of great importance.

In Dirac notation, the Schrödinger equation is written as an abstract equation in the Hilbert space $\mathcal{H}$

$$i\hbar\partial_t|\psi(t)\rangle = \widehat{H}|\psi(t)\rangle \qquad (2.1.63)$$

with the initial vector $|\psi(0)\rangle$ (we put $t_0 = 0$, for simplicity). Its *formal* solution satisfying the given initial condition is

$$|\psi(t)\rangle = e^{-it\widehat{H}/\hbar}|\psi(0)\rangle \qquad t > 0 \qquad (2.1.64)$$

where the exponent is defined in the standard way as the series expansion

$$e^{-it\widehat{H}/\hbar} = \sum_{n=0}^{\infty} \frac{(-it/\hbar)^n}{n!} \widehat{H}^n = 1 - \frac{it}{\hbar}\widehat{H} + \cdots. \qquad (2.1.65)$$

We refer to solution (2.1.64) as a formal one, since the calculation of the action of $\widehat{H}^n$ on $|\psi(0)\rangle$ for an arbitrary n is a non-trivial problem (if the eigenstates of $\widehat{H}$ are not known).

◇ **Transition amplitudes (propagators) in quantum mechanics**

To any operator acting on a Hilbert space, we can put in correspondence its *integral kernel*, which is nothing other than matrix elements of the operator in some basis of $\mathcal{H}$ formed by the eigenfunctions of a self-adjoint operator (or a set of mutually commuting operators). Let $\{\psi_\lambda\}$ be an orthonormal basis of $\mathcal{H}$ (in general, λ stands for the *set* of eigenvalues of a complete set of commuting operators). Then the integral kernel $K_A(\lambda', \lambda)$ of an operator $\widehat{A}$ is

$$K_A(\lambda', \lambda) = \langle\psi_{\lambda'}|\widehat{A}|\psi_\lambda\rangle$$

so that the action of the operator $\widehat{A}$ on an arbitrary vector $|\varphi\rangle \in \mathcal{H}$ in the λ-representation, i.e. the wavefunction $\varphi(\lambda) \stackrel{\text{def}}{\equiv} \langle\psi_\lambda|\varphi\rangle$, is expressed through $K_A(\lambda', \lambda)$ as follows:

$$(\widehat{A}\varphi)(\lambda) = \langle\psi_\lambda|\widehat{A}|\varphi\rangle = \int d\lambda' \, \langle\psi_\lambda|\widehat{A}|\psi_{\lambda'}\rangle\langle\psi_{\lambda'}|\varphi\rangle$$

$$= \int d\lambda' \, K_A(\lambda, \lambda')\varphi(\lambda'). \qquad (2.1.66)$$

The latter expression justifies the name, *integral* kernel. It is easy to verify that the kernel K_A of the product $\widehat{A} = \widehat{A}_1\widehat{A}_2$ of two operators $\widehat{A}_1$ and $\widehat{A}_2$ is expressed via the convolution of the kernels K_{A_1} and K_{A_2}:

$$K_A(\lambda'', \lambda) = \int d\lambda' \, K_{A_1}(\lambda'', \lambda')K_{A_2}(\lambda', \lambda). \qquad (2.1.67)$$

Thus, in the x-representation (i.e. in the basis of eigenfunctions of the coordinate operator), the integral kernel of the evolution operator is defined as

$$K(x, t|x_0, t_0) = \langle x|e^{-i(t-t_0)\widehat{H}/\hbar}|x_0\rangle. \qquad (2.1.68)$$

This is a basic object allowing us to express the solution of the Schrödinger equation (2.1.9), satisfying initial condition (2.1.57), as follows:

$$\psi(x, t) = \int_{-\infty}^{\infty} dx_0 \, K(x, t|x_0, t_0)\psi(x_0, t_0) \qquad t > t_0. \qquad (2.1.69)$$

To obtain this relation, we multiply (2.1.64) from the left by $\langle x|$ and insert the unit operator

$$\mathbb{1} = \int_{-\infty}^{\infty} dx_0 \, |x_0\rangle\langle x_0|$$

on the right-hand side.

Knowledge of the Hamiltonian eigenfunctions (2.1.59) allows us to write the kernel in the following convenient form:

$$K(x,t|x_0,t_0) = \sum_n e^{-iE_n(t-t_0)/\hbar}\varphi_n(x)\varphi_n^*(x_0). \qquad (2.1.70)$$

Here the summation sign implies a summation over discrete eigenvalues as well as an integration over the continuous eigenvalues.

The reader may easily recognize in (2.1.68) an analog of the transition probability in stochastic physics, discussed in chapter 1. However, we must always remember that in quantum mechanics we deal with probability *amplitudes* as basic objects. Thus, the kernel (2.1.68) is called the *transition amplitude* or *propagator*.

The composition law for the evolution operators (2.1.10) and the completeness of the (coordinate) basis in $\mathcal{H}$ give an important relation for the propagators

$$K(x_2,t_2|x_0,0) = \int_{-\infty}^{\infty} dx_1\, K(x_2,t_2|x_1,t_1)K(x_1,t_1|x_0,0) \qquad 0 \le t_1 \le t_2 \qquad (2.1.71)$$

which is a close analog of the ESKC relation (1.1.45). To prove (2.1.71), we must take the matrix element of (2.1.10) between $\langle x|$ and $|x_0\rangle$ and insert the unit operator

$$\mathbb{1} = \int_{-\infty}^{\infty} dx'\, |x'\rangle\langle x'|$$

on the right-hand side. Then

$$\langle x|\widehat{U}(t,t_0)|x_0\rangle = \int_{-\infty}^{\infty} dx'\, \langle x|\widehat{U}(t,t')|x'\rangle\langle x'|U(t',t_0)|x_0\rangle.$$

◇ **Example: the propagator of a free particle**

To illustrate the meaning of propagators (transition amplitudes) in quantum mechanics presented here, let us calculate, as a simplest example, the propagator of a free particle with the Hamiltonian

$$\widehat{H}_0 = \frac{1}{2m}\widehat{p}^2 = -\frac{\hbar^2}{2m}\partial_x^2. \qquad (2.1.72)$$

For this aim, we need the generalization of the Gaussian integral formula for complex coefficients in the exponent of the integrand:

$$\int_{-\infty}^{\infty} dx\, e^{-\alpha x^2 + \beta x}\, dx = \sqrt{\frac{\pi}{\alpha}}e^{\beta^2/(4\alpha)} \qquad x \in \mathbb{R},\ \alpha,\beta \in \mathbb{C} \qquad (2.1.73)$$

where

$$\operatorname{Re}\alpha > 0 \qquad \beta \text{ is arbitrary} \qquad (2.1.74)$$

or

$$\operatorname{Re}\alpha = \operatorname{Re}\beta = 0. \qquad (2.1.75)$$

Note that for purely imaginary constants (the case (2.1.75)) the integral exists, from the mathematically rigorous point of view, only in the sense of distributions (generalized functions) because of the oscillating (not decreasing at infinity) behaviour of the integrand. This means, according to the general theory of distributions, that the rigorous sense can be attributed to this integral only if the purely imaginary exponent $\exp\{-i(ax^2 + bx)\}$ ($a, b \in \mathbb{R}$) is integrated together with an appropriate test function (with a suitable behaviour at infinity). In fact, case (2.1.74) can be treated in that sense, too: the real part $\exp\{-\eta x^2\}$ ($\eta \in \mathbb{R}$) of the exponent can be considered as a test function,

$$\int_{-\infty}^{\infty} dx\, e^{-\alpha x^2 + \beta x}\, dx = \int_{-\infty}^{\infty} dx\, e^{-\eta x^2} e^{-i a x^2 + \beta x}\, dx \qquad \alpha \equiv \eta + ia. \tag{2.1.76}$$

Sometimes, to be careful with the integration, we introduce an infinitesimal positive real part for α even in the case (2.1.75) and sets $\eta \equiv \operatorname{Re}\alpha$ equal to zero at the very end of the calculations. This procedure is usually called the $i\eta$-*prescription*. Essentially, the result (2.1.73), (2.1.75) for purely imaginary α can be obtained from the genuine Gaussian formula for real α by analytic continuation and the $i\eta$-prescription defines the bypass rule around the singularity (branch point) at $\operatorname{Im}\alpha = 0$, to produce the proper phase:

$$\int_{-\infty}^{\infty} dx\, \exp\{-i(ax^2 + bx)\} = \sqrt{\frac{\pi}{ia}} e^{ib^2/(4a)}. \tag{2.1.77}$$

Using (2.1.73), the evolution operator for the free Hamiltonian $\widehat{H}_0$ (2.1.72) can be derived directly (problem 2.1.3, page 150):

$$K(x, t|x_0, 0) = \langle x | e^{-it\widehat{H}_0/\hbar} | x_0 \rangle$$

$$= \left(\frac{m}{2\pi i\hbar t}\right)^{1/2} \exp\left\{i\frac{m}{2\hbar t}(x - x_0)^2\right\}. \tag{2.1.78}$$

Again, we see a close analogy with the transition probability for a Brownian particle discussed in chapter 1, except for the basic difference that the exponent in (2.1.78) is purely imaginary.

◇ **Energy-dependent propagator**

As seen from (2.1.68), the propagator K depends only on the difference of times, i.e.

$$K(x, t|x_0, t_0) = K(x, x_0; t - t_0). \tag{2.1.79}$$

It is often simpler to use the energy-dependent Green function $G(x, x_0; E)$ instead of the propagator $K(x, x_0; t) = K(x, t|x_0, 0)$. The two are related by the Fourier transform:

$$G(x, x_0; E) = \frac{1}{i\hbar} \int_{-\infty}^{\infty} dt\, e^{iEt/\hbar} K(x, x_0; t)$$

$$= \frac{1}{i\hbar} \int_0^{\infty} dt\, e^{iEt/\hbar} K(x, x_0; t) \tag{2.1.80}$$

(since $K(x, x_0; t) = 0$ for $t < 0$). The integral (2.1.80) can always be made to converge if the energy E becomes slightly complex with a small positive imaginary part (the $i\eta$-prescription).

The transform (2.1.80) of the series (2.1.70) gives the expansion for the energy-dependent Green function

$$G(x, x_0; E) = \sum_n \frac{\varphi_n(x)\varphi_n^*(x_0)}{E - E_n}. \tag{2.1.81}$$

Since the spectrum of $\widehat{H}$ is real, singularities arise when $\operatorname{Im} E = 0$. For a discrete spectrum, we identify the poles of G with the location of bound states and the corresponding residues with the bound state wavefunctions. For a continuous spectrum of $\widehat{H}$, the function G has a cut.

2.1.2 Feynman–Kac formula in quantum mechanics

The basic observation underlying path integrals for time evolution of quantum-mechanical transition amplitudes was made by Dirac (1933) on the role of Lagrangians in quantum mechanics. He observed that the short-time propagator is the exponential of $\{iS/\hbar\}$, where S is the classical action of the system under consideration (see also Dirac (1947)). This eventually led Feynman to the invention of quantum-mechanical path integrals in configuration space (Feynman 1942, 1948) (see also Feynman and Hibbs (1965) and references therein). This complex-valued path integral, which now bears Feynman's name, gives a convenient expression for the propagator of the Schrödinger equation

$$K(x_t, t|x_0, 0) = \int_{\mathcal{C}\{x_0,0;x_t,t\}} \prod_{\tau=0}^{t} \frac{dx(\tau)}{\sqrt{\frac{2i\pi\hbar}{m} d\tau}} \exp\left\{ \frac{i}{\hbar} \int_0^t d\tau \left(\frac{1}{2} m\dot{x}^2 - V(x) \right) \right\}. \qquad (2.1.82)$$

Later, Kac (1949) (see also Kac (1959) and references therein) realized that the analog of Feynman's formula with a *real* exponent can be mathematically rigorously justified on the basis of the Wiener functional measure and path integral and that it can be applied to problems in statistical physics and the theory of stochastic processes. That is how formula (2.1.82) and its statistical counterpart (1.2.94) acquired the name '*Feynman–Kac formula*'.

In this book, we do not follow the historical development. Instead, we introduced and discussed the Feynman–Kac formula in the preceding chapter in the framework of the theory of stochastic processes and Wiener integrals, and now we can use these results for quantum-mechanical problems.

◇ **Relation between the Schrödinger equation for a free particle and the diffusion equation**

If a particle is free, i.e. $V(x) = 0$, the Schrödinger equation (2.1.9) with the Hamiltonian (2.1.72) becomes

$$\frac{\partial \psi}{\partial t} = \frac{i\hbar}{2m} \frac{\partial^2 \psi}{\partial x^2} \qquad (2.1.83)$$

and closely resembles the diffusion equation (1.1.18) with the *formal* diffusion coefficient

$$D_{\text{QM}} = \frac{i\hbar}{2m}. \qquad (2.1.84)$$

Therefore, we can immediately write the solution of (2.1.83):

$$\psi(x, t) = \frac{1}{\sqrt{4\pi D_{\text{QM}} t}} \exp\left\{ -\frac{x^2}{4 D_{\text{QM}} t} \right\}. \qquad (2.1.85)$$

However, it differs fundamentally from the genuine diffusion equation: now the 'diffusion coefficient' is purely imaginary and thus the function ψ can not be normalized: as we discussed in the preceding subsection, the integral

$$\int_{-\infty}^{\infty} dx \, \frac{1}{\sqrt{4\pi D_{\text{QM}} t}} \exp\left\{ -\frac{x^2}{4 D_{\text{QM}} t} \right\}$$

exists only in the sense of generalized functions. This is related to the fact that in quantum mechanics the function ψ has no direct interpretation as a probability distribution (recall that the meaning of probability distribution is attributed to the squared absolute values of wavefunctions, $|\psi|^2$). In addition, the solutions of the free Schrödinger equation, i.e. plane waves, are non-normalizable even after squaring:

we should make an appropriate superposition of this state (a wave packet) to obtain a localized, square-integrable function. Using the formal analogy with the diffusion equation we can immediately write the representation of the solution of the free Schrödinger equation in terms of the path integral (i.e. write the Feynman–Kac formula for this particular case):

$$K_0(x,t|x_0,0) = \int_{\mathcal{C}\{x_0,0;x_t,t\}} \prod_{\tau=0}^{t} \frac{dx(\tau)}{\sqrt{4D_{QM}\pi\, d\tau}} \exp\left\{-\frac{1}{4D_{QM}}\int_0^t d\tau\, \dot{x}^2\right\} \qquad (2.1.86)$$

which, after substituting (2.1.84), takes the form

$$K_0(x,t|x_0,0) = \int_{\mathcal{C}\{x_0,0;x_t,t\}} \prod_{\tau=0}^{t} \frac{dx(\tau)}{\sqrt{\frac{2i\hbar}{m}\pi\, d\tau}} \exp\left\{\frac{i}{\hbar}\int_0^t d\tau\, \frac{m}{2}\dot{x}^2\right\}. \qquad (2.1.87)$$

Thus the formal *Feynman integration 'measure'*

$$d_F x(t) = \prod_{\tau=0}^{t} \frac{dx(\tau)}{\sqrt{\frac{2i\hbar}{m}\pi\, d\tau}} \exp\left\{\frac{i}{\hbar}\int_0^t d\tau\, \frac{m}{2}\dot{x}^2\right\} \qquad (2.1.88)$$

is *complex valued* and contains an oscillating factor instead of a decreasing exponent as in the Wiener measure. Thus (2.1.88) is certainly not a true measure on some set of trajectories. We shall discuss the reasons for this in more detail in section 2.2.1. In the present section, we shall try to use for quantum-mechanical problems our knowledge about path integrals borrowed from the theory of stochastic processes (i.e. from the preceding chapter). This is possible due to Kac's discovery that if we consider the operator $e^{-t\widehat{H}/\hbar}$ instead of $e^{-it\widehat{H}/\hbar}$, we arrive at the Wiener measure and path integral.

◇ Analytical continuation to purely imaginary time in quantum mechanics

Of course, using the Wiener–Kac approach, we obtain information about $e^{-t\widehat{H}/\hbar}$ and not about the fundamental object of quantum dynamics, the evolution operator $e^{-it\widehat{H}/\hbar}$. However, if we want to study eigenfunctions, it is hard to claim that $e^{-it\widehat{H}/\hbar}$ is any more basic than $e^{-t\widehat{H}/\hbar}$. Indeed, as we shall see in this section, the path-integral consideration of the latter operator allows us to deduce general properties of the Hamiltonian spectrum. Besides, this operator gives the lowest eigenvalue λ_0 of a Hamiltonian:

$$\lambda_0 = -\lim_{t\to 0} t^{-1} \ln\langle\psi|e^{-t\widehat{H}/\hbar}\psi\rangle \qquad (2.1.89)$$

where ψ is an arbitrary vector with non-zero projection on the eigenstate ψ_0 with lowest eigenvalue:

$$\widehat{H}\psi_0 = \lambda_0\psi_0.$$

Moreover, if λ_0 is a non-degenerate eigenvalue and $\langle\psi|\psi_0\rangle > 0$, then

$$\psi_0 = \lim_{t\to\infty} \langle\psi|e^{-2t\widehat{H}/\hbar}\psi\rangle^{-\frac{1}{2}} e^{-t\widehat{H}/\hbar}\psi \qquad (2.1.90)$$

(see, e.g., Simon (1979)).

The consideration of the operator $e^{-t\widehat{H}/\hbar}$ in quantum mechanics can be treated as the transition to imaginary time:

$$t \to -it. \qquad (2.1.91)$$

Such a transition has a strict mathematical background, because the solutions of the Schrödinger equation can be *analytically* continued in the time variable t to the lower complex half-plane. The analyticity, in turn, is the consequence of the assumption about the positivity of spectra of physical Hamiltonians (this provides the good behaviour of the operator $\mathrm{e}^{-iz\widehat{H}}$ in the lower half-plane of the variable $z \in \mathbb{C}$). Note that, from a physical point of view, the positivity of the Hamiltonian spectrum is necessary for the stability of the corresponding system.

This fact is especially important for quantum field theory and is basic for *Euclidean quantum field theory* (see, e.g., Simon (1974)). We shall deal with it in the chapters 3 and 4. The central object of this theory is the Euclidean Green functions which, under the so-called *Osterwalder–Schrader conditions*, prove to be equivalent to the Green functions of the corresponding field theory in ordinary Minkowski spacetime (i.e. to theory with real time). It is essential that according to the *Wightman reconstruction theorem* (see, e.g., Simon (1974) and Glimm and Jaffe (1987)) all theory is defined by a set of Euclidean Green functions satisfying the Osterwalder–Schrader conditions (or, equivalently, by the corresponding Green functions with real time). The great advantage of Euclidean field theory is the possibility to interpret quantum fields as generalized Markovian random fields and, hence, to apply the methods of classical statistical physics.

In this section, we have a much more modest goal: using the possibility of a transition to imaginary time, to consider path integrals for operators of the form $\mathrm{e}^{-t\widehat{H}/\hbar}$ and to derive from this path integral some properties of quantum Hamiltonians and Bohr–Sommerfeld semiclassical conditions.

Before proceeding to this consideration, it is necessary to stress that the previously mentioned possibility of substitution $t \to -it$ by no means implies that we may just forget about real-time formulation and work with Wiener integrals only. For many quantum-mechanical (including field theoretical) problems, this way (transition to imaginary time Green functions and then reconstruction of the results in real-time formulation) is redundant and highly inconvenient. Thus, together with the Wiener–Kac path-integral approach to quantum-mechanical problems, the real-time Feynman path integral (2.1.82) and more general phase-space path integrals (see next section) are widely used.

◇ Feynman's proof of the Feynman–Kac formula

As a result of its close relation with the Feynman–Kac formula for the Bloch equation and Wiener integrals we can readily use the result of section 1.2.5 to prove that the path integral (2.1.82) gives the solution of the Schrödinger equation (2.1.9) with the Hamiltonian (2.1.56) (i.e. the Feynman–Kac formula). However, since this result is of great importance and for reader convenience, we also present here independently the less rigorous but illustrative genuine Feynman proof.

The starting point of the proof is the quantum-mechanical analog of the ESKC equation for the propagator K:

$$\int_{-\infty}^{\infty} dx_\tau \, K(x_\tau, \tau | x_0, 0) K(x_t, t | x_\tau, \tau) = K(x_t, t | x_0, 0). \qquad (2.1.92)$$

According to Wiener's theorem from chapter 1 and the previously mentioned analogy between Feynman and Wiener integrals, we must integrate over the set of continuous functions. It is known that the set of such functions can be approximated uniformly on the set by piecewise linear functions, the accuracy depending only on the number of linear pieces. This means that the integrand in the Feynman–Kac formula for short time intervals can be approximated as follows:

$$\exp\left\{-\frac{i}{\hbar}\int_\tau^t ds\, V(x(s))\right\} = \exp\left\{-\frac{i}{\hbar}(t-\tau)V(x_t)\right\} + \widetilde{\delta} \qquad (2.1.93)$$

where $\widetilde{\delta} \xrightarrow[\tau \to t]{} 0$, and this approximation error does not depend at small $(t - \tau)$ on a chosen trajectory (due to the uniform convergence). Thus we obtain the following approximation of the *propagator for a short time interval*:

$$
K(x_t, t | x_\tau, \tau) = \int_{C\{x_\tau,\tau;x_t,t\}} d_F x(s) \left[\exp\left\{ -\frac{i}{\hbar}(t-\tau) V(x_t) \right\} + \widetilde{\delta} \right]
$$

$$
= \frac{\exp\left\{ -\frac{m(x_t - x_\tau)^2}{2i\hbar(t-\tau)} - \frac{i}{\hbar}(t-\tau) V(x_t) \right\}}{\sqrt{\pi \frac{2i\hbar(t-\tau)}{m}}} + \delta \qquad (2.1.94)
$$

where $\delta \xrightarrow[\tau \to t]{} 0$.

Substituting this approximation in the ESKC equation (2.1.92), we get

$$
K(x_t, t | x_0, 0) = \int_{-\infty}^{\infty} dx_\tau \, K(x_\tau, \tau | x_0, 0) \left[\frac{\exp\left\{ -\frac{m(x_t - x_\tau)^2}{2i\hbar(t-\tau)} - \frac{i}{\hbar}(t-\tau) V(x_t) \right\}}{\sqrt{\pi \frac{2i\hbar(t-\tau)}{m}}} + \delta \right]. \qquad (2.1.95)
$$

For $K(x_\tau, \tau | x_0, 0)$, we can use the Taylor series expansion around x_t:

$$
K(x_\tau, \tau | x_0, 0) = K(x_t, \tau | x_0, 0) + \frac{\partial K(x_t, \tau | x_0, 0)}{\partial x_t}(x_\tau - x_t)
$$

$$
+ \frac{1}{2} \frac{\partial^2 K(x_t, \tau | x_0, 0)}{\partial x_t^2}(x_\tau - x_t)^2 + \cdots \qquad (2.1.96)
$$

so that using the Gaussian integrals (see supplement I, volume II), we convert the integral ESKC equation into a differential one:

$$
K(x_t, t | x_0, 0) = K(x_t, \tau | x_0, 0) \left(1 - \frac{i}{\hbar}(t-\tau) V(x_t) \right)
$$

$$
+ \frac{i\hbar}{2m} \frac{\partial^2 K(x_t, \tau | x_0, 0)}{\partial x_t^2}(t-\tau) + \mathcal{O}((t-\tau)^2). \qquad (2.1.97)
$$

Rewriting (2.1.97) in the form

$$
\frac{K(x_t, t | x_0, 0) - K(x_t, \tau | x_0, 0)}{t - \tau} = \frac{i\hbar}{2m} \frac{\partial^2 K(x_t, \tau | x_0, 0)}{\partial x_t^2}
$$

$$
- \frac{i}{\hbar} V(x_t) K(x_\tau, \tau | x_0, 0) + \mathcal{O}((t-\tau))
$$

and taking the limit $\tau \to t$, we obtain that K satisfies the equation

$$
\frac{\partial K(x_t, t | x_0, 0)}{\partial t} = \frac{i\hbar}{2m} \frac{\partial^2 K(x_t, t | x_0, 0)}{\partial x_t^2} - \frac{i}{\hbar} V(x_t) K(x_\tau, \tau | x_0, 0) \qquad (2.1.98)
$$

i.e. the Schrödinger equation (2.1.9) with the Hamiltonian (2.1.56).

◇ **Feynman path integral and action functional**

A very important remark about the Feynman–Kac formula is that the exponent in the integrand of (2.1.82) is nothing but the physical *action functional*

$$
S[x(\tau)] \stackrel{\text{def}}{\equiv} \int_0^t d\tau \, L(x(\tau)) \qquad (2.1.99)
$$

where the *Lagrangian L* for a particle has the form

$$L(x(\tau)) = \frac{m\dot{x}^2(\tau)}{2} - V(x(\tau)) \qquad (2.1.100)$$

so that the Feynman–Kac formula can be rewritten as follows:

$$K(x_t, t|x_0, 0) = \int_{\mathcal{C}\{x_0, 0; x_t, t\}} \prod_{\tau=0}^{t} \frac{dx(\tau)}{\sqrt{\frac{2i\pi\hbar}{m} d\tau}} e^{\frac{i}{\hbar} S[(x(\tau)]}. \qquad (2.1.101)$$

Therefore, from this representation for the propagator in terms of path integrals (Feynman–Kac formula), it is revealed that in quantum mechanics not only the classical trajectory, given by the extremum of the action,

$$\frac{\delta S[x(t)]}{\delta x(t)} = 0 \qquad (2.1.102)$$

is present, but all the continuous trajectories which bring the particle from the point x_0 to the point x_t are essential and each path gives its own contribution.

Recall that in the case of stochastic classical processes, the limit of zero temperature $T \to 0$ (or zero diffusion coefficient D) gives the deterministic behaviour of a system. In quantum mechanics, the contribution of a path is defined by the factor

$$e^{\frac{i}{\hbar} S[(x(\tau)]} \qquad (2.1.103)$$

with the Planck constant $\hbar$ playing a role analogous to that of temperature in statistical physics. In the limit $\hbar \to 0$, the integrand in (2.1.101) becomes rapidly oscillating and the only essential contribution comes from the extremal (classical) trajectory $x_c(\tau)$ defined by (2.1.102) which, in the case under consideration (i.e. for trajectories with fixed initial and final points), is equivalent to the *Euler–Lagrange equation*

$$\frac{d}{dt} \frac{\partial L}{\partial \dot{x}} - \frac{\partial L}{\partial x} = 0. \qquad (2.1.104)$$

We shall consider this semiclassical limit in detail in section 2.2.3.

2.1.3 Properties of Hamiltonian operators from the Feynman–Kac formula

The aim of this subsection is to illustrate the applicability of path integrals to the investigation of the general properties of quantum systems. Namely, we shall study the spectrum of the Hamiltonian (2.1.56) with the potential energy satisfying the conditions

$$V(x) > 0 \qquad V(x) \xrightarrow[|x| \to \infty]{} +\infty \qquad (2.1.105)$$

and we shall show, following Ray (1954) and using the path-integral technique, that this spectrum is *purely discrete*:

$$\widehat{H}\psi_n = \lambda_n \psi_n \qquad \lambda_n \xrightarrow[n \to \infty]{} \infty. \qquad (2.1.106)$$

As we have discussed in the preceding subsection, in considering a Hamiltonian spectrum we may convert the Feynman path integral into the Wiener one via analytical continuation to purely imaginary time. After this transition, we convert the Schrödinger equation into the Bloch equation considered in section 1.2.5, with its fundamental solution given by the Feynman–Kac formula (1.2.94). On the other hand, the same fundamental solution can be represented, similarly to expression (2.1.70), as

$$K_B(x_t, t|x_0, 0) = \sum_n e^{-\lambda_n t} \psi_n(x_0)\psi_n(x_t) \qquad (2.1.107)$$

where the completeness

$$\sum_n \psi_n(x_0)\psi_n(x_t) = \delta(x_t - x_0) \qquad (2.1.108)$$

of the set of eigenfunctions of a self-adjoint operator has been used (the transition to imaginary time converts a quantum-mechanical Hilbert space into a *real* Hilbert space, and thus we have dropped the complex conjugation sign in (2.1.107) and (2.1.108)). Although we work in imaginary time and, hence, mathematically deal with the Bloch equation, we have in mind an application in quantum mechanics and, therefore, use for the propagator the notation $K_B(x_t, t|x_0, 0) \stackrel{\text{def}}{\equiv} K(x_t, it|x_0, 0)$ instead of $W_B(x_t, t|x_0, 0)$ as in section 1.2.5. Note also that, since we do not know in advance the properties of the spectrum, the sum in (2.1.107) implies in general summation over the discrete part of the spectrum and integration over its continuous part.

◇ Relation between discreteness of Hamiltonian spectrum and compactness of the evolution operator

In fact, the property of the operator $\widehat{U}(t) \equiv \widehat{U}(t, 0)$ which we are going to prove with the help of the path-integral representation (Feynman–Kac formula) is that it transforms a unit ball in the Hilbert space $\mathcal{H}$ into a *compact* subset of $\mathcal{H}$, in other words, $\widehat{U}(t)$ is a *compact (fully continuous) operator*. The compactness of the subset is understood here in the sense of the norm $\|\psi\| \stackrel{\text{def}}{\equiv} \sqrt{\langle\psi|\psi\rangle}$ in $\mathcal{H}$, induced by the scalar product. The point is that, according to the Hilbert–Schmidt theorem (Reed and Simon 1972), such an operator has the sought type of spectrum, i.e. a purely discrete one.

To prove the compactness of $\widehat{U}(t)$, we use, without a proof, the following criterion (see, e.g., Reed and Simon (1972)):

Proposition 2.1. The subset

$$\int_{-\infty}^{\infty} dx_0 \, K_B(x_t, t|x_0, 0)\psi_0(x_0)$$

is compact if, for any $\psi_0(x_0)$, the following conditions are fulfilled:

$$\int_{-\infty}^{\infty} dx_t \, (\psi_t(x_t))^2 = \int_{-\infty}^{\infty} dx_t \left(\int_{-\infty}^{\infty} dx_0 \, K_B(x_t, t|x_0, 0)\psi_0(x_0)\right)^2 < R \qquad (2.1.109)$$

$$\int_{|x_t|>X(\varepsilon)} dx_t \left(\int_{-\infty}^{\infty} dx_0 \, K_B(x_t, t|x_0, 0)\psi_0(x_0)\right)^2 < \varepsilon. \qquad (2.1.110)$$

Here the initial wavefunction is square integrable:

$$\int_{-\infty}^{\infty} dx_0 \, \psi_0^2(x_0) = 1$$

and the quantities R and $X(\varepsilon)$ do not depend on $\psi_0(x_0)$.

Thus to prove that the evolution operator defined by the integral kernel (2.1.82) is compact, we must prove the conditions (2.1.109) and (2.1.110).

◇ **Proof of the compactness of the evolution operator**

To simplify the formulae, we have chosen the time and/or coordinate units so that

$$\frac{2\hbar}{m} = 1$$

(cf (1.1.89)).

First, we shall show that (2.1.109) is correct, i.e. that the function $\psi(x_t)$ obtained with the help of the propagator $K_B(x_t, t|x_0, 0)$ as a result of the evolution of a system, belongs to $\mathcal{L}^2(\mathbb{R})$ (in other words, $\psi(x_t)$ is square integrable). Indeed, a wavefunction at the moment t can be written as the Wiener integral with *unconditional* measure (in the present case, this means integration over trajectories with *arbitrary* initial points and fixed final point):

$$\psi_t(x_t) = \int_{-\infty}^{\infty} dx_0\, \psi_0(x_0) \int_{C\{x_0,0;x_t,t\}} d_W x(s) \exp\left\{-\int_0^t ds\, V(x(s))\right\}$$

$$= \int_{C\{0;x_t,t\}} d_W x(s)\, \psi_0(x_0) \exp\left\{-\int_0^t ds\, V(x(s))\right\} \qquad (2.1.111)$$

where the set of trajectories in the second path integral has arbitrary initial points. In the space of functionals integrated with Wiener measure, we can introduce the scalar product (see problem 2.1.7, page 152)

$$\langle A[x(s)] | B[x(s)] \rangle = \int_{C\{0;x_t,t\}} d_W x(s)\, A[x(s)] B[x(s)]. \qquad (2.1.112)$$

This means that expression (2.1.111) can be considered as the scalar product of the functional $\psi_0(x_0)$ (here $\psi_0(x_0)$ is considered as a type of degenerate functional: $\psi_0[x(\tau)] = \psi_0(x_0)$, defined by a value of the function ψ_0 at the point x_0) and the functional $\exp\{-\int_0^t ds\, V(x(s))\}$. Thus the well-known Cauchy–Schwarz–Bunyakowskii inequality for vectors of any Hilbert space, i.e.

$$|\langle \psi | \varphi \rangle|^2 \leq \langle \psi | \psi \rangle \langle \varphi | \varphi \rangle \qquad (2.1.113)$$

can be applied:

$$\psi_t^2(x_t) \leq \int_{C\{0;x_t,t\}} d_W x(s)\, \psi_0^2(x_0) \int_{C\{x_t,t;0\}} d_W x(s) \exp\left\{-2 \int_0^t ds\, V(x(s))\right\}. \qquad (2.1.114)$$

Taking into account that

$$\exp\left\{-2 \int_0^t ds\, V(x(s))\right\} < 1 \qquad (2.1.115)$$

for any $x(t)$, due to condition (2.1.105) (positiveness of the potential), the inequality (2.1.114) can be made stronger:

$$\psi^2(x_t) < \int_{C\{0;x_t,t\}} d_W x(s)\, \psi_0^2(x_0) \int_{C\{0;x_t,t\}} d_W x(s)$$

$$= \int_{-\infty}^{\infty} dx_0\, \psi_0^2(x_0) \frac{\exp\left\{-\frac{(x_t-x_0)^2}{t}\right\}}{\sqrt{\pi t}} \qquad (2.1.116)$$

so that

$$\int_{-\infty}^{\infty} dx_t\, \psi_t^2(x_t) \leq \int_{-\infty}^{\infty} dx_0\, \psi_0^2(x_0) = 1 \qquad (2.1.117)$$

and condition (2.1.109) has been proved. Relation (2.1.117) also means that the evolution operator defined by the integral kernel $K_B(x_t, t|x_0, 0)$ is a *contraction operator*. The correctness of the condition (2.1.110) is again proved with the help of the Cauchy–Schwarz–Bunyakowskii inequality (2.1.114) but the necessary estimations are technically more involved and cumbersome. Interested readers may find them in appendix B.

Thus the operator with the integral kernel given by the Feynman–Kac formula (1.2.94), i.e. the evolution operator

$$\widehat{U}(t) = e^{-\frac{\widehat{H}t}{\hbar}}$$

is a compact (fully continuous) operator in the Hilbert space $\mathcal{H}$. Such an operator satisfies the conditions of the Hilbert–Schmidt theorem (Reed and Simon 1972), according to which the spectrum of this operator is purely discrete, with finite multiplicity, and its eigenvalues form a sequence tending to zero. Therefore, the Hamiltonian operator $\widehat{H}$ has the same properties, except that its eigenvalues tend to infinity:

$$\lambda_n \to +\infty. \tag{2.1.118}$$

The generalization of this consideration to the case of a Hamiltonian in a higher-dimensional space (as well as the generalization of the actual Feynman–Kac formula) is straightforward.

2.1.4 Bohr–Sommerfeld (semiclassical) quantization condition from path integrals

Let us remember some basic facts from the standard formulation of the *semiclassical approximation* (for extensive consideration, see textbooks on quantum mechanics, e.g., Schiff (1955), Davydov (1976), Landau and Lifshitz (1981) or special monographs, e.g., Maslov and Fedoriuk (1982)).

The basic object of the WKB approximation is the *local classical momentum $p(x)$* of the particle

$$p(x) = \sqrt{2m(E - V(x))} \tag{2.1.119}$$

which we have already introduced in section 2.1.4 (cf (2.1.137)). In terms of this quantity, the stationary Schrödinger equation (2.1.12) acquires the form

$$[-\hbar^2 \partial_x^2 - p^2(x)]\psi(x) = 0. \tag{2.1.120}$$

To develop the WKB expansion, the solution of this equation is sought in the form

$$\psi(x) = e^{i\mathcal{S}(x)/\hbar} \tag{2.1.121}$$

where the exponent $\mathcal{S}(x)$, called the *eikonal*, after substitution into (2.1.120), proves to satisfy the Riccati equation

$$-i\hbar \partial_x^2 \mathcal{S}(x) + [\partial_x \mathcal{S}(x)]^2 - p^2(x) = 0. \tag{2.1.122}$$

If $p(x) = p = $ constant, i.e. $V(x) = 0$, this equation is easy to solve:

$$\mathcal{S}(x) = px + \text{constant} \tag{2.1.123}$$

and (2.1.121) becomes a plane wave. In this case, the first term in equation (2.1.122) exactly vanishes. For slowly varying functions $p(x)$ we can develop an approximate expansion with respect to the first term. Since the Planck constant $\hbar$ appears as a factor in this term, it may be used to count the powers of the smallness parameter δ_{WKB}, the exact expression for which turns out to be the following (see, e.g., the previously cited textbooks):

$$\delta_{\text{WKB}} \stackrel{\text{def}}{\equiv} \frac{2\pi \hbar}{p(x)} \left| \frac{\partial_x p(x)}{p(x)} \right|. \tag{2.1.124}$$

Note that the first factor in δ_{WKB} is nothing but the de Broglie wavelength λ_{Br} corresponding to the momentum $p(x)$ and the whole quantity δ_{WKB} measures the relative variation of the effective momentum $p(x)$ over the de Broglie distance λ_{Br}. As usual, the WKB expansion is valid only if the parameter δ_{WKB} is small enough:

$$\delta_{\text{WKB}} \ll 1. \qquad (2.1.125)$$

Although δ_{WKB} is the true physical dimensionless parameter of the expansion, in mathematical manipulations it is convenient to imagine $\hbar$ as a small quantity (note that δ_{WKB} is proportional to $\hbar$) and to expand the eikonal around $\mathcal{S}_0(x) \stackrel{\text{def}}{\equiv} \mathcal{S}(x)|_{\hbar=0}$ in a power series in $\hbar$:

$$\mathcal{S} = \mathcal{S}_0 - i\hbar \mathcal{S}_1 + (-i\hbar)^2 \mathcal{S}_2 + \cdots.$$

Inserting this *semiclassical expansion* into (2.1.122), we obtain the sequence of WKB equations

$$\partial_x^2 \mathcal{S}_k + \sum_{m=0}^{k+1} \partial_x \mathcal{S}_m \partial_x \mathcal{S}_{k+1-m} = 0 \qquad k = 0, 1, 2, 3, \ldots \qquad (2.1.126)$$

from which $\mathcal{S}_0, \mathcal{S}_1, \mathcal{S}_2, \ldots$ can be successively found. Keeping only the first-order terms $\mathcal{S}_0$ and $\mathcal{S}_1$, we arrive at the (not yet normalized) WKB wavefunction

$$\psi_{\text{WKB}}(x) = \frac{1}{\sqrt{p(x)}} e^{\pm \frac{i}{\hbar} \int^x dy\, p(y)}. \qquad (2.1.127)$$

This is the simplest generalization of a plane wave to the case of slowly varying $p(x)$. The explicit solution of (2.1.126) shows that the contribution of $\mathcal{S}_2, \mathcal{S}_3, \ldots$ is indeed small if the WKB condition (2.1.125) is fulfilled. In the classically accessible region, where $V(x) \leq E$ and, hence, $p(x)$ is real, ψ_{WKB} is an oscillating function; in the inaccessible region with $V(x) > E$, it decreases exponentially. Sewing together the function in both regions and the condition of normalization, we arrive at the *Bohr–Sommerfeld quantization condition*

$$\int_a^b dx\, p(x) \approx (n + \tfrac{1}{2})\pi\hbar \qquad (2.1.128)$$

where n is a positive integer, $n \in \mathbb{Z}_+$. We have used the sign of an approximate equality because the relation (2.1.128) is obtained only in the semiclassical approximation. Note also that the condition (2.1.125) is fulfilled only for large integers, $n \gg 1$. Thus, in fact, we can truly rewrite (2.1.128) as

$$\int_a^b dx\, p(x) \approx n\pi\hbar. \qquad (2.1.129)$$

We are going to prove this approximate equality using the path-integral technique.

◇ **Reformulation of the Bohr–Sommerfeld condition to be suitable for the proof by path integration**

We shall assume that the potential energy in the Hamiltonians under consideration satisfies the conditions

(1) $V(x) > 0$

(2) $\text{Meas}\{x : V(x) < \lambda\}|_{\lambda \to \infty} \sim C\lambda^\alpha$ $\qquad (2.1.130)$

(3) $\text{Meas}\{x : \sup_{|y|<\varepsilon} V(x+y) < \lambda\}|_{\lambda \to \infty} \sim C\lambda^\alpha$ $\qquad \alpha > 0$, C are some constants.

146   *Path integrals in quantum mechanics*

The measure ('volume'; cf the explanation of this term in the proof of the Wiener theorem, section 1.1.3)

$$\text{Meas}\{x : f(x) < a\}$$

of the values of a variable x satisfying the condition $f(x) < a$, $a \in \mathbb{R}$ for some continuous function $f(x)$, can be calculated with the help of the integral

$$\text{Meas}\{x : f(x) < a\} = \int_{-\infty}^{\infty} dx\, \theta(a - f(x)) \tag{2.1.131}$$

where $\theta(x)$ is the step-function. Conditions (2) and (3) mean that the two measures have the same asymptotics in λ.

We shall study the properties of the eigenfunctions of the operator

$$\widehat{H} = -\frac{1}{2}\frac{\partial^2}{\partial x^2} + V(x). \tag{2.1.132}$$

That is, we shall prove that the number of its eigenvalues λ_n satisfying the condition $\lambda_n < \lambda$ for a given constant λ can be estimated via the properties of the potential (Ray 1954). More precisely,

$$\sum_{\lambda_n < \lambda} 1 \bigg|_{\lambda \gg 1} \simeq \frac{1}{2\pi}\, \text{Meas}\left\{x, y : V(x) + \frac{y^2}{2} < \lambda\right\}. \tag{2.1.133}$$

In fact, the relation (2.1.133) coincides with the Bohr–Sommerfeld quantization condition (2.1.129). Indeed, let us rescale the coordinate:

$$x \longrightarrow \frac{\sqrt{m}}{\hbar} x$$

after which the operator (2.1.132) becomes the usual energy operator, i.e. the Hamiltonian,

$$\widehat{H} = -\frac{\hbar^2}{2m}\frac{\partial^2}{\partial x^2} + \widetilde{V}(x) \tag{2.1.134}$$

where

$$\widetilde{V}(x) = V\left(\frac{\sqrt{m}}{\hbar} x\right).$$

In terms of this new variable, condition (2.1.133) becomes

$$\frac{1}{\pi\hbar}\int_a^b dx\, \sqrt{2m(E - \widetilde{V}(x))} \simeq \sum_{E_n < E} 1 \bigg|_{E/E_0 \gg 1} \tag{2.1.135}$$

where we have denoted, as usual, the eigenvalues of the Hamiltonian by E_n (instead of λ_n, as in (2.1.133)) and expressed

$$\text{Meas}\left\{x, y : \widetilde{V}(x) + \frac{y^2}{2} < E\right\}$$

as the integral over the coordinate x (see figure 2.1). The condition (2.1.135) is nothing but the Bohr–Sommerfeld quantization condition:

$$\int_a^b dx\, \sqrt{2m(E - \widetilde{V}(x))} \approx n\pi\hbar \qquad n \gg 1 \tag{2.1.136}$$

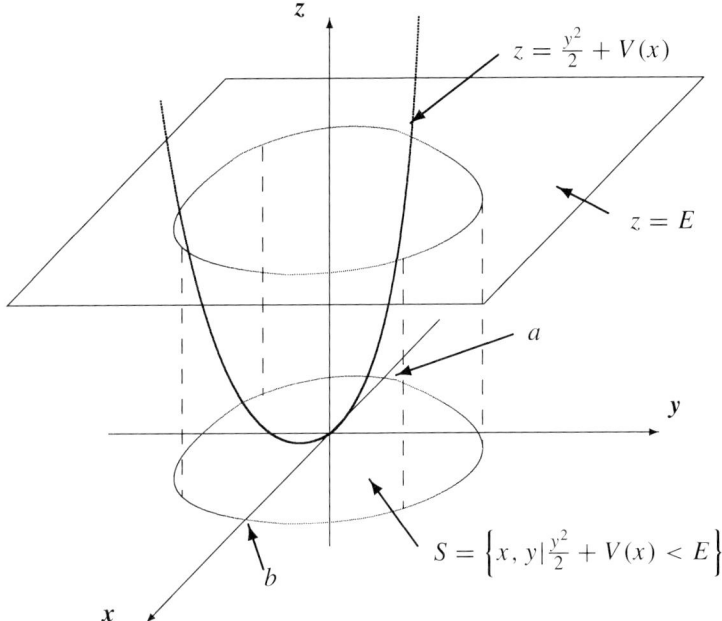

Figure 2.1. Visualization of the domain S, where the variables x and y satisfy the condition $\frac{y^2}{2} + V(x) < E$.

which can be rewritten also as

$$\oint_{\substack{\text{classical} \\ \text{region}}} p\, dx = n 2\pi \hbar \qquad n \gg 1$$

where p is the classical momentum of the particle corresponding to the energy E:

$$p = \sqrt{2m(E - \widetilde{V}(x))} \tag{2.1.137}$$

and n is the number of eigenvalues satisfying the condition $E_n < E$.

◇ **Proof of the Bohr–Sommerfeld condition by a path-integral method**

Now we turn to the proof of the relation (2.1.133).

As usual in this section, we consider the Schrödinger equation in imaginary time, i.e. the Bloch equation (1.2.95) with fundamental solution (1.2.94), which we denote in this chapter by K_B to stress its relation to quantum mechanics. As we have pointed out in the preceding subsection, we can also present the latter as follows

$$K_B(x_t, t | x_0, 0) = \sum_n e^{-\lambda_n t} \psi_n(x_t) \psi_n(x_0) \tag{2.1.138}$$

where λ_n are the eigenvalues of the Hamiltonian (2.1.132):

$$\widehat{H}\psi_n = \lambda_n \psi_n$$

and ψ_n are the normalized eigenfunctions:

$$\int_{-\infty}^{\infty} dx\, \psi_n(x)\psi_k(x) = \delta_{nk}.$$

To approach the Bohr–Sommerfeld condition, we need the following lemma.

Lemma 2.1. The propagator K_B has the following asymptotic representation

$$\int_{-\infty}^{\infty} dx\, K_B(x,t|x,0)\bigg|_{t\to 0} \simeq \frac{1}{\sqrt{\pi t}} \int_{-\infty}^{\infty} dx\, e^{-tV(x)}. \qquad (2.1.139)$$

The proof of this lemma is technically rather cumbersome and we have placed it in appendix C. Here, in the main text, we give only heuristic arguments for the correctness of the lemma:

(i) Since K_B is the transition amplitude for the Bloch equation, the left-hand side of (2.1.139) can be presented with the help of the Feynman–Kac formula as follows:

$$\int_{-\infty}^{\infty} dx\, K_B(x,t|x,0) = \int_{-\infty}^{\infty} dx \int_{\mathcal{C}\{x,0;x,t\}} d_W x(\tau) \exp\left\{-\int_0^t d\tau\, V(x(\tau))\right\} \qquad (2.1.140)$$

where $\mathcal{C}\{x,0;x,t\}$ is the set of *closed* trajectories with the initial and final points at x.

(ii) The Wiener path integral in (2.1.140) is defined via the discrete-time approximation and in the limit $t \to 0$ we can use for the approximation just *one* time interval:

$$\int_{\mathcal{C}\{x,0;x,t\}} d_W x(\tau) \exp\left\{-\int_0^t d\tau\, V(x(\tau))\right\} \xrightarrow[t\to 0]{} \frac{1}{\sqrt{2\pi t}} e^{-tV(x)}$$

so that after the integration over x we obtain the right-hand side of the relation (2.1.139).

The relation (2.1.138) gives (with a normalization of the eigenfunctions)

$$\int_{-\infty}^{\infty} dx\, K_B(x,t|x,0) = \sum_n e^{-\lambda_n t} = \int_0^{\infty} e^{-ut} du \left(\sum_{\lambda_n < u} 1\right). \qquad (2.1.141)$$

The function $g(u) \equiv \sum_{\lambda_n < u} 1$ is a piecewise constant, steplike function (and, hence, its derivative gives the sum of δ-functions; this proves the latter equality in (2.1.141), see problem 2.1.9, page 153). Let us write the right-hand side of (2.1.139) in a similar form:

$$\frac{1}{\sqrt{2\pi t}} \int_{-\infty}^{\infty} dx\, e^{-tV(x)} = \frac{1}{2\pi} \int_{-\infty}^{\infty} dx\, dy\, e^{-t(V(x)+y^2)}$$

$$= \int_0^{\infty} e^{-tu} du \left(\text{Meas}\left\{x,y : V(x) + \frac{y^2}{2} < u\right\}\right)$$

so that (2.1.139) becomes

$$\int_0^{\infty} e^{-ut} du \left(\sum_{\lambda_n < u} 1\right) \simeq \frac{1}{2\pi} \int_0^{\infty} e^{-tu} du \left(\text{Meas}\left\{x,y : V(x) + \frac{y^2}{2} < u\right\}\right). \qquad (2.1.142)$$

Now the Bohr–Sommerfeld condition in the form (2.1.133) follows from (2.1.142) and the *Tauberian theorem* (see, e.g., Widder (1971)) which links the asymptotic behaviour of a Laplace–Stiltjes transform $\widetilde{f}(t)$,

$$\widetilde{f}(t) = \int_0^\infty e^{-tu} \, df(u)$$

and the original $f(u)$. Loosely speaking, this theorem states that functions $f_1(u)$ and $f_2(u)$ have the same asymptotical behaviour at $u \to \infty$, provided that their images $\widetilde{f}_1(t)$ and $\widetilde{f}_2(t)$ have the same asymptotics at $t \to 0$. For the reader's convenience, we present a simplified version of the Tauberian theorem together with a sketch of its proof in appendix D.

Thus relation (2.1.142) and the Tauberian theorem give

$$\sum_{\lambda_n < u} 1 \bigg|_{u \to \infty} \simeq \frac{1}{2\pi} \, \text{Meas} \left\{ x, y : V(x) + \frac{y^2}{2} < u \right\} \tag{2.1.143}$$

which is equivalent to the relation (2.1.133) and, hence, to the Bohr–Sommerfeld quantization condition (2.1.136).

The generalization to higher dimensions is straightforward: we might just substitute the coordinates x, y by d-dimensional vectors and repeat all the process. In particular, (2.1.143) takes the form

$$\sum_{\lambda_n < u} 1 \bigg|_{u \to \infty} \simeq \frac{1}{(2\pi)^d} \, \text{Meas} \left\{ \boldsymbol{x}, \boldsymbol{y} : V(\boldsymbol{x}) + \frac{\boldsymbol{y}^2}{2} < u \right\}.$$

2.1.5 Problems

Problem 2.1.1. Find the expression for the evolution operator of a quantum-mechanical system in the case of a time-dependent Hamiltonian.

Hint. In the time-sliced approximation, when an interval (t, t_0) is sliced into a large number $N+1$ of small intervals of length $\varepsilon = (t - t_0)/(N + 1)$, the solution of the Schrödinger equation with a time-dependent Hamiltonian has the form

$$\psi(t) = \left(1 - \frac{i}{\hbar} \int_{t_N}^{t} d\tau_{N+1} \, \widehat{H}(\tau_{N+1}) \right) \left(1 - \frac{i}{\hbar} \int_{t_{N-1}}^{t_N} d\tau_N \, \widehat{H}(\tau_N) \right)$$
$$\times \cdots \times \left(1 - \frac{i}{\hbar} \int_{t_0}^{t_1} d\tau_1 \, \widehat{H}(\tau_1) \right) \psi(t_0).$$

Thus the evolution operator is given approximately by the product

$$\widehat{U}(t, t_0) \approx \left(1 - \frac{i}{\hbar} \int_{t_N}^{t} d\tau_{N+1} \, \widehat{H}(\tau_{N+1}) \right) \cdots \left(1 - \frac{i}{\hbar} \int_{t_0}^{t_1} d\tau_1 \, \widehat{H}(\tau_1) \right).$$

In the limit $N \to \infty$, this gives

$$\widehat{U}(t, t_0) \approx 1 - \frac{i}{\hbar} \int_{t_0}^{t} d\tau_1 \, \widehat{H}(\tau_1) + \left(-\frac{i}{\hbar}\right)^2 \int_{t_0}^{t} d\tau_2 \int_{t_0}^{\tau_2} d\tau_1 \, \widehat{H}(\tau_2)\widehat{H}(\tau_1)$$
$$+ \left(-\frac{i}{\hbar}\right)^3 \int_{t_0}^{t} d\tau_3 \int_{t_0}^{\tau_3} d\tau_2 \int_{t_0}^{\tau_2} d\tau_1 \, \widehat{H}(\tau_3)\widehat{H}(\tau_2)\widehat{H}(\tau_1) + \cdots \tag{2.1.144}$$

150   Path integrals in quantum mechanics

which is called the *Dyson series*.

The time-ordering operator **T** allows us to write the Dyson series in the compact form (2.1.14).

Problem 2.1.2. Show that the quantum propagator defined by (2.1.68) satisfies the Schrödinger equation with the boundary condition
$$K(x, t|x_0, t_0)|_{t=t_0} = \delta(x - x_0)$$
and that the retarded propagator $K^{(\text{ret})}(x, t|x_0, t_0) \stackrel{\text{def}}{=} \theta(t - t_0) K(x, t|x_0, t_0)$ is the Green function of the Schrödinger equation.

Hint. The first statement follows from the operator equation (2.1.15) and the orthonormality condition for coordinate eigenvectors (2.1.18):
$$\langle x|e^{-i(t-t_0)\widehat{H}}|x_0\rangle|_{t=t_0} = \langle x|x_0\rangle = \delta(x - x_0).$$

The retarded propagator satisfies the inhomogeneous Schrödinger equation
$$[i\hbar\partial_t - \widehat{H}(-i\hbar\partial_x, x; t)] K^{(\text{ret})}(x, t|x_0, t_0) = i\hbar\delta(t - t_0)\delta(x - x_0)$$
and hence, by definition, is a Green function. The non-zero right-hand side arises from the term (cf the discussion on transition probability in the preceding chapter)
$$[\partial_t\theta(t - t_0)]\langle x|\widehat{U}(t, t_0)|x_0\rangle = \delta(t - t_0)\langle x|x_0\rangle.$$

Problem 2.1.3. Derive the propagator (2.1.78) for a free particle using the operator method.

Solution. Using the completeness of the momentum basis and the explicit form of the matrix element $\langle x|p\rangle$, we have
$$\begin{aligned}
K_0(x, t|x_0, 0) &= \langle x|e^{-it\widehat{H}/\hbar}|x_0\rangle \\
&= \int_{-\infty}^{\infty} dp\, \langle x|\exp\left\{-\frac{it}{2m\hbar}\widehat{p}^2\right\}|p\rangle\langle p|x_0\rangle \\
&= \int_{-\infty}^{\infty} dp\, \exp\left\{-\frac{it}{2m\hbar}p^2\right\}\langle x|p\rangle\langle p|x_0\rangle \\
&= \frac{1}{2\pi\hbar}\int_{-\infty}^{\infty} dp\, \exp\left\{-\frac{it}{2m\hbar}p^2 + ip(x - x_0)\right\} \\
&= \left(\frac{m}{2\pi i\hbar t}\right)^{1/2}\exp\left\{i\frac{m}{2\hbar t}(x - x_0)^2\right\}.
\end{aligned} \quad (2.1.145)$$

Problem 2.1.4. Using the explicit expression (2.1.78) for the propagator of a free particle, find the wavefunction $\psi(x, t)$, given that
$$\psi(x_0, t_0) = \sqrt{\frac{1}{\sigma\sqrt{2\pi}}}\exp\left\{-\frac{x^2}{4\sigma^2}\right\}.$$

Hint.
$$\psi(x, t) = \int_{-\infty}^{\infty} dx_0\, K_0(x, t|x_0, t_0)\psi(x_0, t_0)$$

$$= \sqrt{\frac{m}{2\pi i\hbar t \sigma \sqrt{2\pi}}} \int_{-\infty}^{\infty} dx_0 \exp\left\{-\left(\frac{1}{4\sigma^2} - i\frac{m}{2\hbar t}\right)x_0^2 - \frac{imx}{\hbar t}x_0 + \frac{im}{2\hbar t}x^2\right\}$$

$$= \sqrt{\frac{1}{\sigma\sqrt{2\pi}}} \sqrt{\frac{1 - i\frac{\hbar}{2m\sigma^2}t}{1 + \left(\frac{\hbar t}{4m\sigma^2}\right)^2}} \exp\left\{-\frac{x^2}{4\sigma^2}\frac{1}{\left[1 + \left(\frac{\hbar t}{2m\sigma^2}\right)^2\right]} + \frac{i\hbar t x}{8m\sigma^2\left[1 + \left(\frac{\hbar t}{2m\sigma^2}\right)^2\right]}\right\}.$$

Note that

$$\psi(x, t) \xrightarrow[t \to t_0]{} \psi(x_0, t_0).$$

Problem 2.1.5. Give one more *heuristic* (non-rigorous) proof that

$$\psi(x, t) = \int_{-\infty}^{\infty} dx_0 \, K(x, t|x_0, t_0)\psi(x_0, t_0)$$

satisfies the Schrödinger equation, using *direct* differentiation of the propagator represented by the Feynman path integral.

Hint. Taking into account that we do not integrate over $x(t)$ in the path integral representing $K(x, t|x_0, t_0)$, we obtain

$$\frac{\partial}{\partial t}\psi(x, t) = \int_{\mathcal{C}\{x_0, 0; x_t, t\}} \prod_{\tau=t_0}^{t} \frac{dx(\tau)}{\sqrt{\frac{2i\pi\hbar}{m}d\tau}} \exp\left\{\frac{i}{\hbar}\int_{t_0}^{t} d\tau \left(\frac{1}{2}m\dot{x}^2 - V(x)\right)\right\}\psi(x_0, t_0)$$

$$= \frac{i}{\hbar}\left(\frac{m}{2}\dot{x}^2(t) - V(x(t))\right)\psi(x, t).$$

The coordinate differentiation is performed using the discrete-time approximation:

$$\frac{\partial}{\partial x}\int_{t_0}^{t} dt \left(\frac{m}{2}\dot{x}^2(t) - V(x(t))\right) = \lim_{N\to\infty} \frac{\partial}{\partial x_{N+1}} \sum_{j=0}^{N}\left[\frac{m}{2}\frac{(x_{j+1} - x_j)^2}{(\Delta t)^2} - V(x_j)\right]\Delta t$$

$$= \lim_{N\to\infty} \frac{m}{2}\frac{2(x_N - x_{N-1})}{\Delta t} = m\dot{x}(t)$$

(note that here we have chosen a particular way of discretization). This gives

$$\frac{\partial^2}{\partial x^2}\psi(x, t) = \frac{\partial}{\partial x}\int_{\mathcal{C}\{x_0, 0; x_t, t\}} d_F x(\tau) \, e^{-\frac{i}{\hbar}\int_{t_0}^{t} d\tau V(x)} \frac{\partial}{\partial x}\left(\frac{i}{\hbar}\int_{t_0}^{t} dt \left(\frac{m}{2}\dot{x}^2(t) - V(x(t))\right)\right)$$

$$= -\left(\frac{m\dot{x}}{\hbar}\right)^2 \psi(x, t)$$

where we have assumed that the final coordinate $x = x(t)$ and velocity $\dot{x}(t)$ are independent: $\frac{\partial \dot{x}}{\partial x} = 0$. Substitution of these results into the Schrödinger equation confirm once again that $K(x, t|x_0, t_0)$ given by the Feynman path integral defines time evolution of a quantum-mechanical state. Of course, this proof is not mathematically rigorous, but has the advantage of being direct and instructive.

Problem 2.1.6. Using the Feynman path integral show that the transition amplitude $K(x, t|x_0, t_0)$ can be written as the series

$$K(x, t|x_0, t_0) = K_0(x, t|x_0, t_0) - \frac{i}{\hbar}\int_{-\infty}^{\infty} dx' \int_{t_0}^{t} dt' \, K_0(x, t|x', t')V(x', t')K_0(x', t'|x_0, t_0)$$

152   Path integrals in quantum mechanics

$$+ \left(-\frac{i}{\hbar}\right)^2 \int_{-\infty}^{\infty} dx'' dx' \int_{t_0}^{t} dt'' \int_{t_0}^{t''} dt' \, K_0(x,t|x'',t'') V(x'',t'') K_0(x'',t''|x',t')$$
$$\times V(x',t') K_0(x',t'|x_0,t_0) + \cdots$$

where $K_0(x,t|x',t')$ is the propagator for a free particle. Interpret the result.

Hint. Expand the exponent in the Feynman–Kac formula:

$$K(x,t|x_0,t_0) = \int_{\mathcal{C}\{x_0,t_0;x,t\}} d_F x(\tau) \sum_{n=0}^{\infty} \frac{1}{n!} \left(-\frac{i}{\hbar} \int_{t_0}^{t} d\tau \, V(x,\tau)\right)^n$$

and use the Markovian property of a propagator to convert each term in this series to the required form.
  If the potential energy is small compared with the kinetic one (e.g., contains a small constant factor), the series can be considered as the *perturbation expansion*.

Problem 2.1.7. Show that the formula (2.1.112)

$$\langle A[x(t)] | B[x(t)] \rangle = \int_{\mathcal{C}\{x_t,t;0\}} d_W x(s) \, A[x(t)] B[x(t)]$$

for functionals integrated with (unconditional) Wiener measure satisfies all the axioms for a scalar product in Hilbert spaces.

Hint. Verify *explicitly* that all the axioms for a scalar product are indeed satisfied for the defined composition of functionals.

Problem 2.1.8. Starting from the inequality

$$K_B(x,t|x,0) = \int_{\mathcal{C}\{x,t;x,0\}} d_W x(s) \exp\left\{-\int_0^t ds \, V(x(s))\right\}$$
$$\leq \frac{1}{t} \int_{\mathcal{C}\{x,t;x,0\}} d_W x(s) \int_0^t ds \, \exp\{-tV(x(s))\}$$

derive the inequality

$$\int_{-\infty}^{\infty} dx \, K_B(x,t|x,0) \leq \frac{1}{t} \int_{-\infty}^{\infty} d\xi \int_0^t ds \, \exp\{-tV(\xi)\} \frac{1}{\sqrt{\pi t}}$$
$$= \frac{1}{\sqrt{\pi t}} \int_{-\infty}^{\infty} d\xi \, \exp\{-tV(\xi)\}.$$

These inequalities are used for the path-integral derivation of the Bohr–Sommerfeld quantization condition (see appendix C, equations (C.7) and (C.8)).

Hint. The integral on the right-hand side of (C.7) obviously exists and hence we can change the order of the integrations:

$$\frac{1}{t} \int_{\mathcal{C}\{x,0;x,t\}} d_W x(s) \int_0^t ds \, \exp\{-tV(x(s))\} = \frac{1}{t} \int_0^t ds \int_{\mathcal{C}\{x,0;x,t\}} d_W x(\tau) \exp\{-tV(x(s))\}. \quad (2.1.146)$$

The path integral obtained contains a degenerated functional and can be calculated:

$$\int_{C\{x,0;x,t\}} d_W x(\tau) \exp\{-tV(x(s))\} = \int_{-\infty}^{\infty} d\xi \int_{C\{x,0;\xi,s;x,t\}} d_W x(\tau) \exp\{-tV(\xi)\}$$

$$= \int_{-\infty}^{\infty} d\xi \exp\{-tV(\xi)\} \int_{C\{x,t;\xi,s\}} d_W x(\tau) \int_{C\{\xi,s;x,0\}} d_W x(\tau)$$

$$= \int_{-\infty}^{\infty} d\xi \exp\{-tV(\xi)\} \frac{\exp\left\{-\frac{(\xi-x)^2}{s}\right\}}{\sqrt{\pi s}} \frac{\exp\left\{-\frac{(\xi-x)^2}{t-s}\right\}}{\sqrt{\pi(t-s)}}$$

(recall that $C\{x,t;\xi,s;x,0\}$ denotes the set of trajectories passing through the point ξ at the moment s).
Integrating (C.7) over x and using the ESKC relation, we obtain (C.8).

Problem 2.1.9. Prove the relation (cf (2.1.141))

$$\int_0^\infty e^{-ut} du \sum_{\lambda_n < u} 1 = \sum_n e^{-\lambda_n t}.$$

Hint. Write the sum on the left-hand side as follows:

$$\sum_{\lambda_n < u} 1 = \sum_{\text{all } \lambda_n} \theta(u - \lambda_n)$$

and use the fact that the derivative of the θ-function gives the δ-function.

Problem 2.1.10. Find the energy levels E_n ($n \gg 1$) for a harmonic oscillator potential $V(x) = m\omega^2 x^2/2$, using the Bohr–Sommerfeld quantization condition in one dimension.

Hint. Calculate $\oint p\, dx$ as a function of the energy and use the Bohr–Sommerfeld condition.

2.2 Path integrals in the Hamiltonian formalism

In the preceding section, we reduced the problem of path-integral construction in quantum mechanics to that in the theory of stochastic processes, using the possibility of transition to imaginary time which converts the Schrödinger equation into the Bloch equation. This allows us to apply the Wiener path integral, constructed in chapter 1, to quantum-mechanical problems. However, this possibility by no means denies the necessity of construction and use of the evolution operator and its path-integral representation in real time. In the preceding section we have introduced, after Feynman, such an integral, in analogy with the Wiener path integral and proved that the Feynman integral satisfies the Schrödinger equation. As we shall show later (especially in section 2.3), this way gives an important, but restricted type of quantum-mechanical path integral. Thus, the general problem of construction and applications of path integrals in quantum mechanics requires its derivation directly in the real-time formalism and from the standard operator approach.

In this section, we start from such a derivation and discuss why the quantum-mechanical real-time path integral cannot be based on some integration measure (in contrast with the case of the Wiener integral, cf chapter 1). Then, to get some preliminary practice with the quantum-mechanical path integrals, we calculate them for the simplest systems: a free particle and the harmonic oscillator. The next subsection is devoted to the important semiclassical (WKB) approximation in quantum mechanics formulated in terms of path integrals. We have already met this approximation in the preceding section, when deriving the

Bohr–Sommerfeld quantization condition. In section 2.2.3, we shall essentially expand this consideration with many technical details, while in section 2.2.4, we shall discuss the relation between the eigenvalues of a quantum Hamiltonian and the classical periodic orbits (the so-called *periodic orbit theory*). In addition, in this subsection we shortly present some peculiarities of quantum systems with chaotic behaviour in the classical limit.

In section 2.2.5, we consider a particle in a magnetic field facing, for the first time, a Hamiltonian with operator ordering ambiguity, and discuss the influence of the latter on the path-integral construction.

In the last subsection, we illustrate the general fact that the path-integral methods derived in quantum mechanics can be successfully applied in other areas of physics. Namely, we shall briefly discuss the formal similarity of quantum-mechanical problems with a specific type of *optical* problem.

2.2.1 Derivation of path integrals from operator formalism in quantum mechanics

For simplicity, we start from the consideration of a point particle moving in a one-dimensional space and our aim is to derive a path-integral representation for the Green function (2.1.68).

Feynman realized that due to the fundamental composition law (2.1.10) of the time evolution operator, the amplitude (2.1.68) can be sliced into a large number of time evolution operators, each shifting a state for a small time interval

$$\varepsilon = t_j - t_{j-1} = \frac{t - t_0}{N + 1} \qquad j = 0, \ldots, N + 1$$
$$t_{N+1} = t$$

where N is a large positive integer. Explicitly, (2.1.68) can be rewritten as follows:

$$K(x, t | x_0, t_0) = \langle x | \widehat{U}(t, t_N) \widehat{U}(t_N, t_{N-1}) \cdots$$
$$\cdots \widehat{U}(t_j, t_{j-1}) \cdots \widehat{U}(t_2, t_1) \widehat{U}(t_1, t_0) | x_0 \rangle. \qquad (2.2.1)$$

When inserting a complete set of states between each pair of the operators $\widehat{U}$:

$$\int_{-\infty}^{\infty} dx_j \, |x_j\rangle\langle x_j| = 1 \qquad j = 1, \ldots, N$$

the Green function becomes a product of N integrals:

$$K(x, t | x_0, t_0) = \prod_{j=1}^{N} \left[\int_{-\infty}^{\infty} dx_j \right] \prod_{j=1}^{N+1} K(x_j, t_j | x_{j-1}, t_{j-1}) \qquad (2.2.2)$$

where

$$x \equiv x_{N+1} \qquad t \equiv t_{N+1}.$$

The integrand is the product of the amplitudes for infinitesimal time intervals:

$$K(x_j, t_j | x_{j-1}, t_{j-1}) = \langle x_j | e^{-i\varepsilon \widehat{H}(t_j)/\hbar} | x_{j-1}\rangle \qquad (2.2.3)$$

with the Hamiltonian operator $\widehat{H}$, which in general can be time dependent.

Further derivation of the path-integral representation for the Green function crucially depends on the properties of a chosen Hamiltonian operator.

◇ **Derivation of the phase-space path integral for Hamiltonians without terms containing products of non-commuting operators**

The simplest case corresponds to Hamiltonians having the standard form of the sum of a kinetic $T(p,t)$ and potential $V(x,t)$ energy:
$$H(p,x;t) = T(p,t) + V(x,t) \qquad (2.2.4)$$
with the potential $V(x,t)$ being *a smooth* function of the coordinate x. In this situation, the time displacement operator,
$$\exp\left\{-\mathrm{i}\frac{\varepsilon}{\hbar}\widehat{H}\right\} = \exp\left\{-\mathrm{i}\frac{\varepsilon}{\hbar}(\widehat{T} + \widehat{V})\right\} \qquad (2.2.5)$$
is factorizable for a sufficiently small ε, according to the *Baker–Campbell–Hausdorff formula* (see supplement IV, volume II)
$$\exp\left\{-\mathrm{i}\frac{\varepsilon}{\hbar}(\widehat{T}+\widehat{V})\right\} = \exp\left\{-\mathrm{i}\frac{\varepsilon}{\hbar}\widehat{V}\right\}\exp\left\{-\mathrm{i}\frac{\varepsilon}{\hbar}\widehat{T}\right\}\exp\left\{-\mathrm{i}\frac{\varepsilon^2}{\hbar^2}\widehat{R}\right\} \qquad (2.2.6)$$
where the operator $\widehat{R}$ is defined by the series
$$\widehat{R} \stackrel{\text{def}}{\equiv} \frac{\mathrm{i}}{2}[\widehat{V},\widehat{T}] - \frac{\varepsilon}{\hbar}\left(\frac{1}{6}[\widehat{V},[\widehat{V},\widehat{T}]] - \frac{1}{3}[[\widehat{V},\widehat{T}],\widehat{T}]\right) + \mathcal{O}(\varepsilon^2). \qquad (2.2.7)$$
The omitted terms of order $\varepsilon^2, \varepsilon^3, \ldots$ contain higher commutators of $\widehat{V}$ and $\widehat{T}$.

The $\widehat{R}$-term is suppressed by a factor ε^2 and we may expect that in the limit $\varepsilon^2 \to 0$ of infinitesimally small time slices it has a vanishing contribution. We shall clarify this point later; for the moment, let us just drop this term and see the resulting expression for the time evolution amplitude (2.2.1). First, we calculate in this approximation the amplitude for an infinitesimal time displacement:
$$\begin{aligned}
\langle x_j|\mathrm{e}^{-\mathrm{i}\varepsilon\widehat{H}(\widehat{p},\widehat{x},t_j)/\hbar}|x_{j-1}\rangle &\approx \int_{-\infty}^{\infty} dx\, \langle x_j|\mathrm{e}^{-\mathrm{i}\varepsilon\widehat{V}(\widehat{x},t_j)/\hbar}|x\rangle\langle x|\mathrm{e}^{-\mathrm{i}\varepsilon\widehat{T}(\widehat{p},t_j)/\hbar}|x_{j-1}\rangle \\
&= \int_{-\infty}^{\infty} dx\,\langle x_j|\mathrm{e}^{-\mathrm{i}\varepsilon\widehat{V}(\widehat{x},t_j)/\hbar}|x\rangle \int_{-\infty}^{\infty}\frac{dp_j}{2\pi\hbar}\mathrm{e}^{\mathrm{i}p_j(x-x_{j-1})/\hbar}\mathrm{e}^{-\mathrm{i}\varepsilon\widehat{T}(p_j,t_j)/\hbar} \\
&= \int_{-\infty}^{\infty} dx\,\delta(x_j - x)\mathrm{e}^{-\mathrm{i}\varepsilon\widehat{V}(x_j,t_j)/\hbar}\int_{-\infty}^{\infty}\frac{dp_j}{2\pi\hbar}\mathrm{e}^{\mathrm{i}p_j(x-x_{j-1})/\hbar}\mathrm{e}^{-\mathrm{i}\varepsilon\widehat{T}(p_j,t_j)/\hbar} \\
&= \int_{-\infty}^{\infty}\frac{dp_j}{2\pi\hbar}\exp\left\{\frac{\mathrm{i}}{\hbar}p_j(x-x_{j-1}) - \frac{\mathrm{i}}{\hbar}\varepsilon[\widehat{T}(p_j,t_j)+\widehat{V}(x_j,t_j)]\right\}. \quad (2.2.8)
\end{aligned}$$

Inserting this into (2.2.1), we obtain the finite approximation for a *path integral in a phase space*
$$K(x,t|x_0,t_0) = \prod_{j=1}^{N}\left[\int_{-\infty}^{\infty}dx_j\right]\prod_{j=1}^{N+1}\left[\int_{-\infty}^{\infty}\frac{dp_j}{2\pi\hbar}\right]\exp\left\{\frac{\mathrm{i}}{\hbar}S_N(p_1,\ldots,p_{N+1};x_0,\ldots,x_{N+1})\right\} \qquad (2.2.9)$$
where S_N is the sum
$$S_N = \sum_{j=1}^{N+1}\varepsilon\left[p_j\frac{x_j - x_{j-1}}{\varepsilon} - H(p_j,x_j,t_j)\right] \qquad (2.2.10)$$
in which we can easily recognize the finite approximation for the action of the system under consideration (a point particle with the Hamiltonian $H(p,x,t)$).

Another way of heuristic derivation (that is, based on neglecting terms similar to the operator $\widehat{R}$ in (2.2.7), without proper mathematical justification) uses the Taylor expansion of the exponentials for small ε. We suggest the reader should carry out such a derivation as a useful exercise (problem 2.2.1, page 190).

◇ Justification of the path-integral derivation: the Trotter product formula

To clarify the approximation which we made by neglecting the $\widehat{R}$-term in (2.2.8), let us consider, at first, the time-independent Hamiltonians and smooth potentials. In this case, we can show that in the limit $\varepsilon \to 0$, the approximate relation (2.2.8) becomes the exact equality. This follows from the so-called *Trotter product formula*, which reads as

$$\exp\left\{-\frac{i}{\hbar}(t-t_0)\widehat{H}\right\} = \lim_{N\to\infty}\left(\exp\left\{-\frac{i}{\hbar}\varepsilon\widehat{V}(\widehat{x})\right\}\exp\left\{-\frac{i}{\hbar}\varepsilon\widehat{T}(\widehat{p})\right\}\right)^N. \qquad (2.2.11)$$

The Trotter formula implies that the commutator term $\widehat{R}$ from (2.2.6) which is proportional to ε^2 does not contribute in the limit $N \to \infty$.

The heuristic proof of the formula is as follows. We note that

$$\exp\left\{-\frac{i}{\hbar}(t-t_0)\widehat{H}\right\} = \left(\exp\left\{-\frac{i}{\hbar}\varepsilon\widehat{H}\right\}\right)^N$$

and write the difference between the operators with and without $\widehat{R}$-terms in a peculiar way (denoting, for brevity, $\lambda \stackrel{\text{def}}{\equiv} \frac{it}{\hbar}$):

$$\begin{aligned}
&(e^{-\lambda\widehat{T}/N}e^{-\lambda\widehat{V}/N})^N - (e^{-\lambda(\widehat{T}+\widehat{V})/N})^N \\
&= [e^{-\lambda\widehat{T}/N}e^{-\lambda\widehat{V}/N} - e^{-\lambda(\widehat{T}+\widehat{V})/N}](e^{-\lambda(\widehat{T}+\widehat{V})/N})^{N-1} \\
&\quad + e^{-\lambda\widehat{T}/N}e^{-\lambda\widehat{V}/N}[e^{-\lambda\widehat{T}/N}e^{-\lambda\widehat{V}/N} - e^{-\lambda(\widehat{T}+\widehat{V})/N}]e^{-\lambda(\widehat{T}+\widehat{V})(N-2)/N} \\
&\quad + \cdots + (e^{-\lambda\widehat{T}/N}e^{-\lambda\widehat{V}/N})^{N-1}[e^{-\lambda\widehat{T}/N}e^{-\lambda\widehat{V}/N} - e^{-\lambda(\widehat{T}+\widehat{V})/N}].
\end{aligned} \qquad (2.2.12)$$

The right-hand side of the identity (2.2.12) contains the factor

$$[e^{-\lambda\widehat{T}/N}e^{-\lambda\widehat{V}/N} - e^{-\lambda(\widehat{T}+\widehat{V})/N}] \qquad (2.2.13)$$

which by (2.2.6) is of order ε^2. Hence, in the limit, the difference is zero.

In effect, we have given a non-strict proof of the Trotter formula. However, the difference in (2.2.13) contains *operators* and hence a correct consideration must use a more precise and clear meaning of words like 'of order ε^2', etc. In other words, we have to operate with some sort of norm, to speak about the order of a quantity. Recall that the norm $\|\cdot\|$ of an operator $\widehat{A}$ acting on a Hilbert space $\mathcal{H}$ is defined as follows:

$$\|\widehat{A}\| \stackrel{\text{def}}{\equiv} \sup_{\psi\in\mathcal{H}} \frac{\|\widehat{A}\psi\|}{\|\psi\|}. \qquad (2.2.14)$$

Here $\|\varphi\|$ is the Hilbert space norm for any $\varphi \in \mathcal{H}$:

$$\|\varphi\| \stackrel{\text{def}}{\equiv} \sqrt{\langle\varphi|\varphi\rangle}. \qquad (2.2.15)$$

If the Hilbert space $\mathcal{H}$ in (2.2.14) is finite dimensional, the norm of a self-adjoint operator $\widehat{A}$ is just equal to its largest eigenvalue:

$$\|\widehat{A}\| = \max_i\{\lambda_i\} \qquad (2.2.16)$$

where $\{\lambda_i\}$ is the set of eigenvalues: $\widehat{A}\psi = \lambda_i\psi_i$, $i = 1,\ldots,d$ (problem 2.2.2, page 191). For an infinite-dimensional Hilbert space, calculations of the norm are not reduced, in general, to the determination of the maximal eigenvalue of an operator and the limiting procedure (supremum) in the definition is required.

Path integrals in the Hamiltonian formalism 157

To apply this operator norm for the proof of the Trotter formula, we note that, by assumption, the free Hamiltonian $\widehat{H}_0 = \widehat{T}(\widehat{p}, t)$, the total Hamiltonian $\widehat{H} = \widehat{T} + \widehat{V}$ and the potential energy $\widehat{V}$ are *self-adjoint* operators. Hence the corresponding operators $e^{it\widehat{T}/\hbar}$, $e^{it\widehat{V}/\hbar}$ and $e^{it\widehat{H}/\hbar}$ are unitary and, therefore, their norms are equal to unity,

$$\|e^{it\widehat{T}/\hbar}\| = \|e^{it\widehat{V}/\hbar}\| = \|e^{it\widehat{H}/\hbar}\| = 1 \qquad (2.2.17)$$

because any unitary operator $\widehat{A}$ ($\widehat{A}^\dagger \widehat{A} = \widehat{A}\widehat{A}^\dagger = 1$) is norm preserving: $\|\widehat{A}\psi\| = \|\psi\|$. Thus, using (2.2.17) and the operator identity (2.2.12), we obtain

$$\begin{aligned}
&\|[(e^{-\lambda\widehat{T}/N}e^{-\lambda\widehat{V}/N})^N - (e^{-\lambda(\widehat{T}+\widehat{V})/N})^N]\psi\| \\
&\leq \|[e^{-\lambda\widehat{T}/N}e^{-\lambda\widehat{V}/N} - e^{-\lambda(\widehat{T}+\widehat{V})/N}](e^{-\lambda(\widehat{T}+\widehat{V})/N})^{N-1}\psi\| \\
&\quad + \|e^{-\lambda\widehat{T}/N}e^{-\lambda\widehat{V}/N}[e^{-\lambda\widehat{T}/N}e^{-\lambda\widehat{V}/N} - e^{-\lambda(\widehat{T}+\widehat{V})/N}]e^{-\lambda(\widehat{T}+\widehat{V})(N-2)/N}\psi\| \\
&\quad + \cdots + \|(e^{-\lambda\widehat{T}/N}e^{-\lambda\widehat{V}/N})^{N-1}[e^{-\lambda\widehat{T}/N}e^{-\lambda\widehat{V}/N} - e^{-\lambda(\widehat{T}+\widehat{V})/N}]\psi\| \\
&= N\mathcal{O}\left(\frac{t}{N^2}\right) \xrightarrow[N\to\infty]{} 0
\end{aligned} \qquad (2.2.18)$$

where the limit is uniform in $\mathcal{H}$ and we have taken into account that $(\exp\{-\lambda(\widehat{T}+\widehat{V})/N\})^\tau \psi$ for $0 \leq \tau \leq t$ is a compact set in $\mathcal{H}$.

This completes the sketch (mathematically rather superficial) of the following theorem (Trotter 1959):

Theorem 2.1 (Trotter product formula). Let A and B be self-adjoint operators on a separable Hilbert space so that $A + B$, defined on $D(A) \cap D(B)$, is self-adjoint. Then

$$e^{it(\widehat{A}+\widehat{B})} = \lim_{N\to\infty}(e^{it\widehat{A}/N}e^{it\widehat{B}/N})^N. \qquad (2.2.19)$$

Note that, if we consider real exponentials, $e^{-t(\widehat{A}+\widehat{B})}$, $e^{-t\widehat{A}}$, $e^{-t\widehat{B}}$, the Trotter formula is proven under the additional condition that the operators $\widehat{A}$ and $\widehat{B}$ are *bounded from below*. For further details, see, e.g., Nelson (1964) and Simon (1979).

◇ **Continuous limit for the phase-space path integral**

Under the conditions of the Trotter product formula, the discrete approximation (2.2.9) is correct and we can pass to the continuum limit. The continuum limit for the action is without subtleties (the approximation (2.2.10) is an ordinary Darboux sum):

$$S[p, x] = \lim_{\varepsilon\to 0} S_N = \int_{t_0}^{t} d\tau\, (p(\tau)\dot{x}(\tau) - H(p(\tau), x(\tau))). \qquad (2.2.20)$$

Evidently, this is the classical canonical action for the paths $x(t)$, $p(t)$ in the phase space, with the boundary condition $x(t_0) = x_0$, $x(t) = x$.

In the same limit, i.e. $\varepsilon \to 0$, $N \to \infty$, the product of infinitely many integrals in (2.2.9) will be called a path integral over the phase space and will be denoted as

$$\int_{\mathcal{C}\{x_0,t_0;x,t\}} \mathcal{D}x(\tau) \frac{\mathcal{D}p(\tau)}{2\pi\hbar} F[p(\tau), x(\tau)] \stackrel{\text{def}}{\equiv} \lim_{N\to\infty} \prod_{j=1}^{N}\left[\int_{-\infty}^{\infty} dx_j\right] \prod_{j=1}^{N+1}\left[\int_{-\infty}^{\infty} \frac{dp_j}{2\pi\hbar}\right]$$
$$\times F(p_1, \ldots, p_{N+1}; x_0, \ldots, x_{N+1}) \qquad (2.2.21)$$

where $F[\ldots]$ is an arbitrary functional and $x_{N+1} = x$. Here we have introduced the conventional notation $\mathcal{D}x(\tau)$, $\mathcal{D}p(\tau)$ for the infinite-dimensional 'volume elements' which we shall extensively use in what follows. Note that, by definition, in the finite-N expression, there is always one more p_j-integral than the x_j-integrals in the product. While x_0 and x_{N+1} are held fixed and the x_j-integrals are performed for $j = 1, \ldots, N$, each pair (x_j, x_{j-1}) is accompanied by one p_j-integral for $j = 1, \ldots, N+1$.

This asymmetry in the integration over x and p is a consequence of keeping the endpoints fixed in the *position* (configuration) space. There exists the possibility of proceeding in a conjugate way, keeping the initial and final *momenta* p_0 and p fixed (see problem 2.2.4, page 192). In this case, the finitely approximated path integral has the opposite asymmetry—there is now one more x_j-integral than the p_j-integrals—and the resulting path integral reads as

$$\langle p|\widehat{U}(t,t_0)|p_0\rangle \approx \prod_{j=1}^{N}\left[\int_{-\infty}^{\infty}\frac{dp_j}{2\pi\hbar}\right]\prod_{j=1}^{N+1}\left[\int_{-\infty}^{\infty}dx_j\right]$$
$$\times \exp\left\{\frac{i}{\hbar}\sum_{j=1}^{N+1}\varepsilon\left[-x_j\frac{p_j - p_{j-1}}{\varepsilon} - H(p_j, x_j, t_j)\right]\right\}. \quad (2.2.22)$$

A path-integral *symmetric* in p and x arises when considering the quantum-mechanical partition function defined by the trace (cf chapter 4)

$$\mathcal{Z}(t, t_0) \stackrel{\text{def}}{\equiv} \text{Tr}(e^{-i(t-t_0)\widehat{H}/\hbar}). \quad (2.2.23)$$

In the coordinate basis of the Hilbert space, the trace becomes an integral over the amplitude $\langle x|\widehat{U}(t, t_0)|x_0\rangle$ with $x = x_0$:

$$\mathcal{Z}(t, t_0) = \int_{-\infty}^{\infty} dx\, \langle x|\widehat{U}(t, t_0)|x\rangle. \quad (2.2.24)$$

The additional trace integral over $x_{N+1} = x_0$ makes the path integral for Z symmetric in p_j and x_j:

$$\int_{-\infty}^{\infty} dx_{N+1} \prod_{j=1}^{N}\left[\int_{-\infty}^{\infty}dx_j\right]\prod_{j=1}^{N+1}\left[\int_{-\infty}^{\infty}\frac{dp_j}{2\pi\hbar}\right] = \prod_{j=1}^{N+1}\left[\int\int_{-\infty}^{\infty}\frac{dx_j dp_j}{2\pi\hbar}\right]. \quad (2.2.25)$$

It should be noted that, starting from the momentum representation, we would obtain the same expression for the partition function (2.2.24), though the time-slice approximation for the actions in (2.2.10) and (2.2.22) are different (problem 2.2.2, page 154).

With definition (2.2.21), the time amplitude (Green function) can be written in the short form

$$K(x,t|x_0,t_0) = \int_{\mathcal{C}\{x_0,t_0;x,t\}} \mathcal{D}x\frac{\mathcal{D}p}{2\pi\hbar}e^{iS[p(\tau),x(\tau)]/\hbar}. \quad (2.2.26)$$

Here $\mathcal{C}\{x_0, t_0; x, t\}$ is a path in the *phase space* with fixed initial and final *positions* (and arbitrary *momenta*).

What is the mathematical meaning of this 'integral'? Or, formulating the question in a more practical fashion, how do we calculate it?

Attempts to attribute to this expression a traditional meaning, in the spirit of the Wiener continuous integral, as an integral defined by a measure in a functional space, have met too many difficult problems in this case. Another extremal point of view stating that expression (2.2.26) is nothing but 'a hieroglyph' for encoding the (combinatorial) properties of perturbation theory in quantum mechanics seems to be very restrictive. We shall understand the *path integral* (2.2.26) as a *limit of finite-dimensional approximations*. However, it turns out that for more general types of Hamiltonian than those considered in this section, these path integrals are sensitive to a choice of the discrete approximation, the ambiguity in this dependence being of the same nature as the general ambiguity in the quantization of classical systems (operator ordering problem). In section 2.3, we shall discuss and substantiate the limiting procedure (on the rigorously 'physical' level). We shall clarify the set of trajectories in the phase space which are essential for the integral (2.2.26) (supporting the phase path integral). It turns out that these trajectories are necessarily discontinuous (in sharp distinction to the Wiener path integral, cf section 1.1.3). Moreover, the set of functions which contributes to the phase-space path integrals is not uniquely defined: we can vary the sets of trajectories increasing the smoothness of the coordinates for an account of smoothness of momenta and *vice versa*. However, it is impossible to make both coordinates and momenta continuous. This fact is in accordance with the fundamental *uncertainty principle* of quantum mechanics.

◇ **Reduction of the phase-space path integrals to the configuration Feynman path integrals for Hamiltonians with quadratic dependence on momenta**

In the preceding section, we considered another type of path integral in quantum mechanics, in which integration is carried out over trajectories in configuration space only. As we have already noted, in his original paper, Feynman discussed only this quantum-mechanical path integral. The relation between the two types of integral can be easily established for the case of Hamiltonians of the form (2.2.4), with the standard kinetic term

$$H = \frac{p^2}{2m} + V(x,t) \qquad (2.2.27)$$

for which the discrete action (2.2.10) takes the form

$$S_N = \sum_{j=1}^{N+1} \varepsilon \left[p_j \frac{x_j - x_{j-1}}{\varepsilon} - \frac{p_j^2}{2m} - V(x_j, t_j) \right]. \qquad (2.2.28)$$

Since this expression is quadratic in the momenta, we can integrate them out. The standard method for doing this is a *square completion*, i.e. first we rewrite (2.2.28) as follows:

$$S_N = \sum_{j=1}^{N+1} \varepsilon \left[-\frac{1}{2m}\left(p_j - \frac{x_j - x_{j-1}}{\varepsilon} m \right)^2 + \frac{m}{2}\left(\frac{x_j - x_{j-1}}{\varepsilon} \right)^2 - V(x_j, t_j) \right]. \qquad (2.2.29)$$

Integrating over momentum variables in the time-sliced phase-space path integral we arrive at an alternative representation of the Green function:

$$K(x,t|x_0,t_0) = \lim_{N\to\infty} \frac{1}{\sqrt{2\pi\hbar i\varepsilon/m}} \prod_{j=1}^{N} \left[\int_{-\infty}^{\infty} \frac{dx_j}{\sqrt{2\pi\hbar i\varepsilon/m}} \right] \exp\left\{ \frac{i}{\hbar} S_N \right\} \qquad (2.2.30)$$

where S_N now reads as

$$S_N = \sum_{j=1}^{N+1} \varepsilon \left[\frac{m}{2}\left(\frac{x_j - x_{j-1}}{\varepsilon} \right)^2 - V(x_j, t_j) \right] \qquad (2.2.31)$$

with $x_{N+1} = x$. In the continuum limit, the sum (2.2.31) converges toward the action in the Lagrangian form

$$S[x] = \int_{t_0}^{t} dt\, L(x, \dot{x}) \qquad (2.2.32)$$

with the Lagrangian

$$L(x, \dot{x}) = \frac{m}{2}\dot{x}^2 - V(x, t). \tag{2.2.33}$$

Thus, in the continuum limit, the functional integral takes the form of the Feynman path integral considered in the preceding section:

$$K(x, t|x_0, t_0) = \int_{\mathcal{C}\{x_0,t_0;x,t\}} d_F x(\tau) \, e^{\frac{i}{\hbar} \int_{t_0}^{t} d\tau \, V(x(\tau))}$$

$$\equiv \int_{\mathcal{C}\{x_0,t_0;x,t\}} \prod_{\tau=0}^{t} \frac{dx(\tau)}{\sqrt{\frac{2\pi i \hbar}{m} d\tau}} e^{\frac{i}{\hbar} S[x(\tau)]}$$

$$\equiv \int_{\mathcal{C}\{x_0,t_0;x,t\}} \mathcal{D}_{d\tau} x(\tau) e^{\frac{i}{\hbar} S[x(\tau)]}. \tag{2.2.34}$$

Here $\mathcal{D}_{d\tau} x(\tau)$ stands for the 'measure'

$$\mathcal{D}_{d\tau} x(\tau) \stackrel{\text{def}}{\equiv} \prod_{\tau=0}^{t} \frac{dx(\tau)}{\sqrt{\frac{2\pi i \hbar}{m} d\tau}}. \tag{2.2.35}$$

Correspondingly, in the discrete approximation with time slices of thickness ε, we shall use, for brevity, the notation

$$\mathcal{D}_{\varepsilon} x(\tau) \stackrel{\text{def}}{\equiv} \frac{1}{\sqrt{\frac{2\pi i \hbar}{m}\varepsilon}} \prod_{j=1}^{N} \frac{dx_j}{\sqrt{\frac{2\pi i \hbar}{m}\varepsilon}}. \tag{2.2.36}$$

◇ Why the quantum-mechanical path integrals are not defined within integration measure theory

Expressions (2.2.34) are Feynman's original formulae for the quantum-mechanical amplitude (2.1.68). They consist of a sum over all paths in configuration space with a phase factor being the action $S[x]$. It is necessary to mention that various attempts to define this integral outside the framework of a discrete approximation have been made. They require rather involved mathematics, and although definitely of general interest, have not thus far turned out to be analytically powerful and with impressive practical results. Thus we shall not discuss them here, referring the reader to the reviews and papers by Albeverio and Høegh-Krohn (1976), Mizrahi (1976), Truman (1978), DeWitt-Morette *et al* (1979) and Cartier and DeWitt-Morette (1996). Very roughly, we might say that these attempts aim to construct a mathematically profound generalization of the $i\eta$-prescription, used for the correct definition of finite-dimensional complex Gaussian integrals (cf (2.1.77)).

The general reason for the impossibility of constructing a measure for the Feynman integral has its origin in the fundamental properties of quantum mechanics and its essential distinction from classical statistical physics and the theory of stochastic processes. Indeed, as we pointed out in chapter 1, a measure appears inevitably in any theory which contains *probabilities*. Thus we might expect to encounter a measure if considering an expression for some probability. But in quantum mechanics we deal with the more fundamental notion of a probability *amplitude*! By 'more fundamental notion' we mean that knowledge of an amplitude immediately gives the corresponding probability, but the former contains more information about a quantum-mechanical system than the latter. The distinction between the notions is the root of the difficulties with measure definition in the case of the Feynman integral.

It is worth recalling that speaking about the movement of a quantum-mechanical particle in terms of a path it took from here to there easily leads to contradiction and confusion. The standard counterexample to such an idea is the well-known two slit experiment (see, e.g., Dirac (1947) and Schiff (1955)). An electron passes through either (or both) of two slits on its way to a screen. The screen is a detector for position and an interference pattern can be seen. Any attempt to verify that the electron went through one slit or the other by a localization of the electron destroys the interference pattern. If we tried to describe this experiment in terms of the classical notion of paths as in the theory of stochastic processes, we would use conditional (transition) probabilities $W(x, t|x', t')$, satisfying the fundamental Markovian property

$$W(x, t|x_0, t_0) = \int_{-\infty}^{\infty} dx' \, W(x, t|x', t') W(x', t'|x_0, t_0) \qquad t' \in [t_0, t]. \qquad (2.2.37)$$

Such a description does not give any interference pattern. In quantum mechanics, there is a different rule: we work with probability amplitudes $K(x, t|x_0, t_0)$ which also depends on two states and satisfy the similar relation

$$K(x, t|x_0, t_0) = \int_{-\infty}^{\infty} dx' \, K(x, t|x', t') K(x', t'|x_0, t_0) \qquad t' \in [t_0, t] \qquad (2.2.38)$$

but only the square of the absolute value $|K|^2$ is interpreted as a probability. Due to this distinction, although there is the integral in (2.2.38), we cannot say that the system was either in x_1 or x_2 or ... at the time t', since then we should have a relation similar to (2.2.37) for $|K|^2$, but this contradicts (2.2.38).

The same reasoning is applicable to the Feynman path integral. Although it represents a sum over paths, there is no assertion that the system followed definite paths with certain probability (and, hence, no measure appears in the path-integral construction). Rather, we compute probability amplitudes for the paths and sums the amplitudes.

As we have learned in the preceding chapter, it is the Markov property of transition amplitudes that is the cornerstone for their (Wiener) path-integral representation. In the case of quantum mechanics the similar Markov-like relation (2.2.38), this time for transition *amplitudes*, lies at the root of path-integral construction (recall that the Markov-like relation (2.2.38) is the direct consequence of the completeness axiom in quantum mechanics, cf section 2.1.1). Using (2.2.38), we can reduce any transition amplitude to the composition of those for infinitesimal time displacements (i.e. to the discrete approximation for the path integral). Then the only question is what to take for the Green function $K(x, t+\varepsilon|y, t)$ corresponding to this infinitesimal displacement. It was Dirac (1933) who first observed (see also Dirac (1947)) that the exponent

$$\exp\left\{i\frac{S}{\hbar}\right\}$$

is a good approximation for the Green function in the case $S \ll \hbar$ (S is the action of a quantum-mechanical system). In particular, this happens when the action is small because of the small time interval over which K is supposed to propagate. Consequently, for a short time ε, the propagator from y to x is approximated by $K(x, t+\varepsilon|y, t) \approx \exp\{iS/\hbar\}$. This is the heuristic ground for the Feynman path-integral construction in quantum mechanics.

2.2.2 Calculation of path integrals for the simplest quantum-mechanical systems: a free particle and a harmonic oscillator

Before we proceed to a more profound discussion of the properties of the phase-space integral (the analog of the Wiener theorem: determination of a set of essential trajectories which we must integrate over; the relation to the quantum-mechanical operator ordering problem), let us acquire some practical experience with this path integral by considering a few examples of quantum-mechanical systems and the semiclassical approximation for its calculation.

◇ Path-integral calculation of the transition amplitude for a free particle

According to (2.2.26),

$$K_0(x,t|x_0,t_0) = \int_{C\{x_0,t_0;x,t\}} \mathcal{D}x \frac{\mathcal{D}p}{2\pi\hbar} \exp\left\{\frac{i}{\hbar}\int_{t_0}^{t} dt\left(p\dot{x} - \frac{p^2}{2m}\right)\right\} \qquad (2.2.39)$$

and, after integration over momenta, the Feynman integral takes the form

$$K_0(x,t|x_0,t_0) = \int_{C\{x_0,t_0;x,t\}} \mathcal{D}x \exp\left\{\frac{i}{\hbar}\int_{t_0}^{t} dt \frac{m}{2}\dot{x}^2\right\}. \qquad (2.2.40)$$

The time-sliced expression to be integrated is given by (2.2.30) and (2.2.31), where we have to set $V(x) = 0$. The calculations literally repeat the ones we carried out in section 1.1.4, with only the change that we must use the imaginary integration formula (2.1.77) instead of integration of real Gaussians. The calculations yield the free-particle amplitude

$$K_0(x,t|x_0,t_0) = \frac{1}{\sqrt{2\pi i\hbar(t-t_0)/m}} \exp\left\{\frac{i}{\hbar}\frac{m}{2}\frac{(x-x_0)^2}{t-t_0}\right\}. \qquad (2.2.41)$$

This amplitude agrees with the result in the operator approach which we obtained in section 2.1.1 (cf (2.1.78) and problem 2.1.3, page 150).

Another calculation method for this simple integral is the mode expansion, as discussed in section 1.2.7. This method is more involved and seems to be superfluous for the simple case under consideration, but it turns out to be useful for the treatment of a certain class of non-trivial path integrals, after a suitable generalization. Therefore, we also present it here.

Recall that when calculating Wiener integrals in chapter 1, we split all paths into the classical trajectory $x_c(\tau)$ plus deviations $X(\tau)$:

$$x(\tau) = x_c(\tau) + X(\tau). \qquad (2.2.42)$$

The classical trajectory for a free particle has, of course, the same form (1.2.118) as for a free Brownian particle:

$$x_c(\tau) = x_0 + \frac{x-x_0}{t-t_0}(\tau-t_0) \qquad (2.2.43)$$

and is defined by the same equation of motion

$$\ddot{x}_c = 0. \qquad (2.2.44)$$

The deviations vanishing at the endpoints

$$X(t_0) = X(t) = 0 \qquad (2.2.45)$$

are now called the *quantum fluctuations* of the particle trajectory.

When inserting the decomposition (2.2.33) into the action, we observe that, due to the equation of motion (2.2.44) for the classical trajectory, the action separates into a sum of the classical and purely quadratic fluctuation terms:

$$\frac{m}{2}\int_{t_0}^{t} d\tau\,[\dot{x}_c^2(\tau) + 2\dot{x}_c(\tau)\dot{X}(\tau) + \dot{X}^2(\tau)] = \frac{m}{2}\int_{t_0}^{t} d\tau\,\dot{x}_c^2 + m\dot{x}_c X\bigg|_{t_0}^{t} - m\int_{t_0}^{t} d\tau\,\ddot{x}_c X + \frac{m}{2}\int_{t_0}^{t} d\tau\,\dot{X}^2$$

$$= \frac{m}{2}\left[\int_{t_0}^{t} d\tau\,(\dot{x}_c^2 + \dot{X}^2)\right].$$

The absence of a mixed term is a general consequence of the extremality property of the classical path,

$$\delta S|_{x(\tau)=x_c(\tau)} = 0. \tag{2.2.46}$$

This implies that the fluctuation expansion around the classical action

$$S_c \stackrel{\text{def}}{\equiv} S[x_c] \tag{2.2.47}$$

can have no linear term in $X(\tau)$, i.e. it must start as

$$S = S_c + \frac{1}{2}\int_{t_0}^{t} d\tau \int_{t_0}^{t} d\tau' \left.\frac{\delta^2 S}{\delta x(\tau)\delta x(\tau')}\right|_{x(\tau)=x_c(\tau)} X(\tau)X'(\tau') + \cdots. \tag{2.2.48}$$

The amplitude now factorizes into the product of a classical amplitude $e^{iS_c/\hbar}$ and a fluctuation factor $\phi(t-t_0)$:

$$K_0(x,t|x_0,t_0) = e^{iS_c/\hbar}\phi(t-t_0). \tag{2.2.49}$$

For the free particle, the classical action is

$$S_c = \int_{t_0}^{t} dt\,\frac{m}{2}\dot{x}_c^2 = \frac{m}{2}\frac{(x-x_0)^2}{t-t_0}. \tag{2.2.50}$$

The fluctuation factor is given by the path integral

$$\phi(t-t_0) = \int_{\mathcal{C}\{0,t;0,t_0\}} \mathcal{D}_{d\tau}X(\tau)\exp\left\{\frac{i}{\hbar}\int_{t_0}^{t}\frac{m}{2}\dot{X}^2\right\} \tag{2.2.51}$$

Since $X(\tau)$ vanishes at the endpoints and due to time translational invariance, the fluctuation factor ϕ can be determined from the normalization condition for the amplitude as in the calculation of the transition amplitude in section 1.2.6. However, it is instructive and useful for more involved cases to calculate this factor explicitly.

Explicit evaluation of this factor requires the calculation of the multiple integral (i.e. the path integral (2.2.51) in the discrete-time approximation)

$$\phi^{(N)}(t-t_0) = \frac{1}{\sqrt{2\pi i\hbar\varepsilon/m}}\prod_{j=1}^{N}\left[\int_{-\infty}^{\infty}\frac{dX_j}{\sqrt{2\pi i\hbar\varepsilon/m}}\right]\exp\left\{\frac{i}{\hbar}S_N^{(\text{fl})}\right\} \tag{2.2.52}$$

where $S_N^{(\text{fl})}$ is the time-sliced fluctuation action

$$S_N^{(\text{fl})} = \frac{m}{2}\varepsilon\sum_{j=1}^{N+1}\left(\frac{X_j - X_{j-1}}{\varepsilon}\right)^2. \tag{2.2.53}$$

At the end, we have to take the continuum limit

$$n \to \infty \qquad \varepsilon = \frac{t-t_0}{N+1} \to 0.$$

Making use of the finite-difference operators (1.2.195), the action (2.2.53) can be rewritten in the shorter form

$$S_N^{(\text{fl})} = \frac{m}{2}\varepsilon\sum_{j=1}^{N+1}(\bar{\partial}^{(\varepsilon)}X_j)^2. \tag{2.2.54}$$

164   Path integrals in quantum mechanics

Inserting now the expansion (1.2.209) into the time-sliced fluctuation action (2.2.53), the orthogonality relation (1.2.210) results in

$$S_N^{(\text{fl})} = \frac{m}{2}\varepsilon \sum_{j=0}^{N} (\bar{\partial}^{(\varepsilon)} X_j)^2 = \frac{m}{2}\varepsilon \sum_{n=1}^{N+1} \Omega_m \bar{\Omega}_m a_m^2 \qquad (2.2.55)$$

and the fluctuation factor (2.2.52) becomes

$$\phi^{(N)}(t-t_0) = \frac{1}{\sqrt{2\pi i\hbar\varepsilon/m}} \prod_{j=1}^{N}\left[\int_{-\infty}^{\infty} \frac{dX_j}{\sqrt{2\pi i\hbar\varepsilon/m}}\right] \prod_{n=1}^{N} \exp\left\{\frac{i}{\hbar}\frac{m}{2}\Omega_m\bar{\Omega}_m a_m^2\right\}. \qquad (2.2.56)$$

In full analogy with the calculations in section 1.2.7, after the change of the integration variables (1.2.209) and (1.2.212), the integration of (2.2.56) can be performed with the help of the Gaussian formula (2.1.77) with the result:

$$\phi^{(N)}(t-t_0) = \frac{1}{\sqrt{2\pi i\hbar(N+1)\varepsilon/m}}$$

$$= \frac{1}{\sqrt{2\pi i\hbar(t-t_0)/m}} = \phi(t-t_0). \qquad (2.2.57)$$

Combining this factor with the contribution from the classical path, we again obtain for the amplitude of a free particle the formula (2.2.41).

It is straightforward to generalize this result to a point particle moving through any number d of Cartesian space dimensions. If $\boldsymbol{x} = (x_1, \ldots, x_d)$ denotes the spatial coordinates, the action is

$$S[\boldsymbol{x}] = \frac{m}{2}\int_{t_0}^{t} dt\, \dot{\boldsymbol{x}}^2. \qquad (2.2.58)$$

Being quadratic in $\boldsymbol{x}$, the action is the sum of the actions for each component. Hence, the amplitude factorizes and can easily be found from the one-dimensional case (2.2.41)

$$K_0(\boldsymbol{x},t|\boldsymbol{x}_0,t_0) = \frac{1}{(2\pi i\hbar(t-t_0)/m)^{d/2}} \exp\left\{\frac{i}{\hbar}\frac{m}{2}\frac{(\boldsymbol{x}-\boldsymbol{x}_0)^2}{t-t_0}\right\}. \qquad (2.2.59)$$

◇ **Transition amplitude for the harmonic oscillator**

A further problem which can be solved along similar lines is the time evolution amplitude of the linear oscillator:

$$K(x,t|x_0,t_0) = \int_{\mathcal{C}\{x_0,t_0;x,t\}} \mathcal{D}x \frac{\mathcal{D}p}{2\pi\hbar} \exp\left\{\frac{i}{\hbar} S[p,x]\right\}$$

$$= \int_{\mathcal{C}\{x_0,t_0;x,t\}} \mathcal{D}_{d\tau} x \exp\left\{\frac{i}{\hbar} S[x]\right\}$$

$$= \exp\left\{\frac{i}{\hbar} S_c[x]\right\} \phi_\omega(t-t_0). \qquad (2.2.60)$$

Here the canonical action $S[p,x]$ for a harmonic oscillator is

$$S[p,x] = \int_{t_0}^{t} dt\left(p\dot{x} - \frac{1}{2m}p^2 - \frac{m\omega^2}{2}x^2\right) \qquad (2.2.61)$$

Path integrals in the Hamiltonian formalism 165

and the Lagrangian action $S[x]$ reads correspondingly

$$S[x] = \int_{t_0}^{t} dt \, \frac{m}{2}(\dot{x}^2 - \omega^2 x^2). \qquad (2.2.62)$$

After reducing the phase-space path integral to the Feynman (configuration) one, the calculations are again very similar to those in section 1.2.7. The classical part of the action has the same form as (2.2.62):

$$S_c[x] = \int_{t_0}^{t} dt \, \frac{m}{2}(\dot{x}_c^2 - \omega^2 x_c^2) \qquad (2.2.63)$$

and the solution of the corresponding boundary-value problem

$$\ddot{x}_c = -\omega^2 x_c \qquad (2.2.64)$$
$$x(t_0) = x_0 \qquad x(t) = x$$

reads

$$x_c(\tau) = \frac{x \sin \omega(\tau - t_0) + x_0 \sin \omega(t - \tau)}{\sin \omega(t - t_0)}. \qquad (2.2.65)$$

After a simple calculation (using an integration by parts), we can rewrite the classical action in the form

$$S_c = \frac{m\omega}{2\sin\omega(t - t_0)} [(x^2 + x_0^2)\cos\omega(t - t_0) - 2xx_0]. \qquad (2.2.66)$$

The fluctuation factor is defined by the fluctuation part of the action

$$S^{(\text{fl})}[X] = \int_{t_0}^{t} dt \, \frac{m}{2}(\dot{X}^2 - \omega^2 X^2) \qquad (2.2.67)$$

and to find it explicitly, we have to calculate the multiple integral (path integral in the time-sliced approximation):

$$\phi^{(N)}(t - t_0) = \frac{1}{\sqrt{2\pi i\hbar\varepsilon/m}} \prod_{j=1}^{N} \left[\int_{-\infty}^{\infty} \frac{dX_j}{\sqrt{2\pi i\hbar\varepsilon/m}} \right] \exp\left\{ \frac{i}{\hbar} \frac{m}{2} \varepsilon \sum_{j,k=0}^{N+1} X_j [-\partial^{(\varepsilon)}\bar{\partial}^{(\varepsilon)} - \omega^2]_{jk} X_k \right\}$$

$$= \frac{1}{\sqrt{2\pi i\hbar\varepsilon/m}} \prod_{n=1}^{N} \left[\int_{-\infty}^{\infty} \frac{da_n}{\sqrt{2\pi i\hbar\varepsilon^2/m}} \right] \prod_{n=1}^{N} \exp\left\{ \frac{i}{\hbar} \frac{m}{2} (\Omega_n \bar{\Omega}_n - \omega^2) a_n^2 \right\} \qquad (2.2.68)$$

where a_n are the coefficients of the Fourier expansion of $X(\tau)$ (cf the case of a free particle and the calculation in section 1.2.7).

Calculation of this integral is similar to that of the free particle but with the important distinction that the eigenvalues

$$\Omega_n \bar{\Omega}_n - \omega^2 = \frac{1}{\varepsilon^2}\left[2 - 2\cos\left(\frac{\pi n}{t - t_0}\varepsilon\right)\right] - \omega^2 \qquad (2.2.69)$$

of the operator in the exponent of the integrand (2.2.68) may be negative for a large enough time interval $(t - t_0)$.

At first, let us consider the case of strictly positive eigenvalues. Then the Gaussian formula gives

$$\phi^{(N)}(t - t_0) = \frac{1}{\sqrt{2\pi i\hbar\varepsilon/m}} \prod_{n=1}^{N} \frac{1}{\sqrt{\varepsilon^2(\Omega_n \bar{\Omega}_n - \omega^2)}}. \qquad (2.2.70)$$

By introducing an auxiliary frequency $\widetilde{\omega}$, satisfying

$$\sin\frac{\varepsilon\widetilde{\omega}}{2} = \frac{\varepsilon\omega}{2}$$

the product of the eigenvalues can be decomposed into the known result for the free-particle fluctuation factor and the standard product formula (Gradshteyn and Ryzhik (1980), formula 1.391.1)

$$\prod_{n=1}^{N}\left(1 - \frac{\sin^2 b}{\sin^2 \frac{n\pi}{2(N+1)}}\right) = \frac{1}{\sin 2b}\frac{\sin(2(N+1)b)}{(N+1)}. \tag{2.2.71}$$

Indeed, separating the free-particle determinant as a factor (cf section 1.2.7), we obtain

$$\prod_{n=1}^{N}\varepsilon^2(\Omega_n\bar{\Omega}_n - \omega^2) = \prod_{m=1}^{N}\varepsilon^2\Omega_m\bar{\Omega}_m \prod_{n=1}^{N}\left[\frac{\varepsilon^2(\Omega_n\bar{\Omega}_n - \omega^2)}{\varepsilon^2\Omega_n\bar{\Omega}_n}\right]$$

$$= \prod_{m=1}^{N}\varepsilon^2\Omega_m\bar{\Omega}_m \prod_{n=1}^{N}\left(1 - \frac{\sin^2(\varepsilon\widetilde{\omega}/2)}{\sin^2\frac{n\pi}{2(N+1)}}\right). \tag{2.2.72}$$

The first factor is equal, as we know, to $(N+1)$, while the second is found from (2.2.71), with $b = \widetilde{\omega}\varepsilon/2$. As a result, we arrive at the fluctuation determinant

$$\det_{N}(-\varepsilon^2(\partial^{(\varepsilon)}\bar{\partial}^{(\varepsilon)} - \omega^2)) = \prod_{n=1}^{N}[\varepsilon^2(\Omega_n\bar{\Omega}_n - \omega)] = \frac{\sin\widetilde{\omega}(t-t_0)}{\sin\varepsilon\widetilde{\omega}} \tag{2.2.73}$$

and the total fluctuation factor

$$\phi^{(N)}(t - t_0) = \frac{1}{\sqrt{2\pi i\hbar/m}}\sqrt{\frac{\sin\varepsilon\widetilde{\omega}}{\varepsilon\sin\widetilde{\omega}(t-t_0)}}. \tag{2.2.74}$$

Here $\sqrt{i}$ is assumed to mean $e^{i\pi/4}$. The result (2.2.74) is valid for positive eigenvalues and, in particular, as easy to check using (2.2.69), for the time interval satisfying the inequality

$$t - t_0 < \frac{\pi}{\widetilde{\omega}}. \tag{2.2.75}$$

If $t - t_0$ grows larger than $\pi/\widetilde{\omega}$, the smallest eigenvalue $\Omega_1\bar{\Omega}_1 - \omega^2$ becomes negative and the resulting amplitude carries an extra phase factor $e^{-i\pi/2}$ and remains valid until $t - t_0$ becomes larger than $2\pi/\widetilde{\omega}$, where the second eigenvalue becomes negative and introduces a further phase factor $e^{-i\pi/2}$.

To automate the appearance of correct phases, we might use the $i\eta$-*prescription* (cf section 2.1.1). In the present case, it is convenient to realize it as an infinitesimal shift in the frequency ω to the complex plane, $\omega \to \omega - i\eta$. The oscillator amplitude with such an addition has a damped behaviour, $e^{-i\omega t - \eta t}$, as if energy dissipation existed in the system. We have to note, however, that the quantum description of non-conservative systems with energy dissipation is an involved and subtle subject and we will not pursue it further here. Instead, we shall consider the $i\eta$-prescription as a formal trick. With $\eta \neq 0$, we can use the universal Gauss formula (2.1.73), valid for complex numbers with a positive imaginary part. Now, each time that $t - t_0$ passes a multiple of $\pi/\widetilde{\omega}$, the square root of $\sin\widetilde{\omega}(t - t_0)$ in (2.2.74) passes a singularity in a specific way, defined by the $i\eta$-prescription which ensures the proper phase. In such a way, we can consider any value of $t - t_0$.

In the continuum limit $\varepsilon \to 0$, the fluctuation factor becomes

$$\phi(t - t_0) = \frac{1}{\sqrt{2\pi i\hbar/m}}\sqrt{\frac{\omega}{\sin\omega(t - t_0)}}. \qquad (2.2.76)$$

Combining the classical and fluctuation factors, we obtain for the time evolution amplitude of a harmonic oscillator the expression

$$K(x, t|x_0, t_0) = \frac{1}{\sqrt{2\pi i\hbar/m}}\sqrt{\frac{\omega}{\sin\omega(t - t_0)}}$$
$$\times \exp\left\{\frac{1}{2}\frac{i}{\hbar}\frac{m\omega}{\sin\omega(t - t_0)}[(x^2 + x_0^2)\cos\omega(t - t_0) - 2xx_0]\right\}. \qquad (2.2.77)$$

The result can easily be extended to any number d of dimensions, where the action is

$$S[\boldsymbol{x}] = \int_{t_0}^{t} dt\,\frac{m}{2}(\dot{\boldsymbol{x}}^2 - \omega^2 \boldsymbol{x}^2). \qquad (2.2.78)$$

Being quadratic in $\boldsymbol{x}$, the action is the sum of the actions for each component leading to the factorized amplitude

$$K(\boldsymbol{x}, t|\boldsymbol{x}_0, t_0) = \frac{1}{(2\pi i\hbar/m)^{d/2}}\left(\frac{\omega}{\sin\omega(t - t_0)}\right)^{d/2}$$
$$\times \exp\left\{\frac{1}{2}\frac{i}{\hbar}\frac{m\omega}{2\sin\omega(t - t_0)}[(\boldsymbol{x}^2 + \boldsymbol{x}_0^2)\cos\omega(t - t_0) - 2\boldsymbol{x}\boldsymbol{x}_0]\right\}. \qquad (2.2.79)$$

We want to stress once again the following point concerning calculations using the Fourier mode expansion. At first sight, the fluctuation factor in the continuum limit may be calculated via the continuum determinant

$$\phi(t - t_0) \xrightarrow[\varepsilon \to 0]{} \frac{1}{\sqrt{2\pi i\hbar\varepsilon/m}}\frac{1}{\sqrt{\det(-\partial_t^2 - \omega^2)}}. \qquad (2.2.80)$$

But the point is that the determinant $\det(-\partial_t^2 - \omega^2)$ is strongly divergent (because its eigenvalues grow quadratically) and only a combination of the two singular factors in (2.2.80) may result in a finite expression. For its accurate calculation, we need what is called *a regularization*, the most natural and physically transparent being discrete time-slicing which we have just used in the actual calculation of the fluctuation factor and, as we have found, the fluctuation factor can be written in terms of the ratio (cf section 1.2.7) of the determinants for a harmonic oscillator and a free particle

$$\phi(t - t_0) \approx \frac{1}{\sqrt{2\pi i\hbar(t - t_0)/m}}\left[\frac{\det_N(-\varepsilon^2\partial^{(\varepsilon)}\bar{\partial}^{(\varepsilon)} - \varepsilon^2\omega^2)}{\det_N(-\varepsilon^2\partial^{(\varepsilon)}\bar{\partial}^{(\varepsilon)})}\right]^{-\frac{1}{2}} \qquad (2.2.81)$$

$$\xrightarrow[\varepsilon \to 0]{} \frac{1}{\sqrt{2\pi i\hbar(t - t_0)/m}}\left[\frac{\det(-\partial_t^2 - \omega^2)}{\det(-\partial_t^2)}\right]^{-\frac{1}{2}}$$

$$= \frac{1}{\sqrt{2\pi i\hbar(t - t_0)/m}}\prod_{n=1}^{\infty}\left[\frac{\omega_n^2 - \omega^2}{\omega_n^2}\right]^{-\frac{1}{2}}$$

$$= \frac{1}{\sqrt{2\pi i\hbar/m}}\sqrt{\frac{\omega(t - t_0)}{\sin\omega(t - t_0)}}. \qquad (2.2.82)$$

168   *Path integrals in quantum mechanics*

◇ **The harmonic oscillator with time-dependent frequency: calculation by the Gelfand–Yaglom method**

Now let us consider a slight generalization of the oscillator problem in terms of path integrals, namely, the harmonic oscillator with *time-dependent frequency* $\omega(t)$. The associated fluctuation action is still quadratic:

$$S_{\text{fl}}[X] = \int_{t_0}^{t} dt\, \frac{m}{2}(\dot{X}^2 - \omega^2(t)X^2) \tag{2.2.83}$$

and the fluctuation factor is

$$\phi(t, t_0) = \int_{\mathcal{C}\{0,t;0,t_0\}} \mathcal{D}_{d\tau} X(\tau) \exp\left\{\frac{i}{\hbar} S_{\text{fl}}[X]\right\}. \tag{2.2.84}$$

Since $\omega(t)$ is not, in general, translationally invariant in time, the fluctuation factor now depends on both the initial and final time and not only on their difference. The time-sliced approximation for the fluctuation factor is similar to (2.2.81):

$$\phi^{(N)}(t, t_0) = \frac{1}{\sqrt{2\pi i\hbar(t - t_0)/m}} \left[\frac{\det_N(-\varepsilon^2 \partial^{(\varepsilon)} \bar{\partial}^{(\varepsilon)} - \varepsilon^2 \hat{\omega}^2)}{\det_N(-\varepsilon^2 \partial^{(\varepsilon)} \bar{\partial}^{(\varepsilon)})}\right]^{-\frac{1}{2}}. \tag{2.2.85}$$

Here ω^2 denotes the diagonal matrix

$$\omega^2 = \begin{pmatrix} \omega_N^2 & & \\ & \ddots & \\ & & \omega_1^2 \end{pmatrix} \tag{2.2.86}$$

with the matrix elements $\omega_j^2 = \omega^2(t_j)$.

The calculation of the determinant for the matrix $\partial^{(\varepsilon)}\bar{\partial}^{(\varepsilon)} - \varepsilon^2\omega^2$ is quite a difficult problem both in the discrete and continuous cases. But we can reduce the calculation to the solution of the difference or differential equation by the *Gelfand–Yaglom method* (Gelfand and Yaglom 1960) which we have already used in section 1.1.4 (cf example 1.4). The determinant (in the time-sliced approximation) we have to calculate is (cf (1.1.104) and (1.1.106))

$$D^{(N)} \equiv \det_N(-\varepsilon^2 \bar{\partial}^{(\varepsilon)} \partial^{(\varepsilon)} - \varepsilon^2\omega^2) \tag{2.2.87}$$

$$\equiv \begin{vmatrix} 2 - \varepsilon^2 \omega_N^2 & -1 & 0 & 0 & \cdots & \cdots & 0 \\ -1 & 2 - \varepsilon^2 \omega_{N-1}^2 & -1 & 0 & \cdots & \cdots & 0 \\ 0 & -1 & 2 - \varepsilon^2 \omega_{N-2}^2 & -1 & 0 & \cdots & 0 \\ \vdots & & & \ddots & & & \vdots \\ 0 & \cdots & 0 & -1 & 2 - \varepsilon^2 \omega_2^2 & -1 & 0 \\ 0 & \cdots & \cdots & 0 & -1 & 2 - \varepsilon^2 \omega_1^2 & -1 \end{vmatrix}.$$

Proceeding in the same way as in section 1.1.4, we can obtain for $D^{(N)}$ the difference equation

$$(\partial_N^{(\varepsilon)} \bar{\partial}_N^{(\varepsilon)} + \omega_{N+1}^2) D^{(N)} = 0 \tag{2.2.88}$$

or, in a more explicit form,

$$\frac{D^{(N+1)} - 2D^{(N)} + D^{(N-1)}}{\varepsilon^2} + \omega_{N+1}^2 D^{(N)} = 0. \tag{2.2.89}$$

The initial conditions are
$$D^{(1)} = (2 - \varepsilon^2\omega_1^2)$$
$$D^{(2)} = (2 - \varepsilon^2\omega_1^2)(2 - \varepsilon^2\omega_2^2) - 1. \tag{2.2.90}$$

In the continuum limit, the difference equation transforms into the differential equation
$$(\partial_t^2 + \omega^2(t))\widetilde{D}(t) = 0 \tag{2.2.91}$$

for the renormalized function
$$\widetilde{D}(t_j) \stackrel{\text{def}}{\equiv} \varepsilon D_j \tag{2.2.92}$$

with the initial conditions
$$\widetilde{D}(t_0) = 0 \qquad \dot{\widetilde{D}}(t_0) = 1. \tag{2.2.93}$$

The sought ratio of the determinants is equal to the value of $\widetilde{D}(\tau)$ at t:
$$\frac{\det(-\partial_t^2 - \omega^2(t))}{\det(-\partial_t^2)} = \widetilde{D}(t). \tag{2.2.94}$$

Of course, equations (2.2.91) and (2.2.93) cannot be solved for the general time-dependent frequency $\omega^2(t)$.

2.2.3 Semiclassical (WKB) approximation in quantum mechanics and the stationary-phase method

In chapter 1, we learned that in the special case of a small diffusion constant, the stochastic behaviour of a Brownian particle becomes very close to the deterministic behaviour governed by the laws of the ordinary classical mechanics. As we have mentioned in section 2.1.1, in some special situations, a quantum system also behaves similarly to its classical analog. In fact, we have already dealt with such a situation in section 2.1.4, when deriving the semiclassical Bohr–Sommerfeld quantization conditions from path integrals. In that case, we were interested in the characteristics of a system at higher eigenstates (large eigenvalues) and found that a (classical) potential function essentially defines the properties of the corresponding quantum system. This reflects, in particular, the well-known physical fact: if an electron in an atom is highly excited, its wave packet encircles the nucleus in almost the same way as a point particle in classical mechanics. The regular method for calculation of quantum-mechanical amplitudes by expanding them around classical expressions in powers of contributions from quantum fluctuations is known as the *semiclassical*, *eikonal* or *Wentzel–Kramers–Brillouin* (WKB) *approximation* (see, e.g., Davydov (1976) and Maslov and Fedoriuk (1982)); recall that section 2.1.4 contains a short collection of basic facts on this approximation).

As usual, we start and, as a matter of fact, confine ourselves to the one-dimensional case. The multi-dimensional variant of the WKB approximation is essentially more complicated from the technical point of view and we shall discuss only the results and make comments on the qualitative distinctions from the one-dimensional case.

◇ **Stationary-phase approximation for finite-dimensional integrals**

Assume that the typical fluctuations of the action $S[x(\tau)]$ in the time evolution amplitude
$$K(x,t|x_0,t_0) = \int \mathcal{D}_{d\tau} x(\tau) e^{iS[x(\tau)]/\hbar} \tag{2.2.95}$$

are large compared to the Planck constant $\hbar$. If we consider an analogous finite-dimensional integral of the type

$$\int dx\, e^{if(x)/\hbar} \tag{2.2.96}$$

the *Riemann–Lebesgue lemma* (see, e.g., Reed and Simon (1975)) guarantees that for a good enough function $f(x)$, it converges to zero for $\hbar \to 0$. The heuristic reason for this is that in the latter limit, the integral becomes a sum of rapidly oscillating terms canceling each other. For small but non-zero $\hbar$, the evaluation of (2.2.96) proceeds by using the *stationary-phase approximation*. Quite often, this method is also called the *saddle-point approximation*. Canonically, the latter name is attributed to the calculation at $\lambda \to \infty$ of the integrals of the form

$$\int_C dz\, e^{\lambda f(z)} \qquad z \in \mathbb{C}$$

where $f(z)$ is a holomorphic function on the *complex* plane $\mathbb{C}$ and C is some contour. This method implies the deformation of the contour C to a new one, passing through the stationary point (saddle point) of $f(x)$ in the direction of the steepest descent of $\mathrm{Re}\, f(z)$ (that is why the method is also called the *steepest descent method*). Since we are interested in the integrals (2.2.96), where x and $f(x)$ are real quantities, we prefer to refrain from using the last two names and to use the first of the three. For small $\hbar$, the phase $e^{if(x)/\hbar}$ in (2.2.96) varies rapidly, except the point (or points) x_c where

$$\left.\frac{df(x)}{dx}\right|_{x=x_c} = 0.$$

Thus the Riemann–Lebesgue lemma implies that the dominant contribution to the integral comes from regions of x where $f'(x)$ vanishes and it is reasonable to expand $f(x)$ about the point of stationarity, x_c (we suppose, for simplicity, that $f(x)$ has only one such point):

$$f(x) = f(x_c) + \frac{1}{2}(x - x_c)^2 f''(x_c) + \frac{1}{3!}(x - x_c)^3 f'''(x_c) + \cdots. \tag{2.2.97}$$

Neglecting the contributions from the cubic and higher terms, the integral (2.2.96) can be evaluated using the complex Gaussian formula (2.1.77)

$$\int dx\, e^{if(x)/\hbar} \approx \sqrt{\frac{2\pi i\hbar}{f''(x_c)}}\, e^{if(x_c)/\hbar}. \tag{2.2.98}$$

This result is called the *stationary-phase approximation*. It can be shown that the higher terms in the expansion (2.2.97) give higher-order $\hbar$-terms in the expansion of the whole integral. Indeed, denoting $X = x - x_c$, we have

$$\exp\left\{\frac{i}{\hbar}\left[\frac{1}{3!}\frac{d^3 f(x_c)}{dx^3}X^3 + \frac{1}{4!}\frac{d^4 f(x_c)}{dx^4}X^4 + \cdots\right]\right\}$$

$$= 1 + \frac{i}{\hbar}\left[\frac{1}{3!}\frac{d^3 f(x_c)}{dx^3}X^3 + \frac{1}{4!}\frac{d^4 f(x_c)}{dx^4}X^4 + \cdots\right]$$

$$+ \left(\frac{i}{\hbar}\right)^2\left[\frac{1}{2\cdot 3!}\left(\frac{d^3 f(x_c)}{dx^3}\right)^2 X^6 + \cdots\right] + \cdots. \tag{2.2.99}$$

Differentiation of the basic Gaussian integral over the parameter a gives

$$\int_{-\infty}^{\infty} dx\, x^n e^{iax^2/\hbar} = \begin{cases} \sqrt{2\pi i\hbar} \dfrac{(n-1)!!}{a^{(n+1)/2}} (i\hbar)^{n/2} & \text{if } n \text{ is even} \\ 0 & \text{if } n \text{ is odd.} \end{cases} \quad (2.2.100)$$

Using this result, we can see that the corrections due to the cubic, fourth-order and higher terms contain increasing powers of $\hbar$ which is supposed to be small. Thus the stationary-phase expansion can be used for small $\hbar$ to calculate the integral with increasing accuracy. However, the resulting series is, in general, *divergent* and proves to be only an *asymptotic series*. Recall that a formal series

$$\sum_{j=0}^{\infty} c_j x^j \quad (2.2.101)$$

is said to be an *asymptotic series* for the function $f(x)$ at $x \to 0$, if for each N

$$\frac{\left| f(x) - \sum_{j=0}^{N} c_j x^j \right|}{x^N} \xrightarrow[x \to 0]{} 0. \quad (2.2.102)$$

To distinguish clearly the asymptotic series from the more familiar *convergent series*, let us compare their basic properties:

Convergent series	Asymptotic series
For any δ and x, there is N such that $\left\| f(x) - \sum_{j=0}^{K} c_j x^j \right\| < \delta$ for all $K > N$	For any δ and N, there is y such that $\left\| f(x) - \sum_{j=0}^{N} c_j x^j \right\| < \delta$ for all $x < y$

Thus, when approximating a function by an asymptotic series, we should restrict the approximation to a reasonable number of terms, depending on the value of the (small) parameter of the expansion (for further details see textbooks on asymptotic analysis, e.g., Jeffreys (1962)). To obtain some experience in asymptotic expansion, the reader is invited to consider the integral

$$g(\lambda) \stackrel{\text{def}}{\equiv} \int_0^{\infty} dx\, e^{-x^2 - \lambda x^4}$$

which can be approximated by partial sums of the asymptotic series

$$\sum_{n=0}^{\infty} \frac{(-1)^n}{n!} \lambda^n \int_0^{\infty} dx\, x^{4n} e^{-x^2}$$

(see problem 2.2.6, page 192).

◇ **Generalization of the stationary-phase approximation to path integrals**

Now we return to the path integral (2.2.95) and, with the experience from the finite-dimensional case, assume that the dominant contribution is produced by the domain with the smallest oscillations, i.e. around the extremum of the action

$$\delta S[x(\tau)] = 0 \quad (2.2.103)$$

with the solution $x_c(\tau)$ being the classical (extremal) trajectory of the particle. Let $X(\tau)$ describe fluctuations around the classical trajectory:

$$x(\tau) = x_c(\tau) + X(\tau). \tag{2.2.104}$$

Then the expansion of the action reads as

$$S[x] = S[x_c] + \int_{t_0}^{t} d\tau \left.\frac{\delta S}{\delta X(\tau)}\right|_{x=x_c} X(\tau) + \frac{1}{2!}\int_{t_0}^{t} d\tau\, d\tau' \left.\frac{\delta^2 S}{\delta x(\tau)\delta x(\tau')}\right|_{x=x_c} X(\tau)X(\tau')$$

$$+ \frac{1}{3!}\int_{t_0}^{t} d\tau\, d\tau'\, d\tau'' \left.\frac{\delta^3 S}{\delta x(\tau)\delta x(\tau')\delta x(\tau'')}\right|_{x=x_c} X(\tau)X(\tau')X(\tau'') + \cdots \tag{2.2.105}$$

(recall that the linear term is absent due to the extremality condition (2.2.103)). For a point particle with the action (2.1.99), (2.1.100), we have

$$\frac{1}{2!}\int_{t_0}^{t} d\tau\, d\tau' \left.\frac{\delta^2 S}{\delta x(\tau)\delta x(\tau')}\right|_{x=x_c} X(\tau)X(\tau') = \int_{t_0}^{t} d\tau \left[\frac{m}{2}\dot{X}^2 + \frac{1}{2}\frac{d^2 V(x_c)}{dx^2}X^2\right] \tag{2.2.106}$$

and the time evolution amplitude in the leading WKB approximation (i.e. if we include only quadratic terms in the expansion (2.2.105)) becomes

$$\langle x, t|x_0, t_0\rangle = K(x, t|x_0, t_0)$$
$$= \Phi(x, x_c; t, t_0) e^{iS[x_c]/\hbar} \tag{2.2.107}$$

where (cf preceding subsection)

$$\Phi(x, x_c; t, t_0) = \int \mathcal{D}_{d\tau} X(\tau) \exp\left\{\frac{i}{\hbar}\int_{t_0}^{t} d\tau \frac{m}{2}(\dot{X}^2 - \omega^2(\tau)X^2)\right\}$$

$$= \frac{1}{\sqrt{2\pi i\hbar(t-t_0)/m}} \sqrt{\frac{\det(-\partial_\tau^2)}{\det(-\partial_\tau^2 - \omega^2(\tau))}} \tag{2.2.108}$$

$$\omega^2(\tau) = \frac{1}{m}\left.\frac{\partial^2 V(x)}{\partial x^2}\right|_{x=x_c}. \tag{2.2.109}$$

This is the fluctuation factor for a harmonic oscillator with time-dependent frequency and it is natural to solve it by the Gelfand–Yaglom method which we have discussed in chapter 1 and in the preceding subsection. The distinction from the general case is that the frequency (2.2.109) itself depends on the classical trajectory and this induces the dependence of the fluctuation factor (2.2.108) on the initial x_0 and final x positions of the particle. On the other hand, the specific form (2.2.109) of the frequency allows us to present the general result of the Gelfand–Yaglom method in a more convenient form.

For the special case (2.2.109), the solution of the Gelfand–Yaglom equation (2.2.91), (2.2.93) can be written as follows (problem 2.2.7, page 193):

$$\widetilde{D} = \dot{x}_c(t)\dot{x}_c(t_0)\int_{t_0}^{t}\frac{d\tau}{\dot{x}_c^2(\tau)}. \tag{2.2.110}$$

This is the required ratio of the determinants in (2.2.108). But (2.2.110) can still be transformed into a more convenient form.

First, we note that due to the energy conservation law

$$E = \frac{m}{2}\dot{x}_c^2 + V(x_c) = \text{constant}$$

the time dependence of the classical orbit $x_c(\tau)$ is given by

$$t - t_0 = \int_{x_0}^{x_f} dx \, \frac{m}{\sqrt{2m(E - V(x))}} \equiv \int_{x_0}^{x_f} dx \, \frac{m}{p(x)} \qquad (2.2.111)$$

where we have again used the local momentum $p(x) = \sqrt{2m(E - V(x))}$ (cf (2.1.119)) and denoted the final point by x_f. This equation defines the energy as a function of the initial x_0 and final x_f positions and time:

$$E = E(x_f, x_0; t - t_0). \qquad (2.2.112)$$

The integral in (2.2.110) can be expressed through this energy as follows:

$$\int_{t_0}^{t} \frac{d\tau}{\dot{x}_c^2(\tau)} = m^2 \int_{t_0}^{t} \frac{d\tau}{p^2(\tau)} = m^3 \int_{t_0}^{t} \frac{\dot{x} d\tau}{p^3(\tau)}$$

$$= m^3 \int_{x_0}^{x_f} \frac{dx}{p^3} = m^2 \int_{x_0}^{x_f} dx \, \frac{1}{p^2} \frac{\partial p}{\partial E}$$

$$= -m \frac{\partial t}{\partial E} = -m \left(\frac{\partial E}{\partial t}\right)^{-1}. \qquad (2.2.113)$$

To obtain the last line, we have used (2.2.111). The time derivative of the energy can be expressed, in turn, via the spatial derivatives of the classical action (problem 2.2.8, page 193):

$$\frac{\partial E}{\partial t} = \dot{x}(t)\dot{x}(t_0) \frac{\partial}{\partial x_f} \frac{\partial}{\partial x_0} S(x_f, x_0; t - t_0) \qquad (2.2.114)$$

where

$$S(x_f, x_0; t - t_0) \stackrel{\text{def}}{\equiv} S[x_c(t)]. \qquad (2.2.115)$$

The equalities (2.2.110), (2.2.113) and (2.2.114) yield the compact formula

$$\widetilde{D} = m[-\partial_{x_f} \partial_{x_0} S(x_f, x_0; t - t_0)]^{-1}. \qquad (2.2.116)$$

Thus the fluctuation factor (2.2.108) in the WKB approximation receives the form

$$\Phi(x, x_c; t - t_0) = \frac{1}{\sqrt{2\pi i\hbar}} \sqrt{-\partial_x \partial_{x_0} S(x, x_0; t - t_0)}. \qquad (2.2.117)$$

In a d-dimensional space, the fluctuation factors are expressed via the determinant

$$\mathcal{V}_{\text{VPM}} \stackrel{\text{def}}{\equiv} \det\left[-\frac{\partial}{\partial x_f^i} \frac{\partial}{\partial x_0^j} S(x_f, x_0; t - t_0)\right] \qquad (2.2.118)$$

$$i, j = 1, \ldots, d.$$

This determinant of a $d \times d$ matrix is called the *Van Vleck–Pauli–Morette (VPM) determinant*.

In the same way as in the case of ordinary integrals, it is possible to calculate higher corrections in $\hbar$. To this aim we must expand the exponential of third and higher terms in $X(\tau)$ and calculate the corresponding path integrals (in analogy with (2.2.99) and (2.2.100)). Of course, a rigorous control of errors in the expansion is very difficult in the functional case.

◇ Fluctuation factor and geometry of classical trajectories: Maslov–Morse index and Morse theorem

We conclude this subsection with some important remarks about the phase factor of the semiclassical amplitude and the complications arising in the multidimensional case.

As we have learned in the preceding subsection, the fluctuation factor ϕ_{osc} for a harmonic oscillator acquires an additional phase when part of the Fourier modes used for its calculation correspond to negative eigenvalues. The general formula (2.2.117) with the VPM determinant reproduces the expression for ϕ_{osc} as follows:

$$S[x_c] \equiv S(x, x_0; t - t_0)$$
$$= \frac{m\omega}{2\sin\omega(t-t_0)}[(x^2 + x_0^2)\cos\omega(t-t_0) - 2xx_0]$$
$$-\partial_x \partial_{x_0} S = \frac{m\omega}{\sin[\omega(t-t_0)]} \qquad (2.2.119)$$

and the combination of (2.2.119) and (2.2.117) again gives (2.2.76). However, we have noted that the singular points of (2.2.119) at $(t - t_0) = \pi k/\omega$, $k = 1, 2, \ldots$ correspond to the appearance of new negative eigenvalues in the Fourier mode expansions and, hence, an additional phase factor $e^{-i\pi/2}$. In the formula (2.2.76), this phase is hidden due to the implied $i\eta$-prescription, i.e. due to an infinitesimal shift in the frequency ω to the complex plane, $\omega \to \omega - i\eta$, $\eta > 0$.

Analogous situation with negative eigenvalues takes place in the WKB path-integral calculations of the evolution amplitude for any system. The expression for the fluctuation factor of a d-dimensional system with an explicit phase factor looks as follows:

$$\Phi(x, x_c; t - t_0) = \frac{e^{i\pi\nu/2}}{(2\pi i\hbar)^{d/2}} \det\left[-\frac{\partial}{\partial x^i}\frac{\partial}{\partial x_0^j} S(x_f, x_0; t - t_0)\right]. \qquad (2.2.120)$$

Here ν is the *Maslov–Morse index* which states how often the VPM determinant passes through singularities along the orbit. This happens at the so-called *conjugate points* which are the generalization of the turning points in one-dimensional systems.

We shall show, first, that the fluctuation factor acquires indeed the phase $(-\pi/2)$ when the determinant passes through a singularity and then comment on the Maslov–Morse index and its connection with the singularities.

Let us use again the genuine Gelfand–Yaglom expression for the fluctuation factor, writing it in the form

$$\Phi(t) = \frac{c(t)}{|\widetilde{D}(t)|^{1/2}}$$

where $\widetilde{D}(t)$ is the solution of (2.2.91) and (2.2.93) (for brevity, we have chosen $t_0 = 0$ and dropped the indication of dependence on x, x_0). The quantity $c(t)$ is constant for $\widetilde{D}(t) \neq 0$ but may change as t passes through zeros of $\widetilde{D}(t)$ (and, hence, through singularities of the VPM determinant, cf (2.2.116)). Since we are interested in the WKB approximation, we can restrict our attention to quadratic Lagrangians with time-dependent frequencies. Note that equation (2.2.91) for $\widetilde{D}(t)$ coincides with the classical equation of motion for such Lagrangians. Hence, taking into account the boundary conditions (2.2.93), the classical trajectory from the origin at $\tau = 0$ to a point x at a time $\tau = t$ can be written in the form

$$x_c(\tau) = x\frac{\widetilde{D}(\tau)}{\widetilde{D}(t)}$$

the classical action reads as

$$S[x_c(\tau)] = \frac{1}{2} \frac{mx^2}{\widetilde{D}(t)} \frac{d\widetilde{D}(\tau)}{d\tau}\bigg|_{\tau=t}$$

and the corresponding evolution amplitude is

$$K(x,t|0,0) = \frac{c(t)}{|\widetilde{D}(t)|^{1/2}} \exp\left\{\frac{imx^2}{2\hbar\widetilde{D}(t)} \frac{d\widetilde{D}(\tau)}{d\tau}\bigg|_{\tau=t}\right\}. \quad (2.2.121)$$

Let $\widetilde{D}(\tau)$ have a simple zero at $\tau = \tau_1$. Substitution of (2.2.121) in the composition law (2.1.71) for propagators

$$K(0,\tau_1+\varepsilon|0,0) = \int_{-\infty}^{\infty} dx \, K(0,\tau_1+\varepsilon|x,\tau_1-\delta) K(x,\tau_1-\delta|0,0)$$

where ε and δ are small, and the Gaussian integration yield

$$\frac{c(\tau_1+\varepsilon)}{|\widetilde{D}(\tau_1+\varepsilon)|^{1/2}} = \frac{c(\tau_1-\delta)}{|\widetilde{D}(\tau_1-\delta)|^{1/2}} \left[1 + \frac{(\varepsilon+\delta)}{\widetilde{D}(\tau_1-\delta)} \frac{d\widetilde{D}(\tau)}{d\tau}\bigg|_{\tau=(\tau_1-\delta)}\right]^{-1/2}.$$

In the limit $\varepsilon, \delta \to 0$, from this relation together with the assumption that $\widetilde{D}(\tau)$ has a simple zero at τ_1, we obtain the required statement:

$$c(\tau_1+0) = c(\tau_1-0) e^{-i\pi/2}. \quad (2.2.122)$$

Detailed analysis shows that if $\widetilde{D}(\tau)$ has zeros of higher order $k > 1$, the phase factor in (2.2.122) becomes $e^{-i\pi k/2}$, so that the total phase in (2.2.120) is the sum of orders k_i of all zeros of $\widetilde{D}(\tau)$ (or singularities of the VPM determinant):

$$\nu = \sum_{\substack{\text{all}\\\text{zeros}}} k_i.$$

This number can be related to properties of the extremal trajectories of a system, using the notion of an *index* of a bilinear functional and the *Morse theorem*.

By definition, the *index* of a bilinear functional is the dimension of the space on which it is negative definite. In order to illustrate this definition, recall that the fluctuation factor $\phi(t)$ is expressed via the determinant of the quadratic variational derivatives:

$$\frac{\delta^2 S[x_c(\tau)]}{\delta x^i(\tau')\delta x^j(\tau'')}.$$

From the consideration of the harmonic oscillator in the preceding subsection, we know that each negative eigenvalue of this matrix of derivatives produces an additional phase $(-\pi/2)$. Thus the number ν in (2.2.120) is equal to the index of $\delta^2 S$ (considered as a bilinear functional in the space of fluctuations).

Now we want to relate the index to the notion of *conjugate points* in the d-dimensional space where a particle moves. To this aim, consider the families $x(\tau; \boldsymbol{p})$ of extremal paths leaving the initial point x_0 at the time t_0 without specification of the endpoint but defined by their corresponding momenta $\boldsymbol{p}$. Define the so-called Jacobi field as

$$J_{ik}(t,\boldsymbol{p}) = \frac{\partial x^i(\tau;\boldsymbol{p})}{\partial p^k} \quad (2.2.123)$$

which measures the deviations of trajectories with different momenta. For example, in one dimension and with momenta close to each other, we have

$$x(\tau, p + \delta p) - x(\tau, p) = \delta J(\tau, p) + \mathcal{O}((\delta p)^2).$$

The points τ^* where $\det J_{ik}(\tau^*, \boldsymbol{p}) = 0$ are called *conjugate* to $x(t_0)$ or *focal points*. In one dimension, the trajectories merely reconverge at focal points; in the d-dimensional case ($d > 1$), the d-vectors $\boldsymbol{J}_k$ with the components J_{ik} (k is considered to label vectors, i labels components of the vectors) no longer span the d-dimensional tangent space $\mathbb{R}^d$ at $\tau = \tau^*$.

The Jacobi field is closely related to the determinant $\widetilde{D}(\tau)$ and is nothing but the inverse of the VPM matrix:

$$\frac{\partial^2 S[\boldsymbol{x}_c(\tau)]}{\partial x_f^i \partial x_0^j}.$$

In particular, in the one-dimensional case, we have

$$J = \left[\frac{\partial p}{\partial x_f}\right]^{-1} = \left[\frac{\partial^2 S}{\partial x_f \partial x_0}\right]^{-1}.$$

Finally, we present, without proof, the *Morse theorem* (see Morse (1973)):

Theorem 2.2 (Morse). *Let $\boldsymbol{x}_c(\tau)|_{\tau=t_0}^t$ be an extremum of an action S. The index of $\delta^2 S$ is equal to the number of conjugate points to $x(t_0)$ along the extremal curve $\boldsymbol{x}_c(\tau)|_{\tau=t_0}^t$. Each such conjugate point is counted with its multiplicity.*

The surfaces in the x-space, on which the Jacobi field is degenerate (has zero determinant), are called *caustics*. The conjugate points are, in fact, the places where the trajectories touch the caustics.

Thus, the Morse theorem establishes the direct relation between the phase factor of semiclassical transition amplitudes and the geometrical properties of the corresponding classical trajectories.

In the next subsection, we shall continue to study the relations between the characteristics of quantum systems and the behaviour of classical trajectories.

2.2.4 Derivation of the Bohr–Sommerfeld condition via the phase-space path integral, periodic orbit theory and quantization of systems with chaotic classical dynamics

The prime goal of this subsection is to discuss a relation between the eigenvalues of a quantum Hamiltonian (energy levels) and the *periodic orbits* in the classical phase space of the system. As a by-product, we shall rederive the *Bohr–Sommerfeld condition*, obtained in the framework of the Feynman path integral in section 2.1.4, and this time we shall prove it on the basis of the phase-space path integral. In the last part of the subsection, we shall briefly discuss the peculiarities of quantum systems with chaotic behaviour in the classical limit.

◇ Path-integral representation for the trace of the Green function for the stationary Schrödinger equation

Following Gutzwiller (1990 and references therein), a connection between the periodic orbits and the quantum properties of a system can be derived from the stationary-phase evaluation of the path integral. The most convenient object to start with is the trace of the Green operator for the stationary Schrödinger equation.

Consider the stationary problem, the time-independent Schrödinger equation (2.1.12) with the Hamiltonian of the general form (2.1.56). The corresponding *resolvent* of the Hamiltonian operator is defined as follows:

$$\widehat{G}(E) = \lim_{\epsilon \to 0} \frac{1}{E - \widehat{H} + i\epsilon}. \qquad (2.2.124)$$

In the basis of the energy eigenfunctions (2.1.12), this operator is also called the *Green function of the stationary Schrödinger equation* $G(x, x_0; E)$:

$$G(x, x'; E) = \lim_{\epsilon \to 0} \sum_n \frac{\psi_n(x)\psi_n^*(x')}{E - E_n + i\epsilon} \qquad (2.2.125)$$

(the addition of $i\epsilon$ defines the rule of bypassing the poles). Once the Green function has been obtained, we can find the eigenvalues E_n, the spectral density $\rho(E)$ of the energy levels and the eigenfunctions $\psi_n(x)$. In fact,

$$\mathcal{G}(E) \stackrel{\text{def}}{\equiv} \text{Tr}\,\widehat{G}(E) = \int dx\, G(x, x; E) = \lim_{\epsilon \to 0} \sum_n \frac{1}{E - E_n + i\epsilon} \qquad (2.2.126)$$

so that the eigenvalues are given by the poles of $\mathcal{G}(E)$:

$$\rho(E) \equiv \sum_n \delta(E - E_n) = -\frac{1}{\pi} \text{Im}(\text{Tr}\,\widehat{G}(E)) \qquad (2.2.127)$$

$$\psi_n(x)\psi_n^*(x') = \text{Res}\{G(x, x'; E = E_n)\}$$

($\text{Res}\{G(x, x'; E = E_n)\}$ means a residue at the pole $E = E_n$). Thus, even the trace $\mathcal{G}(E)$ of the resolvent operator contains absolutely essential information about quantum systems and we are going to evaluate it by the stationary-phase method.

As explained at the end of section 2.1.1 (see (2.1.81)), the energy-dependent Green function (2.2.125) is nothing but the Fourier transform of the propagator $K(x, x'; t)$ (the $i\epsilon$-rule of the pole bypassing is related to the $i\eta$-prescription for making the Fourier transform (2.1.80) convergent; see the remark after (2.1.80)). Thus in order to derive the semiclassical expression for $\mathcal{G}(E)$, we have to carry out the following steps:

(i) calculate a semiclassical approximation to the propagator;
(ii) perform the Fourier transform to find the energy-dependent Green function;
(iii) take the trace.

The path-integral representation for $\mathcal{G}(E) \equiv \text{Tr}\,\widehat{G}$ follows from (2.1.81) and the path-integral expression for the diagonal elements $K(x, x; t - t_0)$ of the propagator $K(x, x'; t - t_0)$:

$$\begin{aligned}\mathcal{G}(E) &= \frac{i}{\hbar} \int_0^T dT\, \exp\left\{\frac{i}{\hbar}ET\right\} \widetilde{\mathcal{G}}(T) \\ &= \frac{i}{\hbar} \int_0^T dT\, \exp\left\{\frac{i}{\hbar}ET\right\} \int dx_0\, K(x_0, x_0; T) \\ &= \frac{i}{\hbar} \int_0^T dT\, \exp\left\{\frac{i}{\hbar}ET\right\} \int dx_0 \int_{\mathcal{C}\{x_0, t_0; x_0, t\}} \mathcal{D}x(\tau) \exp\left\{\frac{i}{\hbar}S[x(\tau)]\right\} \end{aligned} \qquad (2.2.128)$$

(here $T = t - t_0$). Then, using the WKB approximation discussed in the preceding subsection (cf equations (2.2.103)–(2.2.117)), we find

$$\widetilde{\mathcal{G}}(T) = \int dx_0\, K(x_0, x_0; T)$$

178   *Path integrals in quantum mechanics*

$$= \sum_\alpha \int dx_0 \, \Phi^{(\alpha)}(x_0, x_0; T) \exp\left\{\frac{i}{\hbar} S[x_c^{(\alpha)}(\tau); x_0, x_0; T]\right\} \tag{2.2.129}$$

where we have taken into account that there may be more than one classical solution (orbit) for a given T satisfying the periodic boundary condition and distinguished them by the label α; we assume these solutions are sufficiently separated and just sum over all of them. The quantity $\Phi^{(\alpha)}(x_0, x_0; T)$ is the fluctuation factor (see (2.2.117) for its explicit form) for each solution and $S[x_c^{(\alpha)}(\tau); x_0, x_0; T]$ is the action for the αth classical solution.

◇ **Stationary-phase evaluation of the integral over initial points (calculation of the trace): appearance of periodic orbits**

Let us apply the stationary-phase approximation again, this time to the x_0-integration in (2.2.129) (of course, this is an ordinary, not a *path* integration). The stationary exponent must satisfy

$$\begin{aligned} 0 &= \frac{\partial}{\partial x_0} S[x_c^{(\alpha)}(\tau); x_0, x_0; T] \\ &= \left(\frac{\partial}{\partial x_2} S[x_c^{(\alpha)}(\tau); x_2, x_1; T] + \frac{\partial}{\partial x_1} S[x_c^{(\alpha)}(\tau); x_2, x_1; T]\right)\bigg|_{x_1=x_2=x_0} \\ &= -p_1 + p_2. \end{aligned} \tag{2.2.130}$$

This last equality, where p_1 and p_2 are the momenta at the beginning and the end of the path, follows from classical mechanics (see, e.g., ter Haar (1971)). Equation (2.2.130) shows that the path must not only have the same position x_0 at its two ends, but also the same momentum, i.e. the path must be a *periodic orbit*. In the phase space of a system, such periodic orbits correspond to tori, see figure 2.2. Thus (2.2.129) reduces to a sum over all periodic orbits. Of course, the basic period (per cycle) of the orbit need not be T. It could be T/n, where n is any integer characterizing the number of traverses of the same basic orbit. For a given T, $\mathcal{G}(T)$ will receive contributions from all periodic orbits whose basic period is an integral fraction of T. Assuming that the values of T are in a one-to-one correspondence with tori in the phase space, we replace the sum over the label α in (2.2.129) by the sum over n:

$$\widetilde{\mathcal{G}}(T) = \sum_n \phi^{(n)}(T/n) \exp\left\{\frac{i}{\hbar} n S[x_c(\tau); T/n]\right\} \tag{2.2.131}$$

where the factor $\phi^{(n)}(T/n)$ has come from the integration over x_0 in the stationary-phase approximation (analog of the fluctuation factor Φ) and its calculation yields (problem 2.2.10, page 194):

$$\phi^{(n)}(T/n) = \frac{T}{n} \left(\frac{-i}{2\pi\hbar}\right)^{1/2} \left(-\frac{dE_{cl}}{dT}\right)^{1/2} e^{-\pi\hbar} \tag{2.2.132}$$

where E_{cl} is the energy of the periodic orbit with period T/n. Thus we finally obtain

$$\begin{aligned} \mathcal{G}(E) &= \frac{i}{\hbar}\sum_n \int_0^\infty dT \, \exp\left\{\frac{i}{\hbar}(ET + nS[x_c(\tau); T/n])\right\} \frac{T}{n}\left(\frac{-i}{2\pi\hbar}\right)^{1/2}\left(-\frac{dE_{cl}}{dT}\right)^{1/2} e^{-\pi\hbar} \\ &= \frac{i}{\hbar}\left(\frac{-i}{2\pi\hbar}\right)^{1/2} \sum_n (-1)^n \int_0^\infty ds \, s\sqrt{n}\left(-\frac{dE_{cl}}{dT}\right)^{1/2} \exp\left\{n\frac{i}{\hbar}(Es + S[x_c(\tau); s])\right\} \end{aligned} \tag{2.2.133}$$

where $s = T/n$ is the basic period per cycle.

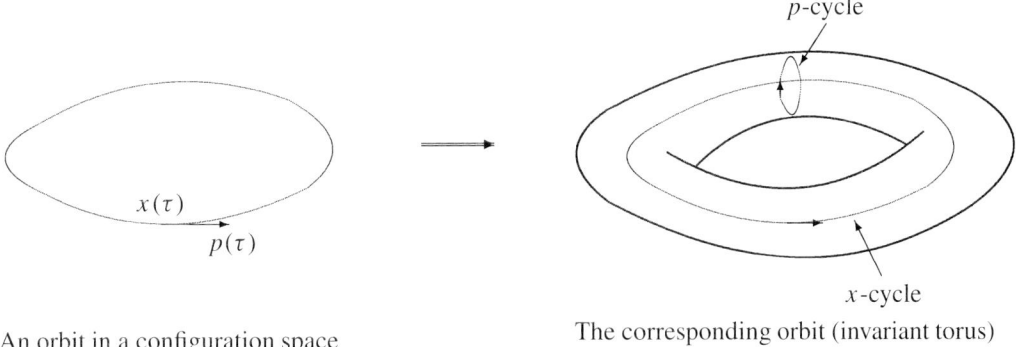

Figure 2.2. A periodic orbit in the configuration space of a dynamical system and its counterpart (invariant torus) in the phase space of the system

◇ **Fourier transform via the stationary-phase approximation method**

The integration over s in (2.2.133) is performed, once again, by the stationary-phase approximation method. The stationary-phase point now satisfies the condition:

$$E = -\frac{\partial S[x_c(\tau); s]}{\partial s} = E_{cl} \qquad (2.2.134)$$

where the last equality is the standard relation in classical mechanics. On the other hand, the variable E is the argument of the function $\mathcal{G}(E)$. Thus for a given E, the stationary-phase approximation picks out the periodic classical orbit with $E_{cl} = E$. We still assume that for any s, there is only one orbit with the basic period s, so that this period, as well as the action S, are fixed by the value of the energy and the set of orbits is parametrized by this basic period $s = s(E)$ or, equivalently, by the energy $E_{cl}(s)$.

A simple calculation of the integral in (2.2.133) by the stationary-phase approximation yields

$$\mathcal{G}(E) = \frac{i}{\hbar}\left(\frac{-i}{2\pi\hbar}\right)^{1/2}\sum_n(-1)^n\left(\frac{2\pi\hbar}{-in}\right)^{1/2}\left(\frac{\partial^2 S[x_c]}{\partial s^2}\right)^{-1/2} s(E)\sqrt{n}$$
$$\times \left(-\frac{dE_{cl}}{ds}\right)^{1/2} \exp\left\{n\frac{i}{\hbar}(Es(E) + S[x_c(\tau); s])\right\}. \qquad (2.2.135)$$

Making use of the relation

$$\frac{\partial^2 S[x_c(\tau); s]}{\partial s^2} = -\frac{\partial E}{\partial s}$$

(obtained by differentiation of (2.2.134)), the expression (2.2.135) for the trace of the resolvent can be cast into the form:

$$\mathcal{G}(E) = \frac{i}{\hbar}\sum_{n=1}^{\infty}(-1)^n s(E)\exp\left\{n\frac{i}{\hbar}I(E)\right\}$$
$$= -\frac{is(E)}{\hbar}\frac{\exp\{iI(E)/\hbar\}}{1 + \exp\{iI(E)/\hbar\}} \qquad (2.2.136)$$

where

$$I(E) \stackrel{\text{def}}{\equiv} S[s(E)] + Es(E) \tag{2.2.137}$$

is the Legendre transformation of S.

More generally, we can consider the trace of the Green function multiplied by some observable $\widehat{A}$: $\mathcal{G}_A(E) \stackrel{\text{def}}{\equiv} \operatorname{Tr} \widehat{G}\widehat{A}$ and evaluate it by the same stationary-phase method based on consideration of periodic orbits (Eckhardt *et al* 1992). By using the energy eigenvectors $|\psi_n\rangle$, this more general trace takes the form

$$\mathcal{G}_A(E) \stackrel{\text{def}}{\equiv} \operatorname{Tr} \widehat{G}(E)\widehat{A} = \lim_{\epsilon \to 0} \sum_n \frac{\langle \psi_n|\widehat{A}|\psi_n\rangle}{E - E_n + i\epsilon} \tag{2.2.138}$$

so that now

$$\rho_A(E) = -\frac{1}{\pi} \operatorname{Im} \mathcal{G}_A(E) = \sum_n \langle \psi_n|\widehat{A}|\psi_n\rangle \delta(E - E_n). \tag{2.2.139}$$

Thus $\operatorname{Tr} \widehat{G}(E)\widehat{A}$ has poles at the quantum eigenvalues, with residues given by the *matrix elements* of the chosen operator $\widehat{A}$.

◇ **Energy levels and the Bohr–Sommerfeld condition**

The relation (2.2.136) shows that $\mathcal{G}(E)$ has poles at the points $E = E_m$ which satisfy

$$I(E) = (2m+1)\pi\hbar \tag{2.2.140}$$

where m is an integer. The residue at each pole is unity, since $dI(E)/dE = s$. Thus the energy levels are given in the stationary-phase approximation (that is, the path-integral variant of the WKB approximation) by E_m which satisfy (2.2.140). This is just the Bohr–Sommerfeld quantization condition which we have already discussed in section 2.1.4 because

$$I(E) = S[s(E)] + Es(E) = \int_0^s d\tau \left(\frac{m}{2}\dot{x}^2 - V + E\right)$$

$$= m\int_0^s d\tau\, \dot{x}^2 = 2\int_{x_1}^{x_2} p\, dx \tag{2.2.141}$$

(x_1, x_2 are the turning points of the classical orbit). In particular, for the harmonic oscillator, the calculation of the last integral in (2.2.141) with $p = \sqrt{2m(E - V(x))} = \sqrt{m(2E - m\omega^2 x^2)}$ together with the use of condition (2.2.140) give the exact values of the energy levels: $E_m = \hbar\omega(m + 1/2)$; see section 2.1.4 and problem 2.1.10, page 153.

Note that if the system is put in a box with periodic boundary conditions, then all energy levels become discrete, and all orbits bounded. Thus the previous method can be used for all levels of the system (even if initially it contained a continuous spectrum) and later the size of the box can tend to infinity.

We would like to stress again that in this derivation it has been assumed that there is only one continuous family of periodic orbits, parametrized by the basic period or energy. In particular, this helped us to replace $\sum_\alpha \int dT$ by $\sum_n n \int ds$. For relatively simple single-well potentials in one dimension, this assumption is satisfied. But even in the one-dimensional space, if the potential had several wells near different minima, there may be several families of orbits, i.e. different orbits with the same basic periods and energy. In higher-dimensional spaces, such a situation is even more plausible. If these families of orbits are sufficiently separated, then the stationary-phase approximation gives an additive contribution to $\mathcal{G}(E)$ from each family. An energy level appears if a member of any of the families satisfies the condition (2.2.140) (up to this approximation). For further details on the WKB approximation and the periodic orbit theory, see Gutzwiller (1990) and references therein.

◇ Generalization to systems with classically chaotic dynamics

In classical mechanics, a trajectory $p(t), x(t)$ in the phase space is called a *chaotic* one if its maximal *Lyapunov exponent* λ is positive. The Lyapunov exponent is defined as

$$\lambda \stackrel{\text{def}}{\equiv} \lim_{t \to \infty} \frac{1}{t} \ln |\boldsymbol{\omega}(t)| \qquad (2.2.142)$$

where $\boldsymbol{\omega}(t)$ is a tangent vector to $p(t), x(t)$ with the condition that $|\boldsymbol{\omega}(0)| = 1$. The exponential instability of the chaotic trajectories implies a continuous frequency spectrum of (classical) motion. The continuous spectrum, in turn, implies correlation decay; this property, which is called *mixing* in ergodic theory, is the most important property of dynamical motion for the validity of a statistical description (see e.g., Lichtenberg and Lieberman (1983)).

While the meaning of classical chaos is beyond question, there is no universally accepted definition of *quantum chaos* (see e.g., de Almeida (1988), Gutzwiller (1990) and Nakamura (1993)). The problem of a definition for quantum chaos arose because the previously mentioned condition of continuous spectrum for classical chaos is violated in quantum mechanics. Indeed, the energy and the frequency spectrum of any quantum motion, bounded in the phase space, are always discrete due to the non-commutativity of the phase-space variables. According to the theory of dynamical systems, such motion corresponds to the limiting case of regular motion. This means that there is no classical-like chaos at all in quantum mechanics. In order to approach an understanding of the properties of quantum systems whose classical motion exhibits chaos, it is natural to study them on the *semiclassical* level. Such a study can be based on *periodic orbit theory* (Gutzwiller (1990) and references therein), basic elements of which for regular (integrable) systems we have presented earlier. The idea is to connect the behaviour of the eigenvalues and eigenfunctions of a quantum system to the structure of the phase space of the corresponding classical system *both* in the *regular* and *chaotic* region. One can again look for a correspondence between the semiclassical 'irregular' spectra and chaotic orbits in non-integrable systems. Noting that even in the absence of invariant tori (the existence of which is the characteristic property of *integrable* non-chaotic systems), the x_0-integration, as in (2.2.129), for the calculation of the trace of the resolvent operator (in the stationary-phase approximation) leads to the claim that the initial and final momenta of the system must be equal to each other (cf (2.2.130)). This, in turn, implies a patching of all (now unstable) periodic orbits in the corresponding path integral.

Calculations in the case of chaotic systems in higher-dimensional spaces and for a more general trace of the form $\mathcal{G}_A \stackrel{\text{def}}{\equiv} \operatorname{Tr} \widehat{G}\widehat{A}$, for some operator $\widehat{A}$, follow essentially the same way as that outlined earlier for the particular case of a one-dimensional space and $\widehat{A} = 1$ but technically they are much more complicated. Therefore, we present only the final result (see Gutzwiller (1990), Eckhardt and Wintgen (1990) and references therein). The contribution from the periodic orbits to the trace can be written in the following general form:

$$\mathcal{G}_{A,\text{period}} = \frac{-i}{\hbar} \sum_\alpha A_\alpha \sum_{n=1}^{\infty} \frac{\exp\left\{n(\frac{i}{\hbar} S_{\text{cl}} - \frac{i}{2}\mu_\alpha \pi)\right\}}{|\det(M_\alpha^n - 1)|^{1/2}} \qquad (2.2.143)$$

where A_α is the integral of the classical counterpart $A(p, x)$ of the operator $\widehat{A}$ along the orbit:

$$A_\alpha = \int_0^{T_\alpha} dt\, A(\boldsymbol{q}_\alpha(t), \boldsymbol{p}_\alpha(t)) \qquad (2.2.144)$$

the phase shift ν_α is the *Maslov–Morse index* of the periodic orbit (cf preceding subsection, equation (2.2.120)) and M_α is the stability matrix around the orbit. The latter is a linearized *Poincaré map*, describing the time evolution of the transverse displacement from the orbit α; M_α is expressed in terms of the Lyapunov exponent λ_α of the orbit. The latter depends on the type of fixed point of the orbit (e.g., for so-called monoclinic orbits with hyperbolic fixed points, $\det(M_\alpha^n - 1) = 4\sinh^2(n\lambda_\alpha/2)$).

It is necessary to note that contributions to the trace $\text{Tr}\,\widehat{G}\widehat{A}$ actually come from two sources: from the periodic paths which give (2.2.143) and from the very short paths, where the propagator turns into a delta function. To evaluate the latter part, we use a Taylor series expansion of the trajectory in powers of the time t and exploits the smallness of t in evaluating the integrals. This then gives a smoothly varying contribution $\mathcal{G}_{A,0}$ to $\mathcal{G}_A(E)$:

$$\mathcal{G}_{A,0} = \int \frac{d^N p\, d^N q}{(2\pi\hbar)^N} \delta(E - H(\boldsymbol{p},\boldsymbol{q})) A(\boldsymbol{p},\boldsymbol{q}) \qquad (2.2.145)$$

($A(\boldsymbol{p},\boldsymbol{q})$ is the classical counterpart of the operator $\widehat{A}$; N is the number of degrees of freedom), i.e. the average of the observable over the energy shell (Berry and Mount (1972), Voros and Grammaticos (1979) and references therein).

Thus periodic orbit theory allows us to establish a correspondence between the classical properties (orbits) of a system and its quantum spectrum (at semiclassical level). Among other factors, this opens the possibility to approach a general definition of quantum chaos. The point is that to discriminate classically chaotic systems at the quantum level, the spectral statistics of the energy levels are of great importance. In particular, it has been found that the spectral statistics of systems with underlying classical chaotic behaviour agree with the predictions of a *Gaussian ensemble*, whereas quantum analogs of classically integrable systems display the characteristics of *Poisson statistics*. The most used spectral statistical characteristics of the energy levels are $P(s)$ and $\Delta_3(L)$. $P(s)$ is the distribution of nearest-neighbour spacings $s_i = (E_{i+1} - E_i)$ of the unfolded levels E_i. This is obtained by accumulating the number of spacings that lie within the bin $(s, s + \Delta s)$ and then normalizing $P(s)$ to unity. For quantum systems whose classical analogs are integrable (regular), $P(s)$ is expected to follow the Poisson distribution

$$P(s) = \exp(-s). \qquad (2.2.146)$$

On the other hand, quantum analogs of chaotic systems exhibit the spectral properties of the Gaussian ensemble (Berry and Tabor 1977):

$$P(s) = \frac{\pi}{2} s \exp\left(-\frac{\pi}{4} s^2\right) \qquad (2.2.147)$$

(for systems with time-reversal symmetry) or

$$P(s) = (32/\pi^2) s^2 \exp(-4s^2/\pi)$$

(for systems without time-reversal symmetry).

The statistic $\Delta_3(L)$ is defined, for a fixed interval $(-L/2, L/2)$, as the least-square deviation of the staircase function $N(E)$ from the best straight line fitting it:

$$\Delta_3(L) = \frac{1}{L} \min_{A,B} \int_{-L/2}^{L/2} [N(E) - AE - B]^2\, dE$$

where $N(E)$ is the number of levels between E and zero for positive energy and between $-E$ and zero for negative energy. For this statistic the Poissonian prediction is

$$\Delta_3(L) = \frac{L}{15} \qquad (2.2.148)$$

while the Gaussian ensemble predicts the same behaviour for $L \ll 1$, but for $L \gg 1$ it gives

$$\Delta_3(L) = \frac{1}{\pi^2} \log L. \qquad (2.2.149)$$

Thus the spectra of regular and (classically) chaotic systems indeed exhibit quite different behaviour. Since the trace $\mathcal{G}(E)$ of the resolvent contains the information about energy levels, its calculation by the periodic orbits is very helpful for discriminating between quantum chaotic and integrable systems.

From (2.2.143) we might understand that the semiclassical quantum eigenvalues are constructed through complicated interference between a set of periodic orbits. When applied to integrable systems, (2.2.143) recovers the Bohr–Sommerfeld quantization conditions. In chaotic systems, however, no invariant torus exists and α covers arbitrarily long periodic orbits. The number of orbits with a period less than $T \gg 1$ proves to be $N(T) \approx \exp\{CT\}$ ($C > 0$ is some positive constant), i.e. exponentially proliferating. In the trace formula (2.2.143), the real part of terms with a given period $T \gg 1$ is of the order $\sim T \exp\{-CT/2\}$. Multiplying this factor by $N(T)$ gives a contribution $\exp\{CT/2\}$, which brings about a serious problem of non-convergence in (2.2.143). Thus, periodic orbits are much too numerous to provide a one-to-one connection between individual paths and quantum eigenvalues and, mathematically, this is reflected in the inherent divergence of the formal Gutzwiller trace formula (2.2.143). Methods to overcome such divergences have been developed in the context of general dynamical systems, where invariant sets can be characterized by their periodic points. The key observation is that classical periodic orbits are strictly organized, both topologically and metrically, and that this organization can be exploited to rewrite ill-behaved sums over periodic orbits in a convergent form. Further details on periodic orbit theory and its applications to the analysis of chaotic systems can be found, e.g., in Gutzwiller (1990) and Nakamura (1993).

2.2.5 Particles in a magnetic field: the Ito integral, midpoint prescription and gauge invariance

In this subsection, we shall consider the construction of the path integral for a particle moving in a magnetic field and approach the important problem of ambiguity in discrete approximations of path integrals. As we shall see in the next section, where this problem is considered in a detailed manner, this ambiguity is a reflection of a general lack of uniqueness in the quantization procedure for a wide class of classical systems. In the path-integral approach, this ambiguity originates in the peculiar properties of stochastic integrals, in particular the *Ito integral* (see problem 1.2.17, page 120). In the usual operator approach, this reveals itself in the known *operator ordering problem* (see the next section).

◇ **Lagrangian of a particle in a magnetic field and its discrete approximation**

The Lagrangian for a charged particle interacting with a magnetic field $\boldsymbol{B}$, derivable from a vector potential $\boldsymbol{A}$, i.e.

$$\boldsymbol{B} = \nabla \times \boldsymbol{A} \qquad (2.2.150)$$

reads as

$$L = \frac{1}{2} m \left(\frac{d\boldsymbol{x}}{d\tau}\right)^2 + \frac{e}{c} \frac{d\boldsymbol{x}}{d\tau} \boldsymbol{A} - V(\boldsymbol{x}) \qquad (2.2.151)$$

where $\boldsymbol{B}$, $\boldsymbol{A}$, $\boldsymbol{x}$ are three-dimensional vectors, c is the speed of light, e is the electric charge. In the action, the additional field-dependent term

$$\frac{e}{c} \int_{t_0}^{t} d\tau \, \frac{d\boldsymbol{x}}{d\tau} \boldsymbol{A} = \frac{e}{c} \int_{x_0}^{x} d\boldsymbol{x} \, \boldsymbol{A} \qquad (2.2.152)$$

184   *Path integrals in quantum mechanics*

in the discrete approximation takes the form

$$\frac{e}{c}\sum_{j=0}^{N}(x_{j+1}-x_j)A(\widetilde{x}_j) \qquad (2.2.153)$$

$$\widetilde{x}_j^k \in [x_j^k, x_{j+1}^k] \qquad k=1,2,3,\ j=0,\ldots,N.$$

Until now we have not taken care of the precise definition of the point at which we evaluate the potential term $\int d\tau\, V(x(\tau))$ in the discrete approximation:

$$\varepsilon \sum_{j=0}^{N} V(\widetilde{x}_j) \qquad (2.2.154)$$

for the action. The point $\widetilde{x}_j$ may be x_j or x_{j+1} or be situated somewhere in between, say at $(x_j+x_{j+1})/2$. This carelessness is based on our experience with Riemann integrals: the essential point in their definition is that they do not depend on the choice of the points $\widetilde{y}$ in the Darboux sum

$$\int_a^b dy\, f(y) = \lim_{N\to\infty}\sum_{j=0}^{N}(y_{j+1}-y_j)f(\widetilde{y}) \qquad \widetilde{y}\in[y_j,y_{j+1}].$$

However, we must remember that $x(\tau)$ in (2.2.151) and (2.2.153) and in all analogous formulae is not an ordinary variable: in the Wiener integral, it would be a *stochastic* Brownian process and the corresponding stochastic integrals have very peculiar properties. We learned this, in particular, from the example of the *Ito integral* (problem 1.2.17, page 120).

Nevertheless, all the formulae we have previously derived are correct! As we shall see very soon, the reason for this is the ε-factor in (2.2.154), due to which any possible difference, caused by different choices of the points $\widetilde{x}_j$, disappears. This is not the case for the term (2.2.153) describing the interaction with a magnetic field.

◇ Discrete-time approximation for a path integral with a potential term and a magnetic field

To check that the V-term is insensitive to the choice of $\widetilde{x}_j$ and to make a correct choice for the A-term, we follow again the Feynman way (Feynman 1948) and compare the evolution of a state vector defined by the path integral and by the Schrödinger equation. In fact, we must repeat the calculations of section 2.1.2 (equations (2.1.92)–(2.1.98)) with an A-term, but for the present purpose it is convenient to do this in a slightly different form. The techniques which we consider here will be useful for a number of problems we shall deal with later in this book.

Using the short-time propagator (2.1.94), we can write the infinitesimal evolution $\psi(t)\to\psi(t+\varepsilon)$ of a state vector as follows:

$$\psi(x,t+\varepsilon) = \int d^3y\,\left(\frac{m}{2\pi i\hbar\varepsilon}\right)^{3/2}\exp\left\{\frac{i}{\hbar}\varepsilon\left[\frac{m}{2}\left(\frac{x-y}{\varepsilon}\right)^2 - V(x+\lambda'(y-x))\right]\right.$$
$$\left. -\frac{i}{\hbar}\frac{e}{c}(y-x)A(x+\lambda(y-x))\right\}\psi(y,t). \qquad (2.2.155)$$

The parameters λ,λ' ($0\leq\lambda,\lambda'\leq 1$) define the choice of $\widetilde{x}_j$. Now let us introduce, for convenience, the new integration variable

$$u\equiv(y-x) \qquad (2.2.156)$$

and expand the exponent and $\psi(y,t)$ in (2.2.155) about $u = 0$:

$$\psi(x, t+\varepsilon) = \exp\left\{-\frac{i\varepsilon V(x)}{\hbar}\right\} \left(\frac{m}{2\pi i\hbar\varepsilon}\right)^{3/2} \int d^3u \, \exp\left\{\frac{imu^2}{2\varepsilon\hbar}\right\}$$
$$\times \exp\left\{-\frac{i}{\hbar}\varepsilon\lambda' u \nabla V(x) - \frac{ie}{\hbar c}\left[uA(x) + \lambda \sum_{m=1}^{3} u_m(u\nabla)A_m(x)\right] + \cdots\right\}$$
$$\times \left[\psi(x,t) + u\nabla\psi + \frac{1}{2}\sum_{m,n=1}^{3} u_m u_n \frac{\partial^2 \psi}{\partial x_m \partial x_n} + \cdots\right]. \quad (2.2.157)$$

The integral in (2.2.157) can be considered as a type of averaging and, denoting for brevity

$$\langle\!\langle f(u) \rangle\!\rangle \stackrel{\text{def}}{\equiv} \left(\frac{m}{2\pi i\hbar\varepsilon}\right)^{3/2} \int d^3u \, \exp\left\{\frac{imu^2}{2\varepsilon\hbar}\right\} f(u)$$

we obtain with the aid of the Gaussian integrals (cf supplement I, volume II):

$$\begin{aligned}
\langle\!\langle 1 \rangle\!\rangle &\sim \mathcal{O}(1) \\
\langle\!\langle u_m \rangle\!\rangle &= 0 \qquad m = 1, 2, 3 \\
\langle\!\langle u_m u_n \rangle\!\rangle &\sim \mathcal{O}(\varepsilon) \\
\langle\!\langle u_{m_1} u_{m_2} \cdots u_{m_{2k+1}} \rangle\!\rangle &= 0 \\
\langle\!\langle u_{m_1} u_{m_2} \cdots u_{m_{2k}} \rangle\!\rangle &\sim \mathcal{O}(\varepsilon^k) \qquad k = 1, 2, \ldots.
\end{aligned} \quad (2.2.158)$$

To derive the Schrödinger equation, we need the expression for the right-hand side of (2.2.157) up to the first order in ε. Relations (2.2.158) show that the term with the parameter λ' and the derivative of the potential disappear in this approximation. This is the reason for the insensitivity of the potential terms to the choice of the points $\widetilde{x}_j$ when they are evaluated in the discrete-time approximation. Thus, all our previous results indeed prove to be correct. In contrast to this, the parameter λ survives in the first-order approximation and equation (2.2.157) takes the form

$$\psi(x, t+\varepsilon) \approx \psi(x,t) - \frac{i}{\hbar}\varepsilon V(x)\psi(x,t) + \lambda\frac{\varepsilon e}{mc}\psi(x,t)\nabla A(x)$$
$$- \frac{i\varepsilon e^2}{2\hbar mc}A^2(x)\psi(x,t) + \frac{i\varepsilon\hbar}{2m}\nabla^2\psi(x,t) + \frac{\varepsilon e}{mc}(A(x)\nabla)\psi(x,t). \quad (2.2.159)$$

In the continuum limit $\varepsilon \to 0$, equation (2.2.159) becomes a Schrödinger-like equation:

$$i\hbar\frac{\partial\psi}{\partial t} = -\frac{\hbar^2}{2m}\nabla^2\psi + \frac{i\hbar e}{mc}(A(x)\nabla)\psi + \lambda\frac{i\hbar e}{mc}\psi\nabla A + \frac{e^2}{2mc^2}A^2\psi + V\psi. \quad (2.2.160)$$

Recall, however, that the *correct* Schrödinger equation for a particle in a magnetic field reads as

$$i\hbar\frac{\partial\psi}{\partial t} = \widehat{H}_A \psi \quad (2.2.161)$$

$$\widehat{H}_A \stackrel{\text{def}}{\equiv} \frac{1}{2m}\left(\widehat{p} - \frac{e}{c}\widehat{A}\right)^2 + \widehat{V} \quad (2.2.162)$$

where

$$\widehat{p} = -i\hbar\nabla. \quad (2.2.163)$$

The reader may easily check that (2.2.160) coincides with (2.2.161) only for the value $\lambda = 1/2$. The latter corresponds to the so-called *midpoint prescription*, i.e. evaluation of $A(x)$ in the discrete approximation at the points

$$\widetilde{x}_j = \tfrac{1}{2}(x_j + x_{j+1}). \tag{2.2.164}$$

◇ **The midpoint prescription, gauge invariance and operator ordering problems**

The specific form (2.2.162) of the Hamiltonian for a particle in a magnetic field is strongly dictated by the requirement of *gauge invariance*. The point is that a physical magnetic field $B = \nabla \times A$ proves to be the same for a class of potentials A related by the *gauge transformations*

$$A(x) \rightarrow A'(x) = A(x) + \nabla \alpha(x) \tag{2.2.165}$$

where $\alpha(x)$ is an arbitrary scalar function. This implies that any physical quantity, in particular, the energy operator (Hamiltonian), must be insensitive to the transformations (2.2.165). The special combination

$$\left(\widehat{p} - \frac{e}{c}\widehat{A}\right)\psi \tag{2.2.166}$$

which appeared in (2.2.161) is indeed invariant with respect to (2.2.165), provided the state vector ψ is also subject to the corresponding gauge transformation:

$$\psi(x) \rightarrow \psi'(x) = e^{-ie\alpha(x)/\hbar c}\psi(x) \tag{2.2.167}$$

(recall that the overall phase of state vectors has no physical meaning; we can detect only the *relative* phase of different vectors). The generalization of this principle of gauge invariance plays a very important role in the construction of modern realistic quantum field theory of fundamental interactions which we shall discuss in the context of the path-integral approach in chapter 3.

Thus the correct choice (midpoint prescription) of points $\widetilde{x}_j$ at which a magnetic field is evaluated in the discrete approximation is crucial for reconstructing the correct Schrödinger equation (2.2.161) and its physical properties (in particular, gauge invariance). This means, in turn, that the integral (2.2.152) is not a Riemann but a stochastic one. If we chose $\lambda = 0$, it would be the *Ito integral* (see problem 1.2.17, page 120). But this choice does not lead to the gauge-invariant result. Let us see how the requirement of gauge invariance distinguishes between different stochastic integrals. In the continuum limit, the path integral for the evolution amplitudes with a magnetic field is

$$K(x, t|x_0, t_0) = \int_{\mathcal{C}\{x_0, t_0; x, t\}} \mathcal{D}_{d\tau} x(\tau) \exp\left\{\frac{i}{\hbar}\int_{t_0}^{t} d\tau \left[\frac{m}{2}\dot{x}^2 - V(x) + \frac{e}{c}\dot{x}A\right]\right\}. \tag{2.2.168}$$

After the gauge transformation (2.2.165), the exponent acquires the additional term

$$\frac{ie}{\hbar c}\int_{t_0}^{t} d\tau \, \dot{x}\nabla\alpha = \frac{ie}{\hbar c}\int_{x_0}^{x} dx \nabla\alpha. \tag{2.2.169}$$

If we use the midpoint prescription $\lambda = 1/2$, then, as follows from (1.2.302), the term (2.2.169) is reduced to the difference of α at the initial and final points:

$$\int_{x_0}^{x} dx \, \nabla\alpha = \alpha(x) - \alpha(x_0). \tag{2.2.170}$$

This quantity is the same for all the paths. We have thus found that the gauge transformation (2.2.165) induces the transformation on K:

$$\begin{aligned}K'(\boldsymbol{x},t|\boldsymbol{x}_0,t_0) &= \mathrm{e}^{\mathrm{i}e\alpha(\boldsymbol{x})/\hbar c}K(\boldsymbol{x},t|\boldsymbol{x}_0,t_0)\mathrm{e}^{-\mathrm{i}e\alpha(\boldsymbol{x})/\hbar c}\\ &= \mathrm{e}^{\mathrm{i}e\alpha(\boldsymbol{x})/\hbar c}\langle\psi(\boldsymbol{x},t)|\psi(\boldsymbol{x}_0,t_0)\rangle\mathrm{e}^{-\mathrm{i}e\alpha(\boldsymbol{x})/\hbar c}\end{aligned} \qquad (2.2.171)$$

which is compensated by the state vector transformation (2.2.167). If we treated (2.2.169) as the Ito integral (1.2.299), we would obtain an extra integral term

$$\int dt\,\nabla^2\alpha$$

(similar to that in (1.2.299)) and, hence, we would obtain a violation of the gauge invariance.

Does this problem with the choice of a correct discrete approximation bring something essentially new into quantum mechanics? The answer is 'no'. In other words, is there an analog of this problem in the usual operator formalism? The answer is 'yes'. In order to see this, pay attention to the fact that the ambiguity concerns the term containing the product of the function $A(\boldsymbol{x})$ and the time derivative $\dot{\boldsymbol{x}}$. In the operator approach, the latter is related to the momentum operator $\hat{\boldsymbol{p}}$ which *does not commute* with $A(\hat{\boldsymbol{x}})$. Thus, if $A_m(\boldsymbol{x})$ were arbitrary functions, we would not know which operator to put in correspondence with the classical product $\boldsymbol{p}A(\boldsymbol{x})$: for example, $\hat{\boldsymbol{p}}\hat{A}$, $\hat{A}\hat{\boldsymbol{p}}$ or something else. This is a particular case of the *operator ordering problem* in quantum mechanics. In the specific situation with a vector magnetic potential, the gauge invariance dictates the specific combination and, hence, the specific ordering rule (2.2.162). Correspondingly, in the path-integral approach, the gauge invariance distinguishes the specific midpoint prescription.

In section 2.3, we shall discuss the general operator ordering problem and its influence on the construction of path integrals.

2.2.6 Applications of path integrals to optical problems based on a formal analogy with quantum mechanics

The topic of this concluding subsection is somehow aside from the main line of this section and even of the chapter. Its main goal is to illustrate the fact that the path-integral methods developed in the framework of quantum mechanics can be applied in quite different areas of physics. Namely, we shall briefly discuss applications of the path-integral technique to some specific problems in (classical) optics.

The spacetime evolution of the electromagnetic field is governed by the Maxwell vector partial differential wave equations of second order in time and space which are of the *hyperbolic type*. At first glance, this nature of the wave equations seems to prevent the use of the path-integral formalism for their solution. However, many optical problems can be studied in the scalar approximation where a single complex scalar field $u(\boldsymbol{r})$, called the *physical–optical disturbance*, is sufficient (see, e.g., Born and Wolf (1959), Foch (1965) and Eichmann (1971)). This is true when the characteristic sizes of the problem, in particular the variation of the *index of refraction*, are small over distances of the order of the wavelength. This approximation, called *quasi-geometric optics*, results in the substitution of the vector wave equation by the d'Alembert wave equation:

$$\frac{n^2(\boldsymbol{r})}{c^2}\frac{\partial^2 u}{\partial t^2} - \nabla^2 u = 0. \qquad (2.2.172)$$

In order to transform the latter into an equation which can be analyzed by path-integral methods, we have to make the following simplifications:

(i) In order to eliminate the time dependence (together with the second-order time derivative) we assume waves of a definite frequency. In this way the d'Alembert equation is converted into the Helmholtz equation which is of second order in the space coordinates (and without time derivatives).

(ii) Many optical problems can be characterized by a well-defined direction of propagation. In this case, the propagating field is described by the so-called *paraxial wave equation* which is of first order in the coordinate along the propagation direction and of second order in the others. The paraxial wave equation strictly resembles the time-dependent Schrödinger equation and direct use of path integrals is now possible.

Let us discuss this approach in somewhat more detail.

◇ Quasi-geometric optics and path integration

The standard geometric optics of inhomogeneous media is based on the *Fermat principle*: a ray propagation is defined by the extremum of the optical *path length* S:

$$S = \int_{z_0}^{z} dz \, L(x, y; \dot{x}, \dot{y}; z) \tag{2.2.173}$$

where

$$L = n(r)(1 + \dot{x}^2 + \dot{y}^2)^{1/2} \tag{2.2.174}$$

is the optical Lagrangian and the dots represent differentiation with respect to z. The z-axis is chosen to be the direction of the propagation and the z-coordinate plays the role of 'time' while the path length S plays the role of an action functional. One can define the canonical momenta:

$$p_x = \frac{n\dot{x}}{\sqrt{1 + \dot{x}^2 + \dot{y}^2}} \qquad p_y = \frac{n\dot{y}}{\sqrt{1 + \dot{x}^2 + \dot{y}^2}} \tag{2.2.175}$$

and the corresponding 'Hamiltonian':

$$H \equiv \dot{x} p_x + \dot{y} p_y - L = -(n^2 - p_x^2 - p_y^2)^{1/2}. \tag{2.2.176}$$

Let us now establish a relation between the ray variables x, y, p_x, p_y and the physical–optical disturbance $u(r)$. For simplicity, we shall consider the case in which $u(r) \equiv u(x, z)$ is independent of y (the generalization is straightforward). If the physical–optical disturbance $u(x, 0)$ is known at the transverse plane at $z = 0$, its value at a small distance ε along the z-direction is given by the integral (Eichmann 1971):

$$u(x, \varepsilon) = \frac{1}{A} \int dx' \, u(x', 0) \exp\{-ikS(x, \dot{x}, x', \dot{x}'; \varepsilon)\} \tag{2.2.177}$$

where A is an appropriate normalization constant and $k = 2\pi/\lambda$ is the wave number in the medium. This integral can be interpreted physically as follows. At any point x' on the initial transverse plane at $z = 0$, rays will emanate in all directions. Some of these rays will terminate at the observation point x on the next transverse plane, at $z = \varepsilon$. Each ray path collected at the observation point has some finite value $S((x, \dot{x}, x', \dot{x}'; \varepsilon)$ for its optical path length. Thus all ray paths contribute to the final value $u(x, \varepsilon)$. However, for large but finite k, most of the continuum rays, with rapidly oscillating phases, add incoherently. Therefore, their contribution to the final value $u(x, \varepsilon)$ is negligible. In contrast, the classical ray path and its neighbours result in a slow phase variation, yielding a coherent contribution to the integral. Thus, the linear superposition of all classical rays collected at the observation point will give the dominant character of the physical–optical disturbance $u(x, \varepsilon)$. The coherence and dominance of the classical ray paths provide a connection between geometric and quasi-geometric optics (similar to the case of quantum and classical mechanics).

The next steps are obvious: at any arbitrary distance z, the disturbance $u(x, z)$ is obtained by iterating shifts by small distances ε along the z-direction. This results in the following path-integral representation for the physical–optical disturbance:

$$u(x, \varepsilon) = \int dx'\, K(x, x'; z) u(x', 0)$$

$$K(x, x'; z) = \mathfrak{N}^{-1} \int \mathcal{D}x(\zeta) \exp\{-ik S(x, x'; z)\} \qquad (2.2.178)$$

($\mathfrak{N}^{-1}$ is a normalization constant).

The problem now reduces to calculating the path integral (2.2.178) for the kernel function K. It is seen that, from the mathematical point view, this integral is quite similar to the quantum-mechanical one. Therefore, for its calculation, we can use the exact or approximate methods developed in the framework of quantum mechanics which we have considered in this chapter (see also problem 2.2.17, page 199). It is clear that similarly to the case of the Feynman–Kac theorem, we can prove that the optical disturbance $u(\boldsymbol{r})$ satisfies a Schrödinger-like equation (in which the z-coordinate plays the role of time), obtained by reduction of the Maxwell equations.

⋄ **Transformation of the optical equation into a Schrödinger-like one via the introduction of a fictitious timelike variable**

If there is no specific direction of ray propagation but the medium still has weakly varying index of refraction, we can start directly from equation (2.2.172) which is, in this case, a good approximation (neglecting the terms $\sim \nabla n/n$) of the Maxwell equations (Born and Wolf 1959). Considering only waves of a definite frequency $\omega = 2\pi c/\lambda = ck$ (λ is the wavelength, k is the wavevector), we can write

$$u(t, \boldsymbol{r}) = u(\boldsymbol{r}) e^{-i\omega t}$$

so that equation (2.2.172) becomes

$$k^2 n^2(\boldsymbol{r}) u + \nabla^2 u = 0. \qquad (2.2.179)$$

Now we can introduce an analog of the potential term:

$$V(\boldsymbol{r}) = E - n^2(\boldsymbol{r}) \qquad (2.2.180)$$

where E is an arbitrary constant. It is convenient to choose the *dimensionless* quantity E equal to the limiting value of $n^2(\boldsymbol{r})$ when $\boldsymbol{r}$ goes to infinity following a given direction. Then $V(\boldsymbol{r})$ behaves as a scattering potential, going to zero as $r \to \infty$. Thus equation (2.2.179) for the optical disturbance u takes the form of the stationary Schrödinger equation:

$$-\frac{1}{k^2} \nabla^2 u + V(\boldsymbol{r}) u = E u \qquad (2.2.181)$$

and the equation for its Fourier transform

$$\widetilde{u}(t, \boldsymbol{r}) = \int dE\, e^{-ikEt} u(E, \boldsymbol{r}) \qquad (2.2.182)$$

is the full analog of the Schrödinger equation. This allows us to present the solution for $\widetilde{u}(t, \boldsymbol{r})$ and, eventually, for the physical–optical disturbance $u(\boldsymbol{r})$, in terms of path integrals, similarly to the case of quantum-mechanical amplitudes. Further details of calculations and discussion of specific optical systems described in this way may be found in Garrod (1966), Gomez-Reino and Liñares (1987) and Ranfagni *et al* (1990).

◇ Remark on the application of path integrals in coupled-wave problems

Some optical problems can be described by the so-called *coupled wave theory*, that is, by a set of linear coupled differential first-order equations derived from the Maxwell equations using certain approximations (see, e.g., Solymar and Cook (1981) and Ranfagni *et al* (1990)). Some examples of phenomena which can be studied within the coupled wave theory are ultrasonic light diffraction, propagation of light in thick holographic gratings, diffraction of x-rays by crystals and fiber optics. This theory assumes that the electromagnetic field propagates as a set of distinct interacting waves. The evolution of the wave amplitudes is described by means of a continuous interchange of energy among the waves, as they propagate through the system. This interchange is mediated by the interaction of the propagating fields with the material of the device.

The system of linear differential equations derived with the coupled wave theory can be written in the following matrix form:

$$\frac{\partial A}{\partial z} = \mathsf{M} A$$

where A is the vector whose components are the N wave amplitudes and M is an $N \times N$ matrix that takes into account the interaction between the propagating waves and the material. There is a strict resemblance between this set of equations and the time-dependent Schrödinger equation:

$$i\frac{\partial}{\partial t}|\psi\rangle = \widehat{H}|\psi\rangle$$

in which the time is replaced by a spatial coordinate z, representing the direction of propagation of the electromagnetic field. This similarity again opens the possibility for using all the power of the path-integral methods (see Ranfagni *et al* (1990) and references therein).

2.2.7 Problems

Problem 2.2.1. Derive the phase-space path-integral representation for the transition amplitude $K(x, t|x_0, t_0)$ in the case of Hamiltonians of the type (2.2.4), using product formula (2.2.2) and the Taylor expansion for exponentials in (2.2.3).

Hint. Use the formulae

$$\langle x''|e^{-i\varepsilon\widehat{H}}|x'\rangle \approx \langle x''|1 - i\varepsilon\widehat{H}|x'\rangle = \langle x''|x'\rangle - i\varepsilon\langle x''|\widehat{H}|x'\rangle$$

$$\langle x''|x'\rangle = \int dp\,\langle x''|p\rangle\langle p|x'\rangle = \frac{1}{2\pi}\int dp\,e^{ip(x''-x')}$$

$$\langle x''|\widehat{H}|x'\rangle = \int dp\,\langle x''|p\rangle\langle p|\widehat{H}|x'\rangle$$

$$= \int dp\,\frac{1}{2\pi}e^{ipx''}H(p,x)\langle p|x'\rangle$$

$$= \frac{1}{2\pi}\int dp\,e^{ip(x''-x')}H(p,x)$$

to convert the product formula (2.2.2) into the form

$$\int dx_1 \cdots dx_N \frac{dp_1 \cdots dp_N}{(2\pi)^{N+1}} \exp\{ip_{N+1}(x - x_N) + ip_N(x_N - x_{N-1}) + \cdots + ip_1(x_1 - x_0)$$
$$- i\varepsilon[H(p_{N+1}, x_N) + H(p_N, x_{N-1}) + \cdots + H(p_1, x_0)]\}$$

and then into the discrete version (2.2.9) of the phase-space path integral.

Problem 2.2.2. Prove the formula(2.2.16) for a self-adjoint operator in a finite-dimensional Hilbert space $\mathcal{H}$.

Hint. Let $\{\psi_i\}_{i=1}^d$ be an orthogonal basis in the d-dimensional Hilbert space of eigenvectors of $\widehat{A}$: $\widehat{A}\psi_i = \lambda_i \psi_i$. Then, according to the definition (2.2.14), we have

$$\|\widehat{A}\| = \sup_{\psi \in \mathcal{H}} \frac{\|\widehat{A}\psi\|}{\|\psi\|}$$

$$= \sup_{c_i} \frac{\sqrt{\sum_i \lambda_i |c_i|^2}}{\sqrt{\sum_k |c_k|^2}} = \max_{k=1,\ldots,d} \{\lambda_k\} \qquad (2.2.183)$$

where $\psi = \sum_i c_i \psi_i$, $c_i \in \mathbb{C}$, $i = 1, \ldots, d$. To prove the latter equality in (2.2.183), it is convenient to use the polar d-dimensional coordinates (see, e.g., the formulae (2.5.80)).

Problem 2.2.3. Show that the multiplication operator

$$\widehat{V}(x)\psi(x) = e^{-x^2}\psi(x)$$

has a unit norm, though $\lambda = 1$ is not an eigenvalue of this operator.

Hint. Obviously, $\|\widehat{V}(x)\psi\| < \|\psi\|$ and, hence,

$$\|\widehat{V}(x)\| \leq 1. \qquad (2.2.184)$$

On the other hand, let us consider the specific set of vectors

$$\psi_R(x) = \frac{1}{2R}\theta(R - |x|) \qquad R \in \mathbb{R} \qquad (2.2.185)$$

so that $\|\psi_R\| = 1$ ($\theta(y)$ is the step-function). Using the inequality

$$e^{-x^2} > 1 - x^2$$

for $x \neq 0$ (which follows, in turn, from the inequality

$$\frac{d}{dx}[e^{-x^2} - (1 - x^2)] = 2x(1 - e^{-x^2}) > 0$$

for $x \neq 0$), we obtain

$$\frac{\|\widehat{V}(x)\psi_R\|}{\|\psi_R\|^2} = \frac{1}{2R}\int_{-R}^{R} dx\, e^{-x^2}$$

$$> \frac{1}{2R}\int dx\, (1 - x^2) = \left(1 - \frac{R^2}{3}\right).$$

As R approaches zero, the right-hand side becomes arbitrarily close to unity and, together with (2.2.184), this means that

$$\|\widehat{V}(x)\| = \sup_{\psi} \frac{\|\widehat{V}\psi\|}{\|\psi\|} = 1.$$

Note that there is no $\psi \in \mathcal{H}$, such that $\widehat{V}\psi = \psi$ and $\lambda = 1$ is not an eigenvalue of $\widehat{V}$. It is clear that, according to (2.2.185), $\lim_{R \to 0} \psi_R(x) = \delta(|x|)$, because this limit is nothing but the derivative of the θ-function. The δ-function is not square integrable and, hence, does not belong to $\mathcal{H} = \mathcal{L}^2(\mathbb{R})$.

Problem 2.2.4. Derive the transition amplitude (Green function) in the phase-space path-integral representation for fixed initial and final *momenta*, cf (2.2.22).

Hint. Follow the same steps as in the case of fixed initial and final coordinates (see section 2.2.1) but work in the momentum space representation of the Hilbert space, starting from the matrix elements of the time evolution operator $\langle p|\widehat{U}(t, t_0)|p_0\rangle$ and assuming that the states $|p\rangle$ have the normalization

$$\langle p_2|p_1\rangle = \delta(p_2 - p_1)$$

and satisfy the completeness relation

$$\int_{-\infty}^{\infty} dp\, |p\rangle\langle p| = 1.$$

Problem 2.2.5. Show that the path integral for the quantum-mechanical partition function $\mathcal{Z}$ (cf (2.2.23)–(2.2.25)) in the finite time-slice approximation has the same form in the coordinate and momentum representation of the canonical commutation relations.

Hint. Use the boundary condition $x_{N+1} = x_0$, rearrange the terms in the discrete approximation for action (e.g., in coordinate representation) appropriately and make the substitution $p_j \to p_{j-1}$.

Problem 2.2.6. Show that the function

$$g(\lambda) \stackrel{\text{def}}{\equiv} \int_0^\infty dx\, e^{-x^2 - \lambda x^4}$$

can be approximated at small λ by a few of the first terms of the series

$$g_{\text{asympt}}(\lambda) \stackrel{\text{def}}{\equiv} \sum_{n=0}^{\infty} \frac{(-1)^n}{n!} \lambda^n \int_0^\infty dx\, x^{4n} e^{-x^2}. \qquad (2.2.186)$$

Try to argue that the latter is an asymptotic series.

Hint. Use the Γ-function integral representation

$$\Gamma(z) = \int_0^\infty dt\, e^{-t} t^{z-1}$$

to calculate the coefficients of the series (2.2.186) explicitly

$$g_{\text{asympt}}(\lambda) = \sum_{n=0}^{\infty} (-1)^n \frac{\Gamma(2n + 1/2)}{2n!} \lambda^n. \qquad (2.2.187)$$

Since $\Gamma(2n + 1/2)|_{n \gg 1} \approx (2n)!$, this series is divergent. On the other hand, the exact function $g(\lambda)$ is proportional to the *parabolic cylinder function* $U(a, z)$

$$g(\lambda) = \frac{\sqrt{\pi}\, \exp\{1/(8\lambda)\}}{2(2\lambda)^{1/4}} U\left(0, \frac{1}{\sqrt{2\lambda}}\right)$$

because $U(a, z)$ has the integral representation (Abramowitz and Stegun 1965)

$$U(a,z) = \frac{\exp\{-z^2/4\}}{\Gamma(a+1/2)} \int_0^\infty ds\, e^{-sz-s^2/2} s^{a-1/2}.$$

Using a table of values of the functions $\Gamma(2n+1/2)$ and $U(0, \frac{1}{\sqrt{2\lambda}})$ we can check that a few of the first terms of (2.2.187) give a good approximation for $g(\lambda)$, if $\lambda \ll 1$. To claim that the series is asymptotic we can estimate errors for the partial sums.

Problem 2.2.7. Show that the solution of equation (2.2.91) with the boundary conditions (2.2.93) for the special case

$$\omega^2(t) = \frac{1}{m}\frac{\partial^2 V(x)}{\partial x^2}\bigg|_{x=x_c}$$

where $x_c(\tau)$ is the classical trajectory, can be written in the form

$$\widetilde{D} = \dot{x}_c(t)\dot{x}_c(t_0)\int_{t_0}^{t}\frac{d\tau}{\dot{x}_c^2(\tau)}.$$

Hint. Differentiation of the classical equation of motion:

$$\partial_t(m\ddot{x}_c + V'(x_c)) = (m\partial_t^2 + V''(x_c))\dot{x}_c = 0$$

shows that

$$D_1(\tau) = \dot{x}_c(\tau) \qquad (2.2.188)$$

is a particular solution of (2.2.91). Using the general fact that the *Wronskian*

$$W \stackrel{\text{def}}{\equiv} \begin{vmatrix} D_1 & D_2 \\ \dot{D}_1 & \dot{D}_2 \end{vmatrix} = D_1\dot{D}_2 - \dot{D}_1 D_2$$

of any two linearly independent solutions D_1, D_2 of equation (2.2.91) is constant, we find that

$$D_2(\tau) = cD_1\int^{\tau}\frac{ds}{D_1^2(s)} \qquad c = \text{constant}.$$

Imposing boundary conditions (2.2.93) on the general solution of (2.2.91)

$$D(\tau) = c_1 D_1(\tau) + c_2 D_2(\tau) \qquad c_1, c_2 \text{ constant}$$

we arrive at the desired solution.

Problem 2.2.8. Prove that on a classical trajectory, there exists the equality

$$\frac{\partial E(x_f, x_0; t-t_0)}{\partial t} = \dot{x}(t)\dot{x}(t_0)\frac{\partial}{\partial x_f}\frac{\partial}{\partial x_0}S(x_f, x_0; t-t_0).$$

Solution.

$$S[x_c] = \int_{t_0}^{t} d\tau\, (p_c(\tau)\dot{x}_c(\tau) - H(p_c, x_c))$$

$$= \int_{x_0}^{x_f} dx\, p(x) - (t-t_0)E \qquad (2.2.189)$$

(H is the Hamiltonian of a particle). Hence,

$$\frac{\partial S}{\partial x_f} = p(x_f) + \left\{ \int_{x_0}^{x_f} dx \left[\frac{\partial p(x)}{\partial E} - (t - t_0) \right] \frac{\partial E}{\partial x_f} \right\}$$

and using

$$\frac{\partial p(x)}{\partial E} = \frac{m}{p(x)} = \frac{1}{\dot{x}}$$

we get

$$\frac{\partial S}{\partial x_f} = p(x_f). \qquad (2.2.190)$$

Thus,

$$\frac{\partial}{\partial x_0} \frac{\partial}{\partial x_f} S(x_f, x_0; t - t_0) = \frac{\partial}{\partial x_0} p(x_f)$$

$$= \frac{m}{p(x_f)} \frac{\partial E(x_f, x_0; t - t_0)}{\partial x_0} \qquad (2.2.191)$$

where the latter equality follows from the definition of p: $p = \sqrt{2m(E - V(x))}$. Using again (2.2.189), we can rewrite

$$\frac{\partial E}{\partial x_0} = \frac{\partial}{\partial x_0} \left(-\frac{\partial S}{\partial t} \right) = -\frac{\partial}{\partial t} \frac{\partial S}{\partial x_0}$$

$$= \frac{\partial}{\partial t} p(x_0) = \frac{m}{p(x_0)} \frac{\partial E}{\partial t}. \qquad (2.2.192)$$

Here we have used the equality

$$\frac{\partial S}{\partial x_0} = -p(x_0)$$

which can be deduced similarly to (2.2.190). Combination of (2.2.191) and (2.2.192) gives the required equality.

Problem 2.2.9. Verify that the fluctuation factor expressed in terms of the VPM determinant as in (2.2.117) gives the correct result for a free particle and a harmonic oscillator.

Hint. In the one-dimensional case, the classical action $S[x_c(\tau)]$ for a free particle reads as

$$S_{fp}[x_c(\tau)] = \frac{m}{2} \frac{(x - x_0)^2}{t - t_0}$$

and for a harmonic oscillator:

$$S_{osc}[x_c(\tau)] = \frac{m\omega}{2\sin[\omega(t - t_0)]} [(x^2 + x_0^2) \cos \omega(t - t_0) - 2xx_0].$$

Differentiation of these expressions and substitution into (2.2.117) produce the formulae obtained independently in section 2.2.2.

Problem 2.2.10. Calculate the factor $\phi^{(n)}(T)$ in (2.2.131) which appears after the calculation in the stationary-phase approximation of the trace $\widetilde{\mathcal{G}}(T)$ of the resolvent for the stationary Schrödinger equation.

Hint. Making use of the relation (2.2.114), the fluctuation factor $\Phi^{(\alpha)}(x_0, x_0; T)$ in (2.2.129) (which we found in section 2.2.3, equation (2.2.117)) can be presented in more a convenient form for the current aim:

$$\Phi^{(\alpha)}(x_0, x_0; T) = \left(\frac{-i}{2\pi\hbar}\right)^{1/2} \left(-\frac{dE}{dT}\right)^{1/2} \frac{1}{\dot{x}(0)} \quad (2.2.193)$$

where $x(0)$ represents the initial and final points of the trajectory under consideration. But for a *periodic* orbit, $x(0)$ is an arbitrary point on it. This implies, in turn, that both $S[x_c]$ and dE/dT do not depend on $x(0)$. Thus, in fact, we have to calculate the integral:

$$2\int_{x_1}^{x_2} \frac{dx(0)}{\dot{x}(0)} = \oint \frac{dx(0)}{\dot{x}(0)} = \oint d\tau = T/n \quad (2.2.194)$$

where x_1, x_2 are the turning points of the orbit and we have taken into account that each orbit should be counted twice because of its counterpart with movement in the opposite direction. The reason for the appearance of the phase factor $\exp\{i\pi n\}$ is quite similar to that discussed in section 2.2.3 (see relation (2.2.120) and the discussion before it) and we refer the interested reader to Keller (1958) and Levit and Smilansky (1977) for further details.

Problem 2.2.11. Using gauge invariance, find the path-integral expression for the propagator of a particle in a magnetic field which is not sensitive to the choice of a discrete-time approximation for the corresponding action (i.e. insensitive to the choice of the points $\tilde{x}_j$, cf (2.2.153)).

Hint. Due to gauge invariance, we can use any vector potential from the set of vector functions related by the transformation (2.2.165) (for more on gauge-invariant theories see chapter 3). On the other hand, as we have discussed in section 2.2.5, the dependence of the propagator on the discrete approximations is related to the non-commutativity of the momentum operators and the vector potential:

$$[\hat{p}_k, \hat{A}_l(x)] = -i\hbar \partial_k A_l(x).$$

However, for the Hamiltonian H_A (2.2.162) to be independent for any order of the operators $\hat{p}, \hat{A}$, it is enough to require the field A to satisfy the *Coulomb gauge condition*:

$$\nabla A = 0 \quad (2.2.195)$$

which can be achieved by the transformation (2.2.165) with the appropriate function $\alpha(x)$. If the vector potential is subject to the Coulomb gauge condition (2.2.195), the dangerous λ-dependent term in (2.2.160) disappears and any discrete approximation gives the same result.

Problem 2.2.12. Write the expression for the propagator of a particle in a magnetic field in terms of the *phase-space* path integral. Using the result of the preceding problem 2.2.11, calculate explicitly the propagator for constant homogeneous magnetic field $B = $ constant.

Hint. Let us choose the coordinate basis so that $B = \nabla \times A$ points along the z-direction. Such a field can be described by the vector potential:

$$A_x = A_z = 0 \qquad A_y = Bx. \quad (2.2.196)$$

This vector potential satisfies the Coulomb gauge condition $\nabla A = 0$. The phase-space action reads

$$S[p, x] = \int_{t_0}^{t} d\tau \left[p\dot{x} - \frac{1}{2m}\left(p - \frac{e}{c}A\right)^2 \right]$$

196   Path integrals in quantum mechanics

The transition amplitude (propagator) is

$$\langle x, t | x_0, t_0 \rangle_{\text{magn.f.}} = \int \mathcal{D}x(\tau) \frac{\mathcal{D}p(\tau)}{(2\pi\hbar)^3} e^{\frac{i}{\hbar}S}$$

$$= \lim_{N \to \infty} \prod_{j=1}^{N} \int dx_j \, dy_j \, dz_j \prod_{k=1}^{N+1} \int \frac{d^3 p_k}{(2\pi\hbar)^3} \exp\left\{\frac{i}{\hbar} S^{(N)}\right\} \quad (2.2.197)$$

where $S^{(N)}$ is the discrete-time action

$$S^{(N)} = \sum_{j=1}^{N+1} \left\{ \mathbf{p}_j (\mathbf{x}_j - \mathbf{x}_{j-1}) - \frac{\varepsilon}{2m} \left[p_{xj}^2 + \left(p_{yj} - \frac{e}{c} B x_j\right)^2 + p_{zj}^2 \right] \right\}.$$

Recall that, according to the discussion in the preceding problem 2.2.11, page 195, we can use any discrete approximation (not only the midpoint prescription) because $\mathbf{A}(\mathbf{x})$ satisfies the Coulomb condition. Integration over the y_j, z_j variables in (2.2.197) produces the δ-functions

$$(2\pi\hbar)^{2N} [\delta(p_{y(N+1)} - p_{yN}) \delta(p_{z(N+1)} - p_{zN}) \cdots \delta(p_{y2} - p_{y1}) \delta(p_{z2} - p_{z1})].$$

Here we have used the Fourier representation for the δ-functions:

$$\delta(x) = \frac{1}{2\pi} \int_{-\infty}^{\infty} dy \, e^{ixy}. \quad (2.2.198)$$

By performing the p_{yj}, p_{zj} integrations for $j = 1, \ldots, N$, the transition amplitude can be presented in the form

$$\langle x, t | x_0, t_0 \rangle_{\text{magn.f.}} = \int_{-\infty}^{\infty} \frac{dp_y \, dp_z}{(2\pi\hbar)^2} \prod_{j=1}^{N} \int_{-\infty}^{\infty} dx_j \prod_{k=1}^{N+1} \int_{-\infty}^{\infty} \frac{dp_{xk}}{2\pi\hbar}$$

$$\times \exp\left\{\frac{i}{\hbar} \left[p_y(y - y_0) + p_z(z - z_0) - (t - t_0) \frac{p_z^2}{2m} \right]\right\} \exp\left\{\frac{i}{\hbar} S_{\text{red}}^{(N)}\right\}$$

where $p_y \equiv p_{y(N+1)}$, $p_z \equiv p_{z(N+1)}$ and the reduced action $S_{\text{red}}^{(N)}$ has the form of the action for a one-dimensional harmonic oscillator:

$$S_{\text{red}}^{(N)} = \sum_{j=1}^{N+1} \left[p_{xj}(x_j - x_{j-1}) - \frac{\varepsilon}{2m} p_{xj}^2 - \frac{\varepsilon m}{2} \left(\frac{eB}{mc}\right)^2 \left(x_j - \frac{cp_y}{eB}\right)^2 \right]$$

with frequency

$$\Omega = \frac{e}{mc} B$$

called the *cyclotron frequency*.

The result for a harmonic oscillator we know from section 2.2.2 (equation (2.2.77)). The remaining integrals are standard Gaussian ones and result in the final expression for the transition amplitude:

$$K_{\text{magn.f.}}(x, t | x_0, t_0) \equiv \langle x, t | x_0, t_0 \rangle_{\text{magn.f.}}$$

$$= \left(\frac{m}{2\pi i\hbar(t - t_0)}\right)^{\frac{3}{2}} \frac{\Omega(t - t_0)}{2 \sin(\Omega(t - t_0)/2)}$$

$$\times \exp\left\{\frac{i}{\hbar}\frac{m}{2}\left[\frac{(z-z_0)^2}{t-t_0} + \frac{\Omega}{2}\cot(\Omega(t-t_0)/2)((x-x_0)^2 + (y-y_0)^2)\right.\right.$$
$$\left.\left. + \Omega(x_0 y - xy_0) + \Omega(xy - x_0 y_0)\right]\right\}.$$

Another way to obtain the propagator for a particle in a constant magnetic field is to perform, first, all the integrations over the momentum variables in (2.2.197). This reduces the phase-space path integral to the Feynman configuration integral with the Lagrangian

$$L = \frac{1}{2}m\dot{x}^2 + \frac{e}{c}\dot{x}A(x)$$

(cf (2.2.151)). Then calculation of the Feynman integral can be carried out by the variational method (since the action is quadratic, the semiclassical approximation gives an exact result). The fluctuation factor in this case can be found from the ESKC-like relation (2.1.71).

Problem 2.2.13. Find the transition amplitude for a particle in a linear potential:

$$L = \tfrac{1}{2}m\dot{x}^2 + Fx$$

where F is a constant.

Hint. The transition amplitude, as usual for a quadratic Lagrangian, has the form

$$\langle x, y | x_0, t_0 \rangle_{\text{lin.pot.}} = e^{\frac{i}{\hbar}S_c[x_c(\tau)]} \int \mathcal{D}_{d\tau} X(\tau) e^{\frac{i}{\hbar}S[X(\tau)]}$$

where $S[X(\tau)]$ for a linear potential is just the free particle action

$$S[X(\tau)] = \int_{t_0}^{t} d\tau\, \tfrac{1}{2}m\dot{X}^2(\tau).$$

The classical equation of motion,

$$m\ddot{x} = F$$

has the obvious solution

$$x_c(\tau) = \frac{F}{2m}(\tau - t_0)^2 + v(\tau - t_0) + x_0$$

where v is the velocity of the particle defined by the requirement that the particle should arrive at the point x at the time t:

$$v = \frac{x - x_0}{t - t_0} - \frac{F}{2m}(t - t_0).$$

Hence,

$$S_c[x_c(\tau)] = \frac{m}{2}\frac{(x-x_0)^2}{t-t_0} + \frac{F}{2}(x+x_0)(t-t_0) - \frac{F^2}{24m}(t-t_0)^3.$$

Thus the result is

$$\langle x, t | x_0, t_0 \rangle_{\text{lin.pot.}} = \sqrt{\frac{m}{2\pi i\hbar(t-t_0)}} \exp\left\{\frac{i}{\hbar}\left[\frac{m}{2}\frac{(x-x_0)^2}{t-t_0} + \frac{F}{2}(x+x_0)(t-t_0) - \frac{F^2}{24m}(t-t_0)^3\right]\right\}.$$

198   *Path integrals in quantum mechanics*

Problem 2.2.14. Using as a guide the consideration of a driven oscillator in sections 1.2.7 and 1.2.8, calculate the evolution amplitude for a harmonic oscillator in an external field

$$S^{(j)} = \int_{t_0}^{t} d\tau \left[\frac{m}{2}(\dot{x}^2 - \omega^2 x^2) + j(\tau)x(\tau) \right] \quad (2.2.199)$$

by the method of Fourier expansion. Calculate this amplitude also by the WKB method.

Hint. Since the quite analogous calculations by the Fourier decomposition method have been considered in detail in section 1.2.8, we present as a hint only the result

$$\langle x, t | x_0, t_0 \rangle_{\text{d.osc.}} = \left[\frac{m\omega}{2\pi i \hbar \sin(\omega(t-t_0))} \right]^{\frac{1}{2}} \exp\left\{ \frac{i}{\hbar} \frac{(x^2 + x_0^2)\cos(\omega(t-t_0)) - 2xx_0}{\sin(\omega(t-t_0))} \right\}$$

$$\times \exp\left\{ -\frac{1}{m\omega} \int_{t_0}^{t} d\tau \int d\tau' \, j(\tau) j(\tau') \frac{\sin(\omega(t-\tau))\sin(\omega(\tau'-t_0))}{\sin(\omega(t-t_0))} \right\}$$

$$\times \exp\left\{ \int_{t_0}^{t} d\tau \, j(\tau) \frac{x_0 \sin(\omega(t-\tau)) + x \sin(\omega(\tau-t_0))}{\sin(\omega(t-t_0))} \right\}. \quad (2.2.200)$$

In the WKB method (which gives, of course, the exact result for this problem), the amplitude is represented in the form

$$\langle x, t | x_0, t_0 \rangle_{\text{d.osc.}} = e^{\frac{i}{\hbar} S_c^{(j)}[x_c(\tau)]} \int \mathcal{D}_{d\tau} X(\tau) \exp\left\{ \frac{i}{\hbar} \int_{t_0}^{t} d\tau \frac{m}{2}(\dot{X}^2 - \omega X^2) \right\}$$

where the path integral is the usual one for the purely harmonic oscillator and, hence, is known. Thus, the problem in this method is reduced to calculating $S_c^{(j)}[x_c(\tau)]$. First, we must solve the classical equation of motion

$$\ddot{x}_c(\tau) + \omega^2 x_c(\tau) = \frac{j(\tau)}{m}. \quad (2.2.201)$$

With the help of the Green function $G(\tau - \tau')$, satisfying the equation

$$\frac{d^2}{d\tau^2} G(\tau - \tau') + \omega^2 G(\tau - \tau') = \frac{1}{m} \delta(\tau - \tau') \quad (2.2.202)$$

the solution is written as

$$x_c(\tau) = \int_{t_0}^{t} d\tau' \, G(\tau - \tau') j(\tau').$$

Taking into account the solution for the homogeneous equation (with $j(\tau) = 0$) and that the derivative of the step-function gives the δ-function, the solution of (2.2.202) can be sought in the form

$$G(\tau - \tau') = C \sin \omega(\tau - \tau')[\theta(\tau - \tau') + a\theta(\tau - \tau')].$$

Substitution into (2.2.202) results in $a = -1$, $C = 1/(4m\omega)$. This gives the classical trajectory. Then, after integration, we find that the action $S_c^{(j)}$ and the evolution amplitude (2.2.200).

Problem 2.2.15. Consider a one-dimensional charged particle in the harmonic potential and external monochromatic electric field $E(\tau) = A \sin(\Omega \tau)$ (e.g., an ion in some crystal subject to the external field)

$$L = \frac{m}{2}(\dot{x}^2 - \omega^2 x^2) + E(\tau)x(\tau).$$

Assuming the particle initially in the ground state of the harmonic oscillator Hamiltonian (without an external electric field), find the probability p_n of the nth energy level to be occupied after the lapse of N periods of oscillation of the external field ($N = 1, 2, 3, \ldots$). Use the result of the preceding problem.

Hint. The evolution of the state is defined by the propagator (2.2.200)

$$\psi(x, NT) = \int_{-\infty}^{\infty} dx_0 \, \langle x, NT | x_0, 0 \rangle_{\text{d.osc.}} \psi_0(x_0)$$

where $T = 2\pi/\Omega$ is the period of oscillation of the electric field and ψ_0 is the ground state of the harmonic oscillator. The desired probability is found as

$$p_n = |\langle \psi_n | \psi(x, NT) \rangle|^2$$

where

$$\langle \psi_n | \psi(x, NT) \rangle = \int_{-\infty}^{\infty} dx \, \psi_n(x) \psi(x, NT)$$

is the probability amplitude of the nth energy level to be occupied and ψ_n is the eigenstate of the harmonic oscillator. Recall that (see, e.g., Landau and Lifshitz (1981))

$$\psi_n(x) = (2^n n! \sqrt{\pi} \ell_0)^{-\frac{1}{2}} H_n(x/\ell_0) \exp\left\{-\frac{1}{2} \frac{x^2}{\ell_0^2}\right\}. \qquad (2.2.203)$$

Here

$$\ell_0 \stackrel{\text{def}}{\equiv} \sqrt{\frac{\hbar}{m\omega}}$$

and H_n is the nth Hermite polynomial. As a result of the calculations, we find that the *Poisson* distribution

$$p_n = \frac{1}{n!} \lambda^n e^{-\lambda}$$

where

$$\lambda = \frac{2A^2}{m\hbar\omega^3} \left[\frac{\sin(N\pi\omega/\Omega)}{\Omega/\omega - \omega/\Omega}\right]^2$$

(A is the amplitude of the external field).

Problem 2.2.16. Using the separation of trajectories into their classical and quantum fluctuation parts, find the Green function for the harmonic oscillator with a linear term which has the Lagrangian

$$\mathcal{L}[x, \dot{x}] = \frac{m}{2}\dot{x}^2 - \frac{c(t)}{2}x^2 + b(t)x\dot{x} - e(t)x$$

assuming that the various coefficients may be time dependent.

Hint. Show that in the quantum fluctuation factor the coefficient of the linear term can be absorbed into a redefined time-dependent frequency, so that its calculation can be carried out by the standard Gelfand–Yaglom method.

Problem 2.2.17. Find the kernel function $K(x, x'; z)$ for the physical–optical disturbance $u(\boldsymbol{r})$ in the case of *paraxial* rays which satisfy the conditions

$$\dot{x} \ll 1 \qquad \dot{y} \ll 1 \qquad (2.2.204)$$

and for a *parabolically focusing medium*, i.e. with the index of refraction of the form

$$n = 1 - \tfrac{1}{2}\omega^2 x^2. \qquad (2.2.205)$$

Hint. Taking into account the paraxiality condition (2.2.204), expand the optical Lagrangian (2.2.174) in series to obtain

$$L \approx \tfrac{1}{2}\dot{x}^2 + \tfrac{1}{2}\dot{y}^2 + n$$

where all higher-order terms are neglected. Then use the results of the path-integral calculation for the quantum-mechanical harmonic oscillator. For further details and physical discussion see Eichmann (1971).

2.3 Quantization, the operator ordering problem and path integrals

We have already discussed the basic principles of the quantization procedure (section 2.1.1). To define a quantization means to establish a rule of putting in correspondence to any *classical* observables, i.e. to a function $f(p,x)$ on the phase space of a system, some *quantum* observable $\widehat{f}$, i.e. an operator in some Hilbert space $\mathcal{H}$. The very function $f(p,x)$ in this context is called a *symbol of the operator* $\widehat{f}$. For example, in the case of a system with the phase space $\mathbb{R}^{2d}$, we put in correspondence to the coordinates x_i and momenta p_j ($i,j = 1,\ldots,d$) the operators as in (2.1.29) and (2.1.30). It is important, however, that a quantization is not uniquely defined by this prescription: for instance, to the product $p_j x_j$, we might put in correspondence either the operator $\widehat{p}_j \widehat{x}_j$ or $\widehat{x}_j \widehat{p}_j$ or $(\widehat{p}_j \widehat{x}_j + \widehat{x}_j \widehat{p}_j)/2$. Thus there are essentially *different* ways of quantization. This is the so-called *operator ordering problem* in quantum mechanics.

At first glance, a path integral, dealing only with commutative objects, is insensitive to the choice of the quantization and hence, we might think that the path-integral approach is incomplete and does not reflect all the peculiarities of quantum mechanics. In the main part of the preceding sections we have, in fact, circumvented the problem, considering only Hamiltonians of the specific form $H = T(\widehat{p}) + V(\widehat{x})$, for which the operator ordering problem is not essential. However, considering a particle in a magnetic field in section 2.2.5, we have already met the problem and found that in the path-integral formalism it reveals itself in the special ways of constructing the time-sliced approximation. But there, the ordering rule was dictated by the requirement of gauge invariance. In the present section, we shall consider this topic from a general point of view and show that an accurate derivation of the phase-space path integral in the case of more complicated Hamiltonians (with terms containing products of non-commuting operators) does depend on the way of quantization (choice of operator ordering), so that the phase-space integral carries information about all the details of a given quantum system.

An adequate formalism to take into account all the peculiarities of the quantization procedure essentially uses the notion of the *symbol of an operator*, some binary operations (composition) in the set of the symbols, the relation of a symbol with the operator kernel for a given operator, etc. Therefore, in the first subsection we shall review shortly the main facts about symbols of operators. The next subsection is devoted to the derivation of the phase-space path integral in the case of Hamiltonians with the operator ordering problem and to an analog of the Wiener theorem. In the last subsection, we shall consider the so-called normal symbol for the evolution operator, its relation to what is termed the *coherent state path integral*, the perturbation expansion and path-integral representation for the scattering operator.

2.3.1 Symbols of operators and quantization

The correspondence between a quantum-mechanical operator and a function on the classical phase space of a system, called the *symbol* of this operator, must be subject to some requirements. First,

Quantization, the operator ordering problem and path integrals

this correspondence is supposed to be linear. Second, it must satisfy the fundamental *correspondence principle*:

(i) the mapping $f \to \widehat{f}$ depends on some parameter (usually, but not necessarily, this parameter is the Planck constant, so in what follows we shall for simplicity denote it as $\hbar$);
(ii) in some sense, which we shall clarify soon,

$$f(p, x) = \lim_{\hbar \to 0} \widehat{f}.$$

To clarify these conditions, let us note that if a quantization is defined, there is a bilinear operation $\star : f_1, f_2 \to f_1 \star f_2$ (the so-called *star-operation* or *star-product*) in the set of symbols which copies the operator product: if f, f_1, f_2 are the symbols of the operators $\widehat{f}, \widehat{f}_1, \widehat{f}_2$ and $\widehat{f} = \widehat{f}_1 \widehat{f}_2$, then $f = f_1 \star f_2$. Explicitly, the correspondence (i) and (ii) means the existence of the following relations

$$\lim_{\hbar \to 0} (f_1 \star f_2)(p, x) = f_1(p, x) f_2(p, x) \tag{2.3.1}$$

$$\lim_{\hbar \to 0} \frac{i}{\hbar}(f_1 \star f_2 - f_2 \star f_1) = \{f_1, f_2\} \tag{2.3.2}$$

where $\{f_1, f_2\}$ is the Poisson bracket:

$$\{f_1, f_2\}(p, x) \stackrel{\text{def}}{\equiv} \sum_{j=1}^{d} \left(\frac{\partial f_1}{\partial p_j} \frac{\partial f_2}{\partial x_j} - \frac{\partial f_1}{\partial x_j} \frac{\partial f_2}{\partial p_j} \right). \tag{2.3.3}$$

Note that coordinate $\widehat{x}_j$ and momentum $\widehat{p}_k$ operators with different indices $j \neq k$ commute with each other and, hence, products of *different* coordinates and components of momentum $\widehat{x}_j \widehat{p}_k, j \neq k$ have no ordering problems. Thus, to simplify formulae, we shall confine ourselves in most of this subsection to the one-dimensional case (i.e. to a two-dimensional phase space). The generalization to a d-dimensional space is straightforward.

◇ **Construction of symbols of operators: xp-, px- and Weyl symbols**

The simplest operators on $\mathcal{H}$ have the form

$$\widehat{f} = \sum_{m=0}^{M} \sum_{n=0}^{N} c_{mn} \widehat{x}^m \widehat{p}^n \qquad M, N = 0, 1, 2, \ldots. \tag{2.3.4}$$

Operators of the form (2.3.4) are dense in the set of all operators on $\mathcal{H} = \mathcal{L}^2(\mathbb{R})$. In other words, any operator can be represented as a sum (perhaps infinite) of the operators (2.3.4). Thus a correspondence between the operators (2.3.4) and their symbols defines the quantization (i.e. the ordering rule) for any operator (i.e. fixes the correspondence between any operator and its symbol).

Loosely speaking, to define a symbol we have to choose some ordering of $\widehat{x}$ and $\widehat{p}$ in (2.3.4) and *then* substitute the operators with classical variables x and p. For example, we may put in correspondence to the operator (2.3.4) the xp-symbol:

$$f_{xp}(p, x) - \sum_{m=0}^{M} \sum_{n=0}^{N} c_{mn} x^m p^n. \tag{2.3.5}$$

Another option is to define the *px-symbol*. To this aim, we have to rearrange the operator (2.3.4), using the canonical commutation relations

$$[\widehat{p}, \widehat{x}] = -i\hbar$$

i.e. $\widehat{x}\widehat{p} = \widehat{p}\widehat{x} + i\hbar$, so as to represent it in the form

$$\widehat{f} = \sum_{m=0}^{M}\sum_{n=0}^{N} \widetilde{c}_{mn} \widehat{p}^n \widehat{x}^m \tag{2.3.6}$$

where the new coefficients $\widetilde{c}_{mn}$ are related to c_{mn} in (2.3.4) via the reordering procedure. Then the *px*-symbol of the operator (2.3.4) (or (2.3.6) because this is just another form of the *same* operator) reads as

$$f_{px}(p, x) = \sum_{m=0}^{M}\sum_{n=0}^{N} \widetilde{c}_{mn} x^m p^n. \tag{2.3.7}$$

One more symbol which is especially useful for solving many quantum-mechanical problems is called the *Weyl symbol*. We shall consider it in some more detail.

First, we have to introduce the symmetric operator product $((\widehat{A}_1^{k_1} \cdots \widehat{A}_N^{k_N}))$ of non-commuting operators $\widehat{A}_1^{k_1}, \ldots, \widehat{A}_N^{k_N}$ with the help of the formula

$$(\alpha_1 \widehat{A}_1 + \cdots + \alpha_n \widehat{A}_N)^k = \sum_{k_1+\cdots+k_N=k} \frac{k!}{k_1!\cdots k_N!} \alpha_1^{k_1}\cdots\alpha_N^{k_N} ((\widehat{A}_1^{k_1}\cdots\widehat{A}_N^{k_N})) \tag{2.3.8}$$

where α_i are arbitrary complex numbers and the summation is understood over systems of positive integer numbers $(k_1, k_2\ldots, k_N)$. Thus the symmetric product $((\widehat{A}_1^{k_1}\cdots\widehat{A}_N^{k_N}))$ is, by definition, the coefficient at

$$\frac{k!}{k_1!\cdots k_N!}\alpha_1^{k_1}\cdots\alpha_N^{k_N}$$

in the expansion of the left-hand side of (2.3.8). In the particular case $N = 2$, the definition (2.3.8) reads as

$$(\alpha \widehat{A} + \beta \widehat{B})^k = \sum_{m+l=k} \frac{k!}{m!l!}\alpha^m \beta^l ((\widehat{A}^m \widehat{B}^l)). \tag{2.3.9}$$

For instance,

$$((\widehat{A}\widehat{B})) = \tfrac{1}{2}(\widehat{A}\widehat{B} + \widehat{B}\widehat{A}) \tag{2.3.10}$$

$$((\widehat{A}^2\widehat{B})) = \tfrac{1}{3}(\widehat{A}^2\widehat{B} + \widehat{A}\widehat{B}\widehat{A} + \widehat{B}\widehat{A}^2). \tag{2.3.11}$$

Now we are ready to introduce the Weyl symbol. Let us consider a polynomial of the form (2.3.5) (but, for completeness, in a d-dimensional space)

$$f_{\mathrm{W}}(\boldsymbol{p}, \boldsymbol{x}) = \sum_{\substack{m_1+\cdots+m_d \leq K \\ n_1+\cdots+n_d \leq K}} c_{m_1\ldots m_d n_1\ldots n_d} x_1^{m_1}\cdots x_d^{m_d} p_1^{n_1}\cdots p_d^{n_d} \tag{2.3.12}$$

and put in correspondence to it the operator

$$\widehat{f} = \sum_{\substack{m_1+\cdots+m_d \leq K \\ n_1+\cdots+n_d \leq K}} c_{m_1\ldots m_d n_1\ldots n_d} ((\widehat{x}_1^{m_1}\widehat{p}_1^{n_1}))((\widehat{x}_2^{m_2}\widehat{p}_2^{n_2}))\cdots((\widehat{x}_d^{m_d}\widehat{p}_d^{n_d})). \tag{2.3.13}$$

Then the function $f_w(p, x)$ is called the Weyl symbol of the operator $\hat{f}$.

A more convenient formula connecting the Weyl symbol with the corresponding operator has the following form

$$\hat{f} = \int_{-\infty}^{\infty} d^d s \, d^d r \, e^{i(s\hat{p}+r\hat{x})} \, \widetilde{f}_w(s, r) \tag{2.3.14}$$

where $\widetilde{f}_w(s, r)$ is the Fourier image of a symbol $f_w(p, x)$:

$$\widetilde{f}_w(s, r) = \frac{1}{(2\pi)^{2d}} \int_{-\infty}^{\infty} d^d p \, d^d x \, e^{-i(sp+rx)} f_w(p, x) \tag{2.3.15}$$

$$f_w(p, x) = \int_{-\infty}^{\infty} d^d s \, d^d r \, e^{i(sp+rx)} \, \widetilde{f}_w(s, r). \tag{2.3.16}$$

One may check that (2.3.14) is consistent with the definition of Weyl symbols given earlier (problem 2.3.1, page 226).

◇ **Relation between operator symbols and operator kernels; the star-product for Weyl symbols**

Another way to connect a symbol with the corresponding operator is to find the relation between the former and the kernel of this operator (see the definition of an integral kernel in (2.1.66)). In the case of the Weyl symbol, this relation is defined by the formula (see problem 2.3.2, page 226)

$$K_f(x, y) = \frac{1}{(2\pi \hbar)^d} \int_{-\infty}^{\infty} d^d p \, e^{\frac{i}{\hbar}(x-y)p} f_w\left(p, \frac{x+y}{2}\right) \tag{2.3.17}$$

where $K_f(x, y)$ is the integral kernel of the operator $\hat{f}$ with the Weyl symbol $f_w(x, p)$, so that the action of the operator $\hat{f}$ on a vector ψ from $\mathcal{L}^2(\mathbb{R}^d)$ is given by the expression

$$(\hat{f}\psi)(x) = \int_{-\infty}^{\infty} d^d y \, K_f(x, y) \psi(y)$$

$$= \frac{1}{(2\pi \hbar)^d} \int_{-\infty}^{\infty} d^d p \, d^d y \, e^{\frac{i}{\hbar}(x-y)p} f_w\left(p, \frac{x+y}{2}\right) \psi(y). \tag{2.3.18}$$

In fact, this formula states that the kernel $K_f(x, y)$ is obtained from the Weyl symbol with a shifted argument $f_w(p, (x+y)/2)$, via the Fourier transform. Thus, using the inverse Fourier transform, it is easy to find

$$f_w(p, x) = \int_{-\infty}^{\infty} d^d y \, e^{\frac{i}{\hbar}py} K_f\left(x - \frac{y}{2}, x + \frac{y}{2}\right) \tag{2.3.19}$$

$$= \int_{-\infty}^{\infty} d^d y \, e^{\frac{i}{\hbar}py} \left\langle x - \frac{y}{2} \middle| \hat{f}(\hat{x}, \hat{p}) \middle| x + \frac{y}{2} \right\rangle. \tag{2.3.20}$$

This formula allows the straightforward calculation of Weyl symbols of operators (problem 2.3.3, page 227).

The relations (2.3.17) and (2.3.19) imply that the Weyl symbol $f_w^\dagger$ of the conjugate operator $\hat{f}^\dagger$ is the complex conjugate to that of the operator $\hat{f}$:

$$f_w^\dagger(p, x) = \bar{f}_w(p, x). \tag{2.3.21}$$

As we have discussed in section 2.1.1, the kernel $K(x, y)$ of the operator $\widehat{f} = \widehat{f_1}\widehat{f_2}$ is expressed through the kernels $K_1(x, y)$, $K_2(x, y)$ of the operators $\widehat{f_1}$, $\widehat{f_2}$, according to the composition formula:

$$K(x, y) = \int_{-\infty}^{\infty} d^d z\, K_1(x, z) K_2(z, y). \tag{2.3.22}$$

From this relation and by using the formulae (2.3.17) and (2.3.19), we can find the formula of composition (star-operation) for two Weyl symbols (i.e. for the Weyl symbol of the product of two operators):

$$f_w(p, x) = ((f_1)_w \star (f_2)_w)(p, x)$$
$$= \frac{1}{(2\pi\hbar)^d} \int_{-\infty}^{\infty} d^d p_1\, d^d x_1\, d^d p_2\, d^d x_2\, \exp\left\{\frac{i}{\hbar}(x - x_2)p_1 + (x_1 - x)p_2 + (x_2 - x_1)p\right\}$$
$$\times (f_1)_w(p_1, x_1)(f_2)_w(p_2, x_2) \tag{2.3.23}$$

(see problem 2.3.4, page 227). Note that the exponent in the integrand of (2.3.23) can be written as a sum of determinants:

$$(x - x_2)p_1 + (x_1 - x)p_2 + (x_2 - x_1)p = \sum_{j=1}^{d} \begin{vmatrix} 1 & 1 & 1 \\ x^j & x_1^j & x_2^j \\ p^j & p_1^j & p_2^j \end{vmatrix} \tag{2.3.24}$$

($x^j, x_1^j, \ldots, p_2^j$ are the components of the vectors $x, x_1, \ldots, p_2$).

Relation (2.3.23) can also be rewritten in another form

$$f_w(p, x) = ((f_1)_w \star (f_2)_w)(p, x)$$
$$= (f_1)_w\left(p - \frac{i\hbar}{2}\frac{\partial}{\partial x_2}, x + \frac{i\hbar}{2}\frac{\partial}{\partial p_2}\right)(f_2)_w(p_2, x_2)\bigg|_{\substack{x_2=x \\ p_2=p}}. \tag{2.3.25}$$

To prove the equivalence of (2.3.23) and (2.3.25), it is enough to use the Fourier transform of $(f_2)_w$ which allows us to calculate the action of the differential operators from $(f_1)_w$ in (2.3.25) and then to make a change of variables converting the result into the formula (2.3.23). Expanding $(f_1)_w$ in (2.3.25) in Taylor series, we arrive at the formal equality

$$f_w(p, x) = ((f_1)_w \star (f_2)_w)(p, x)$$
$$= \sum_{\substack{m_1,\ldots,m_d \\ n_1,\ldots,n_d}} \frac{(-1)^{n_1+\cdots+n_d}}{m_1!\cdots m_d!n_1!\cdots n_d!}\left(\frac{i\hbar}{2}\right)^{m_1+\cdots+m_d+n_1+\cdots+n_d}$$
$$\times [\partial_{x_1}^{m_1}\cdots\partial_{x_d}^{m_d}\partial_{p_1}^{n_1}\cdots\partial_{p_d}^{n_d} f_{w1}(p, x)][\partial_{x_1}^{n_1}\cdots\partial_{x_d}^{n_d}\partial_{p_1}^{m_1}\cdots\partial_{p_d}^{m_d} f_{w2}(p, x)]. \tag{2.3.26}$$

This formula clearly shows that the correspondence principle (2.3.1), (2.3.2) is fulfilled for the Weyl quantization.

The following useful formulae for symbols represent traces of operators:

$$\text{Tr}\,\widehat{f} = \int_{-\infty}^{\infty} d^d x\, K(x, x)$$
$$= \frac{1}{(2\pi\hbar)^d} \int_{-\infty}^{\infty} d^d x\, d^d p\, f_w(p, x) \tag{2.3.27}$$

$$\text{Tr}\,\widehat{f_1}\widehat{f_2}^\dagger = \frac{1}{(2\pi\hbar)^d} \int_{-\infty}^{\infty} d^d x\, d^d p\, (f_1)_w(p, x)(f_2)_w^*(p, x). \tag{2.3.28}$$

Quantization, the operator ordering problem and path integrals

Table 2.1. Examples of ordering rules and functions defining the corresponding symbols.

Correspondence rule	$\Omega(u,v)$	Ordering rule
Weyl	1	$((\widehat{x}^n \widehat{p}^m))$
Symmetric	$\cos \dfrac{uv}{2}$	$\tfrac{1}{2}(\widehat{x}^n \widehat{p}^m + \widehat{p}^m \widehat{x}^n)$
xp-rule	$\exp\left\{-\mathrm{i}\dfrac{uv}{2}\right\}$	$\widehat{x}^n \widehat{p}^m$
px-rule	$\exp\left\{\mathrm{i}\dfrac{uv}{2}\right\}$	$\widehat{p}^m \widehat{x}^n$
Born–Jordan	$\sin\dfrac{uv}{2} \Big/ \dfrac{uv}{2}$	$\dfrac{1}{m+1}\displaystyle\sum_{l=0}^{n} \widehat{p}^{m-l}\widehat{x}^n \widehat{p}^l$

◊ The general form of the correspondence mapping for an arbitrary ordering rule

Similar considerations can be carried out for other ordering rules and the general relation between operators and the corresponding symbols can be formulated in a systematic and universal way (e.g., Langouche *et al* (1982) and Balazs and Jennings (1984)).

Let us consider a pair p, x of canonically conjugate variables and the corresponding non-commuting operators $\widehat{p}$, $\widehat{x}$. For an exponential function of p and x an arbitrary correspondence rule, to generate the operators $\widehat{p}$, $\widehat{x}$ from the (commutative) coordinates p, x, can be presented in the following general form:

$$\exp\left[\frac{\mathrm{i}}{\hbar}(ux+vp)\right] \to O_\Omega(u,v;\widehat{x},\widehat{p}) \equiv \Omega(u,v)\exp\left[\frac{\mathrm{i}}{\hbar}(u\widehat{x}+v\widehat{p})\right] \qquad (2.3.29)$$

where $\Omega(u,v)$ is a function of the auxiliary variables u, v which defines the ordering rule. This produces the mapping for arbitrary monomials:

$$M(n,m) = x^{\mu_1}\ldots x^{\mu_n} p^{\nu_1}\ldots p^{\nu_m} \qquad (2.3.30)$$

by means of the differentiation

$$M(n,m) \to \frac{1}{\mathrm{i}^{n+m}}\left.\frac{\partial^{n+m} O_\Omega(u,v;\widehat{q},\widehat{p})}{\partial u_{\mu_1}\ldots \partial u_{\mu_n}\partial v_{\nu_1}\ldots \partial v_{\nu_m}}\right|_{u=v=0}. \qquad (2.3.31)$$

It is seen that the ordering rule is defined by the function $\Omega(u,v)$. Some known examples are displayed in table 2.1.

We can make this correspondence explicit. In order to construct the operator $\widehat{F}_\Omega$, corresponding to a classical function $f(p,x)$, according to the rule defined by some $\Omega(u,v)$, let us consider the Fourier integral

$$f(p,x) = \int du\, dv\, \widetilde{f}(u,v)\exp\left\{\frac{\mathrm{i}}{\hbar}(ux+\mathrm{i}vp)\right\} \qquad (2.3.32)$$

$$\widetilde{f}(u,v) = \frac{1}{(2\pi\hbar)}\int dp\,dx\, f(p,x)\exp\left\{-\frac{\mathrm{i}}{\hbar}(ux+\mathrm{i}vp)\right\}. \qquad (2.3.33)$$

We define the operator $\widehat{f}^\Omega(\widehat{p},\widehat{q})$ via

$$\widehat{f}^\Omega(\widehat{p},\widehat{q}) = \int du\,dv\, \widetilde{f}(u,v)\Omega(u,v)\exp\left\{\frac{\mathrm{i}}{\hbar}(u\widehat{x}+\mathrm{i}v\widehat{p})\right\} \qquad (2.3.34)$$

206   *Path integrals in quantum mechanics*

which gives

$$\widehat{f^\Omega}(\widehat{p},\widehat{x}) = \frac{1}{(2\pi\hbar)} \int du\, dv\, dp\, dx\, f(p,x)\Omega(u,v) \exp\left\{-\frac{i}{\hbar}u(x-\widehat{x}) - \frac{i}{\hbar}v(p-\widehat{p})\right\}. \quad (2.3.35)$$

(We have presented the one-dimensional version of these general formulae but the generalization to the d-dimensional case is trivial: integrations go over $\mathbb{R}^d$ and the pre-integral factor acquires the power d.)

◇ **The normal symbol of an operator**

Now we shall introduce the *normal symbols* (sometimes also called the *Wick symbols*) which are especially important for many-body problems in quantum mechanics, second quantization formalism and quantum field theory.

The normal symbols are constructed with the help of creation and annihilation operators (cf section 2.1.1). For the current aim it is convenient to rescale the operators used in section 2.1.1, so that the multidimensional generalization of (2.1.48) reads as

$$[\widehat{a}_k, \widehat{a}_j] = [\widehat{a}_k^\dagger, \widehat{a}_j^\dagger] = 0 \quad (2.3.36)$$

$$[\widehat{a}_k, \widehat{a}_j^\dagger] = \hbar\delta_{kj}. \quad (2.3.37)$$

The space $\mathcal{L}^2(\mathbb{R}^d)$ has the orthonormal basis

$$\psi_{k_1,k_2,\ldots,k_d} = \frac{(\widehat{a}_1^\dagger)^{k_1}\cdots(\widehat{a}_d^\dagger)^{k_d}}{\sqrt{\hbar^{k_1+\cdots+k_d}k_1!k_2!\cdots k_d!}}\psi_0 \quad (2.3.38)$$

with the normalized ground state, ψ_0: $\widehat{a}_k\psi_0 = 0$ (cf (2.1.53)).

Since the commutation relations between operators with different indices are trivial (i.e. the operators commute), the generalization of the one-dimensional case to the multidimensional is quite simple and in what follows we mainly restrict the discussion to the case of one degree of freedom, making comments on higher-dimensional formulae where necessary.

One possible realization of the commutation relations for the creation and annihilation operators is based on the formulae (2.1.47) and (2.1.29), (2.1.30). Another very convenient and important realization, called the *Bargmann–Fock realization*, is constructed as follows.

Let $\mathcal{F}^2$ be a space of entire anti-holomorphic complex functions $\phi(z^*)$, $z \in \mathbb{C}$. Introducing the scalar product

$$\langle\psi|\phi\rangle \stackrel{\text{def}}{\equiv} \frac{1}{\hbar}\int dz\, dz^*\, e^{-\frac{1}{\hbar}z^*z}\psi^*(z^*)\phi(z^*) \qquad \psi,\phi \in \mathcal{F}^2 \quad (2.3.39)$$

we convert $\mathcal{F}^2$ into a Hilbert space. The integration measure in (2.3.39) is defined by the equality $dz\, dz^* = dx\, dy/\pi$, where $z = x + iy$ and $dx\, dy$ is the usual Lebesgue measure on $\mathbb{R}^2 \cong \mathbb{C}$. The action of the $\widehat{a}, \widehat{a}^\dagger$-operators on functions from $\mathcal{F}^2$ is defined by the formulae

$$(\widehat{a}^\dagger f)(z^*) = z^* f(z^*) \quad (2.3.40)$$

$$(\widehat{a} f)(z^*) = \hbar\frac{\partial}{\partial z^*} f(z^*). \quad (2.3.41)$$

An advantage of this realization can already be seen at this stage: the problem of finding the ground state ψ_0 becomes almost trivial. Indeed, from (2.3.41) it is obvious that $\psi_0 = $ constant and, using (2.3.39), we easily check that

$$\psi_0 = 1 \quad (2.3.42)$$

is the normalized ground state

$$\widehat{a}\psi_0 = \hbar\frac{\partial}{\partial z}1 = 0.$$

Integration by parts shows that the operators $\widehat{a}$ and $\widehat{a}^\dagger$ are conjugate to each other with respect to the scalar product (2.3.39) (problem 2.3.5, page 228). The successive action of the creation operator $\widehat{a}^\dagger$ gives the orthonormal basis of $\mathcal{F}^2$:

$$\psi_k(z^*) = \frac{(z^*)^k}{\sqrt{\hbar^k k!}}. \tag{2.3.43}$$

In the space $\mathcal{F}^2$, there exists another very convenient basis of the so-called *coherent states* $\Upsilon_v(z^*)$, which are defined by the expression

$$\Upsilon_v(z^*) \stackrel{\text{def}}{\equiv} e^{\frac{1}{\hbar}vz^*} \qquad v \in \mathbb{C}. \tag{2.3.44}$$

It is clear that the vector $\Upsilon_v(z^*)$ is an eigenfunction of the operator $\widehat{a}$:

$$\widehat{a}\Upsilon_v(z^*) = v\Upsilon_v(z^*). \tag{2.3.45}$$

Note, however, that although any vector from $\mathcal{F}^2$ can be represented as a superposition (in fact, integral) of the coherent states, the basis (2.3.44) is not orthonormal:

$$\langle\Upsilon_z|\Upsilon_v\rangle = \Upsilon_v(z^*) = e^{\frac{1}{\hbar}z^*v}. \tag{2.3.46}$$

Here $\Upsilon_v(z^*)$ must be understood as a function from $\mathcal{F}^2$ (in the same way as the matrix element $\langle p|x\rangle = \exp\{\frac{i}{\hbar}px\}$ is a wavefunction). It is seen from (2.3.46) that we have chosen to work with non-normalized coherent states:

$$\langle\Upsilon_z|\Upsilon_z\rangle = e^{\frac{1}{\hbar}z^*z}. \tag{2.3.47}$$

Such a choice proves to be technically convenient in many cases. The resolution of unity in terms of the coherent states (i.e. the completeness relation for them) reads as

$$\mathbb{I} = \frac{1}{\pi}\int d^2v\, |\Upsilon_v\rangle e^{-\frac{1}{\hbar}v^*v}\langle\Upsilon_v|. \tag{2.3.48}$$

In fact, the coherent states form an *overcompleted set* of states because of the absence of the orthogonality.

An advantage of the coherent states shows up, in particular, in the calculation of the normal symbols of operators. To introduce the latter, let us consider, in a Fock space with d degrees of freedom, polynomials in $\widehat{a}$, $\widehat{a}^\dagger$. Using the commutation relations (2.3.36) and (2.3.37), these polynomials can be written in the form

$$\widehat{A} = \sum_{\substack{m_1+\cdots+m_d\leq K \\ n_1+\cdots+n_d\leq K}} A_{m_1\ldots m_d n_1\ldots n_d}(\widehat{a}_1^\dagger)^{m_1}\cdots(\widehat{a}_d^\dagger)^{m_d}\widehat{a}_1^{n_1}\cdots\widehat{a}_d^{n_d}. \tag{2.3.49}$$

The corresponding polynomial in commuting variables is called the *normal symbol*:

$$A(z_i^*, z_j) = \sum_{\substack{m_1+\cdots+m_d\leq K \\ n_1+\cdots+n_d\leq K}} A_{m_1\ldots m_d n_1\ldots n_d}(z^*_1)^{m_1}\cdots(z^*_d)^{m_d}z_1^{n_1}\cdots z_d^{n_d}. \tag{2.3.50}$$

Of course, we can also define *anti-normal symbols* which can be put in correspondence to a polynomial with commuting variables, operators with the annihilation operator *on the left* of the creation operators. We shall not use the anti-normal symbols in this book.

As we have mentioned, the coherent states provide the simplest way of calculating the normal symbols. Again we present the calculation for the case of one degree of freedom; generalization to a higher-dimensional phase space is straightforward and can be carried out by the reader as a useful exercise. Using (2.3.49), (2.3.50) and (2.3.45), we obtain

$$\langle \Upsilon_v | \widehat{A} \Upsilon_z \rangle = \sum_{n,m} A_{nm} \langle \Upsilon_v | (\widehat{a}^\dagger)^n \widehat{a}^m \Upsilon_z \rangle$$

$$= \sum_{n,m} A_{nm} \langle \widehat{a}^n \Upsilon_v | \widehat{a}^m \Upsilon_z \rangle$$

$$= \sum_{n,m} A_{nm} v^{*n} z^m \langle \Upsilon_v | \Upsilon_z \rangle$$

$$= A(v^*, z) \langle \Upsilon_v | \Upsilon_z \rangle \qquad (2.3.51)$$

so that the normal symbol is expressed via an operator $\widehat{A}$ by the formula

$$A(z^*, z) = \frac{\langle \Upsilon_z | \widehat{A} \Upsilon_z \rangle}{\langle \Upsilon_z | \Upsilon_z \rangle} \qquad (2.3.52)$$

$$= e^{-\frac{1}{\hbar} z^* z} \langle \Upsilon_z | \widehat{A} \Upsilon_z \rangle \qquad (2.3.53)$$

(the latter equality follows from (2.3.47)). Note that the normal symbol $A^\dagger$ of the Hermitian conjugate operator $\widehat{A}^\dagger$ is obtained by the complex conjugation

$$A^\dagger(z^*, z) = A^*(z^*, z). \qquad (2.3.54)$$

Now let us find the relation of the normal symbol with the corresponding operator kernel $K_A(z^*, z)$ which, taking into account the scalar product (2.3.39), should be defined as follows

$$(\widehat{A}f)(z^*) = \frac{1}{\hbar} \int dz^* \, dz \, e^{-\frac{1}{\hbar} v^* v} K_A(z^*, v) f(v^*). \qquad (2.3.55)$$

Using (2.3.46), (2.3.53) and (2.3.54), we obtain

$$(\widehat{A}f)(z^*) = \langle \widehat{A}^\dagger \Upsilon_z | f \rangle = \frac{1}{\hbar} \int dv^* \, dv \, e^{-\frac{1}{\hbar} v^* v} (\widehat{A}^\dagger \Upsilon_z)^*(v^*) f(v^*)$$

$$= \int dv^* \, dv \, e^{-\frac{1}{\hbar} v^* v} (A^\dagger(v^*, z) \langle \Upsilon_z | \Upsilon_v \rangle)^* f(v^*)$$

$$= \int dv^* \, dv \, e^{-\frac{1}{\hbar}(z^* - v^*)v} A(z^*, v) f(v^*) \qquad (2.3.56)$$

so that

$$K_A(z^*, v) = e^{-\frac{1}{\hbar} z^* v} A(z^*, v). \qquad (2.3.57)$$

The star-operation for the normal symbols reads as

$$A(z^*, z) = (A_1 \star A_2)$$

$$= \frac{1}{\hbar} \int dv^* \, dv \, e^{-\frac{1}{\hbar}(z^* - v^*)(z - v)} A_1(z^*, v) A_2(v^*, z) \qquad (2.3.58)$$

(see problem 2.3.6, page 228). This star-product can be presented in a differential form (similar to expression (2.3.26) for Weyl symbols):

$$A(z^*, z) = (A_1 \star A_2)$$
$$= \sum_m \frac{\hbar^m}{m!} [\partial_z^m A_1(z^*, z)][\partial_{z^*}^m A_2(z^*, z)] \qquad (2.3.59)$$

which clearly shows that the correspondence principle (2.3.1), (2.3.2) is fulfilled for the normal symbols as well as for the Weyl symbols. Of course, we can find the relation between the two types of symbol. We give it without proof (see, e.g., Berezin (1971)):

$$A(\boldsymbol{v}^*, \boldsymbol{v}) = \left(\frac{2}{\hbar}\right)^d \int d^d z^* \, d^d z \, e^{-\frac{2}{\hbar}(z^* - \boldsymbol{v}^*)(z - \boldsymbol{v})} f_{\rm W}(\boldsymbol{p}, \boldsymbol{x}). \qquad (2.3.60)$$

The latter formula is written for a system with d degrees of freedom: $\boldsymbol{z}^*, \boldsymbol{z}, \boldsymbol{v}^*, \boldsymbol{v}$ denote vectors from $\mathbb{C}^d$; $\boldsymbol{v}^* \boldsymbol{z} \stackrel{\rm def}{\equiv} \sum_{j=1}^d v^*_j z_j$ stands for the standard scalar product in $\mathbb{C}^d$. All the preceding formulae for normal symbols can be easily generalized to the multidimensional case by substitutions of the type $v^* z \to \boldsymbol{v}^* \boldsymbol{z}$ and $dz^* dz/\hbar \to d^d z^* d^d z/\hbar^d$.

2.3.2 General concept of path integrals over trajectories in phase space

One of the principal problems of quantum mechanics is the evaluation of transition amplitudes

$$\langle \psi | \widehat{U}(t, t_0) | \varphi \rangle$$

or, in other words, the kernel of the evolution operator of a system. Unfortunately, except for some specific cases (e.g., the free particle and harmonic oscillator), it is impossible to calculate these kernels directly. From the general point of view, the construction of a path integral in quantum mechanics is just a specific way of representing the evolution operator kernels in a form suitable for further (approximate) evaluation. This path-integral representation is based essentially on two facts:

(1) the operator identity
$$\widehat{U}(t) = [\widehat{U}(t/N)]^N$$

(2) the formula for the kernel of products of operators:

$$K_{U(t)}(x, y) = \int_{-\infty}^{\infty} \left[\prod_{j=1}^{N-1} dx_j \right] K_{U(t/N)}(x, x_1) K_{U(t/N)}(x_1, x_2) \cdots K_{U(t/N)}(x_{N-1}, y). \qquad (2.3.61)$$

In fact, the latter expression is nothing but the discrete-time approximation for the Feynman (configuration space) path integral. The advantage of representation (2.3.61) is that evaluating the kernel of a short-time evolution operator (short-time propagator) is much easier to do than that for an arbitrary time. An actual calculation of the short-time propagator requires the use of the complete set of momentum eigenvalues and, hence, integration over momenta (cf, for example, (2.2.8)) which results in the discrete approximation for the *phase-space* path integral.

However, in the general case (apart from Hamiltonians of the form $H(p, x) = T(p) + V(x)$), we have to choose, at first, a way of quantization, i.e. an operator ordering rule for the Hamiltonian under consideration. This, in turn, essentially defines the discrete approximation (2.3.61) for the path integral, exhibiting, in this way, the general quantum-mechanical ambiguity (operator ordering problem) in the path-integral formalism. For example, relation (2.3.17) clearly shows that Weyl ordering results in the *midpoint prescription*: when we substitute the Hamiltonian operator by its classical counterpart (i.e. by the Weyl symbol) we have to evaluate the latter in the midpoints, as indicated on the right-hand side of (2.3.17).

Thus the formalism of operator symbols allows us to take into account different possibilities for the quantization of complicated systems (with inequivalent different orders for the operators in the Hamiltonians). In this approach, the basic objects of the formalism prove to be the *phase-space* path integrals. Specific prescriptions (e.g., the midpoint one) for the discrete approximation of configuration Feynman path integrals *automatically* appear after integration (if possible!) over momentum variables.

It should be stressed that the construction of the phase-space path integrals, outlined in this section, is suitable for *arbitrary Hamiltonians*, while direct construction of path integrals in a configuration space is heavily based on the Feynman–Kac formula and meets serious problems for complicated Hamiltonians (i.e. those more general than the form $H = T(p) + V(x)$). We shall consider such complicated Hamiltonians in section 2.5.

◇ **Discrete approximation for a phase-space path integral as a star-product of symbols of evolution operators**

Let $\widehat{H}$ be some Hamiltonian and H be one of its symbols (for example, Weyl or normal symbol). For a small $t = \varepsilon$, we have

$$\widehat{U}(\varepsilon) = \exp\left\{-\frac{i}{\hbar}\widehat{H}\varepsilon\right\} = 1 - \frac{i}{\hbar}\widehat{H}\varepsilon + \mathcal{O}(\varepsilon^2). \tag{2.3.62}$$

Thus the symbol of the evolution operator is

$$U(\varepsilon) = 1 - \frac{i}{\hbar}H\varepsilon + \mathcal{O}(\varepsilon^2) = \exp\left\{-\frac{i}{\hbar}H\varepsilon\right\} + \mathcal{O}(\varepsilon^2). \tag{2.3.63}$$

Let $\widehat{G}(t)$ be an operator with the symbol $G(t) = \exp\{-\frac{i}{\hbar}Ht\}$. From (2.3.63), we have

$$\widehat{U}(\varepsilon) = \widehat{G}(\varepsilon)(1 + \mathcal{O}(\varepsilon^2)) \tag{2.3.64}$$

and by substituting this approximate equality into the operator identity

$$\widehat{U}(t) = (\widehat{U}(t/N))^N$$

we arrive at an expression for $\widehat{U}(t)$ with arbitrary t:

$$\widehat{U}(t) = \lim_{N\to\infty} \widehat{U}_N(t) \tag{2.3.65}$$

$$\widehat{U}_N(t) = \left(\widehat{G}\left(\frac{t}{N}\right)\right)^N. \tag{2.3.66}$$

It is worth noting that, strictly speaking, the expansions (2.3.62) and (2.3.63) are correct if the Hamiltonians are of the order of unity (in an appropriate sense). In other words, we should expand the consideration presented in the proof of the Trotter formula (section 2.2.1) for Hamiltonians of the form $H(p,x) = T(p) + V(x)$, to the case of arbitrary Hamiltonians. In general, this is a very complicated mathematical problem. But from the physical point of view, violation of the expansions (2.3.62) and (2.3.63) would mean that the Hamiltonian is not truly the generator of time evolution. Essentially, we just leave this exotic situation aside.

Since the symbol of the operator $\widehat{G}(t/N)$ is known, we can write (2.3.65) in terms of the symbols

$$U_N(t) = G\left(\frac{t}{N}\right) \star G\left(\frac{t}{N}\right) \star \cdots G\left(\frac{t}{N}\right). \qquad (2.3.67)$$

Let ξ, for brevity, denote a vector in the $2d$-dimensional phase space:

$$\xi = \{x, p\} = \{x_1, \ldots, x_d; p_1, \ldots, p_d\}. \qquad (2.3.68)$$

Taking into account that the $\star$-operation for some symbols f_1, f_2 is given by the integrals of the form

$$(f_1 \star f_2)(\xi) = \int d^d\xi_1 \, d^d\xi_2 \, \Lambda(\xi; \xi_1, \xi_2) f_1(\xi_1) f_2(\xi_2) \qquad (2.3.69)$$

where $\Lambda(\xi; \xi_1, \xi_2)$ is one of the functions defining the star-product for a given type of symbol (cf (2.3.23) and (2.3.58)), we obtain in this way an expression for $\widehat{U}_N(t)$ in terms of a multidimensional integral, i.e. the time-sliced approximation for the phase-space path integral for the evolution operator (more precisely, for its symbol). At this stage, we can clearly see the implication of the operator ordering problem (in other words, the existence of different ways of quantization of the same classical system) on the derivation of path integrals. Indeed, different ways of quantization (different symbols of operators) imply different functions $\Lambda(\xi; \xi_1, \xi_2)$ and, hence, different expressions for the discretely approximated path integral. Naive passage to the limit of continuous time hides the difference for basic types of symbol giving the same answer (thus the latter is correct only for special Hamiltonians or symbols). Careful calculations lead to peculiarities. Additional details, for example shifted arguments of the type

$$\int_{t_0}^{t} dt \, H(p(\tau+0), x(\tau)) \stackrel{\text{def}}{=} \lim_{\varepsilon \to 0} \int_{t_0}^{t} dt \, H(p(\tau+\varepsilon), x(\tau)) \qquad (2.3.70)$$

or additional non-integral terms appear. The non-triviality of the infinitesimal shifts (2.3.70) follows from the fact that trajectories which must be integrated over in the *phase-space* path integral are *discontinuous* (see later). This is a non-trivial generalization of the Wiener theorem for the path integral in the theory of stochastic processes and this fact reflects the basic principle of quantum mechanics, namely, the uncertainty principle (there are no trajectories of quantum particles in phase spaces!).

For definiteness, let us choose in what follows the *Weyl symbol*, still dropping, for brevity, the index indicating this: $U(t) \equiv U_\text{w}(t)$, $H \equiv H_\text{w}$. Using the explicit form (2.3.23) of the star-operation for the Weyl symbols, the time-sliced approximation (2.3.67) can be written in the form

$$U_N(p, x; t, t_0) = \frac{1}{(\pi \hbar)^{2d(N-1)}} \int \prod_{m=1}^{N-1} d^d p_m \, d^d x_m \, d^d k_m \, d^d q_m$$

$$\times \exp\left\{-2\frac{\mathrm{i}}{\hbar}\left[2\sum_{m=1}^{N-1}[(p_m - k_m)(q_{m+1} - q_m) - (x_m - q_m)(k_{m+1} - k_m)]\right.\right.$$

$$\left.\left. - \varepsilon \sum_{m=0}^{N-1} H(p_m, x_m)\right]\right\}. \qquad (2.3.71)$$

Here, for compactness, we have introduced the auxiliary variables

$$k_1 \equiv p_0 \qquad q_1 \equiv x_0 \qquad k_N \equiv p \qquad q_N \equiv x.$$

In the continuous limit $N \to \infty$, this gives

$$U(p, x; t, t_0) = \int \mathcal{D}p(\tau)\mathcal{D}x(\tau)\mathcal{D}k(\tau)\mathcal{D}q(\tau)$$
$$\times \exp\left\{-2\frac{i}{\hbar}\int_{t_0}^{t} d\tau\, [2((x(\tau) - q(\tau))\dot{k}(\tau) - (p(\tau) - k(\tau))\dot{q}(\tau))\right.$$
$$\left. - H(p(\tau), x(\tau))]\right\} \quad (2.3.72)$$

where
$$k(t_0) \equiv p(t_0) \qquad q(t_0) \equiv x(t_0) \qquad k(t) \equiv p \qquad q(t) \equiv x.$$

Note that, if we did not know the correct discrete approximation (2.3.71) and *started* from (2.3.72), we could obtain the standard expression for the path integrals:

$$U = \int \mathcal{D}p(\tau)\mathcal{D}x(\tau) \exp\left\{\frac{i}{\hbar}S\right\}. \quad (2.3.73)$$

To see this, let us substitute the exponential in (2.3.72) by the following integral sum:

$$2\sum_{m=2}^{N-1}(p_m - k_m)(q_{m+1} - q_m) - \sum_{m=1}^{N-1}(x_{m+1} - q_{m+1})(k_{m+1} - k_m)$$
$$- (x_2 - q_2)k_1 - (x_N - q_N)k_N - \varepsilon\sum_{m=0}^{\infty} H(p_m, x_m)$$

which has the same continuous limit as that in (2.3.72) and then integrate over k_m, q_m; $m = 2, \ldots, N-1$. The continuous limit of the result of the integration yields the path integral (2.3.73). This shows that the naive continuous limit may hide peculiarities in the correct discrete approximations and even lead to a wrong result. It should be stressed, however, that this concerns the *local* terms in the exponentials of path integrals. The latter remark is important for quantum field theories where such local terms become *inessential* due to the *renormalization procedure* (see chapter 3).

◇ Representation for a symbol of the evolution operator as a ratio of two path integrals

The path-integral representation for symbols of operators offers a slight modification which allows us to drop a careful treatment of the overall pre-integral factors. For the sake of generality, let us write this modification in a form independent of a concrete type of finite-dimensional approximation of (phase-space) path integrals.

Let $\mathcal{X}$ be the Hilbert space consisting of trajectories in a phase space, with the scalar product

$$\langle \xi | \xi \rangle \stackrel{\text{def}}{\equiv} \int_{t_0}^{t} d\tau\, \xi^2(\tau) < \infty \quad (2.3.74)$$

where
$$\xi^2(\tau) = \sum_{j=1}^{d}(p_j^2(\tau) + x_j^2(\tau))$$

(this is the phase-space analog of the configuration scalar product (2.1.112)).

For a phase-space path integral the space $\mathcal{X}$ plays the same role as the space of continuous functions for the Wiener integral: we shall argue that it is this set of trajectories in the phase space which must be integrated over.

As we have already discussed, different types of finite approximation of path integrals exist: for example, the time-slicing (discrete-time) approximation or Fourier expansion with a finite number of terms in the series. The general procedure for a finite-dimensional approximation can be described as follows.

Let P_N be a collection of orthogonal projector operators in $\mathcal{X}$ and $\mathcal{X}_N \stackrel{\text{def}}{\equiv} P_N \mathcal{X}$ be spaces of functions used for a finite-dimensional approximation of the paths, with the properties

$$\dim\{\mathcal{X}_N\} = D_N < \infty$$
$$P_N P_{N+1} = P_{N+1} P_N = P_N \qquad (2.3.75)$$
$$\lim_{N \to \infty} P_N = \mathbb{I}$$

where $\mathbb{I}$ is the identity operator in $\mathcal{X}$ (the limit is understood in the strong operator sense). The meaning of these projectors P_N is obvious: in particular, in the case of the time-sliced approximation they substitute paths by piecewise linear functions defined by their values at N points; in the case of the Fourier expansion, they cut the series off after the Nth term.

Now we introduce a general notation, $\widehat{B}_N$, for the operators which, by acting on functions from $\mathcal{X}_N$, produce the exponent, except a Hamiltonian, in the finite-dimensional approximation of the phase-space path integral (2.3.72), so that this part of the exponent now can be written as

$$\langle \boldsymbol{\xi}_{(N)} | \widehat{B}_N | \boldsymbol{\xi}_{(N)} \rangle \qquad \boldsymbol{\xi}_{(N)} \in \mathcal{X}_N. \qquad (2.3.76)$$

An explicit form of this operator depends on the rule of the star-product of the chosen symbol and on a chosen approximation. For the time-sliced approximation, the B_N are the finite-difference operators, while for the finite mode (Fourier) expansion, they are ordinary differential operators. For example, for the Weyl symbol and discrete-time approximation, the explicit form of (2.3.76) is given by the exponent in the second term of (2.3.71). Now the finite approximation of the path integral for the symbol of the evolution operator can be defined as

$$U_N(\boldsymbol{\xi}_{(N)}; t, t_0) \stackrel{\text{def}}{\equiv} \frac{\int d^{D_N} \boldsymbol{\xi}_{(N)} \exp\left\{\frac{i}{\hbar}\left[\langle \boldsymbol{\xi}_{(N)} | \widehat{B}_N | \boldsymbol{\xi}_{(N)} \rangle - \int_{t_0}^{t} d\tau\, H(\boldsymbol{\xi}(\tau), \tau)\right]\right\}}{\int d^{D_N} \boldsymbol{\xi}_{(N)} \exp\left\{\frac{i}{\hbar}\left[\langle \boldsymbol{\xi}_{(N)} | \widehat{B}_N | \boldsymbol{\xi}_{(N)} \rangle\right]\right\}}. \qquad (2.3.77)$$

This formula is inspired by the fact that the Weyl symbol of the unity operator ('evolution' operator with zero Hamiltonian) is equal to ordinary unity and this is obviously true for (2.3.77). The advantage of definition (2.3.77) is that it makes the path integral *insensitive* to a total pre-integral factor of a finite-dimensional approximation.

The fact that the approximation (2.3.77) indeed converges to the symbol of the evolution operator for an arbitrary chosen symbol can be proved by a method similar to that for the Feynman integral which we have discussed in section 2.1.1: we must prove that the corresponding operator satisfies the Schrödinger equation. To this aim, we prove, starting from the approximation (2.3.77), that the symbol $U = \lim_{N \to \infty} U_N$ satisfies the Markov-like condition:

$$U(t, t') \star U(t', t_0) = U(t, t_0) \qquad t_0 \leq t' \leq t. \qquad (2.3.78)$$

Then differentiating it and using the fact that an evolution is generated by a Hamiltonian, i.e.

$$U(t, t_0) = 1 - \frac{i}{\hbar} \int_{t_0}^{t} d\tau\, H(\boldsymbol{\xi}, \tau) + \mathcal{O}((t - t_0)^2) \qquad (2.3.79)$$

we arrive at the equation

$$i\hbar \frac{\partial U(t, t_0)}{\partial t} = H(t) \star U(t, t_0) \qquad (2.3.80)$$

which is equivalent to the Schrödinger equation for the evolution operator.

◇ The Wiener-like theorem for phase-space path integrals

To conclude this subsection, we shall discuss the set of trajectories in a phase space on which the path integral is concentrated. But since the integral is not generated by some measure, we should clarify, first of all, the meaning of the notion 'set of trajectories on which the integral is concentrated'. To do this, let us consider some Hilbert space $\mathcal{X}$ of classical trajectories and a ball $S(r)$ of radius r in this space, defined with respect to the standard norm in $\mathcal{X}$, that is $\langle \boldsymbol{\xi}(\tau)|\boldsymbol{\xi}(\tau)\rangle \leq r^2$. With the help of the projector operators (2.3.75), we also define the collection of spaces $\mathcal{X}_N \equiv P_N \mathcal{X}$ and the corresponding balls $S_N(r)$ and consider the integral

$$J_N(r) = \int_{\boldsymbol{\xi} \in S_N(r)} d^{D_N}\xi \, \exp\left\{-\frac{i}{2\hbar} \langle \boldsymbol{\xi}|\widehat{B}_N|\boldsymbol{\xi}\rangle\right\} \qquad (2.3.81)$$

where the operators $\widehat{B}_N$ are defined in (2.3.76). If the functional

$$\exp\left\{-\frac{i}{2\hbar} \langle \boldsymbol{\zeta}|B|\boldsymbol{\zeta}\rangle\right\}$$

generated some measure in a functional space, the ratio

$$\mathcal{J}_N(r) \stackrel{\text{def}}{\equiv} \frac{J_N(r)}{J_N(\infty)} \qquad (2.3.82)$$

would satisfy the conditions:

(i) for any $r \geq 0$ there exists the limit

$$\mathcal{J}(r) = \lim_{N \to \infty} \mathcal{J}_N(r) \qquad (2.3.83)$$

(ii) there exist the limits $\lim_{r \to \infty} \mathcal{J}_N(r)$ for any N and

$$\lim_{r \to \infty} \mathcal{J}(r) = 1. \qquad (2.3.84)$$

In the case of phase-space path integrals, these conditions can be taken as a definition (Berezin 1981): if (2.3.83) and (2.3.84) are fulfilled for some Hilbert space $\mathcal{X}$, the phase-space integral is said to be concentrated on (or supported by) this space.

Theorem 2.3 (the analog of the Wiener theorem). A phase-space path integral is concentrated on the Hilbert space of trajectories with the scalar product (2.3.74).

Proof. To simplify the notation, let us consider the case of one degree of freedom. As follows from analysis of the integral (2.3.72) (Berezin 1981), after the shift

$$p(\tau) \to p + \tilde{p}(\tau) \qquad x(\tau) \to x + \tilde{x}(\tau)$$

we can consider only trajectories which satisfy the conditions

$$\tilde{p}(t) + \tilde{p}(t_0) = 0 \qquad \tilde{x}(t) + \tilde{x}(t_0) = 0.$$

These functions have the Fourier expansion

$$\tilde{p}(\tau) = \sum_{n=-\infty}^{\infty} \alpha_{2n+1} \exp\left\{(2n+1)\pi i \frac{\tau}{t-t_0}\right\} \qquad (2.3.85)$$

$$\tilde{x}(\tau) = \sum_{n=-\infty}^{\infty} \beta_{2n+1} \exp\left\{(2n+1)\pi i \frac{\tau}{t-t_0}\right\} \qquad (2.3.86)$$

with the expansion coefficients subject to the conditions

$$\alpha^*_{2n+1} = \alpha_{-(2n+1)} \qquad \beta^*_{2n+1} = \beta_{-(2n+1)}.$$

The space $\mathcal{X}_N$ is obtained from $\mathcal{X}$ by cutting off the series (2.3.85), (2.3.86) after the Nth term. For later convenience, instead of the ball $S_N(r)$, we consider the ellipsoid $S_N(\sigma, \mu, r)$:

$$\sum_{|n|<N} (\sigma_{2n+1}|\alpha_{2n+1}|^2 + \mu_{2n+1}|\beta_{2n+1}|^2) \leq r^2. \qquad (2.3.87)$$

The calculation of the ratio (2.3.82) gives (see problem 2.3.9, page 228)

$$\mathcal{J}_N(\sigma, \mu, r) = \frac{J_N(\sigma, mu, r)}{J_N(\sigma, \mu, \infty)} \qquad (2.3.88)$$

$$= \int_{-\infty}^{\infty} ds \, \frac{e^{isr^2} - e^{-isr^2}}{2\pi i s} F_N^{-1}(s) \qquad (2.3.89)$$

where

$$F_N(s) \stackrel{\text{def}}{\equiv} \prod_{n=-N}^{N} \left(1 - \frac{s^2 \hbar^2 \sigma_{2n+1}\mu_{2n+1}}{\pi^2(2n+1)^2}\right) \qquad (2.3.90)$$

s is an auxiliary variable and the contour of integration goes around the poles from below. Thus

$$\mathcal{J}(\sigma, \mu, r) = \lim_{N \to \infty} \frac{J_N(r)}{J_N(\infty)}$$

$$= \int_{-\infty}^{\infty} ds \, \frac{e^{isr^2} - e^{-isr^2}}{2\pi i s} F_\infty^{-1}(s). \qquad (2.3.91)$$

The product (2.3.90) has the limit $N \to \infty$, if the sum

$$\sum_{n=-\infty}^{\infty} \frac{\sigma_{2n+1}\mu_{2n+1}}{(2n+1)^2} \qquad (2.3.92)$$

converges. For the case under consideration (the space $\mathcal{X}$), $\sigma_{2n+1} = \mu_{2n+1} = 1$ and the latter condition is fulfilled. The first factor in the integrand of (2.3.91) in the limit $r \to \infty$ becomes the δ-function $\delta(s)$, so that (2.3.91) in this limit reads as

$$\lim_{r\to\infty} \mathcal{J}(r) \equiv \lim_{r\to\infty} \mathcal{J}(1,1,r) = \int_{-\infty}^{\infty} ds \, \delta(s) F_\infty^{-1}(s)$$

$$= F_\infty^{-1}(0) = 1. \qquad (2.3.93)$$

Corollary 2.1 (the uncertainty principle). *There exists the space of trajectories in a classical phase space supporting the phase-space path integral, in which the smoothness of the coordinate variables can be increased at the expense of the momentum variables and vice versa.*

Proof. Let $\mathcal{X}^{\sigma,\mu}$ be the Hilbert space of trajectories with the new scalar product

$$\langle \xi | \xi \rangle \stackrel{\text{def}}{\equiv} \sum_n (\sigma_{2n+1} |\alpha_{2n+1}|^2 + \mu_{2n+1} |\beta_{2n+1}|^2). \tag{2.3.94}$$

The ellipsoid (2.3.87) becomes a ball in the space $\mathcal{X}^{\sigma,\mu}$. Now we can choose

$$\sigma_{2n+1} = (2n+1)^{1+\varepsilon} \qquad \mu_{2n+1} = (2n+1)^{-2\varepsilon} \tag{2.3.95}$$

where ε is some small positive number. This choice obviously satisfies (2.3.92). With this choice of σ and μ, the scalar product (2.3.94) is well defined if the coefficients α_{2n+1} of the momentum expansion are square summable:

$$\sum_{n=-\infty}^{\infty} |\alpha_{2n+1}|^2 < \infty. \tag{2.3.96}$$

Condition (2.3.96) implies that momenta are continuous functions of time. In contrast to this, the smoothness of coordinates, due to the choice of μ_{2n+1} in (2.3.95), is worse than that of square-integrable functions. For $\varepsilon > \frac{1}{2}$, the coordinates, as functions of time, may turn out to be even distributions (generalized function). Exchanging the properties of σ and μ, we can increase the smoothness of coordinates at the expense of momenta. □

This result is in good accord with the uncertainty principle. Indeed, we can measure the values of a function at some point only if the function is continuous at that point. Hence, according to the uncertainty principle, a coordinate and the corresponding momentum cannot be continuous simultaneously. The better we want to know, say a coordinate, and hence increase its smoothness, the worse the smoothness of the corresponding momentum becomes and vice versa.

In any case, the space $\mathcal{X}$ with the most natural scalar product (2.3.74) is suitable for calculating phase-space path integrals. Sometimes, the spaces $\mathcal{X}^{\sigma,\mu}$ provide better convergence for finite-dimensional approximations of path integrals. A reasonable choice for the functional space depends on a problem: for example, if the Hamiltonian has the standard form $\widehat{H} = \widehat{p}^2/(2m) + V(\widehat{x})$ and we are going to transform the phase-space path integral into the Feynman one (integrating over the momentum), it is natural (according to the Wiener theorem and the close relation of the Feynman and Wiener integrals) to choose a space which guarantees continuity (but not differentiability) of coordinate variables as functions of time.

2.3.3 Normal symbol for the evolution operator, coherent-state path integrals, perturbation expansion and scattering operator

In this subsection, we consider again the case of one degree of freedom. Generalization to d degrees of freedom is quite straightforward (it consists mainly of substitutions of expressions $\widehat{a}^\dagger \widehat{a}, z^*z, zz, \ldots$ by $\sum_{j=1}^d \widehat{a}_j^\dagger \widehat{a}_j, \sum_{j=1}^d z^*_j z_j, \sum_{j=1}^d z_j z_j, \ldots$, respectively).

◇ **Normal symbol for the evolution operator**

Let

$$\widehat{H}(\widehat{a}^\dagger, \widehat{a}) = \sum_{mk} (\widehat{a}^\dagger)^m \widehat{a}^k \tag{2.3.97}$$

be some Hamiltonian constructed of creation and annihilation operators (2.3.36), (2.3.37) and $H(z^*, z)$ be its normal symbol:

$$H(z^*, z) = \sum_{mk} (z^*)^m z^k. \tag{2.3.98}$$

Using the formula (2.3.58) for the $\star$-product of normal symbols and acting as in the derivation of the Weyl symbol for the evolution operator, we can write the discrete-time approximation for the normal symbol of the evolution operator:

$$U_N(z^*, z) = \frac{1}{(2\pi \hbar)^{(N-1)}} \int dz^*_1 \, dz_1 \cdots dz^*_{N-1} \, dz_{N-1} \, e^{\frac{1}{\hbar} \Theta_N}$$
$$\times U_1(z^*, z_N) U_2(z^*_N, z_{N-1}) \cdots U_N(z^*_1, z) \tag{2.3.99}$$

where

$$\Theta_N \stackrel{\text{def}}{=} (z^* - z^*_N) z_N + \sum_{j=2}^{N} (z^*_j - z^*_{j-1}) z_{j-1} + (z^*_1 - z^*) z \tag{2.3.100}$$

$$z_j = z(\tau_j) \qquad z^*_j = z^*(\tau_j)$$

$$U_j(z^*_{j+1}, z_j) = \exp\left\{-\frac{i}{\hbar} \varepsilon H(z^*_{j+1}, z_j; \tau_j)\right\}.$$

Recall that $z^*(\tau), z(\tau)$ are the complex coordinates

$$z(\tau) = \frac{1}{\sqrt{2}} (q(\tau) + ip(\tau))$$

$$z^*(\tau) = \frac{1}{\sqrt{2}} (q(\tau) - ip(\tau)) \tag{2.3.101}$$

$$\tau_j = t_0 + \frac{j}{N+1} (t - t_0) \qquad \varepsilon = \frac{t - t_0}{N+1}.$$

Note that the boundary conditions for trajectories in the case of the normal symbol are quite unusual: we fix an *initial* point for $z(\tau)$ and a *final* one for $z^*(\tau)$,

$$z^*(t) = z^* \qquad z(t_0) = z \tag{2.3.102}$$

while $z^*(t_0), z(t)$ are *variables of integration*. This is in sharp contrast to the phase-space path integrals which we have considered before: while in the latter the boundary conditions are fixed for a single function (coordinate $x(t) = x$, $x(t_0) = x_0$ or momentum $p(t) = p$, $p(t_0) = p_0$), in the path integral (2.3.99) for the normal symbol we have to fix different functions: $z^*(\tau)$ at $\tau = t$ and $z(\tau)$ at $\tau = t_0$.

In the limit $N \to \infty$, the symbol (2.3.99) turns into the path integral

$$U(z^*, z; t, t_0) = \int_{C\{z^*(t) = z^* | z(t_0) = z\}} \mathcal{D}z^*(\tau) \mathcal{D}z(\tau) \exp\left\{-\frac{i}{\hbar} \int_{t_0+0}^{t} d\tau \, H(z^*(\tau), z(\tau - 0); \tau)\right.$$
$$\left. + \frac{1}{\hbar} \int_{t_0+0}^{t} d\tau \left[\frac{dz^*(\tau)}{d\tau} z(\tau - 0)\right] + \frac{1}{\hbar} (z^*(t_0 + 0) - z^*) z\right\}. \tag{2.3.103}$$

The shifted arguments in (2.3.103) reflect the corresponding shifts in the discrete approximation expressions:

$$\sum_{j=0}^{N} \varepsilon H(z^*(\tau_{j+1}), z(\tau_j); \tau_{j+1}) \longrightarrow \int_{t_0+0}^{t} d\tau \, H(z^*(\tau), z(\tau - 0); \tau) \tag{2.3.104}$$

218   *Path integrals in quantum mechanics*

$$\sum_{j=2}^{N+1}(z^*(\tau_j) - z^*(\tau_{j-1}))z(\tau_{j-1}) = \sum \varepsilon \frac{dz^*}{d\tau}\bigg|_{\tau=\tilde{\tau}_j} z(\tau_{j-1})$$

$$\longrightarrow \int_{t_0+0}^{t} d\tau \left[\frac{dz^*(\tau)}{d\tau} z(\tau - 0)\right] \quad (2.3.105)$$

$$(z^*(\tau_1) - z^*)z \longrightarrow (z^*(t_0 + 0) - z^*)z \quad (2.3.106)$$

where the equality in the second line has been obtained using the Lagrange formula and where $\tau_{j-1} \leq \tilde{\tau}_j \leq \tau_j$. Of course, if the functions $z^*(\tau), z(\tau)$ were continuously differentiable functions, we could just drop all the shifts. But as we have learned in the preceding subsection, in the case of the Weyl symbol, functions supporting the path integral are discontinuous, so that the shifts must be taken into account. Moreover, in a rigorous analysis of the path integral, even the inequality of shifts in the formulae (2.3.105) and (2.3.104) turns out to be essential. (The inequality follows from the fact that $\tilde{\tau}_{j+1} - \tau_{j-1} \leq \varepsilon$.)

We shall not go into the details of the rigorous analysis of the finite-dimensional approximation of path integrals for normal symbols (see Berezin (1981)). Essentially, constructing the general approximation and defining the path integral as an appropriate $N \to \infty$ limit of the *ratio* of *two* finite-dimensional integrals follows the same path as in the case of the Weyl symbols (cf the preceding subsection).

◇ **Coherent-state path integrals**

As we have pointed out in section 2.3.1, the normal symbol for an operator is closely related to the coherent states (cf (2.3.51)). Quite often, the path integral (2.3.103) for the normal symbol of the evolution operator is called the *coherent-state path integral*. Similar to the case of the path integral in the coordinate representation (that is, the path-integral representation for Weyl, px- or xp-symbols), cf problem 2.2.1, page 190, the path integral (2.3.103) can be directly derived with the use of these states without exploiting the formalism of operator symbols. (However, it is worth stressing again that the operator symbol formalism allows us to give a clear interpretation for the quantum-mechanical path integral, to treat the operator ordering problem consistently in terms of path integrals and to consider all the variety of different forms of path integrals within a unified approach.)

The coherent states $|\Upsilon_z\rangle$ introduced in section 2.3.1 (see equation (2.3.44)) are *holomorphic* functions of the label z (this property is very helpful for many applications of this basis) but they are not normalized. To derive the path integral (2.3.103) directly in terms of the coherent states, it is more convenient to use *normalized* coherent states $|z\rangle$:

$$|z\rangle \stackrel{\text{def}}{\equiv} \frac{|\Upsilon_z\rangle}{\sqrt{\langle\Upsilon_z|\Upsilon_z\rangle}}. \quad (2.3.107)$$

These normalized states are generated from the ground state vector $|0\rangle$ of the corresponding creation and annihilation operators (cf (2.3.36), (2.3.37), (2.3.40) and (2.3.41)) by the operator $\exp\{z\hat{a}^\dagger - z^*\hat{a}\}$:

$$|z\rangle = \exp\{z\hat{a}^\dagger - z^*\hat{a}\}|0\rangle$$
$$= \exp\left\{-\frac{1}{2}|z|^2\right\} \exp\{z\hat{a}^\dagger\} \exp\{-z^*\hat{a}\}|0\rangle$$
$$= \exp\left\{-\frac{1}{2}|z|^2\right\} \exp\{z\hat{a}^\dagger\}|0\rangle$$
$$= \exp\left\{-\frac{1}{2}|z|^2\right\} \sum_{n=0}^{\infty} \frac{1}{\sqrt{n!}} z^n |n\rangle \quad (2.3.108)$$

($|n\rangle$ is the eigenvector of the number operator: $\widehat{a}^\dagger\widehat{a}|n\rangle = n|n\rangle$). Here we have used the Baker–Campbell–Hausdorff formula

$$e^{\widehat{A}+\widehat{B}} = e^{[\widehat{A},\widehat{B}]/2} e^{\widehat{A}} e^{\widehat{B}} \qquad (2.3.109)$$

which is valid whenever $[\widehat{A}, \widehat{B}]$ commutes with both $\widehat{A}$ and $\widehat{B}$. We denote the normalized vectors as $|z\rangle$ following the general custom but we should remember that now these states are not holomorphic (or anti-holomorphic) functions of z due to the normalization factor. The set of these states, just like the set of $|\Upsilon_z\rangle$, is not orthogonal

$$\langle z_1|z_2\rangle = \exp\{-\tfrac{1}{2}|z_1|^2 - \tfrac{1}{2}|z_2|^2 + z_1^* z_2\} \qquad (2.3.110)$$

(but it is seen that $\langle z|z\rangle = 1$) and satisfies the condition of the resolution of unity:

$$\pi^{-1}\int d^2z\,|z\rangle\langle z| = \pi^{-1}\sum_{n,m}\frac{1}{\sqrt{n!}}\frac{1}{\sqrt{m!}}\int d(\operatorname{Re} z)\,d(\operatorname{Im} z)\,\exp\{-|z|^2\}(z^*)^n z^m |m\rangle\langle n|$$

$$= \sum_n |n\rangle\langle n| = \mathbb{I}. \qquad (2.3.111)$$

As we have mentioned in section 2.3.1, this relation means that the coherent states form an *overcomplete basis*.

Using the standard method of subdividing the action of the evolution operator into infinitesimal shifts (see section 2.2.1) and inserting the resolution of unity (2.3.111) in terms of the coherent states, we find that

$$\langle z|\exp\left\{-\frac{i}{\hbar}T\widehat{H}\right\}|z_0\rangle = \lim_{N\to\infty}\int\left[\prod_{j=1}^N d^2 z_j\right]\prod_{j=0}^N \langle z_{j+1}|(\mathbb{I} - i\varepsilon\widehat{H}/\hbar)|z_j\rangle$$

$$= \lim_{N\to\infty}\int\left[\prod_{j=1}^N d^2 z_n\right]\prod_{j=0}^N \langle z_{j+1}|z_n\rangle\left[1 - \frac{i}{\hbar}\varepsilon H(z_{j+1}, z_j)\right]$$

$$= \lim_{N\to\infty}\int\left[\prod_{j=1}^N d^2 z_n\right]\prod_{j=0}^N \langle z_{j+1}|z_n\rangle \exp\left\{-\frac{i}{\hbar}\varepsilon H(z_{j+1}, z_j)\right\} \qquad (2.3.112)$$

where $T = t - t_0$, $\varepsilon = T/(N+1)$, $z_{N+1} = z$ and $H(z_{j+1}, z_j) = \langle z_{j+1}|\widehat{H}|z_j\rangle$ is the normal symbol of the Hamiltonian operator $\widehat{H}$ (cf (2.3.53)). Using the approximate equality

$$\langle z_{j+1}|z_j\rangle = 1 - \langle z_{j+1}|(|z_{j+1}\rangle - |z_j\rangle)$$
$$\approx \exp\{\langle z_{j+1}|(|z_{j+1}\rangle - |z_j\rangle)\} \qquad (2.3.113)$$

and interchanging (as usual, this operation is not quite justified mathematically) the order of integration and the limit $N\to\infty$, we finally arrive at the formal continuous expression for the path integral:

$$U(z^*, z; t, t_0) = \langle z, t|z_0, t_0\rangle = \langle z|\exp\left\{-\frac{i}{\hbar}(t-t_0)\widehat{H}\right\}|z_0\rangle$$

$$= \int_{\mathcal{C}\{z^*(t)=z^*|z(t_0)=z\}} \mathcal{D}z^*(\tau)\mathcal{D}z(\tau)$$

$$\times \exp\left\{\frac{i}{\hbar}\int_{t_0}^t d\tau\,[i\hbar\langle z(\tau)|\dot{z}(\tau)\rangle - H(z^*(\tau), z(\tau))]\right\} \qquad (2.3.114)$$

where we have introduced
$$|\dot{z}(\tau)\rangle \stackrel{\text{def}}{\equiv} \frac{d}{d\tau}|z(\tau)\rangle.$$

Making use of relation (2.3.110) for the overlap of the coherent states, it is easy to verify that the path integral (2.3.114) coincides with that in equation (2.3.103) including the infinitesimal shifts of the time variable which we have dropped for simplicity in expression (2.3.114).

For the trace formula,
$$\operatorname{Tr} e^{-iT\widehat{H}(a^\dagger,a)} = \int \frac{dz^* dz}{\pi} \langle z|e^{-iT\widehat{H}(a^\dagger,a)}|z\rangle$$
$$= \lim_{N\to\infty} \prod_{j=1}^{N} \int_{z=z_0} \frac{dz_j^* dz_j}{\pi} \exp\{-z_j^*(z_j - z_{j-1}) - i\epsilon H(z_j^*, z_{j-1})\} \quad (2.3.115)$$

the boundary condition becomes a periodic one: $z_N = z_0$.

Note that the coherent-state path integral can be written in another form. To this aim, we introduce the so-called *contravariant symbol* $h(z^*, z)$ of the Hamiltonian operator $\widehat{H}$ (Berezin 1981), which is defined by the relation
$$\widehat{H} = \int d^2z\, h(z^*, z)|z\rangle\langle z| \qquad (2.3.116)$$

(assuming that such a representation does exist for a given operator $\widehat{H}$). Imposing the requirement that
$$\int d^2z_1\, d^2z_2\, |h(z_1^*, z_1)h(z_2^*, z_2)\langle z|z_1\rangle\langle z_1|z_2\rangle\langle z_2|z\rangle| < \infty \qquad (2.3.117)$$

for all $|z\rangle$, we can prove (Chernoff 1968) that the operator
$$\left[\int d^2z\, \exp\left\{-\frac{i}{\hbar}\frac{(t-t_0)}{N}h(z^*, z)\right\}|z\rangle\langle z|\right]^N$$

strongly converges to the evolution operator $\exp\{-i(t-t_0)\widehat{H}/\hbar\}$ as $N \to \infty$. This fact allows us to present the normal symbol for the evolution operator as the following path integral:
$$\langle z|\exp\left\{-\frac{i}{\hbar}(t-t_0)\widehat{H}\right\}|z_0\rangle = \lim_{N\to\infty} \int \left[\prod_{j=1}^{N} d^2z_n \exp\left\{-\frac{i}{\hbar}\varepsilon h(z_{j+1}, z_j)\right\}\right] \prod_{j=0}^{N} \langle z_{j+1}|z_n\rangle. \quad (2.3.118)$$

The continuous-like form of this path integral coincides with (2.3.103) only after the substitution $H(z_{j+1}, z_j) \to h(z_{j+1}, z_j)$. This shows once again a formal nature of the continuous-time form of the phase-space path integral. Note that $H_{\text{cl}} = H(z_{j+1}, z_j) + \mathcal{O}(\hbar) = h(z_{j+1}, z_j)$, so that both forms coincide in the classical limit, in which $\hbar \to 0$.

A very interesting property of the coherent space path integral is that it can be obtained as a limit of the well-defined *Wiener path integral* (Daubechies and Klauder 1985):
$$\langle z|\exp\left\{-\frac{i}{\hbar}t\widehat{H}\right\}|z_0\rangle = \lim_{N\to\infty} \int \left[\prod_{j=1}^{N} d^2z_n \exp\left\{-\frac{i}{\hbar}\varepsilon H(z_{j+1}, z_j)\right\}\right] \prod_{j=0}^{N} \langle z_{j+1}|z_n\rangle$$
$$= \lim_{\nu\to\infty} \int \mathcal{D}p\, \mathcal{D}x\, \exp\left\{-\frac{1}{2\nu}\int d\tau\, (\dot{p}^2 + \dot{x}^2)\right\} \exp\left\{i\int d\tau\, [p\dot{x} - H(p, x)]\right\}$$
$$= \lim_{\nu\to\infty} (2\pi)e^{\nu t/2} \int d\mu_W^\nu(p, x)\, \exp\left\{i\int dt\, [p\dot{x} - H(p, x)]\right\} \qquad (2.3.119)$$

where the phase variables p and x are related to z^*, z by (2.3.101). Thus the coherent-state path integral can be regularized (defined) through the sequence of the well-defined (see chapter 1) Wiener path integrals. The proof of this relation pertains to problem 2.3.10, page 229.

One more important advantage of the coherent state form of the path integral is the possibility of a straightforward generalization to group and homogeneous manifolds, due to the existence of generalized coherent states adjusted to the symmetries of these manifolds. The point is that the coherent states (2.3.107) (or (2.3.44)) are intrinsically related to the (nilpotent) Heisenberg–Weyl Lie group. There exists a generalization of the coherent states to the case of other Lie groups, e.g., $SU(2)$ and $SU(1,1)$ (see the collection of related reprints and the introductory article in Klauder and Skagerstam (1985)). The reader will find some further details on this subject in section 2.5.5.

◇ Derivation of the normal symbol of the evolution operator for an arbitrary Hamiltonian starting from a linear Hamiltonian

For verification that (2.3.103) gives indeed the correct symbol of the evolution operator and for its further elaboration, it is helpful to calculate the symbol for linear Hamiltonians of the form

$$\widehat{H}^{(J)} = J(\tau)\widehat{a}^\dagger + \bar{J}(\tau)\widehat{a} \qquad (2.3.120)$$

with the corresponding normal symbol

$$H^{(J)}(z^*, z) = J(\tau)z^* + \bar{J}(\tau)z. \qquad (2.3.121)$$

Since the path integral with such a symbol of the Hamiltonian has a quadratic polynomial in the exponent of the integrand, we can calculate it by any method which we have discussed in this and the preceding chapters. One should only be careful with the infinitesimal shifts of the time variable. The result reads as

$$U^{(J)}(z^*, z; t, t_0) = \exp\left\{-\frac{i}{\hbar}\int_{t_0}^{t} d\tau\, [z^* J(\tau) + \bar{J}(\tau)z] - \frac{1}{\hbar}\int d\tau\, ds\, \theta(s - \tau - 0)\bar{J}(s)J(\tau)\right\}. \qquad (2.3.122)$$

Of course, the infinitesimal shift $\theta(s - \tau - 0) = \lim_{\varepsilon \to 0} \theta(s - \tau - \varepsilon)$ of the step-function argument is not important for ordinary functions $\bar{J}, J$; but it becomes essential if $\bar{J}, J$ are distributions.

Now the symbol of the evolution operator for an arbitrary Hamiltonian $\widehat{H}$ can be represented in the form of an infinite series, making use of the Fourier transform of the Hamiltonian symbol

$$H(z^*, z; \tau) = \int dv^*\, dv\, e^{-i(z^*v + zv^*)} \widetilde{H}(v^*, v; \tau). \qquad (2.3.123)$$

The exponent $(z^*v + zv^*)$ in this representation for $H(z^*, z; \tau)$ can be interpreted as a particular case of the symbol for the linear Hamiltonian (cf (2.3.121)). This allows us to write

$$U^{(H)}(z^*, z; t, t_0) = \sum_{n=0}^{\infty} \left(-\frac{i}{\hbar}\right) \frac{1}{n!} U^{(n)}(z^*, z; t, t_0) \qquad (2.3.124)$$

where

$$U^{(n)}(z^*, z; t, t_0) = \int d^n v^*\, d^n v\, d^n \tau\, \exp\left\{-i\left[z^*\left(\sum_{k=1}^{n} v_k\right) + \left(\sum_{k=1}^{n} v^*_k\right)z\right]\right.$$
$$\left. - \hbar \sum_{k,l=1}^{n} v^*_k v_l \theta(\tau_k - \tau_l - 0)\right\} \widetilde{H}(v^*_1, v_1; \tau_1) \cdots \widetilde{H}(v^*_n, v_n; \tau_n). \qquad (2.3.125)$$

Here we have used (2.3.122) with $\bar{J}(\tau) = \hbar \sum_1^n v^*_k \delta(\tau - \tau_k)$, $J(\tau) = \hbar \sum_1^n v_k \delta(\tau - \tau_k)$.

These formulae can be represented in a more convenient form with the help of the operator $\widehat{L}$ in the space of functionals:

$$\widehat{L} \stackrel{\text{def}}{=} -\mathrm{i} \int_{t_0 \leq \tau, s \leq t} d\tau\, ds\, \theta(s - \tau - 0) \frac{\delta}{\delta z(s)} \frac{\delta}{\delta z^*(\tau)}. \qquad (2.3.126)$$

The use of this operator is based on the fact that the functional

$$u[z^*(\tau), z(\tau)] \stackrel{\text{def}}{=} \exp\left\{-\frac{\mathrm{i}}{\hbar} \int_{t_0}^t d\tau\, H(z^*, z; \tau)\right\}$$

for the *linear* Hamiltonian $H^{(J)}(z^*, z; \tau) = J(\tau) z^* + \bar{J}(\tau) z$, proves to be the eigenfunction of $\widehat{L}$:

$$\widehat{L} u = \lambda u$$
$$\lambda = \mathrm{i}\hbar^2 \int d\tau\, ds\, \theta(s - \tau - 0) \bar{J}(s) J(\tau). \qquad (2.3.127)$$

Thus we can rewrite (2.3.122) in the form

$$U^{(J)}(z^*, z; t, t_0) = \mathrm{e}^{\mathrm{i}\widehat{L}} \exp\left\{-\frac{\mathrm{i}}{\hbar} \int_{t_0}^d \tau\, H^{(J)}(z^*, z; \tau)\right\}\bigg|_{\substack{z(t_0)=z \\ z^*(t)=z^*}}.$$

This formula looks like a redundant complication of (2.3.122). But using (2.3.125), we can straightforwardly verify that (2.3.128) is correct for an *arbitrary* Hamiltonian:

$$U^{(H)}(z^*, z; t, t_0) = \mathrm{e}^{\mathrm{i}\widehat{L}} \exp\left\{-\frac{\mathrm{i}}{\hbar} \int_{t_0}^t d\tau\, H(z^*, z; \tau)\right\}\bigg|_{\substack{z(t_0)=z \\ z^*(t)=z^*}} \qquad (2.3.128)$$

(we leave this verification to the reader as a useful exercise).

Note that all the shifts of time variables have been reduced in the latter expression to the only shift in formula (2.3.126) and the net result of this shift is the condition

$$LH(z^*, z; \tau) = 0 \qquad (2.3.129)$$

(because $H(z^*, z; \tau)$ depends on $z^*(\tau)$, $z(\tau)$ with *coinciding* time arguments). The meaning of this condition is quite transparent. Indeed, our aim is to find normal symbols of operators. By construction, the Hamiltonian $\widehat{H}$ was taken in the *normal form* from the very beginning (cf (2.3.97)) and $H(z^*, z; \tau)$ is its normal symbol, therefore we do not have to transform it any further. In contrast, the evolution operator $\exp\{-\mathrm{i}Ht/\hbar\}$ does not have a normal form and to find its normal symbol we used the path-integral representation which was then converted into formula (2.3.128). Representation (2.3.128) (together with (2.3.129)) may serve as a background for calculating the evolution and related operators by the perturbation expansion.

◇ Perturbation expansion for the scattering operator

Let us consider an application of the formula (2.3.128) for the derivation of the perturbation series of the *scattering operator*, also called the *S-matrix*.

We start by briefly recalling some basic facts on the scattering operator (for details, see, e.g., Taylor (1972)). The basic assumption for constructing a scattering operator is that the so-called *asymptotic states* $\psi_\pm$, i.e. the states of a system at $t \to \pm\infty$, are defined by some operator $\widehat{H}_0$, called the *free Hamiltonian*, which differs from the exact Hamiltonian $\widehat{H}(t)$, the latter governing the evolution of the system according to relations (2.1.8), (2.1.9) and (2.1.11). These asymptotic states are defined by the relations

$$\lim_{t\to+\infty} \psi(t) = e^{-\frac{i}{\hbar}\widehat{H}_0 t}\psi_+$$

$$\lim_{t\to-\infty} \psi(t) = e^{-\frac{i}{\hbar}\widehat{H}_0 t}\psi_-. \tag{2.3.130}$$

The state $\psi(t)$ is supposed to be the exact evolving state of the system

$$\psi(t) = \widehat{U}(t,0)\varphi \tag{2.3.131}$$

with initial state φ at moment $t_0 = 0$. Relations (2.3.130) imply that $\psi_\pm$ linearly depend on φ:

$$\psi_\pm = \widehat{V}_\pm \varphi \tag{2.3.132}$$

where

$$\widehat{V}_\pm = \lim_{t\to\pm\infty} \widehat{V}(t)$$

$$\widehat{V}(t) = e^{\frac{i}{\hbar}t\widehat{H}_0}\widehat{U}(t,0). \tag{2.3.133}$$

Thus, according to (2.3.132),

$$\psi_+ = \widehat{V}_+ \widehat{V}_-^{-1} \psi_-. \tag{2.3.134}$$

The operator

$$\widehat{S} \stackrel{\text{def}}{\equiv} \widehat{V}_+ \widehat{V}_-^{-1} \tag{2.3.135}$$

is called the *scattering operator*. It can be related to the evolution operator and free Hamiltonian as follows:

$$\widehat{S} = \lim_{\substack{t\to+\infty \\ t'\to-\infty}} \widehat{S}(t,t') \tag{2.3.136}$$

where

$$\widehat{S}(t,t') = \widehat{V}(t)\widehat{V}^{-1}(t')$$
$$= e^{\frac{i}{\hbar}t\widehat{H}_0}\widehat{U}(t,0)\widehat{U}^{-1}(t',0)e^{-\frac{i}{\hbar}t'\widehat{H}_0}$$
$$= e^{\frac{i}{\hbar}t\widehat{H}_0}\widehat{U}(t,t')e^{-\frac{i}{\hbar}t'\widehat{H}_0}. \tag{2.3.137}$$

From (2.3.137), we derive the evolution equation for $\widehat{S}(t,t')$:

$$\left(i\hbar\frac{\partial}{\partial t} - \widehat{H}_1(t)\right)\widehat{S}(t,t') = 0 \tag{2.3.138}$$

$$\widehat{S}(t',t') = \mathbb{I} \tag{2.3.139}$$

where

$$\widehat{H}_1 = e^{\frac{i}{\hbar}t\widehat{H}_0}\widehat{H}_{\text{int}}(t,t')e^{-\frac{i}{\hbar}t'\widehat{H}_0} \tag{2.3.140}$$

$$\widehat{H}_{\text{int}} = \widehat{H}(t) - \widehat{H}_0. \tag{2.3.141}$$

Relations (2.3.137)–(2.3.141) define the *interaction representation* in quantum mechanics for the scattering operator.

224   *Path integrals in quantum mechanics*

In many applications, the interaction Hamiltonian has the form

$$\widehat{H}_{\text{int}} = \widehat{H}_V e^{\alpha|t|} \qquad \alpha > 0 \tag{2.3.142}$$

with $\widehat{H}_V$ being independent of t. In these cases, the scattering operator depends on the parameter α and is called the *adiabatic* scattering operator. In particular, in non-relativistic quantum mechanics, the possible choice is

$$\widehat{H}_0 = \widehat{H}_0^{(\text{f.p.})} \equiv \frac{1}{2m}\widehat{p}^2 \tag{2.3.143}$$

$$\widehat{H}_V = \widehat{V}(\widehat{x}) \tag{2.3.144}$$

(f.p. stands for free particle) and if $V(x)$ is a rapidly decreasing potential, there exists the limit $\alpha \to 0$ of the adiabatic scattering operator.

With the choice (2.3.143) and (2.3.144), it is natural to construct the scattering operator in terms of some symbol based on the canonical variables p, x (e.g., Weyl, px- or xp-symbols). Then, using an analog of formula (2.3.128) for these symbols, we can develop the perturbation expansion for the explicit calculation of this operator (i.e. its symbol or kernel) as a power series in the potential (2.3.144) (or, more precisely, some small parameter entering the potential).

Another option, especially important in many-body problems (quantum field theories, see chapter 3), is to take as $\widehat{H}_0$ the Hamiltonian of a harmonic oscillator, which, in terms of annihilation and creation operators, has the form

$$\widehat{H}_0^{(\text{harm})} = \hbar\omega\widehat{a}^\dagger\widehat{a}. \tag{2.3.145}$$

Then the rest part $\widehat{H}_{\text{int}}$ of the total Hamiltonian $\widehat{H}$ can be written as follows:

$$\widehat{H}_1 = \sum_{mn} C_{mn}(\widehat{a}^\dagger)^m \widehat{a}^n \qquad c_{mn} \in \mathbb{R}. \tag{2.3.146}$$

Since the Hamiltonian $\widehat{H}_0^{(\text{harm})}$ is quadratic in the operators $\widehat{a}^\dagger, \widehat{a}$, the path integral (2.3.103) can be calculated exactly. The equations for an extremal trajectory corresponding to the exponential in (2.3.103) read:

$$\frac{dz(\tau)}{d\tau} + i\omega z(\tau) = 0$$

$$\frac{dz^*(\tau)}{d\tau} - i\omega z(\tau) = 0 \tag{2.3.147}$$

$$z^*(t) = z^* \qquad z(t_0) = z.$$

The solution of these equations is given by the expressions

$$z^*(\tau) = z^* e^{i\omega t} \tag{2.3.148}$$

$$z(\tau) = z e^{-i\omega t}. \tag{2.3.149}$$

The path integral for the Hamiltonian (2.3.156) has a Gaussian form and, according to our previous calculations, it equals the integrand evaluated at the extremal trajectories (2.3.148) and (2.3.149). The result for the symbol of the evolution operator is

$$U^{(\text{harm})}(z^*, z; t, t_0) = \exp\left\{\frac{1}{\hbar}(e^{i\omega(t-t_0)} - 1)z^*z\right\} \tag{2.3.150}$$

and for the integral kernel:

$$K^{(\text{harm})}(z^*, z; t, t_0) = \exp\left\{\frac{1}{\hbar} e^{i\omega(t-t_0)} z^* z\right\}. \tag{2.3.151}$$

From the latter formula, it follows that if $f(z^*)$ is an arbitrary vector from $\mathcal{F}^2$ (cf (2.3.39)), its evolution generated by the Hamiltonian (2.3.145) will be

$$\widehat{U}^{(\text{harm})}(t-t_0) f(z^*) = \frac{1}{2\pi i} \int d\bar{\xi}\, d\xi\, e^{-\bar{\xi}\xi} \exp\{z^*\xi e^{-i\omega t}\} f(\bar{\xi})$$
$$= f(z^* e^{-i\omega t}). \tag{2.3.152}$$

This means, in turn, that if any operator $\widehat{A}$ has a kernel $A(z^*, z)$, the operator

$$\exp\left\{\frac{i}{\hbar}\widehat{H}_0^{(\text{harm})} t\right\} \widehat{A} \exp\left\{-\frac{i}{\hbar}\widehat{H}_0^{(\text{harm})} t_0\right\}$$

has the kernel

$$A(z^* e^{i\omega t}, z e^{-i\omega t_0}). \tag{2.3.153}$$

Applying this formula to the scattering operator and using the interaction representation, we can see that the path integral for $\widehat{S}(t, t')$ in (2.3.138) and (2.3.139) coincides with that for the evolution operator generated by an operator with the normal symbol

$$H_1(z^* e^{i\omega\tau}, z e^{-i\omega\tau}; \tau) \tag{2.3.154}$$

where $H_1(z^*, z; \tau)$ is the symbol of the operator $\widehat{H}_1$. Then using (2.3.128), we obtain

$$S(z^*, z; t, t_0) = e^{i\widehat{L}} \exp\left\{-\frac{i}{\hbar} \int_{t_0}^{t} d\tau\, H_1(z^* e^{i\omega\tau}, z e^{-i\omega\tau}; \tau)\right\}\bigg|_{\substack{z(\tau)=z \\ z^*(\tau)=\zeta}} \tag{2.3.155}$$

where the operator $\widehat{L}$ is defined in (2.3.126). If $\widehat{H}_1$ can be considered as a small perturbation for the Hamiltonian $H_0^{(\text{harm})}$ (for example, H_1 contains a small overall constant factor), this formula can be used as a basis for perturbation theory. In fact, it reproduces the well-known *Wick theorem* (see e.g., Itzykson and Zuber (1980) and more on this subject in chapter 3). For the perturbative calculation, we expand the second exponent in (2.3.155) up to a desirable order in H_1 and find the result of the action of the operator $\exp\{i\widehat{L}\}$.

One more possibility for the development of the perturbation expansion is to calculate the path integral for the quadratic Hamiltonian

$$\widehat{H}_0^{(J)}(t) = \omega \widehat{a}^\dagger \widehat{a} + \bar{J}(t)\widehat{a} + \widehat{a}^\dagger J(t). \tag{2.3.156}$$

with linear terms, so that the corresponding symbol or kernel of the evolution operator depends on the external functions $\bar{J}, J$ and represent the additional terms of the total Hamiltonian using functional derivatives:

$$H_1(z^*(\tau), z(\tau)) = H_1\left(\frac{1}{i}\frac{\delta}{\delta J(\tau)}, \frac{1}{i}\frac{\delta}{\delta \bar{J}(\tau)}\right) \exp\left\{\int_{t_0}^{t} ds\, \bar{J}(s) z(s) + z(s) J(s)\right\}\bigg|_{\bar{J}=J=0}. \tag{2.3.157}$$

This again allows us to calculate the evolution operators order by order in powers of H_1, the perturbation expansion being essentially equivalent to that based on (2.3.155). The perturbation expansion starting from (2.3.157) is especially effective in quantum field theory and we shall consider it in detail in chapter 3.

2.3.4 Problems

Problem 2.3.1. Prove the consistency of formula (2.3.14) with the definition of the Weyl symbol given in (2.3.12) and (2.3.13).

Hint. Using the expansion
$$e^{i(sp+rx)} = \sum_{k=0}^{\infty} \frac{i^k}{k!}(sp+rx)^k$$
we might see that it is enough to check that the operator $(s\widehat{p}+r\widehat{x})^k$ has a Weyl symbol of the form $(sp+rx)^k$. This can be done using the formula
$$(sp+rx)^k = \sum_{k_1+\cdots+k_d=k} \frac{k!}{k_1!\cdots k_d!}(s_1p_1+r_1x_1)^{k_1}\cdots(s_dp_d+r_dx_d)^{k_d}$$
which is true even after the substitution $x_j \to \widehat{x}_j$, $p_j \to \widehat{p}_j$ because the operators $(s_j\widehat{p}_j+r_j\widehat{x}_j)$ and $(s_k\widehat{p}_k+r_k\widehat{x}_k)$ are commutative for $j \neq k$. Now to prove the statement it is enough to show that the operator $(s_j\widehat{p}_j+r_j\widehat{x}_j)^n$ has a Weyl symbol of the form $(s_jp_j+r_jx_j)$, but this follows straight from definitions (2.3.12) and (2.3.13).

Problem 2.3.2. Prove relation (2.3.17) between the kernel and Weyl symbol of an operator in $\mathcal{L}^2(\mathbb{R}^d)$.

Hint. According to the Baker–Campbell–Hausdorff formula (2.2.6),
$$e^{i(s\widehat{p}+r\widehat{x})} = e^{i\hbar rs/2}e^{ir\widehat{x}}e^{is\widehat{p}}.$$
This allows us to calculate the action of the operator on the left-hand side on an arbitrary function from $\mathcal{L}^2(\mathbb{R}^d)$
$$[e^{i(s\widehat{p}+r\widehat{x})}\psi](x) = e^{i(rx+\hbar rs/2)}\psi(x+\hbar s)$$
so that
$$(\widehat{f}\psi)(x) = \int_{-\infty}^{\infty} d^d r \, d^d s \, e^{i(rx+\hbar rs/2)}\widetilde{f}_W(r,s)\psi(x+\hbar s).$$
Substituting the Fourier transform (2.3.15), we obtain
$$(\widehat{f}\psi)(x) = \frac{1}{(2\pi)^{2d}}\int_{-\infty}^{\infty} d^d r \, d^d s \, d^d x' \, d^d p \, e^{i[r(x-x')-s(p-\hbar r/2)]} f(p,x')\psi(x+\hbar s)$$
$$= \frac{1}{(2\pi)^{2d}}\int_{-\infty}^{\infty} d^d r \, d^d s \, d^d x' \, d^d p \, e^{i[r(x-x'+\hbar s/2)-sp]} f(p,x')\psi(x+\hbar s)$$
$$= \frac{1}{(2\pi)^{2d}}\int_{-\infty}^{\infty} d^d s \, d^d x' \, d^d p \, \delta(x-x'+\hbar s/2) e^{-isp} f(p,x')\psi(x+\hbar s)$$
$$= \frac{1}{(2\pi)^{2d}}\int_{-\infty}^{\infty} d^d s \, d^d p \, e^{-isp} f(p, x+\hbar s/2)\psi(x+\hbar s).$$
Introducing the variable $y = x+\hbar s$ instead of s, we arrive at
$$(\widehat{f}\psi)(x) = \frac{1}{(2\pi)^{2d}}\int_{-\infty}^{\infty} d^d y \, d^d p \, e^{\frac{i}{\hbar}(x-y)p} f(p,(x+y)/2)\psi(y).$$
This proves (2.3.17).

Problem 2.3.3. Calculate the Weyl symbols of the following operators:

(1) $f(\hat{p})$, $f(\hat{x})$;
(2) $\hat{p}_i \hat{p}_j f(\hat{x})$, $\hat{p}_i f(\hat{x})\hat{p}_j$, $f(\hat{x})\hat{p}_i\hat{p}_j$;
(3) $f(\hat{x})\hat{p}_i g^{ij}(\hat{x})\hat{p}_j f(\hat{x})$.

Here f, g^{ij} are some smooth functions; $i, j = 1, \ldots, d$.

Hint. Applying formula (2.3.20), we have for the Weyl symbol f_w of the operator $f(\hat{p})$

$$f_w(p) = \int d^d v\, e^{ipv/\hbar} \left\langle x - \frac{v}{2} \middle| f(\hat{p}) \middle| x + \frac{v}{2} \right\rangle$$

$$= \int d^d v\, d^d p'\, e^{ipv/\hbar} \left\langle x - \frac{v}{2} \middle| f(\hat{p}) \middle| p' \right\rangle\left\langle p' \middle| x + \frac{v}{2} \right\rangle$$

$$= \frac{1}{(2\pi\hbar)^d} \int d^d v\, d^d p'\, e^{i(p-p')v/\hbar} f(p') = f(p). \tag{2.3.158}$$

Similarly, for $f(x)$, the correspondence with the Weyl symbol is

$$f(\hat{x}) \Leftrightarrow f(x). \tag{2.3.159}$$

Straightforwardly, we show the following Weyl correspondence:

$$\hat{p}_i \hat{p}_j f(\hat{x}) \Leftrightarrow \left(p_i - \frac{i\hbar}{2}\frac{\partial}{\partial x^i}\right)\left(p_j - \frac{i\hbar}{2}\frac{\partial}{\partial x^j}\right) f(x)$$

$$\hat{p}_i f(\hat{x}) \hat{p}_j \Leftrightarrow \left(p_i - \frac{i\hbar}{2}\frac{\partial}{\partial x^i}\right)\left(p_j + \frac{i\hbar}{2}\frac{\partial}{\partial x^j}\right) f(x) \tag{2.3.160}$$

$$f(\hat{x}) \hat{p}_i \hat{p}_j \Leftrightarrow \left(p_i + \frac{i\hbar}{2}\frac{\partial}{\partial x^i}\right)\left(p_j + \frac{i\hbar}{2}\frac{\partial}{\partial x^j}\right) f(x) \tag{2.3.161}$$

and

$$f(\hat{x})\hat{p}_i g^{ij}(\hat{x})\hat{p}_j f(\hat{x}) \Leftrightarrow p_i p_j f^2(x) g^{ij} + \tfrac{1}{2}(\tfrac{1}{2} f^2(x) g^{ij}{}_{,ij}(x) + f_{,i}(x) f_{,j}(x) g^{ij}(x) - f_{,ij}(x) f(x) g^{ij}(x)). \tag{2.3.162}$$

In the latter expression we have used the shorthand notation for the differentiation of an arbitrary function $F(x)$:

$$F_{,i}(x) \equiv \frac{\partial}{\partial x^i} f(x) \qquad F_{,ij}(x) \equiv \frac{\partial}{\partial x^i}\frac{\partial}{\partial x^j} f(x).$$

Problem 2.3.4. Prove formula (2.3.23) for the star-product of Weyl symbols.

Hint. Substituting into (2.3.22) the expression for the kernels in terms of Weyl symbols (cf (2.3.17)), we obtain

$$f_w(x,y) = \frac{1}{2^d(\pi\hbar)^{2d}} \int_{-\infty}^{\infty} d^d p_1\, d^d p_2\, d^d y\, d^d z\, \exp\left\{\frac{i}{\hbar}(x-y-z)p_1 + (z-x-y)p_2 + 2yp\right\}$$

$$\times (f_1)_w\left(p_1, \frac{x-y+z}{2}\right)(f_2)_w\left(p_2, \frac{z+x+y}{2}\right).$$

Introducing the new variables

$$x_1 = \frac{x-y+z}{2} \qquad x_2 = \frac{z+x+y}{2}$$

and taking into account that the Jacobian of the substitution is

$$\left|\frac{\partial(x_1, x_2)}{\partial(y, z)}\right| = \frac{1}{2^d}$$

we arrive at the formula (2.3.23).

Problem 2.3.5. Verify that the operators (2.3.40) and (2.3.41) are Hermitian conjugate with respect to the scalar product (2.3.39).

Hint. Use integration by parts.

Problem 2.3.6. Prove formula (2.3.58) for the star-product of normal symbols.

Hint. The following chain of equalities prove the required formula:

$$(A_1 \star A_2)(z^*, z) = e^{-\frac{1}{\hbar}z^*z}\langle \Upsilon_z | \widehat{A}_1 \widehat{A}_2 \Upsilon_z \rangle$$
$$= e^{-\frac{1}{\hbar}z^*z}\langle \widehat{A}_1^\dagger \Upsilon_z | \widehat{A}_2 \Upsilon_z \rangle$$
$$= e^{-\frac{1}{\hbar}z^*z} \int dv^* dv \, e^{-\frac{1}{\hbar}v^*v} A_1(z^*, v) e^{\frac{1}{\hbar}z^*v} A_2(v^*, z) e^{-\frac{1}{\hbar}v^*z}.$$

Problem 2.3.7. Calculate the Weyl symbol of the evolution operator for a harmonic oscillator and then obtain an expression for the transition amplitude.

Hint. Start from the discrete approximation (2.3.71) and, using the fact that all variables enter the exponential quadratically, calculate the integral by any method which we have discussed (e.g., by Gaussian integration over the variables one by one or by the stationary-phase method). Then use relation (2.3.17) to find the amplitude (integral kernel).

Problem 2.3.8. The basic formulae for the xp-ordering are the following:

(i) the relation between the xp-symbol $f_{xp}(x, p)$ and the corresponding kernel reads as

$$K_f(x, y) = \frac{1}{(2\pi\hbar)^d} \int_{-\infty}^{\infty} d^d p \, e^{\frac{i}{\hbar}(x-y)p} f_{xp}(x, p) \qquad (2.3.163)$$

(ii) the star-product of xp-symbols can be written as

$$(f_{xp} \star g_{xp})(p, x) = \frac{1}{(2\pi\hbar)^d} \int_{-\infty}^{\infty} d^d p_1 \, d^d x_1 \exp\left\{-\frac{i}{\hbar}(x_1 - x)(p_1 - p)\right\}$$
$$\times f_{xp}(p_1, x) g_{xp}(p, x_1). \qquad (2.3.164)$$

Using these formulae, construct the path-integral representation for the xp-symbol of the evolution operator, calculate it for a harmonic oscillator and compare the resulting expressions for the symbol and kernel with the results of the preceding problem (i.e. with calculations for the Weyl ordering rule).

Problem 2.3.9. Calculate the ratio (2.3.88) for the ellipsoid (2.3.87).

Hint. The integral in the exponent of $J_N(\sigma, \mu, r)$ in terms of Fourier coefficients has the form

$$\int_{t_0}^{t} d\tau \, \xi \omega \dot{\xi} = \pi i \sum_n (2n+1)[\alpha^*_{2n+1}\beta_{2n+1} - \beta^*_{2n+1}\alpha_{2n+1}].$$

Quantization, the operator ordering problem and path integrals

The integral over the ellipse can be written as follows:

$$J_N(\sigma,\mu,r) = \int_{-r^2}^{r^2} dv\, \delta\left(v - \sum_{|n|<N}(\sigma_{2n+1}|\alpha_{2n+1}|^2 + \mu_{2n+1}|\beta_{2n+1}|^2)\right)$$

$$\times \int \prod_{|n|<N} d\alpha^*_{2n+1}\, d\alpha_{2n+1}\, d\beta^*_{2n+1}\, d\beta_{2n+1}\, \exp\left\{\frac{i}{2\hbar}\int_{t_0}^{t} d\tau\, \xi \omega \dot{\xi}\right\}.$$

Then it is convenient to use the representation

$$\delta(u) = \frac{1}{2\pi}\int_{-\infty}^{\infty} ds\, e^{isu}$$

for δ-functions, which allows us to integrate easily over $\alpha^*_{2n+1}, \alpha_{2n+1}, \beta^*_{2n+1}, \beta_{2n+1}$ and v (for further details see Berezin (1981)).

Problem 2.3.10. Show that the limit

$$\lim_{v\to\infty} \int \mathcal{D}p\,\mathcal{D}x\, \exp\left\{-\frac{1}{2v}\int d\tau\,(\dot{p}^2+\dot{x}^2)\right\} \exp\left\{i\int d\tau[p\dot{x} - H(p,x)]\right\}$$

$$= \lim_{v\to\infty}(2\pi)e^{vT/2}\int d\mu^v_W(p,x)\, \exp\left\{i\int dt\,[p\dot{x} - H(p,x)]\right\} \qquad (2.3.165)$$

of the two-dimensional *Wiener* path integrals (cf (2.3.119)) gives the coherent state quantum-mechanical path integral defined by (2.3.112) and (2.3.114).

Hint. Using relation (2.3.101) between the complex and real coordinates of the phase space, and the explicit expression (2.3.110) for the overlap of coherent states, the discrete approximation for the coherent-state path integral can be cast in the form

$$\langle p,x|e^{-iT\hat{H}}|p_0,x_0\rangle \equiv \lim_{N\to\infty}\frac{1}{(2\pi)^N}\int\prod_{l=1}^{N} dp_l\, dx_l\, \exp\left\{\sum_{l=0}^{N}[i\tfrac{1}{2}(p_{l+1}+p_l)(x_{l+1}-x_l)\right.$$

$$-\tfrac{1}{4}[(p_{l+1}-p_l)^2+(x_{l+1}-x_l)^2]$$

$$\left.- i\varepsilon H(\tfrac{1}{2}(p_{l+1}+p_l+ix_{l+1}-ix_l),\tfrac{1}{2}(x_{l+1}+x_l-ip_{l+1}+ip_l))]\right\}.$$

Here the states $|p,x\rangle$ are defined by

$$|p,x\rangle \stackrel{\text{def}}{\equiv} e^{-ix\hat{p}}e^{ip\hat{x}}|0\rangle$$

and they coincide with (2.3.108) up to the notation (i.e. transition to the real coordinates). On the other hand, the Wiener path integral in (2.3.165) (or (2.3.119)) has quite a similar discrete-time approximation. Having in hand these discrete-time approximations for both path integrals it is straightforward to show that in the limit $v \to \infty$ the Wiener integral indeed becomes the coherent-state path integral in quantum mechanics (Daubechies and Klauder 1985).

2.4 Path integrals and quantization in spaces with topological constraints

So far, we have discussed the path-integral description of systems in unbounded Euclidean spaces, where each coordinate runs over the whole real line. However, for many physical systems, appropriate coordinates are confined to some restricted domain of the Euclidean space or correspond to topologically non-trivial compact manifolds. There is no doubt that the geometrical properties of a phase space are important characteristics of the corresponding Hamiltonian systems, so that both classical and quantum dynamics strongly depend on them. Consider, for example, a free particle in a one-dimensional space with the Hamiltonian $H_0 = p^2/(2m)$. If the corresponding phase space is assumed to be a plane $\mathbb{R}^2$, then classical trajectories are straight lines perpendicular to the momentum axis in the phase space and outgoing to infinity. In quantum theory, we have the spectrum of the Hamiltonian $E = p^2/(2m)$, $p \in \mathbb{R}$, and the eigenfunctions are the plane waves $\psi \sim \exp\{ipx\}$.

Contracting the configuration space to a finite size L can be achieved in different ways:

(1) One may identify the boundary points of the configuration space to turn it into a circle $S^{(1)}$ of length L; then the phase space becomes a cylinder $\mathbb{R} \otimes S^{(1)}$.
(2) Another way is to install infinite walls at the boundary points $x = 0, L$ to prevent a particle from penetrating outside the interval; then the phase space is a strip $\mathbb{R} \otimes (0, L)$.

The *classical* motion of the particle now becomes periodical. The system returns to its initial state (a point in the phase spaces) with the period:

(1) $T = L/v_0$, where v_0 is the (conserved) particle velocity, if the particle is on a circle;
(2) $T = 2L/v_0$ for a particle moving between two walls because the particle has to be reflected from both walls to reach the initial point (note that each reflection changes the sign of the particle momentum).

The *quantum* theories are also different. For the cylindrical phase space $\mathbb{R} \otimes S^1$, the simplest boundary conditions for wavefunctions are periodic:

$$\psi(\phi) = \psi(\phi + 2\pi). \tag{2.4.1}$$

This results in the discrete spectrum of a free particle on a circle:

$$E_n = \frac{2\pi^2 n^2 \hbar^2}{mL^2} \tag{2.4.2}$$

where L is the circumference of the circle, and the eigenfunctions are

$$\psi_n(\phi) = \frac{1}{2\pi} e^{in\phi} \qquad n = 1, 2, 3, \ldots. \tag{2.4.3}$$

Recall that two vectors describe the *same* state if they differ from each other by a phase factor (cf postulate 2.1 in section 2.1.1). Thus we have the right to equate wavefunctions in physically equivalent points up to a complex multiplier and impose a more general *quasi-periodic* boundary condition on a circle:

$$\psi(\phi + 2\pi) = e^{i\alpha} \psi(\phi). \tag{2.4.4}$$

We shall consider this possibility later.

For a particle confined inside the interval $[0, L]$ (in a 'box'), the wavefunctions have zero (Dirichlet) boundary conditions

$$\psi(0) = \psi(L) = 0 \tag{2.4.5}$$

and these lead to the following energy levels:

$$E_n = \frac{\pi^2 n^2 \hbar^2}{2mL^2} \qquad (2.4.6)$$

and the eigenfunctions

$$\psi_n(x) = \sqrt{\frac{2}{L}} \sin\left(\pi n \frac{x}{L}\right) \qquad n = 1, 2, 3, \ldots. \qquad (2.4.7)$$

The standard scheme for obtaining the path-integral solution of the Schrödinger equation, described in the preceding sections, cannot be *straightforwardly* applied to the quantum-mechanical problems with non-trivial geometry. This was mentioned by Pauli in his lectures of 1950–51 (see Pauli (1973)). Indeed, for a free particle in a one-dimensional 'box', i.e. inside the interval $[0, L]$ with infinite walls at the boundaries, the straightforward generalization of (2.2.30) (with $V = 0$), i.e.

$$K(x, t | x_0, t_0) = \lim_{N \to \infty} \frac{1}{\sqrt{2\pi \hbar i \varepsilon / m}} \prod_{j=1}^{N} \left[\int_0^L \frac{dx_j}{\sqrt{2\pi \hbar i \varepsilon / m}} \right] \exp\left\{ \frac{i}{\hbar} \sum_{j=1}^{N+1} \varepsilon \left[\frac{m}{2} \left(\frac{x_j - x_{j-1}}{\varepsilon} \right)^2 \right] \right\} \qquad (2.4.8)$$

does not give the correct result because the function $\psi(x, t)$ for the infinitesimally small time $t = \varepsilon \to 0$,

$$\psi(x, \varepsilon) = \int_0^L dx_0 \frac{1}{\sqrt{2\pi i \hbar \varepsilon / m}} \exp\left\{ \frac{im(x - x_0)^2}{2\hbar \varepsilon} \right\} \psi_0(x_0, 0) \qquad (2.4.9)$$

already does not satisfy the boundary condition at $x = 0$ and $x = L$, even if the initial wavefunction does satisfy

$$\psi_0(0, 0) = \psi_0(L, 0) = 0.$$

Note that the kernel and, hence, $\psi(x, t)$ are the solutions of the Schrödinger equation (because *inside* the interval $[0, L]$, the equation does not differ from the ordinary one).

Thus because of the wrong boundary behaviour, formula (2.4.8) does not give a solution to the problem and requires modification.

In the next subsection, we shall derive the correct path-integral representation for a transition amplitude in a box and on a half-line. In the subsequent part of this section we shall expand the consideration to systems with periodic boundary conditions: namely, a free particle on a circle and with a torus-shaped phase space.

2.4.1 Point particles in a box and on a half-line

The correct way for constructing the path integral for a particle in a box had also been described by Pauli (1973). The solution of the problem is based on the *method of images*: to provide the correct boundary values (2.4.5) of the wavefunctions, we have to subtract from the exponent in (2.4.9) an infinite number of analogous terms obtained by the reflection of x_0 with respect to the points $x = 0$ and $x = L$. For example, the subtraction

$$\exp\left\{ \frac{im(x - x_0)^2}{2\hbar \varepsilon} \right\} - \exp\left\{ \frac{im(x + x_0)^2}{2\hbar \varepsilon} \right\}$$

provides the correct behaviour at $x = 0$, while the subtraction

$$\exp\left\{ \frac{im(x - x_0)^2}{2\hbar \varepsilon} \right\} - \exp\left\{ \frac{im(x - 2L + x_0)^2}{2\hbar \varepsilon} \right\}$$

gives the required zero at $x = L$. But the subtracted terms, in turn, also violate the boundary conditions: the first one at $x = L$ and the second at $x = 0$. This requires further subtractions and leads to an infinite series for the integral kernel of the evolution operator:

$$K_{\text{box}}(x - x_0; t, t_0) = \sum_{n=-\infty}^{\infty} [K_0(x - x_0 + 2Ln; t, t_0) - K_0(x + x_0 + 2Ln; t, t_0)] \quad (2.4.10)$$

where $K(x - y; t, t_0)$ is the ordinary evolution kernel (2.2.41) for a free particle. Each term in this series satisfies the Schrödinger equation and the total sum has the correct boundary values:

$$K_{\text{box}}(L, t|x_0, t_0) = K_{\text{box}}(0, t|x_0, t_0) = 0.$$

Thus the problem has been solved. But this path-integral representation meets technical trouble because we have to integrate the Gaussian exponents over the finite interval $[0, L]$ (which cannot be done analytically). To circumvent this difficulty, we can use the following trick. The time evolution of a wavefunction $\psi_0(x_0, t_0)$ can be written as

$$\psi(x, t) = \int_0^L dx_0 \, K_{\text{box}}(x - x_0; t, t_0) \psi_0(x_0, t_0)$$

$$= \sum_{n=-\infty}^{\infty} \left[\int_{2nL}^{(2n+1)L} dx_0 \, K_0(x - x_0; t, t_0) - \int_{2(n-1)L}^{2nL} dx_0 \, K(x - x_0; t, t_0) \right] \psi_0(x_0, t_0)$$

$$(2.4.11)$$

where we have used obvious changes of variables. Now, if we expand the domain of definition of the state vector ψ_0 from the interval $[0, L]$ onto the whole real line $\mathbb{R}$ as a periodical and antisymmetric function

$$\psi_0(-x) = -\psi_0(x)$$
$$\psi_0(x + 2Ln) = \psi_0(x) \qquad n = 0, \pm 1, \pm 2, \ldots \quad (2.4.12)$$

the evolution formula (2.4.11) can be cast again into the standard form

$$\psi(x, t) = \int_{-\infty}^{\infty} dx_0 \, K_0(x - x_0; t, t_0) \psi_0(x_0, t_0). \quad (2.4.13)$$

Thus the appropriate continuation of the initial function converts the evolution formula into the desired form with infinite limits of integration. The fact that $\psi(x, t)$ in (2.4.13) satisfies the Schrödinger equation is obvious. The continuation (2.4.12) provides the correct boundary conditions for $\psi(x, t)$ in (2.4.13):

$$\psi(0, t) = \psi(L, t) = 0.$$

◇ **Direct construction of the path integral for a free particle in a box by the method of images**

This method of continuation to the whole real line $\mathbb{R}$ can be applied to the very construction of the path integral. To this aim, it is helpful to use the so-called *Poisson formula*

$$\sum_{l=-\infty}^{\infty} e^{2\pi i x l} = \sum_{m=-\infty}^{\infty} \delta(x - m) \quad (2.4.14)$$

(see problem 2.4.1, page 243), where $x \in \mathbb{R}$ and the δ-function is defined on a real line (l and m, of course, are integers).

The scalar product between the localized states (2.4.7) is given by the following explicit formula:

$$\langle x|x_0\rangle = \frac{2}{L}\sum_{n=0}^{\infty}\sin(k_n x)\sin(k_n x_0) \qquad (2.4.15)$$

where k_n runs over the discrete positive momenta

$$k_n = \frac{\pi}{L}n \qquad n = 1, 2, 3, \ldots. \qquad (2.4.16)$$

The sum (2.4.15) over positive integers can be converted into the sum over *all* integers:

$$\langle x|x_0\rangle = \frac{1}{2L}\sum_{n=-\infty}^{\infty}[e^{ik_n(x-x_0)} - e^{ik_n(x+x_0)}]$$

and with the help of the Poisson formula (2.4.14) and the θ-functions, the scalar product can be rewritten in the form

$$\langle x|x_0\rangle = \sum_{l=-\infty}^{\infty}\int_{-\infty}^{\infty}\frac{dk}{2\pi}[e^{ik(x-x_0+2Ll)} - e^{ik(x+x_0+2Ll)}]$$

$$= \sum_{y=\pm x}\sum_{l=-\infty}^{\infty}\int_{-\infty}^{\infty}\frac{dk}{2\pi}\exp\{ik(y-x_0+2Ll) + i\pi(\theta(-y) - \theta(-x_0))\}. \qquad (2.4.17)$$

The first sum runs over two values, $y = +x$ and $y = -x$, and the first θ-function in the exponential provides the correct sign of the terms in this sum. The second θ-function, of course, is equal to zero because $x_0 \in [0, L]$ and, hence, $(-x_0) \le 0$. This term has been formally included for later convenience. The main advantage of the last form of the scalar product in (2.4.17) is that the combination of the summations over $l \in [-\infty, \infty]$ and $y = \pm x$, together with the integration over the interval $[0, L]$, is equivalent to the integration over the whole real line $\mathbb{R}$:

$$\int_0^L dx \sum_{y=\pm x}\sum_{l=-\infty}^{\infty} \leftrightarrow \int_{-\infty}^{\infty} dx. \qquad (2.4.18)$$

Thus, acting as in section 2.2.1, but with the scalar product (2.4.17), and making use of the equivalence (2.4.18), we arrive at the representation of the evolution amplitude for a particle in a box via the conventional path integral with the additional summations (problem 2.4.2, page 244):

$$\langle x, t|x_0, t_0\rangle_{\text{box}} = \sum_{l=-\infty}^{\infty}\sum_{y=\pm x+2Ll}\prod_{j=1}^{N}\int_{-\infty}^{\infty}dx_j \prod_{j=1}^{N+1}\int_{-\infty}^{\infty}\frac{dp}{2\pi\hbar}$$

$$\times \exp\left\{\frac{i}{\hbar}\sum_{j=1}^{N+1}[p_j(x_j - x_{j-1}) + \hbar\pi(\theta(-x_j) - \theta(-x_{j-1})) - \varepsilon H(p_j, x_j)]\right\} \qquad (2.4.19)$$

where $x_{N+1} \equiv y$. In the continuous time limit, this can be written as the path integral

$$\langle x, t|x_0, t_0\rangle_{\text{box}} = \sum_{\substack{\pm x+2Ll \\ l=0,\pm 1,\pm 2,\ldots}}\int \mathcal{D}x(\tau)\int \frac{\mathcal{D}p}{2\pi\hbar}\exp\left\{\frac{i}{\hbar}S\right\} \qquad (2.4.20)$$

where the continuous action can be formally written as

$$S = \int_{t_0}^{t} d\tau \, [p\dot{x} - H(p, x) - \hbar\pi \dot{x}\delta(x)] \tag{2.4.21}$$

and the sum in (2.4.20) means that we have to add up the path integrals with final points at $\pm x + 2Ll$ ($l = 0, \pm 1, \pm 2, \ldots$). The last term in S is obtained as follows:

$$\sum_j (\theta(-x_j) - \theta(-x_{j-1})) = \sum_j \varepsilon \frac{x_j - x_{j-1}}{\varepsilon} \cdot \frac{\theta(-x_j) - \theta(-x_{j-1})}{x_j - x_{j-1}}$$

$$\xrightarrow[\varepsilon \to 0]{} \int_{t_0}^{t} d\tau \, \dot{x} \frac{d\theta(-x)}{dx} = -\int_{t_0}^{t} d\tau \, \dot{x}\delta(x).$$

A path bounced off the walls of the box is eliminated by the corresponding (obtained by reflection) paths crossing the borders and which have the same action. These paths receive a negative sign in the path integral from the phase factor $\exp\{i\pi\theta(-y)\}$. Only paths remaining completely *within* the box have no canceling partners (see figure 2.3). Since the new *topological term*

$$S_{\text{topol}}[x] = -\pi\hbar \int_{t_0}^{t} d\tau \, \dot{x}(\tau)\delta(x(\tau)) = \hbar\pi(\theta(-x) - \theta(-x_0)) \tag{2.4.22}$$

is purely a boundary one, all the integral in (2.4.19) can be evaluated in the same way as for a free particle without the infinite walls: the integration over x_j, p_j can be done as usual and we obtain the amplitude

$$\langle x, t | x_0, t_0 \rangle_{\text{box}} = \sum_{l=-\infty}^{\infty} \sum_{y=\pm x+2dl} \frac{1}{\sqrt{2\pi \hbar i(t - t_0)}} \left[\exp\left\{\frac{i}{\hbar} \frac{m}{2} \frac{(x - x_0 + 2Ll)^2}{t - t_0}\right\} \right.$$

$$\left. - \exp\left\{\frac{i}{\hbar} \frac{m}{2} \frac{(x + x_0 + 2Ll)^2}{t - t_0}\right\} \right] \tag{2.4.23}$$

(cf problem 2.4.4, page 244).

Note that if we want to use $\langle x, t | x_0, t_0 \rangle \equiv K(x, t | x_0, t_0)$ as an integral kernel (propagator) of the evolution operator of some Hamiltonian H, and to write the evolution of a wavefunction $\psi_0(x_0, t_0)$ in the form

$$\psi(x, t) = \int_{-\infty}^{\infty} dx_0 \, K(x, t | x_0, t_0) \psi_0(x_0, t_0)$$

using an integral with *infinite* limits, we have to continue the function $\psi_0(x_0, t_0)$ in the non-physical region as was explained at the beginning of this section.

◇ A quantum free particle on a half-line

The situation with a quantum-mechanical particle in a space with one infinite wall is somewhat analogous to the problem of the excluded volume which we have discussed on the basis of the Wiener path integral in section 1.2.7 (see especially problem 1.2.5, page 113): only a half-space, say $x > 0$, is accessible to the quantum-mechanical particle, and the completeness relation reads as

$$\int_0^{\infty} dx \, |x\rangle\langle x| = 1. \tag{2.4.24}$$

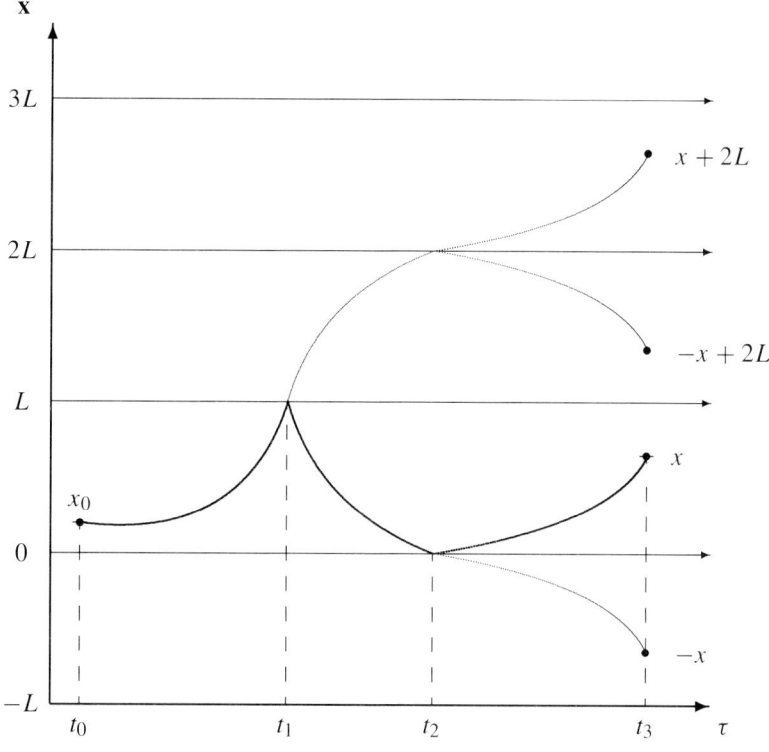

Figure 2.3. The correspondence of a path in a box and the real line.

Consider again a free particle. Then the orthogonality relation can be calculated with the help of the eigenfunction

$$\psi_p(x) = \langle p|x\rangle = \sqrt{\frac{1}{\pi}} \sin kx \qquad k = p/\hbar \qquad (2.4.25)$$

of the free Hamiltonian, with the appropriate boundary condition $\psi_0(0) = 0$. Thus the orthogonality relation has the form

$$\begin{aligned}\langle x|x'\rangle &= \frac{1}{k}\int_{-\infty}^{\infty} dk\ \sin kx \sin kx' \\ &= \frac{1}{2\pi}\int_{-\infty}^{\infty} dk\ [\exp\{ik(x-x')\} - \exp\{ik(x+x')\}] \\ &= \delta(x-x') - \delta(x+x'). \end{aligned} \qquad (2.4.26)$$

The second δ-function seems to be redundant because both x and x' are supposed to be positive. But, on the other hand, this prompts us to extend the domain of the values of x, x' to the whole real line, similarly to the case of a box. This inspires us to present (2.4.25) as

$$\langle x|x'\rangle = \sum_{\sigma=\pm 1}\frac{1}{2\pi\hbar}\int_{-\infty}^{\infty} dp\ \exp\left\{\frac{i}{\hbar}p(\sigma x - x') - i\pi(\theta(-\sigma x) - \theta(-x'))\right\} \qquad (2.4.27)$$

($\theta(x)$ is the step-function). Now we are ready to write down the discrete approximation for the path integral

$$\langle x, t | x_0, t_0 \rangle_{\text{h.-line}} = \prod_{j=1}^{N} \int_0^{\infty} dx_j \prod_{j=1}^{N+1} \sum_{\sigma=\pm 1} \int_{-\infty}^{\infty} \frac{dp_j}{2\pi\hbar}$$

$$\times \exp\left\{ \sum_{j=1}^{N+1} \left[\frac{i}{\hbar} p(\sigma x_j - x_{j-1}) + i\pi(\theta(-\sigma x_j) - \theta(-x_{j-1})) - \frac{i}{\hbar}\varepsilon H(p) \right] \right\}.$$

(2.4.28)

In the same way as for the particle in a box the sum over the 'reflected' points $\sigma x = \pm x$ is now combined, at each j, with the integral $\int_0^{\infty} dx_j$, to form an integral over the whole real line including the non-physical half-space $x < 0$. Only the last sum cannot be accommodated in this way, and we obtain the following path-integral representation for the evolution amplitude:

$$\langle x, t | x_0, t_0 \rangle_{\text{h.-line}} = \sum_{x_{N+1} = \pm x} \prod_{j=1}^{N} \int_{-\infty}^{\infty} dx_j \prod_{j=1}^{N+1} \int_{-\infty}^{\infty} \frac{dp_j}{2\pi\hbar}$$

$$\times \exp\left\{ \sum_{j=1}^{N+1} \left[\frac{i}{\hbar} p(x_j - x_{j-1}) + i\pi(\theta(-x_j) - \theta(-x_{j-1})) - \frac{i}{\hbar}\varepsilon H(p) \right] \right\}.$$

(2.4.29)

This path integral is of the usual type, with integration over all paths in the whole space. The only special features are the additional term in the exponent (with the θ-functions) and the final symmetrization in the final points $\pm x$. Note that having assumed $x_0 > 0$, the initial phase factor $\theta(-x_0)$ can be omitted, but it is convenient to keep it for symmetry reasons and for convenience in the transition to the continuum limit.

In the continuum limit, the exponent corresponds to an action

$$S[p, x] = \int_{t_0}^{t} dt \left[p\dot{x} - \frac{p^2}{2m} + \hbar\pi \partial_t \theta(-x) \right]$$

$$= S_0[p, x] + S_{\text{topol}}.$$

(2.4.30)

The first part S_0 of the action is the conventional action for a free particle. The second part S_{topol} again gives only the boundary term. Hence, we can immediately write down the result:

$$\langle x, t | x_0, t_0 \rangle_{\text{h.-line}} = \sum_{y=\pm x} \frac{1}{\sqrt{2\pi i\hbar(t-t_0)}} \exp\left\{ \frac{i}{\hbar} \frac{m}{2} \frac{(y-x_0)^2}{t-t_0} + i\hbar\pi(\theta(-y) - \theta(-x_0)) \right\}$$

$$= \frac{1}{\sqrt{2\pi i\hbar(t-t_0)}} \left[\exp\left\{ \frac{i}{\hbar} \frac{m}{2} \frac{(x-x_0)^2}{t-t_0} \right\} - \exp\left\{ \frac{i}{\hbar} \frac{m}{2} \frac{(x+x_0)^2}{t-t_0} \right\} \right].$$

(2.4.31)

The correctness of this result can be checked again by the operator method (problem 2.4.4, page 244).

◇ Remarks about path integrals for particles confined within an arbitrary domain of $\mathbb{R}^d$

In the case of a multidimensional space and for a complicated form of the domain $\Omega \subset \mathbb{R}^D$ where a particle is confined, the construction of the corresponding path-integral representation is not a simple problem. However, if we know the complete set of eigenfunctions ψ_λ of the Hamiltonian $\widehat{H}$: $\widehat{H}\psi_\lambda = E_\lambda \psi_\lambda$, with the boundary conditions

$$\psi_\lambda(\boldsymbol{x})|_{x \in \partial \Omega} = 0 \qquad (2.4.32)$$

where $\partial \Omega$ denotes the boundary of the domain $\Omega \subset \mathbb{R}^D$, the appropriate integral kernel can be written easily. To this aim, let us analytically continue the functions $\psi_\lambda(\boldsymbol{x})$ outside the domain Ω (we assume that the Hamiltonian $H(\boldsymbol{p}, \boldsymbol{x})$ is an analytical function of $\boldsymbol{x}$ and, therefore, $\psi_\lambda(\boldsymbol{x})$ can be analytically continued). Then the kernel we seek can be written as

$$K_\Omega(\boldsymbol{x}, t | \boldsymbol{x}_0, t_0) = \int d^d x' \, K(\boldsymbol{x}, t | \boldsymbol{x}', t_0) I(\boldsymbol{x}', \boldsymbol{x}_0) \qquad (2.4.33)$$

where $K(\boldsymbol{x}, t | \boldsymbol{x}', t_0)$ is the usual kernel (represented by the path integral) for a particle with the Hamiltonian $H(\boldsymbol{p}, \boldsymbol{x})$ in the whole space and

$$I(\boldsymbol{x}', \boldsymbol{x}_0) \stackrel{\text{def}}{\equiv} \sum_\lambda \psi_\lambda(\boldsymbol{x}') \psi_\lambda(\boldsymbol{x}_0).$$

The fact that K_Ω satisfies the Schrödinger equation is obvious (because K satisfies it). Thus we need only verify the boundary conditions

$$K_\Omega(\boldsymbol{x}, t | \boldsymbol{x}_0, t_0)|_{x \in \partial \Omega} = 0. \qquad (2.4.34)$$

To do this, remember that $K(\boldsymbol{x}, t | \boldsymbol{x}', t_0)$ is the kernel of the evolution operator $\exp\{\frac{i}{\hbar}(t - t_0)\widehat{H}\}$. Hence, the relation (2.4.33) can be written as

$$K_\Omega(\boldsymbol{x}, t | \boldsymbol{x}_0, t_0) = e^{\frac{i}{\hbar}(t-t_0)\widehat{H}} \sum_\lambda \psi_\lambda(\boldsymbol{x}) \psi_\lambda(\boldsymbol{x}_0)$$

$$= \sum_\lambda e^{\frac{i}{\hbar}(t-t_0)E_\lambda} \psi_\lambda(\boldsymbol{x}) \psi_\lambda(\boldsymbol{x}_0).$$

Therefore, (2.4.34) follows from (2.4.32).

As an example, we can derive by this method the kernel for a particle confined inside a two-dimensional disc of radius R. The eigenfunctions of the free Hamiltonian $\widehat{H}_0 = \widehat{p}^2/(2m)$ in polar coordinates has the form

$$\psi_{km}(r, \phi) = \frac{1}{C_{km}} J_m(\sigma_{km} r) e^{im\phi} \qquad k = 0, 1, 2, \ldots; \quad m = 0, \pm 1, \pm 2, \ldots$$

where $J_m(\sigma_{km} r)$ are the Bessel functions and the constants σ_{km} are defined by the boundary conditions

$$\psi_{km}(R, \phi) = 0 \Rightarrow J_m(\sigma_{km} R) = 0.$$

The Bessel functions can be analytically continued in the region $r > R$ and this provides the construction of the integral kernel for a free particle inside the circle in the form (2.4.33).

We must remark that the way of constructing integral kernels outlined here is not of great practical value. The point is that it assumes knowledge, *in advance*, of the complete set of eigenfunctions of the Hamiltonian under consideration. But such a set contains, in fact, all the information about the corresponding physical system. In particular, the integral kernel is given directly by the series

$$K_\Omega(\boldsymbol{x}, t | \boldsymbol{x}_0, t_0) = \sum_\lambda e^{\frac{i}{\hbar}(t-t_0)E_\lambda} \psi_\lambda(\boldsymbol{x}) \psi_\lambda(\boldsymbol{x}_0).$$

238   *Path integrals in quantum mechanics*

Thus, its representation in the path-integral form (2.4.33) seems to be redundant. Nevertheless, we cannot exclude that, in some cases, representation (2.4.33) may turn out to be more convenient. In some exceptional cases, we succeed in finding $I(x', x_0)$ in (2.4.33) without knowledge of the eigenfunctions, using some natural symmetries of the system. For instance, the continuation of functions in a box (2.4.12) and the corresponding kernel are determined by periodicity and antisymmetry.

2.4.2 Point particles on a circle and with a torus-shaped phase space

The trajectories of a point particle on a circle are parametrized by an angular variable $\phi(t) \in [0, 2\pi]$ subject to the constraint that $\phi = 0$ and $\phi = 2\pi$ be identical points. Recall that the corresponding phase space has the shape of a cylinder.

The first step in the derivation of a path-integral representation for evolution amplitudes is quite usual. The time evolution operator is decomposed into a product of the infinitesimal ones:

$$\langle \phi, t | \phi_0, t_0 \rangle = \langle \phi | \exp\left\{-\frac{i}{\hbar}(t-t_0)\widehat{H}\right\} | \phi_0 \rangle$$

$$= \langle \phi | \prod_{n=1}^{N+1} \exp\left\{-\frac{i}{\hbar}\varepsilon \widehat{H}\right\} | \phi_0 \rangle. \tag{2.4.35}$$

The compactification shows up in the completeness relations to be inserted between factors on the right-hand side for $n = 1, \ldots, N$:

$$\int_0^{2\pi} d\phi_n |\phi_n\rangle\langle\phi_n| = 1. \tag{2.4.36}$$

If the integrand is singular at $\phi = 0$, the integration must end below 2π, at an infinitesimal value before it. Otherwise we would double-count the contributions from the identical points $\phi = 0$ and $\phi = 2\pi$. The orthogonality of the wavefunctions reads as

$$\langle \phi_j | \phi_k \rangle = \delta^{(S)}(\phi_j - \phi_k) \qquad \phi_j, \phi_k \in [0, 2\pi) \tag{2.4.37}$$

where $\delta^{(S)}(\phi)$ is the δ-function *on a circle* $S^{(1)}$ which can be defined as the kernel of the *identity* operator acting on the states $|\phi\rangle$ and can be expanded in the series

$$\delta^{(S)}(\phi) = \frac{1}{2\pi} \sum_{m=-\infty}^{\infty} e^{im\phi}. \tag{2.4.38}$$

This is the circle analog of the usual Fourier decomposition

$$\delta(x) = \frac{1}{2\pi} \int_{-\infty}^{\infty} dk \, e^{ikx}$$

of the δ-function on the whole real line. Due to the compactness of a circle, the integral is substituted by the sum.

◇ **Path integral for a particle on a circle**

Let us consider, for simplicity, a Hamiltonian without ordering ambiguities, $H(p, \phi) = T(p) + V(\phi)$, and calculate the evolution amplitude for the infinitesimal time shift

$$\langle \phi_j, t_j | \phi_{j-1}, t_{j-1} \rangle = \langle \phi | \exp\left\{-\frac{i}{\hbar}\varepsilon \widehat{H}\right\} | \phi_0 \rangle$$

$$= \exp\left\{-\frac{i}{\hbar}\varepsilon \widehat{H}(-i\partial_{\phi_j}, \phi_j)\right\} \langle \phi_j | \phi_{j-1} \rangle \tag{2.4.39}$$

which, with the use of (2.4.38), becomes

$$\langle \phi_j, t_j | \phi_{j-1}, t_{j-1} \rangle = \langle \phi_j | \exp\left\{-\frac{i}{\hbar} \varepsilon \widehat{H}\right\} | \phi_0 \rangle$$

$$= \exp\left\{-\frac{i}{\hbar} \varepsilon \widehat{H}(-i\partial_{\phi_j}, \phi_j)\right\} \frac{1}{2\pi} \sum_{m=-\infty}^{\infty} \exp\{im_j(\phi_j - \phi_{j-1})\}$$

$$= \frac{1}{2\pi} \sum_{m=-\infty}^{\infty} \exp\left\{im_j(\phi_j - \phi_{j-1}) - \frac{i}{\hbar} \varepsilon \widehat{H}(\hbar m_j, \phi_j)\right\}. \quad (2.4.40)$$

The completeness relation (2.4.36) allows us to write the desired finite-dimensional (time-sliced) approximation of the path integral for a system with cyclic coordinates:

$$\langle \phi, t | \phi_0, t_0 \rangle_{\text{circle}} \approx \prod_{j=1}^{N} \int_0^{2\pi} d\phi_j \prod_{j=1}^{N+1} \sum_{m_j=-\infty}^{\infty} \frac{1}{2\pi}$$

$$\times \exp\left\{i \sum_{j=1}^{N+1} \left[m_j(\phi_j - \phi_{j-1}) - \frac{i}{\hbar} \varepsilon \widehat{H}(\hbar m_j, \phi_j)\right]\right\} \quad (2.4.41)$$

$$\phi_{N+1} = \phi.$$

Such an expression looks rather inconvenient for practical calculations. Fortunately, just as in the cases of a box and half-line, it can be turned into a more comfortable equivalent form, involving a proper continuous path integral. This is possible at the expense of a single additional infinite sum which guarantees the cyclic invariance in the variable ϕ. To find the equivalent form, we have to use the Poisson formula (2.4.14) which allows us to write the orthogonality relation (2.4.37) as

$$\langle \phi_j | \phi_k \rangle = \sum_{l=-\infty}^{\infty} \delta(\phi_j - \phi_k + 2\pi l)$$

$$= \sum_{l=-\infty}^{\infty} \int_{-\infty}^{\infty} \frac{dk}{2\pi} \exp\{ik(\phi_j - \phi_k + 2\pi l)\}. \quad (2.4.42)$$

Using this form of the orthogonality relation in (2.4.39), we can convert (2.4.41) into the expression

$$\langle \phi, t | \phi_0, t_0 \rangle_{\text{circle}} \approx \prod_{j=1}^{N} \int_0^{2\pi} d\phi \prod_{j=1}^{N+1} \int_{-\infty}^{\infty} \frac{dk_n}{2\pi} \sum_{l_j=-\infty}^{\infty}$$

$$\times \exp\left\{i \sum_{j=1}^{N+1} \left[k_j(\phi_j - \phi_{j-1} + 2\pi l_j) - \frac{1}{\hbar} \varepsilon H(\hbar k_j, \phi_j)\right]\right\}. \quad (2.4.43)$$

Note that the sums over l_j in (2.4.43) together with the integration over ϕ_j in the range $[0, 2\pi)$ is equivalent to just the integration over ϕ_j over the whole real line, leaving only the last sum (over $l \equiv l_{N+1}$) explicit (because there is no corresponding integration over ϕ_{N+1}). Thus we arrive at the following representation for the propagator on a circle:

$$\langle \phi, t | \phi_0, t_0 \rangle_{\text{circle}} \approx \sum_{l=-\infty}^{\infty} \prod_{j=1}^{N} \int_{-\infty}^{\infty} d\phi \prod_{j=1}^{N+1} \int_{-\infty}^{\infty} \frac{dp_j}{2\pi\hbar}$$

$$\times \exp\left\{\frac{i}{\hbar}\sum_{j=1}^{N+1}[p_j(\phi_j - \phi_{j-1} + 2\pi l\delta_{j,N+1}) - \varepsilon H(p_j, \phi_j)]\right\}. \quad (2.4.44)$$

Comparing this with the time-sliced approximation for the path integral on the whole line $\mathbb{R}$, we see that it differs only by the additional sum over l:

$$\langle \phi, t|\phi_0, t_0\rangle_{\text{circle}} = \sum_{l=-\infty}^{\infty} \langle \phi + 2\pi l, t|\phi_0, t_0\rangle_{\text{line}} \qquad \phi, \phi_0 \in [0, 2\pi]. \quad (2.4.45)$$

Therefore, in the continuum limit, (2.4.44) tends to the path integral

$$\langle \phi, t|\phi_0, t_0\rangle_{\text{circle}} = \sum_{l=-\infty}^{\infty} \int_{C\{\phi+2\pi l, t; \phi_0, t_0\}} \mathcal{D}\phi(\tau) \int \frac{dp(\tau)}{2\pi\hbar} \exp\left\{\frac{i}{\hbar}\int_{t_0}^{t} d\tau\, [p\dot\phi - H(p, \phi)]\right\}. \quad (2.4.46)$$

The substitution of the path integral on a circle by the sum of the path integrals on the whole real line is illustrated in figure 2.4.

As an example, consider a free particle moving on a circle, with the Hamiltonian

$$H(p, \phi) = \frac{p^2}{2m}. \quad (2.4.47)$$

We know that the corresponding path integral for a free particle on a real line is

$$\langle \phi, t|\phi_0, t_0\rangle = \frac{1}{\sqrt{2\pi\hbar i(t - t_0)}} \exp\left\{\frac{i}{\hbar}\frac{m}{2}\frac{(\phi - \phi_0)^2}{t - t_0}\right\}. \quad (2.4.48)$$

Using (2.4.46), the amplitude on a circle is given by the periodic Gaussian

$$\langle \phi, t|\phi_0, t_0\rangle = \sum_{l=-\infty}^{\infty} \frac{1}{\sqrt{2\pi\hbar i(t - t_0)}} \exp\left\{\frac{i}{\hbar}\frac{m}{2}\frac{(\phi - \phi_0 + 2\pi l)^2}{t - t_0}\right\}. \quad (2.4.49)$$

◇ **The propagator on a circle with quasi-periodic boundary conditions**

Thus the transition amplitude on a circle can be written as the linear combination (2.4.45) of the amplitudes on a real line. Note that, if we rewrite the sum (2.4.45) with arbitrary coefficients:

$$\langle \phi, t|\phi_0, t_0\rangle_{\text{circle}} = \sum_{l=-\infty}^{\infty} C_l \langle \phi + 2\pi l, t|\phi_0, t_0\rangle_{\text{line}} \qquad \phi, \phi_0 \in [0, 2\pi] \quad (2.4.50)$$

the amplitude, considered as the propagator $K(\phi, t|\phi_0, t_0) \equiv \langle \phi, t|\phi_0, t_0\rangle$, still satisfies the Schrödinger equation. The reason is that the Schrödinger equation (as any other differential equation) acts only *locally* and thus does not feel the topological properties of the space. Therefore, it is the same for a circle and a line and, if the differential operator corresponding to the Schrödinger equation acts on each term on the right-hand side of (2.4.50), we obtain zero. This means that the left-hand side also satisfies the equation for arbitrary C_l, $l = 0, \pm 1, \pm 2, \ldots$. The condition of physical equivalence of the points ϕ and $\phi + 2\pi$ on a circle means that the amplitude $\langle \phi, t|\phi_0, t_0\rangle_{\text{circle}}$, after the shift $\phi \to \phi + 2\pi$, acquires at most a phase:

$$\langle \phi + 2\pi, t|\phi_0, t_0\rangle_{\text{circle}} = e^{i\alpha}\langle \phi, t|\phi_0, t_0\rangle_{\text{circle}}$$

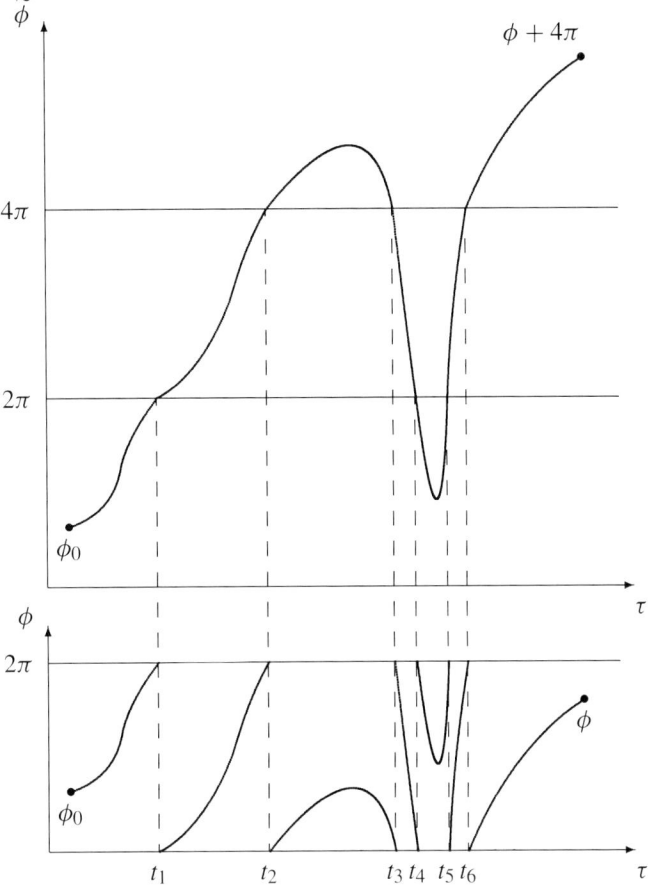

Figure 2.4. The correspondence of a path on a circle and the real line (universal covering of the circle).

(a phase does not change the corresponding probability distribution). Applying this requirement to (2.4.50), we obtain
$$C_{l+1} = e^{i\alpha} C_l$$
so that
$$C_l = e^{i\alpha l} \qquad l = 0, \pm 1, \pm 2, \ldots$$
(the phase C_0 can be taken arbitrarily and we have chosen $C_0 = 1$). Thus the requirement of *physical equivalence* (i.e. up to a phase for amplitudes) of the points ϕ and $\phi + 2\pi$ leads to a more general expression for the transition amplitude:
$$\langle \phi, t | \phi_0, t_0 \rangle_{\text{circle}} = \sum_{l=-\infty}^{\infty} e^{i\alpha l} \langle \phi + 2\pi l, t | \phi_0, t_0 \rangle_{\text{line}} \qquad \phi, \phi_0 \in [0, 2\pi]. \qquad (2.4.51)$$

Of course, this expression can be derived directly if we start from the quasi-periodic boundary condition (2.4.4) (the so-called *α-quantization*) instead of the trivial boundary condition (2.4.1). We suggest the reader does this as an exercise (problem 2.4.5, page 244).

A number of physical interpretations are possible for the phase α. For instance, if we study a periodic crystal, it is possible to reduce it to the consideration of a single cell of the lattice. The purely periodic boundary condition

$$\psi(x+a) = \psi(x)$$

(a is the cell size) is, definitely, too restrictive and inadequate for the problem. The so-called *Bloch wavefunctions* satisfy the more general conditions

$$\psi(x+a) = e^{i2\pi ka}\psi(x)$$

where k is the wavevector. Thus, in this case, the phase α is related to the parameters of the crystal: $\alpha = 2\pi k a$ (we mentioned the one-dimensional variant of the problem; in the realistic three-dimensional case, the phase becomes $\alpha = 2\pi \mathbf{k}\mathbf{a}$, where the vector $\mathbf{a}$ defines the crystal cell).

◇ **Propagator in the case of a torus-like phase space**

The same method of path-integral construction can be applied to a system with a torus-shaped phase space. This means that the system is described by a coordinate and a momentum bounded by intervals: $0 \leq x \leq L$, $0 \leq p \leq \Lambda$ and all quantities are supposed to be *periodic* functions of x and p with periods L and Λ, respectively. Of course, in this case, we can hardly speak about 'a particle' (free or in some external potential). Such a phase space is similar to a toplike system (in the sense that these systems have both *compact* coordinates and *momentum* degrees of freedom) or to a system on a finite lattice (with periodic boundary conditions).

Proceeding in the same way as in the case of a cylindrical phase space (a particle on a circle), we arrive at the following time-sliced approximation of the path integral for the amplitudes:

$$\langle x,t|x_0,t_0\rangle_{\text{torus}} \approx \sum_{l=-\infty}^{\infty} \prod_{j=1}^{N} \int_0^{2\pi} dx \int_0^{\Lambda} \frac{dp_1}{2\pi\hbar} \prod_{j=2}^{N+1} \int_{-\infty}^{\infty} \frac{dp_n}{2\pi\hbar}$$

$$\times \exp\left\{\frac{i}{\hbar}\sum_{j=1}^{N+1}[p_j(x_j - x_{j-1} + 2\pi l\delta_{j,N+1}) - \varepsilon H(p_j, x_j)]\right\}. \quad (2.4.52)$$

The most noticeable distinction from the corresponding expression (2.4.44) is that the integration over p_1 goes over the finite interval $[0, \Lambda]$ and not over the whole real line. If the Hamiltonian in (2.4.52) does not depend on the coordinate variable x (as in the case of a free particle): $H(p,x) \equiv H(p)$, we may integrate over all x_j, $j = 1, \ldots, N$ producing N δ-functions and then integrate over $p_2, \ldots, p_{N+1}$, with the result:

$$\langle x,t|x_0,t_0\rangle_{\text{torus}} = \sum_{l=-\infty}^{\infty} \int_0^{\Lambda} \frac{dp}{2\pi\hbar} \exp\left\{\frac{i}{\hbar}[p(x - x_0 + lL) - H(p)(t - t_0)]\right\}. \quad (2.4.53)$$

Using the Poisson formula (2.4.14), the expression (2.4.53) can be presented in the form

$$\langle x,t|x_0,t_0\rangle_{\text{torus}} = \frac{1}{L}\int_0^{\Lambda} \frac{dp}{2\pi\hbar} \exp\left\{\frac{i}{\hbar}[p(x - x_0) - H(p)(t - t_0)]\right\} \sum_{m=-\infty}^{\infty} \delta\left(p - \frac{2\pi m}{L}\right). \quad (2.4.54)$$

For consistency, the integrand in (2.4.54) must be periodic in p, with the period Λ. This is true *only* if the condition

$$L\Lambda = 2\pi k \qquad k \in \mathbb{Z} \quad (2.4.55)$$

(the so-called *volume quantization* condition) is fulfilled. Then, due to the δ-functions, we can integrate over p and (2.4.54) becomes

$$\langle x, t | x_0, t_0 \rangle_{\text{torus}} = \frac{1}{L} \sum_{m=0}^{k-1} \exp\left\{ i \left[2\pi m \frac{(x - x_0)}{L} - H\left(\frac{m}{k}\Lambda\right)(t - t_0) \right] \right\}. \tag{2.4.56}$$

Thus we conclude that at fixed k, the momentum and coordinate take the following possible values:

$$p_m = \frac{m}{k}\Lambda$$
$$x_m = \frac{m}{k}L \qquad m = 0, 1, \ldots, k-1. \tag{2.4.57}$$

Note that the discrete parameter k plays the role of the Planck constant in the sense that the semiclassical and classical approximations correspond to larger values of k (in other words, are valid in the limit $k \to \infty$) and not to the standard limit $\hbar \to 0$. Sometimes we describe such a situation as the appearance of a *discrete* (quantized) Planck constant. The quantization condition (2.4.55) was obtained for the first time by Berezin (1974) (note that Berezin used angular variables $\phi = 2\pi x/L$, $\theta = 2\pi p/\Lambda$). This is a characteristic property of systems with compact phase spaces: another well-known example is a quantum top where the total angular momentum (spin) of the top with discrete values plays the role of the Planck constant. In this case, the phase space proves to be a two-dimensional sphere (Berezin 1975).

2.4.3 Problems

Problem 2.4.1. Prove the Poisson formula (2.4.14).

Hint. The sum $g(x) \stackrel{\text{def}}{=} \sum_{m=-\infty}^{\infty} \delta(x - m)$ on the right-hand side of (2.4.14) is a periodic function of x with a unit period and has the Fourier series

$$g(x) = \sum_{l=-\infty}^{\infty} g_l e^{2\pi i x l}.$$

The Fourier coefficients are given by

$$g_l = \int_{-1/2}^{1/2} dx\, g(x) e^{-2\pi i x l} = 1.$$

These are precisely the Fourier coefficients on the left-hand side of (2.4.14).

Mathematically minded readers may worry that this proof is somewhat formal because an equation involving δ-functions is valid, rigorously speaking, only when integrated with an appropriate smooth test function. Let $f(x)$ be such a test function. Then the Poisson formula implies:

$$\sum_{m=-\infty}^{\infty} f(m) = \int_{-\infty}^{\infty} dx \sum_{l=-\infty}^{\infty} e^{2\pi i x l} f(x).$$

Now if we truncate the sum on the right-hand side at finite large values $\pm L$, it can easily be calculated (being of the geometric type):

$$\sum_{l=-L}^{L} e^{2\pi i x l} = \frac{\sin(\pi x (2L + 1))}{\sin(\pi x)}.$$

The right-hand side is a well-known approximate expression for the δ-function (i.e. it becomes the δ-function in the limit $L \to \infty$). This proves formula (2.4.14).

Problem 2.4.2. Derive the path integral (2.4.19) for a particle in a box using the standard way of path-integral construction for the scalar product (2.4.17). Show that the summation in (2.4.17) results in integration over the whole real line.

Hint. To simplify calculations, you may, at first, consider the trivial (zero) Hamiltonians. Generalization to a general Hamiltonian brings nothing new compared with the case of an unbounded line.

Problem 2.4.3. Verify that the operator approach gives the same result (2.4.49) for the transition amplitude of a free particle on a circle.

Hint. The wavefunction of a free particle on a circle reads as

$$\psi_n(\phi) = \frac{1}{\sqrt{2\pi}} e^{in\phi} \qquad n \in \mathbb{Z}$$

and the energy eigenvalues are

$$E_n = \frac{\hbar^2}{2mr^2} n^2$$

(r is the radius of the circle). Thus

$$\langle \phi, t | \phi_0, t_0 \rangle_{\text{circle}} = \langle \phi | \exp\left\{ -\frac{i}{\hbar}(t - t_0)\widehat{H} \right\} | \phi_0 \rangle$$

$$= \sum_{n=-\infty}^{\infty} \psi_n(\phi) \bar{\psi}_n(\phi_0) \exp\left\{ -\frac{i}{\hbar} \frac{\hbar^2 n^2}{2mr^2}(t - t_0) \right\}$$

$$= \sum_{n=-\infty}^{\infty} \frac{1}{2\pi} \exp\left\{ in(\phi - \phi_0) - i\frac{\hbar n^2}{2mr^2}(t - t_0) \right\}.$$

Finally, the use of the Poisson formula (2.4.14) allows us to substitute the summation over m by the integration over momentum plus another summation and after the integration, to cast this expression into the form (2.4.49).

Problem 2.4.4. Prove that formula (2.4.31) is correct for the transition amplitude of a free particle in a one-dimensional half-space (with an infinite wall) and (2.4.23) for a particle in a box.

Hint. Inserting into (2.4.31) the Fourier transform of the Gaussian distribution, derive that

$$\langle x, t | x_0, t_0 \rangle_{\text{h.-line}} = \int_{-\infty}^{\infty} \frac{dp}{2\pi\hbar} \left[\exp\left\{ \frac{i}{\hbar} \frac{m}{2} \frac{(x - x_0)^2}{t - t_0} \right\} - \exp\left\{ \frac{i}{\hbar} \frac{m}{2} \frac{(x + x_0)^2}{t - t_0} \right\} \right] \exp\left\{ -\frac{p^2(t - t_0)}{2m\hbar} \right\}$$

$$= 2\int_0^{\infty} \frac{dp}{2\pi\hbar} \sin\left(\frac{px}{\hbar}\right) \sin\left(\frac{px_0}{\hbar}\right) \exp\left\{ -\frac{p^2(t - t_0)}{2m\hbar} \right\}$$

which is the correct spectral representation for the time evolution amplitude (cf (2.1.70)).

The formula for a particle in a box is verified in the same way.

Problem 2.4.5. Derive the path-integral representation for the propagator of a free particle on a circle in the case of the non-trivial boundary conditions (2.4.4), starting from the discrete approximation, the scalar product (2.4.37) and the completeness relation.

Hint. Follow the steps outlined in section 2.4.2. The answer is given by the expression (2.4.51).

2.5 Path integrals in curved spaces, spacetime transformations and the Coulomb problem

This section is devoted to the construction of path integrals in curved configuration spaces. More precisely, we are going to construct a path-integral representation for the propagator of a particle moving in a d-dimensional Riemann space with a non-zero curvature tensor (some basic facts about Riemann geometry are given in supplement V, volume II; the standard textbook on differential geometry is Kobayashi and Nomizu (1969), good introductory books are Isham (1989) and Visconti (1992)). This consideration is important because of two types of application. First, this is the background for direct applications to quantum-mechanical problems with curved configuration space geometry (including the ambitious aim of path-integral formulation of quantum gravity, see section 3.4). Second, this formalism has proved to be necessary for finding the path-integral solution of such a basic quantum-mechanical problem as a particle's motion in the Coulomb potential, as well as the solution of a number of other exactly solvable potentials. The essential idea involved in the exact derivation of the path-integral solution is the mapping of the path integral for the Coulomb problem (or other solvable potentials) onto the path integral for the harmonic oscillator. This is achieved by the reparametrization of paths according to a new space and time variable via the so-called *Kustaanheimo–Stiefel transformation* which was originally invented for solving the quite different problems of celestial mechanics. For the path-integral solution of the Coulomb problem this idea was first exploited by Duru and Kleinert (1979, 1982) (see also Kleinert (1995)) and later expanded to a number of other exactly solvable Hamiltonians (see, e.g., Grosche (1996), Grosche and Steiner (1998) and references therein).

In the last section 2.5.5, we treat a specific case of curved manifolds, namely, the Lie group manifolds (for the basic notions of Lie group theory see supplement IV, volume II). From the practical point of view, this calculation is important for systems possessing some symmetry. In addition, the existence of group transformations (i.e. transitivity of group manifolds) opens the possibility for an *exact* evaluation of the corresponding path integrals.

2.5.1 Path integrals in curved spaces and the ordering problem

We shall not go into all the details of the quantization procedure on arbitrary manifolds (see, e.g., the classical papers by De Witt (1957) and Berezin (1975)). Instead, we shall concentrate only on those points which are necessary for the construction of path integrals and, in particular, for solving the Coulomb problem. As always, the starting point is the corresponding Schrödinger equation. In a d-dimensional Riemannian manifold $\mathcal{M}$, with the coordinates q^a, the metric $g_{ab}(q)$ and the line element (see supplement V, volume II)

$$ds^2 = g_{ab}\, dq^a\, dq^b$$

the Schrödinger equation reads as

$$-i\hbar \frac{\partial}{\partial t}\psi(\boldsymbol{q};t) = \left[-\frac{\hbar^2}{2m}\Delta_{\text{LB}} + V(\boldsymbol{q})\right]\psi(\boldsymbol{q};t). \quad (2.5.1)$$

ψ is some state function, defined in the Hilbert space $\mathcal{L}^2(\mathbb{R}^d)$, the space of all square-integrable functions in the sense of the scalar product

$$\langle\psi_1|\psi_2\rangle = \int_{\mathcal{M}} d^d q\, \sqrt{g}\, \bar{\psi}_1(\boldsymbol{q})\psi_2(\boldsymbol{q}) \quad (2.5.2)$$

$$g(\boldsymbol{q}) \stackrel{\text{def}}{=} \det(g_{ab}(\boldsymbol{q})) \quad (2.5.3)$$

and Δ_{LB} is the *Laplace–Beltrami operator*

$$\Delta_{\text{LB}} \stackrel{\text{def}}{\equiv} g^{-\frac{1}{2}}\partial_a g^{\frac{1}{2}} g^{ab}\partial_b = g^{ab}\partial_a\partial_b + g^{ab}(\partial_a \ln\sqrt{g})\partial_b + g^{ab}{}_{,a}\partial_b. \qquad (2.5.4)$$

Here and in what follows, we use the notation and conventions adopted in theories of curved spaces:

(i) summation over pairs of repeating indices (one upper, one lower) is understood;
(ii) indices are lowered and raised with the help of the metric g_{ab} and its inverse g^{ab} ($g_{ab}g^{bc} = \delta_a^c$);
(iii) for derivatives of a function f, we use the shorthand notation

$$f_{,a} \stackrel{\text{def}}{\equiv} \partial_a f \equiv \frac{\partial f}{\partial q^a}.$$

The Hamiltonian

$$\widehat{H} \stackrel{\text{def}}{\equiv} -\frac{\hbar^2}{2m}\Delta_{\text{LB}} + V(\boldsymbol{q}) \qquad (2.5.5)$$

is usually defined on some dense subset $D(\widehat{H}) \subseteq \mathcal{L}^2(\mathbb{R}^d)$, so that $\widehat{H}$ is self-adjoint. The time evolution of a state $\psi(\boldsymbol{q},t)$, written via the propagator $K(\boldsymbol{q},t,|\boldsymbol{q}_0,t_0)$ of equation (2.5.1), reads as

$$\psi(\boldsymbol{q};t) = \int d^d q_0 \sqrt{g(\boldsymbol{q}_0)}\, K(\boldsymbol{q},t|\boldsymbol{q}_0,t_0)\psi(\boldsymbol{q}_0,t_0). \qquad (2.5.6)$$

The construction of the path-integral representation for the propagator in the case of a curved space with an arbitrary metric $g_{ab}(q)$ is not so easy as in the case of a flat space with an ordering-insensitive Hamiltonian of the form $H = T(p) + V(q)$. The general approach to the construction of path integrals for Hamiltonians containing terms with ordering ambiguity has been described in section 2.3. In the special case of a curved configuration space, the first derivation was by De Witt (1957). His result reads as

$$K(\boldsymbol{q},t,|\boldsymbol{q}_0,t_0) = \int_{C\{q_0,t_0|q,t\}} \mathcal{D}\!\left[\sqrt{g(\tau)}q(\tau)\right] \exp\left\{\frac{i}{\hbar}\int_{t_0}^{t} dt \left[\frac{m}{2} g_{ab}(\boldsymbol{q})\dot{q}^a\dot{q}^b - V(\boldsymbol{q}) + \hbar^2\frac{R}{6m}\right]\right\}$$

$$\stackrel{\text{def}}{\equiv} \lim_{N\to\infty} \left(\frac{m}{2\pi i\varepsilon\hbar}\right)^{Nd/2} \prod_{j=1}^{N-1}\int \sqrt{g(\boldsymbol{q}_j)}\, d^d q_j$$

$$\times \exp\left\{\frac{i}{\hbar}\sum_{j=1}^{N}\left[\frac{m}{2\varepsilon}g_{ab}(\boldsymbol{q}_{j-1})(q_j^a - q_{j-1}^a)(q_j^b - q_{j-1}^b)\right.\right.$$

$$\left.\left. -\varepsilon V(\boldsymbol{q}_{j-1}) + \varepsilon\frac{\hbar^2}{6m}R(\boldsymbol{q}_{j-1})\right]\right\}. \qquad (2.5.7)$$

Here

$$R = g^{ab}(\Gamma^c_{ab,c} - \Gamma^c_{cb,a} + \Gamma^d_{ab}\Gamma^c_{cd} - \Gamma^d_{cb}\Gamma^c_{ad}) \qquad (2.5.8)$$

is the scalar curvature and

$$\Gamma^a_{bc} = \tfrac{1}{2}g^{ad}(g_{bd,c} + g_{dc,b} - g_{bc,d}) \qquad (2.5.9)$$

are the Christoffel symbols. As usual, we identify: $\boldsymbol{q}_N = \boldsymbol{q}$.

The most noticeable difference between (2.5.7) and the path integrals which we have encountered earlier in this book is that in the exponential there is a term containing $\hbar^2$ in addition to the classical action, so that the path integral now has the following general form:

$$K \sim \int_{C\{q_0,t_0|q,t\}} \mathcal{D}\!\left[\sqrt{g(\tau)}q(\tau)\right] \exp\left\{\frac{i}{\hbar}S_{\text{eff}}\right\}$$

Path integrals in curved spaces, spacetime transformations and the Coulomb problem 247

with the *effective* action

$$S_{\text{eff}} = \int dt\, L_{\text{eff}} \equiv \int dt\, (L - \Delta V_{\text{DeW}}) \tag{2.5.10}$$

in the exponential, instead of just the classical one, $S = \int dt\, L$, with the classical Lagrangian of the form

$$L(q, \dot{q}) = \frac{m}{2} g_{ab} \dot{q}^a \dot{q}^b - V(q).$$

The quantum correction $\Delta V_{\text{DeW}} = -\frac{\hbar^2}{6m} R$ is necessary for propagator (2.5.7) to be the solution of the Schrödinger equation (2.5.1).

From the general discussion in section 2.3, we know that such a term, in general, is not indispensable. If we choose some ordering for a Hamiltonian and perform the multiple star-product for the infinitesimal evolution operators as in (2.3.67), the resulting discrete-time approximation for the evolution operator kernel or symbol does not contain any extraordinary terms. The only question is whether the star-operation for this chosen ordering has a suitable form, which would enable us to write the evolution kernel in the continuous limit as a conventional path integral. As we have also learned in section 2.3, the operator-ordering ambiguity in quantum mechanics reveals itself in path integrals as these depend on the choice of different discrete approximations. The order of operators in the Hamiltonian which we consider in this subsection is defined by the form of the Laplace–Beltrami operator (equations (2.5.4) and (2.5.5)). On the other hand, the lattice approximation in the DeWitt path integral (2.5.7) is also fixed: the metric terms in the action are evaluated at the 'prepoint' q_{j-1}. The 'quantum correction' term, of the order $\hbar^2$, just reflects and compensates the discrepancy between the chosen operator ordering and the lattice prescription. Changing the lattice definition, i.e. the evaluation of the metric terms at other points, e.g., the 'postpoint' q_j or the 'midpoint' $\tilde{q}_j \stackrel{\text{def}}{\equiv} \frac{1}{2}(q_j + q_{j-1})$, changes ΔV, because in the Taylor expansion of the relevant terms, *all* terms of the order $\mathcal{O}(\varepsilon)$ contribute to the path integral. This fact is particularly important in the expansion of the kinetic term in the Lagrangian, where we have $(q_j - q_{j-1})^4/\varepsilon \sim \mathcal{O}(\varepsilon)$, because, as we have learned in chapter 1, $(q_j - q_{j-1}) \sim \sqrt{\varepsilon}$.

◇ **Weyl-ordering rule and Weyl symbol for the propagator in a curved space**

A very convenient lattice prescription is the midpoint definition, which is connected to the Weyl-ordering prescription in the Hamiltonian $\widehat{H}$ (cf section 2.3). Let us discuss the construction of the path integral in this prescription in some detail. First, we introduce the modified momentum operators $\widehat{p}_a$:

$$\widehat{p}_a = \frac{\hbar}{i}\left(\frac{\partial}{\partial q^a} + \frac{\Gamma_a}{2}\right) \qquad \Gamma_a \equiv \Gamma^b_{ab} = \frac{1}{2}\frac{\partial \ln\sqrt{g}}{\partial q^a} \tag{2.5.11}$$

which are Hermitian with respect to the scalar product (2.5.2) and obviously have the same canonical commutation relations with $\widehat{q}^a$:

$$[\widehat{q}^a, \widehat{p}_b] = i\hbar \delta^a{}_b.$$

In terms of the momentum operators (2.5.11), we rewrite the quantum Hamiltonian $\widehat{H}$ (2.5.5) by using the Weyl-ordering prescription:

$$\widehat{H}(p,q) = \frac{1}{8m}(g^{ab}(\widehat{q})\widehat{p}_a\widehat{p}_b + 2\widehat{p}_a g^{ab}(\widehat{q})\widehat{p}_b + \widehat{p}_a \widehat{p}_b g^{ab}(\widehat{q})) + \Delta V_{\text{Weyl}}(\widehat{q}) + V(\widehat{q}). \tag{2.5.12}$$

Here a well-defined quantum correction appears which is given by

$$\Delta V_{\text{Weyl}} = \frac{\hbar^2}{8m}(g^{ab}\Gamma^d_{ac}\Gamma^c_{bd} - R) = \frac{\hbar^2}{8m}[g^{ab}\Gamma_a\Gamma_b + 2(g^{ab}\Gamma_a)_{,b} + g^{ab}{}_{,ab}]. \tag{2.5.13}$$

We stress once again that the source of this 'quantum correction' is that we started from the Laplace–Beltrami quantum Hamiltonian (2.5.4), (2.5.5), but in the construction of the path integral we want to use the Weyl-ordering rule and Weyl symbols.

The Weyl-correspondence gives a unique prescription for the construction of the path integral. We have for the integral kernel

$$K(q,t,|q_0,t_0) = \left\langle q \left| \exp\left[-\frac{i}{\hbar}\widehat{H}(\widehat{p},\widehat{q})\right] \right| q_0 \right\rangle$$

$$= \left(\prod_{j=1}^{N-1} \int \sqrt{g^{(j)}}\, dq_j\right) \prod_{j=1}^{N} \left\langle q_j \left| \exp\left[-\frac{i}{\hbar}\frac{t-t_0}{N}\widehat{H}(\widehat{p},\widehat{q})\right] \right| q_{j-1} \right\rangle. \quad (2.5.14)$$

In order to simplify the notation and to make the formulae easier to read, we drop the boldface type for vectors in this section as well as the indices of summation (for example $p_a(q_0^a - q^a) \equiv p(q_0 - q)$) and the explicit indication of dimensionality in the integration measure: $d^d p, d^d q, d^d u, \to dp, dq, du$.

Making use of the basic formulae for Weyl symbols (2.3.17) and (2.3.18) or the equality

$$\int du\, e^{iuq} \left\langle q_0 \left| p - \frac{u}{2} \right\rangle \left\langle p + \frac{u}{2} \right| q \right\rangle = \frac{1}{(2\pi\hbar)^d} \int du\, \exp\left[\frac{i}{\hbar}p(q_0-q) + \frac{i}{\hbar}u\left(q - \frac{q_0+q}{2}\right)\right]$$

$$= e^{ip(q_0-q)/\hbar} \delta\left(q - \frac{q_0+q}{2}\right) \quad (2.5.15)$$

we arrive at the expression for the short-time propagator

$$\langle q_j | \exp[-i\varepsilon \widehat{H}(\widehat{p},\widehat{q})/\hbar] | q_{j-1}\rangle$$

$$= \frac{1}{(2\pi\hbar)^d}\left\langle q_j \left| \int dp\, dq\, e^{-ih(p,q)/\hbar} \int du\, dv\, \exp[i(q-\widehat{q})u + i(p-\widehat{p})v] \right| q_{j-1}\right\rangle$$

$$= \frac{1}{(2\pi\hbar)^d} \int dp\, \exp\left\{\frac{i\varepsilon}{\hbar} p(q_j - q_{j-1}) - \frac{i\varepsilon}{\hbar} H(p,\widetilde{q}_j)\right\} \quad (2.5.16)$$

where $\widetilde{q}_j = \frac{1}{2}(q_j + q_{j-1})$ is the midpoint coordinate.

Inserting this into equation (2.5.14), we obtain the *Hamiltonian path integral*

$$K(q,t,|q_0,t_0) = [g(q_0)g(q)]^{-\frac{1}{4}} \lim_{N\to\infty} \prod_{j=1}^{N-1}\int dq_j \prod_{j=1}^{N}\int \frac{dp_j}{(2\pi\hbar)^d} \exp\left\{\frac{i}{\hbar}\sum_{j=1}^{N}[\Delta q_j \cdot p_j - \varepsilon H_{\text{eff}}(p_j,\widetilde{q}_j)]\right\}. \quad (2.5.17)$$

Here $\Delta q_j \stackrel{\text{def}}{\equiv} q_j - q_{j-1}$ (since we need a careful control of ε-factors, it is more convenient to use this finite difference than the lattice derivatives (1.2.195)). The effective Hamiltonian to be used in path integral (2.5.17) reads as

$$H_{\text{eff}}(p_j,\widetilde{q}_j) = \frac{1}{2m} g_{ab}(\widetilde{q}_j) p_j^a p_j^b + V(\widetilde{q}_j) + \Delta V_{\text{Weyl}}(\widetilde{q}_j). \quad (2.5.18)$$

The Gaussian integration over the momenta yields a Lagrangian path integral of the form

$$K(q,t,|q_0,t_0) = [g(q_0)g(q)]^{-\frac{1}{4}} \int_{\mathcal{C}\{q_0,t_0|q,t\}} \mathcal{D}\left(\sqrt{g(\tau)}q(\tau)\right) \exp\left[\frac{i}{\hbar}\int_{t_0}^{t} dt\, L_{\text{eff}}(q,\dot{q})\right]$$

$$\stackrel{\text{def}}{\equiv} [g(q_0)g(q)]^{-\frac{1}{4}} \lim_{N\to\infty} \left(\frac{m}{2\pi i\varepsilon\hbar}\right)^{\frac{Nd}{2}} \left(\prod_{j=1}^{N-1} \int dq_j\right)$$

$$\times \prod_{j=1}^{N} \sqrt{g(\widetilde{q}_j)} \exp\left\{\frac{i}{\hbar}\left[\frac{m}{2\varepsilon} g_{ab}(\widetilde{q}_j)\Delta q_j^a \Delta q_j^b - \varepsilon V(\widetilde{q}_j) - \varepsilon\Delta V_{\text{Weyl}}(\widetilde{q}_j)\right]\right\}. \quad (2.5.19)$$

Equation (2.5.19) is, of course, equivalent to equation (2.5.7), although its form is different due to another discrete-time approximation. Namely, the midpoint prescription arises here in a natural way, as a consequence of the Weyl-ordering rule.

◇ **Feynman-like proof of correctness of the path-integral representation for the propagator in a curved space**

As usual, to prove the correctness of (2.5.19) directly, we must show that the short-time kernel

$$K(q_j, q_{j-1}; \varepsilon) = \left(\frac{m}{2\pi i\varepsilon\hbar}\right)^{d/2} [g(q_{j-1})g(q_j)]^{-\frac{1}{4}} \sqrt{g(\widetilde{q}_j)}$$

$$\times \exp\left\{\frac{i}{\hbar}\left[\frac{m}{2\varepsilon} g_{ab}(\widetilde{q}_j)\Delta q_j^a \Delta q_j^b - \varepsilon V(\widetilde{q}_j) - \varepsilon\Delta V_{\text{Weyl}}(\widetilde{q}_j)\right]\right\}. \quad (2.5.20)$$

satisfies the Schrödinger equation (2.5.1). We may proceed exactly as in section 2.1.2. But in the more complicated case of a curved phase space, it is simpler to act in a slightly different way. Let us expand $\psi(q_0, t_0)$ in the Taylor series around the point (q, t):

$$\psi(q_0, t_0) = \psi(q, t) - \varepsilon\frac{\partial\psi(q, t)}{\partial t} + (q^a - a_0^a)\frac{\partial\psi(q, t)}{\partial q^a} + \frac{1}{2}(q^a - a_0^a)(q^b - b_0^b)\frac{\partial^2\psi(q, t)}{\partial q_0^b q_0^a} + \cdots$$

and insert the expansion into equation (2.5.6). This gives

$$\psi(q, t) + \varepsilon\frac{\partial\psi(q, t)}{\partial t} = B_0\psi(q, t) + B^b\frac{\partial\psi(q, t)}{\partial q^b} + B^{ab}\frac{\partial^2\psi(q, t)}{\partial q^b \partial q^a} + \cdots \quad (2.5.21)$$

where the coefficients in the expansion are given by

$$B_0 = \int dq_0 \sqrt{g(q_0)} K(q, q_0; \varepsilon)$$

$$\simeq \left(\frac{m}{2\pi i\varepsilon\hbar}\right)^{d/2} g^{-\frac{1}{4}}(q) e^{-i\varepsilon[V(q)+\Delta_{\text{Weyl}}V(q)]/\hbar} \int dq_0\, g^{\frac{1}{2}}(\widetilde{q}) g^{\frac{1}{4}}(q_0) \exp\left\{\frac{im}{2\varepsilon\hbar}\xi^a g_{ab}(\widetilde{q})\xi^b\right\} \quad (2.5.22)$$

$$B^b = \int dq_0 \sqrt{g(q_0)} K(q, q_0; \varepsilon)\xi^b$$

$$\simeq \left(\frac{m}{2\pi i\varepsilon\hbar}\right)^{d/2} g^{-\frac{1}{4}}(q) e^{-i\varepsilon[V(q)+\Delta_{\text{Weyl}}V(q)]/\hbar} \int dq_0\, g^{\frac{1}{2}}(\widetilde{q}) g^{\frac{1}{4}}(q_0) \exp\left\{\frac{im}{2\varepsilon\hbar}\xi^a g_{ac}(\widetilde{q})\xi^c\right\} \xi^b \quad (2.5.23)$$

$$B^{ab} = \int dq_0 \sqrt{g(q_0)} K(q, q_0; \varepsilon)\xi^a\xi^b$$

$$\simeq \left(\frac{m}{2\pi i\hbar c}\right)^{d/2} g^{-\frac{1}{4}}(q) e^{-i\varepsilon[V(q)+\Delta_{\text{Weyl}}V(q)]/\hbar} \int dq_0\, g^{\frac{1}{2}}(\widetilde{q}) g^{\frac{1}{4}}(q_0) \exp\left\{\frac{im}{2\varepsilon\hbar}\xi^c g_{cd}(\widetilde{q})\xi^d\right\} \xi^a\xi^b. \quad (2.5.24)$$

Here we have introduced, for convenience, the special notation for the difference of the particle coordinates: $\xi^a \stackrel{\text{def}}{\equiv} (q^a - q_0^a)$.

The calculation of these coefficients B_0, B_b, B_{ab} is rather tedious and we shall mark only its main steps (for further details see in Grosche (1992)).

First of all, we have to extract the dependence of the metric on q_0^a or, equivalently, on ξ^a. To this aim, we are going to expand the terms containing $g_{ab}(\tilde{q})$, $g_{ab}(q_0)$ into the power series in ξ^a and to show that the higher terms of this expansion give contributions of higher orders in ε.

Let us consider the following class of integrals:

$$\langle\!\langle \xi^{a_1}\xi^{a_2}\cdots\xi^{a_N}\rangle\!\rangle_g \stackrel{\text{def}}{\equiv} \sqrt{g(q)}\left(\frac{m}{2\pi i\hbar\varepsilon}\right)^{d/2}\int dq_0 \exp\left\{-\frac{m}{2i\hbar\varepsilon}(q^a-q_0^a)g_{ab}(q^b-q_0^b)\right\}$$
$$\times (q^{a_1}-q_0^{a_1})(q^{a_2}-q_0^{a_2})\cdots(q^{a_N}-q_0^{a_N}) \qquad a_i = 1,\ldots,d.$$

The Gaussian integrations give, for the lowest powers of ξ^a in the integrands, the following results:

$$\langle\!\langle \xi^a\xi^b\rangle\!\rangle_g = \frac{i\varepsilon\hbar}{m}g^{ab} \qquad (2.5.25)$$

$$\langle\!\langle \xi^a\xi^b\xi^c\xi^d\rangle\!\rangle_g = \left(\frac{i\varepsilon\hbar}{m}\right)^2[g^{ab}g^{cd}+g^{ac}g^{bd}+g^{ad}g^{bc}] \qquad (2.5.26)$$

$$\langle\!\langle \xi^a\xi^b\xi^c\xi^d\xi^e\xi^f\rangle\!\rangle_g = \left(\frac{i\varepsilon\hbar}{m}\right)^3[g^{ab}g^{cd}g^{ef}+g^{ac}g^{bd}g^{ef}+g^{ad}g^{bc}g^{ef}+g^{ab}g^{ce}g^{df}+g^{ab}g^{cf}g^{de}$$
$$+g^{cd}g^{ae}g^{bf}+g^{cd}g^{af}g^{be}+g^{ac}g^{be}g^{df}+g^{ac}g^{bf}g^{de}+g^{bd}g^{ae}g^{cf}$$
$$+g^{bd}g^{af}g^{ce}+g^{ad}g^{be}g^{cf}+g^{ad}g^{bf}g^{ce}+g^{bc}g^{ae}g^{df}+g^{bc}g^{af}g^{de}] \quad (2.5.27)$$

(problem 2.5.2, page 282). This shows that the integration of increasing powers of ξ^a indeed gives expressions of increasing orders in ε. Therefore, we can expand the integrands in (2.5.22)–(2.5.24) in a power series and keep only the first few terms to obtain the result up to the first order in ε. The Taylor expansion of the terms containing the metric at the midpoint yields

$$g^{1/4}(q-\xi)g^{1/2}(q-\xi/2) \simeq g^{3/4}(q)[1-\Gamma_a\xi^a + \tfrac{1}{8}(4\Gamma_a\Gamma_b+3\Gamma_{a,b})\xi^a\xi^b] \qquad (2.5.28)$$

$$\exp\left\{\frac{im}{2\varepsilon\hbar}g_{ab}(q-\xi/2)\xi^a\xi^b\right\} \simeq \exp\left\{\frac{im}{2\varepsilon\hbar}g_{ab}(q)\xi^a\xi^b\right\}$$
$$\times\left[1+\frac{m}{2i\varepsilon\hbar}g_{ab}\Gamma^b_{cd}\xi^a\xi^c\xi^d - \frac{m}{8i\varepsilon\hbar}(g_{av}\Gamma^v_{bc,d}+g_{au}\Gamma^u_{vd}\Gamma^v_{bc}+g_{uv}\Gamma^u_{ad}\Gamma^v_{bc})\xi^a\xi^b\xi^c\xi^d\right.$$
$$\left.+\frac{1}{2}\left(\frac{m}{2i\varepsilon\hbar}\right)^2 g_{av}g_{du}\Gamma^v_{bc}\Gamma^u_{ef}\xi^a\xi^b\xi^c\xi^d\xi^e\xi^f\right] \qquad (2.5.29)$$

and their combination reads as

$$g^{1/4}(q-\xi)g^{1/2}(q-\xi/2)\exp\left\{\frac{im}{2\varepsilon\hbar}g_{ab}(q-\xi/2)\xi^a\xi^b\right\}$$
$$\simeq g^{3/4}(q)\exp\left[\frac{im}{2\varepsilon\hbar}g_{ab}(q)\xi^a\xi^b\right]$$
$$\times\left[1-\Gamma_a\xi^a+\frac{m}{2i\varepsilon\hbar}g_{ad}\Gamma^d_{bc}\xi^a\xi^b\xi^c+\frac{1}{2}\left(\frac{m}{2i\varepsilon\hbar}\right)^2 g_{av}g_{du}\Gamma^v_{bc}\Gamma^u_{ef}\xi^a\xi^b\xi^c\xi^d\xi^e\xi^f\right.$$
$$\left.-\frac{m}{8i\varepsilon\hbar}(g_{av}\Gamma^v_{bc,d}+g_{au}\Gamma^u_{vd}\Gamma^v_{bc}+g_{uv}\Gamma^u_{ad}\Gamma^v_{bc})\xi^a\xi^b\xi^c\xi^d+\frac{1}{8}(4\Gamma_a\Gamma_b+3\Gamma_{a,b})\xi^a\xi^b\right]. \quad (2.5.30)$$

Here the various derivatives of the metric tensor g_{ab} have been expressed via the Christoffel symbols.

Now we are ready to calculate the coefficients of the Taylor expansion in (2.5.21). Up to the first order in ε, the coefficients B_{ab} are derived from (2.5.25):

$$B^{ab} \approx -\frac{i\varepsilon\hbar}{2m} g^{ab}. \tag{2.5.31}$$

Similarly,

$$B^a \approx -\frac{i\varepsilon\hbar}{2m}[\Gamma_a g^{ab} + (\partial_a g^{ab})]. \tag{2.5.32}$$

For the ξ^2- and ξ^4-terms in B_0, we obtain

$$\langle\!\langle \tfrac{1}{8}(4\Gamma_a\Gamma_b + 3\Gamma_{a,b})\xi^a\xi^b \rangle\!\rangle_g = \frac{i\varepsilon\hbar}{8m} g^{ab}(4\Gamma_a\Gamma_b + 3\Gamma_{a,b}) \tag{2.5.33}$$

$$\left\langle\!\!\left\langle -\frac{m}{8i\varepsilon\hbar}(g_{av}\Gamma^v_{bc,d} + g_{au}\Gamma^u_{vd}\Gamma^v_{bc} + g_{uv}\Gamma^u_{ad}\Gamma^v_{bc})\xi^a\xi^b\xi^c\xi^d \right\rangle\!\!\right\rangle_g$$

$$= -\frac{i\varepsilon\hbar}{8m} g^{ab}[8\Gamma_a\Gamma_b + \Gamma^c_{ab,c} + 2\Gamma_{a,b} + g_{uv}g^{cd}(2\Gamma^u_{ad}\Gamma^v_{bc} + \Gamma^u_{ab}\Gamma^v_{cd}) + 5\Gamma^c_{ab}\Gamma^d_{bd} + 2\Gamma^d_{ac}\Gamma^c_{bd}]. \tag{2.5.34}$$

For the ξ^6-terms, equation (2.5.27) yields

$$\left\langle\!\!\left\langle \frac{1}{2}\left(\frac{m}{2i\varepsilon\hbar}\right)^2 g_{av}g_{du}\Gamma^v_{bc}\Gamma^u_{ef}\xi^a\xi^b\xi^c\xi^d\xi^e\xi^f \right\rangle\!\!\right\rangle_g$$

$$= \frac{i\varepsilon\hbar}{8m} g^{ab}\left[4\Gamma_a\Gamma_b + 4\Gamma^c_{ab}\Gamma_c + 4\Gamma^d_{ac}\Gamma^c_{bd} + g_{uv}g^{cd}(2\Gamma^u_{ac}\Gamma^v_{bd} + \Gamma^u_{ab}\Gamma^v_{cd})\right]. \tag{2.5.35}$$

Therefore, the combination of the relevant terms results finally in the required Weyl term:

$$\int g^{1/4}(q-\xi) g^{\frac{1}{2}}(q-\bar\xi) \exp\left[\frac{im}{2\varepsilon\hbar} g_{ab}(q-\bar\xi)\xi^a\xi^b\right] d\xi$$

$$= \left(\frac{m}{2\pi i\varepsilon\hbar}\right)^{-d/2}\left[1 + \frac{i\varepsilon\hbar}{8m}g^{ab}(\Gamma_{a,b} - \Gamma^c_{ab}\Gamma_c + 2\Gamma^d_{ac}\Gamma^c_{bd} - \Gamma^c_{ab,c})\right]$$

$$= \left(\frac{m}{2\pi i\varepsilon\hbar}\right)^{-d/2} \exp\left(\frac{i\varepsilon}{\hbar}\Delta V_{\mathrm{Weyl}}\right). \tag{2.5.36}$$

Inserting all the contributions into equation (2.5.21), we prove that the path integral (2.5.19) defines an evolution which satisfies the Schrödinger equation (2.5.1).

2.5.2 Spacetime transformations of Hamiltonians

Now we return for a moment to the discussion of systems in flat Euclidean spaces and in Cartesian coordinates. Consider a one-dimensional system with a standard Hamiltonian

$$\widehat{H} = \frac{\widehat{p}}{2m} + V(\widehat{x})$$

and the stationary Schrödinger equation

$$\widehat{H}|\psi\rangle = E|\psi\rangle$$

which, in the coordinate representation, has the following explicit form

$$-\frac{\hbar^2}{2m}\frac{d^2\psi(x)}{dx^2} + V(x)\psi(x) = E\psi(x). \qquad (2.5.37)$$

We assume that the potential $V(x)$ is so complicated that to solve equation (2.5.37) straightforwardly is impossible. In this case, the following trick may prove to be helpful. Let us make a coordinate transformation

$$x = F(q). \qquad (2.5.38)$$

In the new variable q, equation (2.5.37) becomes

$$-\frac{\hbar^2}{2mF'^2}\frac{d^2\psi}{dq^2} - \frac{\hbar^2}{2m}\frac{F''}{F'^3}\frac{d\psi}{dq} + V(F(q))\psi = E\psi$$

(recall that $F' \stackrel{\text{def}}{\equiv} \partial F/\partial q$). The substitution

$$\psi(F(q)) = \sqrt{F'}\varphi(q) \qquad (2.5.39)$$

simplifies this equation which for the function $\varphi(q)$ reads as

$$-\frac{\hbar^2}{2m}\frac{d^2\varphi}{dq^2} + (W(q) - EF'^2(q))\varphi = 0 \qquad (2.5.40)$$

where

$$W \stackrel{\text{def}}{\equiv} F'^2 V(F(q)) + \frac{\hbar^2}{4m}\left[\frac{3}{2}\left(\frac{F''}{F'}\right)^2 - \frac{F'''}{F'}\right]. \qquad (2.5.41)$$

Equation (2.5.40) has the form of a Schrödinger equation with the 'potential' $(W(q) - EF'^2(q))$ and zero eigenvalue. If this new potential term allows us to solve exactly the corresponding eigenvalue problem

$$-\frac{\hbar^2}{2m}\frac{d^2\varphi_n}{dq^2} + (W(q) - EF'^2(q))\varphi_n = \mathcal{E}_n\varphi_n$$

the eigenvalues of the initial problem can be found from the equations

$$\mathcal{E}_n(E) = 0 \qquad n = 0, 1, 2, 3, \ldots.$$

The corresponding eigenfunctions are determined by (2.5.39). Knowledge of the eigenfunctions and eigenvalues gives, in fact, complete information about the behaviour of the system under consideration.

This general approach to simplifying the basic relations via changes of variables can be applied to the calculation of complicated path integrals.

◇ **Spacetime transformations in the path-integral formalism**

Consider a one-dimensional path integral

$$K(x, x_0; t - t_0) = \int_{\mathcal{C}\{x_0, t_0 | x, t\}} \mathcal{D}x(\tau) \exp\left[\frac{i}{\hbar}\int_{t_0}^{t}\left(\frac{m}{2}\dot{x}^2 - V(x)\right)dt\right]. \qquad (2.5.42)$$

It is now assumed that the potential V is so complicated that a direct evaluation of the path integral is not possible. Inspired by the discussion of the transformations of variables in the Schrödinger equation, we shall look for a modification of this path integral which permits the explicit calculation of $K(x, x_0; t - t_0)$. This technique was originally developed by Duru and Kleinert (1979, 1982) (see also Kleinert (1995) and references therein). It was further evolved by other authors (see e.g., Grosche (1992, 1996), Grosche and Steiner (1998) and references therein).

Let us introduce a more general Hamiltonian $\widehat{H}_E$ which, similar to (2.5.40), contains the energy variable:

$$\widehat{H}_E = -\frac{\hbar^2}{2m}\frac{d^2}{dx^2} + V(x) - E \tag{2.5.43}$$

with the corresponding path integral

$$K_E(x, x_0; T) = e^{iTE/\hbar} K(x'', x'; T) \tag{2.5.44}$$

where $K(T)$ denotes the path integral of equation (2.5.42) and $T \equiv t - t_0$.

Let us now use the spacetime transformation (2.5.38)

$$x = F(q) \tag{2.5.45}$$

to obtain from (2.5.43)

$$\widehat{H}_E = -\frac{\hbar^2}{2m}\frac{1}{F'^2(q)}\left[\frac{d^2}{dq^2} - \frac{F''(q)}{F'(q)}\frac{d}{dq}\right] + V(F(q)) - E. \tag{2.5.46}$$

In equations (2.5.37)–(2.5.40) we considered this sort of transformation for the *stationary* Schrödinger equation. To remove the q-dependent factor in front of the second derivative in $\widehat{H}_E$ in the *time-evolution* Schrödinger equation, we have to accompany the space transformation (2.5.45) by the transformation of the time variable $t \to s$, so that

$$dt = f(x)\, ds \tag{2.5.47}$$

where the function f satisfies the constraint

$$f[F(q)] = F'^2(q). \tag{2.5.48}$$

This leads to the transformation of the time derivative

$$\frac{\partial}{\partial t} = \frac{1}{f(x)}\frac{\partial}{\partial s}$$

and

$$-i\hbar\frac{\partial}{\partial t}\psi = \widehat{H}_E\psi \longrightarrow -i\hbar\frac{\partial}{\partial s}\psi = f\widehat{H}_E\psi.$$

Thus it is reasonable to introduce a new Hamiltonian $\widetilde{H} = f\widehat{H}_E$:

$$\begin{aligned}\widetilde{H} &= -\frac{\hbar^2}{2m}\left[\frac{d^2}{dq^2} - \frac{F''(q)}{F'(q)}\frac{d}{dq}\right] + f[F(q)][V(F(q)) - E] \\ &= -\frac{\hbar^2}{2m}\left[\frac{d^2}{dq^2} - \Gamma(q)\frac{d}{dq}\right] + f[F(q)][V(F(q)) - E]\end{aligned} \tag{2.5.49}$$

where $\Gamma(q) = F''(q)/F'(q)$. Comparison with (2.5.11) and (2.5.12) shows that $\Gamma(q)$ plays the role of the Christoffel symbol (2.5.9) for the one-dimensional metric $g(q) = (F')^2$. Since we study at present the one-dimensional case, the metric simply becomes a function (the indices take only one value and g_{ab} is not actually a coordinate-dependent *matrix* but a *function*). The corresponding measure in the scalar product and the Hermitian momentum are now defined as

$$(f_1, f_2) = \int dx\, J(x) f_1^*(x) f_2(x) \qquad (2.5.50)$$

$$J(x) \stackrel{\text{def}}{\equiv} \sqrt{g(q)} = e^{\int^x \Gamma(x')\,dx'} \qquad (2.5.51)$$

$$p_q = -i\hbar \left[\frac{d}{dq} + \frac{1}{2}\Gamma(q)\right]. \qquad (2.5.52)$$

The Hamiltonian $\widetilde{H}$, expressed in terms of q and p_q, is

$$\widetilde{H} = \frac{p_q^2}{2m} + f[F(q)][V(F(q))_E] + \Delta V(q) \qquad (2.5.53)$$

where $\Delta V(q)$ is proportional to the squared Planck constant:

$$\Delta V(q) = \frac{\hbar^2}{8m}\left[3\left(\frac{F''(q)}{F'(q)}\right)^2 - 2\frac{F'''(q)}{F'(q)}\right]. \qquad (2.5.54)$$

Note that $\widetilde{H}$ coincides with the Hamiltonian in (2.5.40). The path integral, corresponding to the Hamiltonian $\widetilde{H}$, is

$$\widetilde{K}(q, q_0; S) = \int_{\mathcal{C}\{q_0, t_0 | q, t\}} \mathcal{D}q(\sigma) \exp\left\{\frac{i}{\hbar}\int_{s_0}^{s} d\sigma \left[\frac{m}{2}\dot{q}^2 - f[F(q)][V(F(q)) - E] - \Delta V(q)\right]\right\} \qquad (2.5.55)$$

with $S \equiv (s - s_0)$. In the usual way, from the short-time kernel of equation (2.5.55), we can derive that the time evolution equation

$$\widetilde{\psi}(q; s) = \int dq_0\, \widetilde{K}(q, q_0; S)\widetilde{\psi}(q_0; s_0) \qquad (2.5.56)$$

is equivalent to the time-dependent Schrödinger equation

$$i\hbar\frac{\partial}{\partial s}\widetilde{\psi}(q; s) = \widetilde{H}\widetilde{\psi}(q; s). \qquad (2.5.57)$$

◇ **Justification of spacetime transformations in the path integral**

Of course, a rigorous transformation from $K(x, x_0; T)$ to $\widetilde{K}(q, q_0; S)$ must be based on time-sliced approximation (i.e. on the precise definition) of these path integrals. We shall only outline here the main points of this consideration (for details, see, e.g., Kleinert (1995) and Grosche (1992)).

Let us consider the path integral $K(T)$ in its lattice definition:

$$K(x, x_0; T) = \lim_{N\to\infty}\left(\frac{m}{2\pi i\varepsilon\hbar}\right)^{N/2}\prod_{j=1}^{N-1}\int dx_j\, \exp\left\{\frac{i}{\hbar}\sum_{j=1}^{N}\left[\frac{m}{2\varepsilon}(x_j - x_{j-1})^2 - \varepsilon V(\bar{x}^{(j)})\right]\right\}. \qquad (2.5.58)$$

To convert it into the discrete-time approximation for the desired path integral (2.5.55) which is supposed to be exactly evaluated, we have to make the transformation (2.5.45) and, following the midpoint prescription, to expand all quantities about $\widetilde{q}_j = (q_j + q_{j+1})/2$.

This procedure, together with the transformation of the measure $\prod_{j=1}^{N-1} dx_j$, casts (2.5.58) into the form

$$K(q, q_0; T) = [F'(q_0)F'(q)]^{-\frac{1}{2}} \lim_{N\to\infty} \left(\frac{m}{2\pi i\varepsilon\hbar}\right)^{N/2} \prod_{j=1}^{N-1} \int dq_j$$

$$\times \prod_{j=1}^{N} F'(\widetilde{q}_j) \exp\left\{\frac{i}{\hbar}\left[\frac{m}{2\varepsilon\hbar}\Delta q_j \Delta q_j F'^2(\widetilde{q}_j) - \varepsilon V(\widetilde{q}_j) - \frac{\varepsilon\hbar^2}{8m}\frac{F'^2(\widetilde{q}_j)}{F'^4(\widetilde{q}_j)}\right]\right\}. \quad (2.5.59)$$

Note that this (discretely approximated) path integral has the canonical form (2.5.19) with the one-dimensional metric $g_{ab} = F'^2$.

This remains true for higher-dimensional spaces, for which (2.5.59) reads as

$$\hat{K}(q, q_0; T) = [F_{;q}(q_0)F_{;q}(q)]^{-1/2} \lim_{N\to\infty} \left(\frac{m}{2\pi i\varepsilon\hbar}\right)^{Nd/2} \prod_{j=1}^{N-1} \int dq^{(j)}$$

$$\times \prod_{j=1}^{N} F_{;q}(\widetilde{q}_j) \exp\left\{\frac{i}{\hbar}\left[\frac{m}{2\varepsilon}\Delta q_k^{(j)} \Delta q_n^{(j)} F_{,k}(\widetilde{q}_j) F_{,n}(\widetilde{q}_j) - \varepsilon V(\widetilde{q}_j)\right.\right.$$

$$- \frac{\varepsilon\hbar^2}{8m}(F_{,k}^{(j)} F_{,n}^{(j)})^{-1}\left(\frac{F_{;q,k}(\widetilde{q}_j) F_{;q,n}(\widetilde{q}_j)}{F_{;q}^2(\widetilde{q}_j)} - \frac{F_{;q,kn}(\widetilde{q}_j)}{F_{;q}(\widetilde{q}_j)}\right)$$

$$\left.\left.- \frac{\varepsilon\hbar^2}{8m} F_{,p}^{(j)} F_{,nkl}^{(j)} F_{pnkl}^{-1}(q^{(j)})\right)\right]\right\} \quad (2.5.60)$$

where

$$F_{pnkl}^{-1}(q^{(j)}) \stackrel{\text{def}}{\equiv} (F_{,p}^{(j)} F_{,n}^{(j)})^{-1} (F_{,k}^{(j)} F_{,l}^{(j)})^{-1}$$
$$+ (F_{,p}^{(j)} F_{,k}^{(j)})^{-1} (F_{,l}^{(j)} F_{,n}^{(j)})^{-1} + (F_{,p}^{(j)} F_{,l}^{(j)})^{-1} (F_{,k}^{(j)} F_{,n}^{(j)})^{-1} \quad (2.5.61)$$

and for the d-dimensional transformations $x^a = F^a(q)$ $(a = 1,\ldots,d)$ we have put

$$F_{;q} \stackrel{\text{def}}{\equiv} \det \frac{\partial F^a}{\partial q^n}.$$

Here the derivatives are considered as elements of a $d \times d$ matrix $B^a{}_n = \partial F^a/\partial q^n$. After the transition to the coordinates q^m, the coordinate-dependent metric

$$g_{mn}(q) = \sum_{a=1}^{d} \frac{\partial F^a}{\partial q^m} \frac{\partial F^a}{\partial q^n} \quad (2.5.62)$$

appears. Readers acquainted with Riemann geometry may easily recognize that the derivatives $\partial F^a/\partial q^m$ of the transformation functions F^a in (2.5.62) play the role of orthogonal d-beins. Making use of metric (2.5.62), we can rewrite (2.5.60) in the canonical form (2.5.19) (path integral for coordinates with a non-trivial metric) with the $\hbar^2$-term $\Delta_{\text{Weyl}}V$ looking as follows:

$$\Delta_{\text{Weyl}}V(q) = \frac{\hbar^2}{8m} g^{mn}(q) \Gamma^{k}_{lm}(q) \Gamma^{l}_{km}(q). \quad (2.5.63)$$

This is just the Weyl-ordered quantum potential without a curvature term. In the present case, the curvature vanishes since we started from a *flat* Euclidean space which remains flat after the transformation.

To proceed further, we add the energy constant into the exponential (i.e. we use the Hamiltonian $\hat{H}_E$) and perform the time transformation according to

$$dt = f(s)\,ds \qquad s_0 = s(t_0) = 0 \qquad s = s(t) \tag{2.5.64}$$

(as before we make the choice $f = F'^2$). Translation of this transformation into the discrete notation according to the midpoint prescription requires symmetrization over the interval $(j, j-1)$ to prefer neither of the endpoints over the other, i.e.

$$\Delta t_j = \varepsilon = \Delta s_j F'(q_j) F'(q_{j-1}) \tag{2.5.65}$$
$$\Delta s_j = s_j - s_{j-1} \equiv \delta_j. \tag{2.5.66}$$

The expansion of (2.5.65) about midpoints yields

$$\varepsilon \simeq \delta_j F'^2(\tilde{q}_j) \left\{1 + \frac{\Delta^2 q_j}{4}\left[\frac{F'''(\tilde{q}_j)}{F'(\tilde{q}_j)} - \left(\frac{F''(\tilde{q}_j)}{F'(\tilde{q}_j)}\right)^2\right]\right\}. \tag{2.5.67}$$

Insertion of this expression into the discrete approximation of the initial integrals gives

$$\left(\frac{m}{2\pi i\varepsilon\hbar}\right)^{\frac{N}{2}} \prod_{j=1}^{N-1} \int dx_j \prod_{j=1}^{N} \exp\left\{\frac{i}{\hbar}\left[\frac{m}{2\varepsilon}\Delta^2 x_j - \varepsilon V(x_j) + \varepsilon E\right]\right\}$$

$$= \prod_{j=1}^{N-1} \int dq_j\, F'(q_j) \prod_{j=1}^{N} \left(\frac{m}{2\pi i\varepsilon\hbar}\right)^{\frac{1}{2}}$$

$$\times \exp\left\{\frac{i}{\hbar}\left\{\frac{m}{2\varepsilon}\left[F'^2(\tilde{q}_j)\Delta^2 q_j + \frac{1}{12}F'(\tilde{q}_j)F'''(\tilde{q}_j)\Delta^4 q_j\right] - \varepsilon V(x_j) + \varepsilon E\right\}\right\}$$

$$= [F'(q_0)F'(q)]^{-\frac{1}{2}} \prod_{j=1}^{N-1} \int dq_j \prod_{j=1}^{N} \left(\frac{m}{2\pi i\delta_j\hbar}\right)^{\frac{1}{2}}$$

$$\times \prod_{j=1}^{N} \exp\left\{\frac{i}{\hbar}\left(\frac{m}{2\delta_j}\left[1 + \frac{\Delta^2 q_j}{4}\left(\frac{F'''(\tilde{q}_j)}{F'(\tilde{q}_j)} - \left(\frac{F''(\tilde{q}_j)}{F'(\tilde{q}_j)}\right)^2\right)\right]^{-1}\right.\right.$$

$$\left.\left.\times \left[\Delta^2 q_j + \frac{\Delta^4 q_j}{12}\frac{F'''(\tilde{q}_j)}{F'(\tilde{q}_j)}\right] - \delta_j F'^2(\tilde{q}_j)[V(q_j) - E]\right)\right\}. \tag{2.5.68}$$

Up to $\mathcal{O}(\delta)$-terms, where $\delta = \max_j \delta_j$, the last quantity can be rewritten as follows (cf (2.5.25)–(2.5.27)):

$$[F'(q_0)F'(q)]^{-\frac{1}{2}}\left(\frac{m}{2\pi i\delta\hbar}\right)^{\frac{N}{2}} \prod_{j=1}^{N-1}\int dq_j\, \exp\left\{\frac{i}{\hbar}\left(\frac{m}{2\delta}\Delta^2 q_j - \delta_j F'^2(q_j)[V(q_j) - E]\right.\right.$$

$$\left.\left. - \frac{i\delta\hbar}{8m}\left[3\left(\frac{F''(q_j)}{F'(q_j)}\right)^2 - 2\frac{F'''(q_j)}{F'(q_j)}\right]\right)\right\}. \tag{2.5.69}$$

The expression (2.5.69) is nothing other than the discrete-time approximation of the path integral (2.5.55). Thus integral (2.5.55) (which supposedly allows exact evaluation!) can indeed be obtained from the initial path integral (2.5.42) with the additional energy variable (cf (2.5.44)) by spacetime transformations.

Path integrals in curved spaces, spacetime transformations and the Coulomb problem 257

◇ **Relation between propagators in the initial and transformed spacetime coordinates**

For practical use of these spacetime transformations in calculating $K(x, x_0; T)$, we need a direct relation between K and $\tilde{K}$. We formulate this important relation as follows.

Theorem 2.4. The propagator K is given in terms of $\tilde{K}$ by the equations

$$K(x, x_0; T) = \frac{1}{2\pi i \hbar} \int_{-\infty}^{\infty} dE \, e^{-iTE/\hbar} G(x, x_0; E) \qquad (2.5.70)$$

$$G(x, x_0; E) = i[f(x')f(x'')]^{\frac{1}{4}} \int_0^{\infty} ds \, \tilde{K}(q, q_0; s). \qquad (2.5.71)$$

Sketch of proof. Consider the Green function $G(x, x_0; E)$:

$$G(x, x_0; E) = i \int_0^{\infty} dT \, e^{iTE/\hbar} K(x, x_0; T) \qquad (2.5.72)$$

$$K(x, x_0; T) = \frac{1}{2\pi i \hbar} \int_{-\infty}^{\infty} dE \, e^{-iTE/\hbar} G(x, x_0; E) \qquad (2.5.73)$$

(cf the definitions (2.1.80)). Recall (see the remark below (2.1.80)) that, to make the integral (2.5.72) convergent, we need to add an infinitesimal positive imaginary part to E. Then, using the genuine definition of K as a transition amplitude:

$$K(x, x_0; T) = \langle x| \exp\{iT\widehat{H}/\hbar\}|x_0\rangle$$

we have

$$G(x, x_0; E) = \lim_{\varepsilon \to 0} i \int_0^{\infty} dT \, e^{iTE/\hbar - T\varepsilon} \langle x|e^{iT\widehat{H}/\hbar}|x_0\rangle$$

$$= \langle x|\frac{1}{\widehat{H} - E}|x_0\rangle. \qquad (2.5.74)$$

The latter expression can be written in different equivalent forms:

$$G(x, x_0; E) = \langle x|f(\widehat{x})[(\widehat{H} - E)f(\widehat{x})]^{-1}|x_0\rangle = f(x)\langle x|\frac{1}{(\widehat{H} - E)f(\widehat{x})}|x_0\rangle$$

$$= \langle x|[f(\widehat{x})(\widehat{H} - E)]^{-1} f(\widehat{x})|x_0\rangle = f(x_0)\langle x|\frac{1}{f(\widehat{x})(\widehat{H} - E)}|x_0\rangle$$

$$= \langle x|\sqrt{f(\widehat{x})}[f(\widehat{x})(\widehat{H} - E)]^{-1}\sqrt{f(\widehat{x})}|x_0\rangle$$

$$= \sqrt{f(x)f(x_0)}\langle x|\frac{1}{\sqrt{f(\widehat{x})}(\widehat{H} - E)\sqrt{f(\widehat{x})}}|x_0\rangle$$

where $f(x)$ is an arbitrary function. This, in turn, means that (2.5.72) can be equivalently written in the following ways:

$$G(x, x_0; E) = f(x) \int_0^{\infty} ds \, \langle x|e^{-isf(\widehat{x})(\widehat{H}-E)/\hbar}|x_0\rangle \qquad (2.5.75)$$

$$= f(x_0) \int_0^{\infty} ds \, \langle x|e^{-is(\widehat{H}-E)f(\widehat{x})/\hbar}|x_0\rangle \qquad (2.5.76)$$

$$= \sqrt{f(x)f(x_0)} \int_0^{\infty} ds \, \langle x|e^{-is\sqrt{f(\widehat{x})}(\widehat{H}-E)\sqrt{f(\widehat{x})}/\hbar}|x_0\rangle. \qquad (2.5.77)$$

For the short-time evolution, the matrix element in the last integrand can be explicitly calculated:

$$\langle x_j | e^{-i\delta_j f^{\frac{1}{2}}(\hat{x})(\hat{H}-E) f^{\frac{1}{2}}(\hat{x})/\hbar} | x_{j-1} \rangle$$

$$\simeq \langle x_j | 1 - i\delta_j f^{\frac{1}{2}}(\hat{x})(\hat{H} - E) f^{\frac{1}{2}}(\hat{x})/\hbar | x_{j-1} \rangle$$

$$= \langle x_j | x_{j-1} \rangle - \frac{i\delta_j}{\hbar} \sqrt{f(x_j) f(x_{j-1})} \langle x_j | \hat{H} - E | x_{j-1} \rangle$$

$$= \langle F(q_j) | F(q_{j-1}) \rangle - \frac{i\delta_j}{\hbar} \sqrt{f[F(q_j)] f[F(q_{j-1})]} \langle F(q_j) | \hat{H}_E | F(q_{j-1}) \rangle$$

$$= \langle q_j | q_{j-1} \rangle - \frac{i\delta}{\hbar} \langle q_j | \tilde{H} | q_{j-1} \rangle$$

$$= \frac{1}{2\pi} \int dp_q \exp\left[\frac{i}{\hbar} p_q \Delta q_j - \frac{i\delta}{2m\hbar} p_q^2 - \frac{i\delta}{\hbar} F'^2(\tilde{q}_j)(V(\tilde{q}_j) - E) - \frac{i\delta}{\hbar} \Delta V(\tilde{q}_j) \right]$$

$$= \left(\frac{m}{2\pi i\delta\hbar} \right)^{\frac{1}{2}} \exp\left[\frac{im}{2\delta\hbar} \Delta^2 q_j - \frac{\delta}{\hbar} F'^2(\tilde{q}_j)(V(\tilde{q}_j) - E) - \frac{\delta}{\hbar} \Delta V(\tilde{q}_j) \right] \quad (2.5.78)$$

where the quantum potential ΔV is given by (2.5.54). The product of N such short-time transition amplitudes gives the discrete approximation (2.5.69) and the continuous limit gives the path integral (2.5.55). This proves the statement of the theorem.

2.5.3 Path integrals in polar coordinates

In this subsection, we shall derive the path integral in d-dimensional polar coordinates (Kleinert (1995), Grosche (1992, 1996), Grosche and Steiner (1998) and references therein; see also Böhm and Junker (1987) for a discussion within the group theoretic approach). We shall obtain an expansion in the angular momentum l, where the angle-dependent part can be integrated out and the remaining radius-dependent part is simply called the *radial path integral*. We shall discuss some properties of the radial path integral and show that it is possible to obtain the radial Schrödinger equation from the short-time kernel.

The starting point is the usual path integral in d dimensions

$$K^{(d)}(x, x_0; T) = \int_{C\{x_0, t_0 | x, t\}} \mathcal{D}x(t) \exp\left\{ \frac{i}{\hbar} \int_{t_0}^{t} \left[\frac{m}{2} \dot{x}^2 - V(x) \right] dt \right\}$$

$$= \lim_{N \to \infty} \left(\frac{m}{2\pi i\varepsilon\hbar} \right)^{Nd/2} \int dx_1 \ldots \int dx_{N-1}$$

$$\exp\left\{ \frac{i}{\hbar} \sum_{j=1}^{N} \left[\frac{m}{2\varepsilon\hbar} (x_j - x_{j-1})^2 - V(x_j) \right] \right\}. \quad (2.5.79)$$

Let the potential energy of a particle depend only on the distance from some point but not on the direction: $V(x) = V(|x|)$ (as usual, the origin of the coordinate system is chosen to coincide with that point). It is natural to introduce the d-dimensional polar coordinates

$$
\begin{aligned}
x_1 &= r\cos\theta_1 \\
x_2 &= r\sin\theta_1 \cos\theta_2 \\
x_3 &= r\sin\theta_1 \sin\theta_2 \cos\theta_3 \\
&\ldots \\
x_{d-1} &= r\sin\theta_1 \sin\theta_2 \ldots \sin\theta_{d-2}\cos\phi \\
x_d &= r\sin\theta_1 \sin\theta_2 \ldots \sin\theta_{d-2}\sin\phi
\end{aligned} \tag{2.5.80}
$$

where

$$0 \leq \theta_\nu \leq \pi \qquad \nu = 1,\ldots,d-2 \qquad 0 \leq \phi \leq 2\pi$$

$$r = |x| = \left(\sum_{\nu=1}^{d} x_\nu^2\right)^{1/2}.$$

Therefore, $V(x) = V(r)$. We have to use the addition theorem:

$$\cos\psi^{(1,2)} = \cos\theta_1^{(1)}\cos\theta_1^{(2)} + \sum_{m=1}^{d-2}\cos\theta_{m+1}^{(1)}\cos\theta_{m+1}^{(2)}\prod_{n=1}^{m}\sin\theta_n^{(1)}\sin\theta_n^{(2)} + \prod_{n=1}^{d-1}\sin\theta_n^{(1)}\sin\theta_n^{(2)} \tag{2.5.81}$$

where $\psi^{(1,2)}$ is the angle between two d-dimensional vectors $x^{(1)}$ and $x^{(2)}$ so that $x^{(1)}x^{(2)} = r^{(1)}r^{(2)}\cos\psi^{(1,2)}$. The metric tensor in polar coordinates is

$$(g_{ab}) = \mathrm{diag}(1, r^2, r^2\sin^2\theta_1, \ldots, r^2\sin^2\theta_1 \ldots \sin^2\theta_{d-2}). \tag{2.5.82}$$

If $d = 3$, equation (2.5.81) reduces to

$$\cos\psi^{(1,2)} = \cos\theta^{(1)}\cos\theta^{(2)} + \sin\theta^{(1)}\sin\theta^{(2)}\cos(\phi^{(1)} - \phi^{(2)}).$$

The d-dimensional measure $d^d x$ in polar coordinates takes the form

$$d^d x = r^{d-1}\,dr\,d\Omega = r^{d-1}\prod_{k=1}^{d-1}(\sin\theta_k)^{d-1-k}\,dr\,d\theta_k \tag{2.5.83}$$

$$d\Omega = \prod_{k=1}^{d-1}(\sin\theta_k)^{d-1-k}\,d\theta_k. \tag{2.5.84}$$

Here $d\Omega$ denotes the $(d-1)$-dimensional surface element on the unit sphere S^{d-1} and $\Omega(d) = 2\pi^{d/2}/\Gamma(d/2)$ is the volume of the d-dimensional unit S^{d-1}-sphere. The determinant of the metric tensor is given by

$$g = \det(g_{ab}) = \left(r^{d-1}\prod_{k=1}^{d-1}(\sin\theta_k)^{d-1-k}\right)^2. \tag{2.5.85}$$

◇ **Separation of the radial part of the path integral**

In these polar coordinates, the path integral (2.5.79) reads as

$$K^{(d)}(r, \{\theta\}, r_0, \{\theta_0\}; T) = \lim_{N \to \infty} \left(\frac{m}{2\pi i\varepsilon\hbar}\right)^{Nd/2} \int_0^\infty r_1^{d-1} dr_1 \int d\Omega_1 \ldots \int_0^\infty r_{N-1}^{d-1} dr_{N-1} \int d\Omega_{N-1}$$

$$\times \prod_{j=1}^N \exp\left\{\frac{im}{2\varepsilon\hbar}[r_j^2 + r_{j-1}^2 - 2r_j r_{j-1} \cos\psi^{(j,j-1)}] - \frac{i\varepsilon}{\hbar}V(r_j)\right\} \quad (2.5.86)$$

where $\{\theta\}$ denotes the set of the angular variables. Now we want to separate out the angular-dependent part in (2.5.86). To this aim, we need the following expansion formula (Gradshteyn and Ryzhik (1980), formula 8.534)

$$e^{z \cos \psi} = \left(\frac{z}{2}\right)^{-\nu} \Gamma(\nu) \sum_{l=0}^\infty (l+\nu) I_{l+\nu}(z) C_l^\nu(\cos \psi) \quad (2.5.87)$$

where $\nu = d/2 - 1$, the C_l^ν are the *Gegenbauer polynomials* and the I_μ are the *modified Bessel functions*. Equation (2.5.87) is a generalization of the well-known expansion in three dimensions where $\nu = 1/2$ (Gradshteyn and Ryzhik (1980), formula 8.511.4)

$$e^{z \cos \psi} = \sqrt{\frac{\pi}{2z}} \sum_{l=0}^\infty (2l+1) I_{l+\frac{1}{2}}(z) P_l(\cos \psi) \quad (2.5.88)$$

with the Gegenbauer polynomials turning into the *Legendre polynomials* $P_l(\cos \psi)$ (recall that $C_l^{\frac{1}{2}} = P_l$). Strictly speaking, (2.5.87) is not valid for $\nu = 0$, but it is possible to include the case $d = 2$, i.e. $\nu = 0$, if we use (Gradshteyn and Ryzhik (1980), formula 8.934.4)

$$l \lim_{\lambda \to 0} \Gamma(\lambda) C_l^\lambda = 2 \cos l\psi$$

yielding in the two-dimensional case the simpler formula (Gradshteyn and Ryzhik (1980), formula 8.511.4)

$$e^{z \cos \psi} = \sum_{k=-\infty}^\infty I_k(z) e^{ik\psi}. \quad (2.5.89)$$

Equation (2.5.87) allows us to write the jth term of the product in equation (2.5.86) as follows:

$$\exp\left\{\frac{im}{2\varepsilon\hbar}(r_j^2 + r_{j-1}^2) - \frac{i\varepsilon}{\hbar}V(r_j)\right\} \exp\left\{\frac{m}{2i\varepsilon\hbar} r_j r_{j-1} \cos\psi^{(j,j-1)}\right\}$$

$$= \left(\frac{2i\varepsilon\hbar}{mr_j r_{j-1}}\right)^{\frac{d-2}{2}} \Gamma\left(\frac{d-2}{2}\right) \exp\left\{\frac{im}{2\varepsilon\hbar}(r_j^2 + r_{j-1}^2) - \frac{i\varepsilon}{\hbar}V(r_j)\right\}$$

$$\times \sum_{l_j=0}^\infty \left(l_j + \frac{d}{2} - 1\right) I_{l_j + \frac{d-2}{2}}\left(\frac{m}{i\varepsilon\hbar} r_j r_{j-1}\right) C_{l_j}^{\frac{d-2}{2}}(\cos \psi^{(j,j-1)}). \quad (2.5.90)$$

Therefore, we can separate the *radial part* $k_l^{(d)}(r, r_0; T)$ of the path integral

$$K^{(d)}(r, \{\theta\}, r_0, \{\theta_0\}; T) = \Omega^{-1}(d) \sum_{l=0}^\infty \frac{l+d-2}{d-2} C_l^{\frac{d-2}{2}}(\cos \psi) k_l^{(d)}(r, r_0; T). \quad (2.5.91)$$

Here $\cos \psi$ is defined by the relation (2.5.81) with $\theta^{(1)} = \theta$ and $\theta^{(2)} = \theta_0$ and the radial part reads as

$$k_l^{(d)}(r, r_0; T) = (r_0 r)^{-\frac{d-2}{2}} \lim_{N \to \infty} \left(\frac{m}{2\pi i \varepsilon \hbar}\right)^{N/2} \int_0^\infty dr_1 \ldots \int_0^\infty dr_{N-1}$$
$$\times \prod_{j=1}^N \mu_l[r_j r_{j-1}] \exp\left\{\frac{i}{\hbar} \sum_{j=1}^N \left[\frac{m}{2\varepsilon}(r_j - r_{j-1})^2 - \varepsilon V(r_j)\right]\right\} \quad (2.5.92)$$

where we have denoted

$$\mu_l^{(d)}(z_j) \stackrel{\text{def}}{=} \sqrt{2\pi z_j} e^{-z_j} I_{l+\frac{d-2}{2}}(z_j) \quad (2.5.93)$$
$$z_j = (m/i\varepsilon \hbar) r_j r_{j-1}.$$

Note that it is tempting to use the asymptotic form of the modified Bessel functions:

$$I_\nu(z) \simeq (2\pi z)^{-\frac{1}{2}} e^{z - (\nu^2 - \frac{1}{4})/2z} \quad (|z| \gg 1, \text{Re}(z) > 0) \quad (2.5.94)$$

which results in $k_l^{(d)}$ with a simpler form:

$$k_l^{(d)}(r, r_0; T) = (r_0 r)^{-\frac{d-1}{2}} \lim_{N \to \infty} \left(\frac{m}{2\pi i \varepsilon \hbar}\right)^{N/2} \int_0^\infty dr_1 \ldots \int_0^\infty dr_{N-1}$$
$$\times \exp\left\{\frac{i}{\hbar} \sum_{j=1}^N \left[\frac{m}{2\varepsilon}(r_j - r_{j-1})^2 - \hbar^2 \frac{(l + \frac{d-2}{2})^2 - \frac{1}{4}}{2m r_j r_{j-1}} - \varepsilon V(r_j)\right]\right\}. \quad (2.5.95)$$

However, the asymptotic expansion of I_ν in equation (2.5.94) is problematic because it is valid only for $\text{Re}(z) > 0$, while we have $\text{Re}(z) = 0$. The continuous limit of expression (2.5.92) results in a radial path integral with a non-trivial functional measure $\mu_l^{(d)}$:

$$k_l^{(d)}(r, r_0; T) = \int_{C\{r,t;r_0,t_0\}} \mu_l^{(d)}[r^2(\tau)] \mathcal{D}r(\tau) \exp\left\{\frac{i}{\hbar} \int_{t_0}^t dt \left[\frac{m}{2}\dot{r}^2 - V(r)\right]\right\} \quad (2.5.96)$$

and the whole l-dependence is in μ_l, as defined in equation (2.5.93) in the discrete approximation.

Since the integration measure $\mu_l^{(d)}$ depends on d and l only via the specific combination in the index of the Bessel functions, we conclude that

$$\mu_l^{(d)}[r^2] = \mu_{l+\frac{d-3}{2}}^{(3)}[r^2] \quad (2.5.97)$$

and hence all the dimensional dependence of the path integral (2.5.96) can be deduced from the three-dimensional case:

$$k_l^{(d)}(r, r_0; T) = k_{l+\frac{d-3}{2}}^{(3)}(r, r_0, T). \quad (2.5.98)$$

From now on, we shall study the latter case and denote $k_l^{(3)}(T) = k_l(T)$.

The radial path integral (2.5.96) is reminiscent of a one-dimensional path integral (although with a non-trivial integration measure $\mu_l^{(d)}$) but it is defined on a half-line, $r \geq 0$. Using our experience from section 2.4.1, we can rewrite $k_l(T)$ as a superposition of two one-dimensional path integrals:

$$k_l(r, r_0; T) = k_l^{(1)}(r, r_0; T) - (-1)^l k_l^{(1)}(r, -r_0; T) \quad (2.5.99)$$

262   *Path integrals in quantum mechanics*

with infinite limits of integrations over r:

$$k_l^{(1)}(r,r_0;T) = \lim_{N\to\infty} \left(\frac{m}{2\pi i\varepsilon\hbar}\right)^{\frac{N}{2}} \int_{-\infty}^{\infty} dr_1 \ldots \int_{-\infty}^{\infty} dr_{N-1}$$
$$\times \prod_{j=1}^{N} \mu_l^{(d)}[r_j r_{j-1}] \exp\left\{\frac{i}{\hbar}\sum_{j=1}^{N}\left[\frac{m}{2\varepsilon\hbar}\Delta^2 r_{(j)} - \varepsilon V(|r_j|)\right]\right\}. \quad (2.5.100)$$

◇ **Harmonic oscillator in the polar coordinates**

Let us now discuss the most important application of equation (2.5.92), namely the *harmonic oscillator in polar coordinates* with $V(r) = \frac{1}{2}m\omega^2(t)r^2$ (for generality, with time-dependent frequency). This example has great virtue in solving various path-integral problems.

We have to study

$$k_l(r,r_0;T) = (r_0 r)^{\frac{1-d}{2}} \lim_{N\to\infty} \left(\frac{m}{2\pi i\varepsilon\hbar}\right)^{N/2} \int_0^\infty dr_1 \ldots \int_0^\infty dr_{N-1}$$
$$\times \prod_{j=1}^{N}\left[\mu_l^{(d)}[r_j r_{j-1}] \exp\left\{\frac{im}{2\varepsilon\hbar}(r_j - r_{j-1})^2 - \frac{i\varepsilon}{2\hbar}m\omega^2(t_j)r_j^2\right\}\right]$$
$$= (r_0 r)^{\frac{2-d}{2}} \lim_{N\to\infty} k_l^N(r,r_0;T)$$

$$k_l^N(r,r_0;T) \stackrel{\text{def}}{\equiv} \left(\frac{m}{i\varepsilon\hbar}\right)^{N/2} \int_0^\infty r_1 dr_1 \ldots \int_0^\infty r_{N-1} dr_{N-1}$$
$$\times \prod_{j=1}^{N}\left[\exp\left\{\frac{im}{2\varepsilon\hbar}(r_j^2 + r_{j-1}^2) - \frac{i\varepsilon}{2\hbar}m\omega^2(t_j)r_j^2\right\} I_{l+\frac{d-2}{2}}\left(\frac{m}{i\varepsilon\hbar}r_j r_{j-1}\right)\right]$$
$$= (r_0 r)^{\frac{2-d}{2}} \lim_{N\to\infty} \left(\frac{\alpha}{i}\right)^{N/2} e^{i\alpha(r_0^2+r^2)/2} \int_0^\infty r_1 dr_1 \ldots \int_0^\infty r_{N-1} dr_{N-1}$$
$$\times \exp[i(\beta_1 r_1^2 + \beta_2 r_2^2 + \cdots + \beta_{N-1} r_{N-1}^2)]$$
$$\times [I_{l+\frac{d-2}{2}}(-i\alpha r_0 r_1) \ldots I_{l+\frac{d-2}{2}}(-i\alpha r_{N-1} r_N)]. \quad (2.5.101)$$

Here we have introduced $\alpha = m/\varepsilon\hbar$ and $\beta_j = \alpha[1 - \varepsilon^2 m\omega^2(t_j)/2]$. To work out the integration in (2.5.101), we need an analog of the Gaussian integration formula for the case of non-trivial integration measure $\mu_l^{(d)}$; in other words, in the presence of additional factors of the form $\sim rI_{l+(d-2)/2}$. Such an analog is provided by the following formula (Gradshteyn and Ryzhik (1980), formula 6.633.2):

$$\int_0^\infty dx\, x e^{-Cx^2} I_\nu(Ax) I_\nu(Bx) = \frac{1}{2C} e^{-(A^2+B^2)/4C} I_\nu\left(\frac{AB}{2C}\right) \quad (2.5.102)$$

which is valid for $\text{Re}(\nu) > -1$, $|\arg\sqrt{C}| < \pi/4$ and $A, B > 0$. By analytic continuation we can show that

$$\int_0^\infty dr\, r e^{i\alpha r^2} I_\nu(-iar) I_\nu(-ibr) = \frac{i}{2\alpha} e^{(a^2+b^2)/4\alpha i} I_\nu\left(\frac{ab}{2\alpha i}\right) \quad (2.5.103)$$

is valid for $\nu > -1$ and $\text{Re}\,\alpha > 0$. By means of relation (2.5.103), we obtain for $k_l^N(T)$:

$$k_l^N(T) = \left(\frac{\alpha}{i}\right)^{\frac{N}{2}} \exp\left(\frac{i\beta}{2}(r_0^2 + r_0^2)\right) \int_0^\infty r_1\, dr_1 \ldots \int_0^\infty r_{N-1}\, dr_{N-1}$$

Path integrals in curved spaces, spacetime transformations and the Coulomb problem

$$\times \exp[i(\beta_1 r_1^2 + \beta_2 r_2^2 + \cdots + \beta_{N-1} r_{N-1}^2)][I_{l+\frac{d-2}{2}}(-i\alpha r_0 r_1) \ldots I_{l+\frac{d-2}{2}}(-i\alpha r_{N-1} r_N)]$$

$$= \frac{\alpha_N}{i} \exp(i p_N r_0^2 + i q_N r_0^2) I_{l+\frac{d-2}{2}}(-i\alpha_N r_0 r) \qquad (2.5.104)$$

where the coefficients α_N, p_N and q_N are given by

$$\alpha_N = \alpha \prod_{k=1}^{N-1} \frac{\alpha}{2\gamma_k}$$

$$p_N = \frac{\alpha}{2} - \sum_{k=1}^{N-1} \frac{\alpha_k^2}{4\gamma_k}$$

$$q_N = \frac{\alpha}{2} - \frac{\alpha^2}{4\gamma_{N-1}} \qquad (2.5.105)$$

and α_k, γ_k are defined by the recursion equations:

$$\alpha_1 = \alpha \qquad \alpha_{k+1} = \alpha \prod_{j=1}^{k} \frac{\alpha}{2\gamma_k} \qquad (k \geq 1)$$

$$\gamma_1 = \beta_1 \qquad \gamma_{k+1} = \beta_{k+1} - \frac{\alpha^2}{4\gamma_k}. \qquad (2.5.106)$$

We can now determine these quantities by a version of the *Gelfand–Yaglom method*. Let us start with the evaluation of γ_k. Putting

$$\frac{2\gamma_k}{\alpha} = \frac{y_{k+1}}{y_k} \qquad (2.5.107)$$

we obtain from (2.5.106) the difference equation

$$\frac{y_{k+1} - 2y_k + y_{k-1}}{\varepsilon^2} + \omega_k^2 y_k = 0. \qquad (2.5.108)$$

In the limit $N \to \infty$, this gives a differential equation for y:

$$\ddot{y} + \omega^2(t) y = 0. \qquad (2.5.109)$$

The boundary condition for this equation can be found as follows. The discrepancy

$$y_1 \simeq y_0 + \varepsilon \dot{y}_0 \to y_0 \qquad (\varepsilon \to 0)$$

$$y_1 \simeq \frac{2}{\alpha} y_0 \gamma_1 \to 2 y_0 \qquad (\varepsilon \to 0)$$

shows that $y(0) = y_0 = 0$. This, in turn, gives

$$\dot{y}(t)|_{t=0} \approx \frac{y_1 - y_0}{\varepsilon} = \frac{1}{\varepsilon} \left(\frac{2}{\alpha} \gamma_1 - 1 \right) y_0 \to 1 \qquad (\varepsilon \to 0).$$

Thus the boundary conditions equation (2.5.109) are

$$y(0) = 0 \qquad \dot{y}(0) = 1. \qquad (2.5.110)$$

Now, for α_N in the continuous limit, we obtain

$$\lim_{N\to\infty} \alpha_N = \lim_{N\to\infty} \frac{m}{\varepsilon\hbar} \prod_{j=1}^{N-1} \frac{y(j\varepsilon)}{y[(j+1)\varepsilon]}$$

$$= \lim_{N\to\infty} \frac{m}{\varepsilon\hbar} \frac{y(\varepsilon)}{y(\varepsilon N)} = \frac{m}{\hbar y(T)}. \quad (2.5.111)$$

Similarly,

$$\lim_{N\to\infty} q_N = \lim_{N\to\infty} \frac{m}{2\varepsilon\hbar}\left(1 - \frac{y[(N-1)\varepsilon]}{y[\varepsilon N]}\right)$$

$$= \lim_{N\to\infty} \frac{m}{2\hbar} \frac{y[N\varepsilon] - y[t' + (N-1)\varepsilon]}{\varepsilon y[N\varepsilon]} = \frac{m\dot{y}(T)}{2\hbar y(T)}. \quad (2.5.112)$$

Finally, we shall calculate $\lim_{N\to\infty} p_N$. First, we represent it in the form:

$$\lim_{N\to\infty} p_N = \lim_{N\to\infty}\left(\frac{\alpha}{2} - \sum_{k=1}^{N-1} \frac{\alpha_k^2}{4\gamma_k}\right)$$

$$= \frac{m}{2\hbar} \lim_{N\to\infty}\left(\frac{1}{\varepsilon} - \sum_{k=1}^{N-1} \frac{\varepsilon}{y^2[(k+1)\varepsilon]}\right)$$

$$= \frac{m}{2\hbar} \lim_{N\to\infty}\left(\frac{1}{\varepsilon} - \int_\varepsilon^T \frac{dt}{y^2(t)}\right)$$

$$= \frac{m}{2\hbar y(T)} \lim_{N\to\infty}\left[\frac{y(T)}{\varepsilon} - y(T)\int_\varepsilon^T \frac{dt}{y^2(t)}\right]. \quad (2.5.113)$$

Now let us introduce

$$\xi(t) \stackrel{\text{def}}{\equiv} \left[\frac{y(t)}{\varepsilon} - y(t)\int_\varepsilon^t \frac{d\tau}{y^2(\tau)}\right].$$

One can easily check that

$$\ddot{\xi} = \left(\frac{1-\varepsilon}{\varepsilon}\right) y(t).$$

Hence, $\xi(t)$ satisfies the same equation as $y(t)$:

$$\ddot{\xi} + \omega^2(t)\xi = 0. \quad (2.5.114)$$

Furthermore, we find

$$\lim_{N\to\infty} \xi(\varepsilon) = \lim_{N\to\infty} \frac{y(\varepsilon)}{\varepsilon} - y(\varepsilon)\int_\varepsilon^\varepsilon \frac{dt}{y^2(t)} = 1$$

$$\lim_{\varepsilon\to 0} \dot{\xi}(\varepsilon) = \lim_{\varepsilon\to 0}\left(\frac{\dot{y}(\varepsilon)}{\varepsilon} - \frac{1}{y(\varepsilon)}\right)$$

$$= \lim_{\varepsilon\to 0} \frac{\dot{y}(0)[y(0) + \varepsilon\dot{y}(0)] - \varepsilon}{\varepsilon[y(0) + \varepsilon\dot{y}(0)]} = 0$$

and therefore $\xi(t)$ satisfies the boundary conditions

$$\xi(0) = 1 \qquad \dot{\xi}(0) = 0 \quad (2.5.115)$$

so that we have
$$\lim_{N \to \infty} p_N = \frac{m}{2\hbar} \frac{\xi(T)}{y(T)}. \qquad (2.5.116)$$

Combining the expression (2.5.104) with (2.5.111), (2.5.112) and (2.5.115), we have, in the limit $N \to \infty$,

$$k_l(r, r_0; T) = (r_0 r)^{\frac{2-d}{2}} \frac{m}{i\hbar y(T)} \exp\left[\frac{im}{2\hbar}\left(\frac{\xi(T)}{y(T)}r_0^2 + \frac{\dot{y}(T)}{y(T)}r_0^2\right)\right] I_{l+\frac{d-2}{2}}\left(\frac{mr_0 r}{i\hbar y(T)}\right). \qquad (2.5.117)$$

The functions $y(\tau), \xi(\tau)$ are defined by equations (2.5.109), (2.5.110) and (2.5.114), (2.5.115). In particular, for $\omega(t) = \omega = $ constant

$$y(T) = \frac{1}{\omega} \sin \omega(T)$$
$$\xi(T) = \cos \omega(T)$$

which yield the radial path-integral solution for the radial harmonic oscillator with time-independent frequency:

$$k_l(r, r_0; T) = (r_0 r)^{\frac{2-d}{2}} \frac{m\omega}{i\hbar \sin \omega T} \exp\left\{\frac{im\omega}{2\hbar}(r_0^2 + r_0^2) \cot \omega T\right\} I_{l+\frac{d-2}{2}}\left(\frac{m\omega r_0 r}{i\hbar \sin \omega T}\right). \qquad (2.5.118)$$

Now we can calculate, with the help of equation (2.5.118), the energy levels and state functions, using the expression (2.1.70). For this purpose we use the Hille–Hardy formula (Gradshteyn and Ryzhik (1980), formula 8.976.1)

$$\frac{t^{-\alpha/2}}{1-t} \exp\left[-\frac{1}{2}(x+y)\frac{1+t}{1-t}\right] I_\alpha\left(\frac{2\sqrt{xyt}}{1-t}\right) = \sum_{n=0}^{\infty} \frac{t^n n! e^{-\frac{1}{2}(x+y)}}{\Gamma(n+\alpha+1)} (xy)^{\alpha/2} L_n^{(\alpha)}(x) L_n^{(\alpha)}(y). \qquad (2.5.119)$$

With the substitution $t = e^{-2i\omega T}$, $x = m\omega r_0^2/\hbar$ and $y = m\omega r_0^2/\hbar$, the Hille–Hardy formula allows us to rewrite (2.5.118) in the form

$$k_l(r, r_0; T) = \sum_{N=0}^{\infty} e^{-iTE_N/\hbar} R_N^l(r_0) R_N^l(r) \qquad (2.5.120)$$

where

$$E_N = \omega\hbar\left(N + \frac{d}{2}\right) \qquad (2.5.121)$$

$$R_N^l(r) \stackrel{\text{def}}{=} \sqrt{\frac{2m\omega}{\hbar r^{d-2}} \frac{\Gamma(\frac{N-l}{2}+1)}{\Gamma(\frac{N+l+d}{2})}} \left(\frac{m\omega}{\hbar}r^2\right)^{l+\frac{d-2}{2}} \exp\left\{-\frac{m\omega}{\hbar}r^2\right\} L_{\frac{N-l}{2}}^{(l+\frac{d-2}{2})}\left(\frac{m\omega}{\hbar}r^2\right). \qquad (2.5.122)$$

In the limit $\omega \to 0$, we obtain from equation (2.5.118) the radial part of the propagator for a *free particle*:

$$k_l^{(\omega=0)}(r, r_0; T) = (r_0 r)^{\frac{2-d}{2}} \frac{m}{i\hbar T} \exp\left[\frac{im}{2\hbar T}(r_0^2 + r_0^2)\right] I_{l+\frac{d-2}{2}}\left(\frac{mr_0 r}{i\hbar T}\right)$$

$$= (r_0 r)^{\frac{2-d}{2}} \int_0^\infty dp\, p \exp\left(-\frac{i\hbar T p^2}{2m}\right) J_{l+\frac{d-2}{2}}(pr_0) J_{l+\frac{d-2}{2}}(pr) \qquad (2.5.123)$$

with the wavefunctions and energy spectrum of a free particle as they should be

$$\psi_p(r) = r^{\frac{2-d}{2}} \sqrt{p} J_{l+\frac{d-2}{2}}(pr) \qquad E_p = \frac{p^2}{2m}. \qquad (2.5.124)$$

The one-dimensional case coincides with the motion on a half-line considered in section 2.4.1. Indeed, for $d=1$, the wavefunctions reads as

$$\psi_p(r) = r\sqrt{p} J_{\frac{1}{2}}(pr) = \sqrt{\frac{2}{\pi r}} \sin pr \qquad E_p = \frac{p^2}{2m} \qquad (2.5.125)$$

with the correct boundary condition $\psi_p(0) = 0$.

2.5.4 Path integral for the hydrogen atom: the Coulomb problem

The Coulomb problem is, undoubtedly, one of the most important subjects in quantum physics. It was Pauli who solved it: using the specific symmetries of the Coulomb problem he found the correct result even before the final formulation of quantum mechanics was developed independently in 1925 by Schrödinger and Heisenberg. Pauli did not use a differential equation nor did he solve the corresponding eigenvalue problem as Schrödinger was to do later, instead he exploited the 'hidden' $SO(4)$ symmetry of the Kepler–Coulomb problem. This symmetry classically gives rise to a conserved quantity, called the Runge–Lenz vector.

Calculation of the wavefunctions and energy levels is a relatively simple task in operator language, but this important physical system could not be treated by path integrals and constructing the propagator in an explicit form (not as the series (2.1.70) over eigenstates and eigenvalues) was impossible for a long time. In 1979, Duru and Kleinert applied a spacetime transformation known in astronomy (Kustaanheimo–Stiefel transformation) to the path integral of the Coulomb problem and succeeded in solving this problem (Duru and Kleinert 1979). In fact, their idea of space and time transformations, which we have already discussed in section 2.5.2, opens new possibilities for solving many unsolved path-integral problems (see Kleinert (1995), Grosche (1996), Grosche and Steiner (1998) and references therein). Closely related to the original Coulomb problem is the $1/r$-potential discussion in $\mathbb{R}^d$ for an arbitrary dimension d. For a simpler exposition, we shall start from the $d=2$ case, followed by the original Coulomb problem in $\mathbb{R}^3$.

◇ Coulomb problem in a two-dimensional space

Let us consider the Euclidean *two-dimensional* space with the singular potential $V(r) = -Ze^2/r$ ($r = |x|$, $x \in \mathbb{R}^2$). Here, e^2 denotes the squared charge of an electron and Z defines the actual charge of a particle (multiple of e). Note, however, that this potential is not the potential of a point charge in $\mathbb{R}^2$ (in particular, it violates the Gauss law; a true potential of a charged particle in two-dimensional space is proportional to the logarithm: $V_{\text{ch. part.}}(r) \sim \ln r$). The classical Lagrangian for the Coulomb-like system now has the form

$$L(x, \dot{x}) = \frac{m}{2}\dot{x}^2 + \frac{Ze^2}{r}. \qquad (2.5.126)$$

The path integral is given by

$$K(x, x_0; T) = \int_{\mathcal{C}\{x_0, t_0 | x, t\}} \mathcal{D}x(t) \exp\left\{\frac{i}{\hbar} \int_{t_0}^{t} dt \left(\frac{m}{2}\dot{x}^2 + \frac{Ze^2}{r}\right)\right\}. \qquad (2.5.127)$$

However, the discrete approximation is not trivial for the $1/r$ term. In fact, this is too singular for a path integral, and therefore some regularization must be found. This fact can be understood from different points of view. In particular, as we noted after the Trotter formula (cf (2.2.19)), the transition to imaginary time in the Trotter formula requires the additional condition that the operators entering the exponentials be bounded from below. On the other hand, we have learned that the very definition of a Gaussian-like integral with purely imaginary exponents is heavily based on the possibility of the continuation of the time variable into the complex plane. Thus, in fact, a correct application of the Trotter formula for path-integral construction is guaranteed only for potentials *bounded from below*. One more problem, also related to the singularity of the Coulomb potential, is that the corresponding classical action

$$S = \int_{t_0}^{t} d\tau \left[\frac{m}{2} \dot{x}^2 + \frac{Ze^2}{r} \right] \qquad (2.5.128)$$

can be infinitely lowered by paths which correspond to a slow particle falling into a Ze^2/r-type abyss. In fact, general arguments show that in nature this catastrophe is prevented by quantum fluctuations. But this mechanism can work only for *exact* path integrals with an *infinite* number of time slices or *complete* (infinite series) Fourier decomposition. One can show (see, e.g., Kleinert (1995)), that integrating out infinitely high-frequency Fourier components smooths the singularity of the Coulomb potential and results in the desired stability. But straightforward discrete approximation (which does not contain those high-frequency components) proves to be ill defined. Fortunately, the Duru–Kleinert method of space and time transformations allows us to circumvent this unpleasant situation.

As it turns out, we must perform the following space transformations

$$x_1 = \xi^2 - \eta^2 \qquad x_2 = 2\xi\eta \qquad (2.5.129)$$

which cast the original Lagrangian into the form

$$L(u, \dot{u}) = 2m(\xi^2 + \eta^2)(\dot{\xi}^2 + \dot{\eta}^2) + \frac{Ze^2}{\xi^2 + \eta^2} \qquad (2.5.130)$$

where u is a shorthand notation for both of the new coordinates ξ and η: $u = \{\xi, \eta\} \in \mathbb{R}$. The metric tensor (g_{ab}) and its inverse (g^{ab}), in the new coordinates, are given by

$$(g_{ab}) = 4(\xi^2 + \eta^2) \begin{pmatrix} 1 & 0 \\ 0 & 1 \end{pmatrix} \qquad (g^{ab}) = \frac{1}{4(\xi^2 + \eta^2)} \begin{pmatrix} 1 & 0 \\ 0 & 1 \end{pmatrix} \qquad (2.5.131)$$

with the determinant $g = \det(g_{ab}) = 16(\xi^2 + \eta^2)^2$, so that $dx_1 \, dx_2 = 4(\xi^2 + \eta^2) \, d\xi \, d\eta$. Note that $r = \xi^2 + \eta^2$. The Hermitian momenta corresponding to the scalar product

$$\langle \psi_1, \psi_2 \rangle = 4 \int_{-\infty}^{\infty} \int_{-\infty}^{\infty} d\xi \, d\eta \, (\xi^2 + \eta^2) \psi_1^*(\xi, \eta) \psi_2(\xi, \eta) \qquad (2.5.132)$$

are given by

$$p_\xi = -i\hbar \left(\frac{\partial}{\partial \xi} + \frac{\xi}{\xi^2 + \eta^2} \right) \qquad p_\eta = -i\hbar \left(\frac{\partial}{\partial \eta} + \frac{\eta}{\xi^2 + \eta^2} \right). \qquad (2.5.133)$$

Following the general theory discussed in sections 2.5.1 and 2.5.2 we start by considering the Hamiltonian with the added energy variable E

$$\widehat{H}_E = -\frac{\hbar^2}{2m} \left(\frac{\partial^2}{\partial x_1^2} + \frac{\partial^2}{\partial x_2^2} \right) - \frac{Ze^2}{r} - E \qquad (2.5.134)$$

which in the variables ξ, η reads as

$$\widehat{H}_E = -\frac{\hbar^2}{2m} \frac{1}{4(\xi^2 + \eta^2)} \left(\frac{\partial^2}{\partial \xi^2} + \frac{\partial^2}{\partial \eta^2}\right) - \frac{Ze^2}{\xi^2 + \eta^2} - E. \qquad (2.5.135)$$

The time transformation (cf (2.5.48)) is given by $\epsilon = f(\xi, \eta)\delta$, with $f(\xi, \eta) = 4(\xi^2 + \eta^2)$, and the spacetime-transformed Hamiltonian has the form

$$\widetilde{H} = -\frac{\hbar^2}{2m}\left(\frac{\partial^2}{\partial \xi^2} + \frac{\partial^2}{\partial \eta^2}\right) - 4Ze^2 - 4E(\xi^2 + \eta^2)$$

$$= \frac{1}{2m}(p_\xi^2 + p_\eta^2) - 4Ze^2 - 4E(\xi^2 + \eta^2) \qquad (2.5.136)$$

with the momentum operators and vanishing 'quantum' potential ΔV:

$$p_\xi = -i\hbar \frac{\partial}{\partial \xi} \qquad p_\eta = -i\hbar \frac{\partial}{\partial \eta} \qquad \Delta V = 0. \qquad (2.5.137)$$

To incorporate the time transformation

$$s(\tau) = \int_{t_0}^\tau \frac{d\sigma}{4r(\sigma)} \qquad s = s(t) \qquad (2.5.138)$$

and its lattice definition, $\varepsilon \to 4\widetilde{r}_j \Delta s_j = 4\widetilde{u}_j^2 \delta_j$, into the path integral (2.5.127), we shall use theorem 2.4 and the relations (2.5.70) and (2.5.71):

$$K(x, x_0; T) = \frac{1}{2\pi i \hbar} \int_{-\infty}^\infty dE\, e^{-iTE/\hbar} G(x, x_0; E) \qquad (2.5.139)$$

$$G(x, x_0; E) = i \int_0^\infty ds'' [\widetilde{K}(u, u_0; \mathcal{S}) + \widetilde{K}(-u, u_0; \mathcal{S})] \qquad (2.5.140)$$

where the spacetime-transformed path integral $\widetilde{K}$ is given by

$$\widetilde{K}(u, u_0; \mathcal{S}) = e^{4iZe^2 \mathcal{S}/\hbar} \int \mathcal{D}\xi(s) \int \mathcal{D}\eta(s) \exp\left\{\frac{i}{\hbar}\int_0^{\mathcal{S}} ds \left[\frac{m}{2}(\dot{\xi}^2 + \dot{\eta}^2) + 4E(\xi^2 + \eta^2)\right]\right\} \qquad (2.5.141)$$

(we recall that $u = \{\xi, \eta\}$).

Relation (2.5.141) is slightly modified with respect to the standard one, (2.5.71). The reason is that the square-root mapping (2.5.129) gives rise to a sign ambiguity. Thus, if we consider all paths in the complex plane $x = x_1 + ix_2$ from x_0 to x, they will be mapped into two different classes of paths in the complex u-plane: those which go from u_0 to u and those going from u_0 to $-u$. Here $u \in \mathbb{C}$ is understood to be a complex coordinate: $u = \xi + i\eta$, so that the two real transformations (2.5.129) can be presented as one complex transformation: $x = u^2$. In the complex x-plane with a cut for the function $u = \sqrt{x}$, these are the paths passing an even or odd number of times through the square root from $x = 0$ and $x = -\infty$. We may choose the u_0 corresponding to the initial x_0 to lie on the first sheet (i.e. in the right half u-plane). The final u can be in the right as well as the left half-plane and all paths on the x-plane go over into paths from u_0 to u and paths from u_0 to $-u$. Thus the two contributions arise in equation (2.5.140). Equation (2.5.141) is interpreted as the path integral of a two-dimensional isotropic harmonic oscillator with the frequency $\omega = \sqrt{-8E/m}$. Therefore, we get (cf (2.2.77))

$$\widetilde{K}(u, u_0; \mathcal{S}) = \frac{m\omega}{2i\hbar \sin \omega \mathcal{S}} \exp\left\{\frac{4iZe^2\mathcal{S}}{\hbar} + \frac{im\omega}{2\pi\hbar\sin\omega\mathcal{S}}[(u^2 + u_0^2)\cos\omega\mathcal{S} - 2uu_0]\right\}. \qquad (2.5.142)$$

Path integrals in curved spaces, spacetime transformations and the Coulomb problem   269

◇ **The genuine Coulomb problem: three-dimensional space**

Now we are ready to consider the Euclidean three-dimensional space with the singular potential $V(r) = -Ze^2/r$ ($r = |x|$, $x \in \mathbb{R}^3$), i.e. the genuine Coulomb problem. Of course, this potential does correspond in the space $\mathbb{R}^3$ to the potential of a point charge (in contrast to the two-dimensional case). The classical Lagrangian again has the form

$$\mathcal{L}(x, \dot{x}) = \frac{m}{2}\dot{x}^2 + \frac{Ze^2}{r}. \qquad (2.5.143)$$

The path integral is given by

$$K(x, x_0; T) = \int_{\mathcal{C}\{x,t|x_0,t_0\}} \mathcal{D}x(\tau)\exp\left\{\frac{i}{\hbar}\int_{t_0}^{t} d\tau \left(\frac{m}{2}\dot{x}^2 + \frac{Ze^2}{r}\right)\right\}. \qquad (2.5.144)$$

The transformation which we used in the two-dimensional case corresponds to the two-dimensional realization of the Kustaanheimo–Stiefel transformation. Its three-dimensional variant is more tricky. The point is that the three coordinates x_1, x_2, x_3 are substituted in this case by *four* variables u_1, u_2, u_3, u_4, according to the matrix relation

$$\begin{pmatrix} x_1 \\ x_2 \\ x_3 \end{pmatrix} = \begin{pmatrix} u_3 & u_4 & u_1 & u_2 \\ -u_2 & -u_1 & u_4 & u_3 \\ -u_1 & u_2 & u_3 & -u_4 \end{pmatrix} \begin{pmatrix} u_1 \\ u_2 \\ u_3 \\ u_4 \end{pmatrix}. \qquad (2.5.145)$$

It is clear that this transformation is degenerate (has a zero Jacobian) and can not be used straightforwardly for the change of variables in an integral. To overcome this problem, we use the following method. Let us insert a unit factor in the form

$$1 = \lim_{N\to\infty}\left(\frac{m}{2\pi i\varepsilon\hbar}\right)^{\frac{N}{2}} \prod_{j=1}^{N}\int_{-\infty}^{\infty} d\xi_j \exp\left\{\frac{im}{2\hbar\varepsilon}\Delta^2\xi_j\right\} \qquad (2.5.146)$$

into the discrete approximation of (2.5.144) which, after this, reads as

$$K(x, x_0; T) = \int_{-\infty}^{\infty} d\xi_N \lim_{N\to\infty}\left(\frac{m}{2\pi i\varepsilon\hbar}\right)^{2N} \prod_{j=1}^{N-1}\int d^3 x_j \int_{-\infty}^{\infty} d\xi_j$$

$$\times \exp\left\{\frac{i}{\hbar}\sum_{j=1}^{N}\left[\frac{m}{2\varepsilon}(\Delta^2 x^{(j)} + \Delta^2 \xi^{(j)}) + \varepsilon\frac{Ze^2}{r_j}\right]\right\}. \qquad (2.5.147)$$

Now we have four variables in the exponential of the path integral and it is natural to denote

$$x_4 \stackrel{\text{def}}{\equiv} \xi$$

simultaneously extending the matrix in (2.5.145) up to the square 4×4 matrix. To this aim, define

$$x = \mathsf{A}(u)u$$

where

$$\mathsf{A}(u) \stackrel{\text{def}}{\equiv} \begin{pmatrix} u_3 & u_4 & u_1 & u_2 \\ -u_2 & -u_1 & u_4 & u_3 \\ -u_1 & u_2 & u_3 & -u_4 \\ u_4 & -u_3 & u_2 & -u_1 \end{pmatrix}.$$

We must clarify, however, a prescription for a discrete approximation of the path integral (2.5.147). As usual, we choose the midpoint prescription, so that the transformation of the differences reads as

$$\Delta x_a^{(j)} = 2 \sum_{b=1}^{4} A^{ab}(\tilde{u}^{(j)}) \Delta u_b^{(j)} \qquad (2.5.148)$$

where

$$\tilde{u}_a^{(j)} = \tfrac{1}{2}(u_a^{(j)} + u_a^{(j-1)}).$$

The transformation of the measure has a similar form:

$$dx_{aj} = 2 \sum_{b=1}^{4} A_{ab}(\tilde{u}_j) du_{bj}$$

and the Jacobian is non-zero and given by

$$\frac{\partial(x_1, x_2, x_3, \xi)}{\partial(u_1, u_2, u_3, u_4)} = 2^4 \tilde{r}^{(j)2} \qquad (2.5.149)$$

where

$$\tilde{r}_j \stackrel{\text{def}}{\equiv} A^\top(\tilde{u}_j) \cdot A(\tilde{u}_j). \qquad (2.5.150)$$

Note that if the matrix A depends on a single point u, we have

$$A^\top A = u_1^2 + u_2^2 + u_3^2 + u_4^2 = r = |x|$$

so that (2.5.150) is a reasonable generalization of this relation to the case of the midpoint prescription. According to the latter, we shall use $\tilde{r}$, as defined in (2.5.150), in the potential term in (2.5.147). This Kustaanheimo–Stiefel transformation produces the following classical Lagrangian (which corresponds to the Hamiltonian $H_E = H + E$ with added energy variable):

$$L(u, \dot{u}) = 2m u^2 \dot{u}^2 + \frac{Ze^2}{u^2} - E. \qquad (2.5.151)$$

Thus, it is heuristically clear that if we now, in addition, transform the time variable, defined by the relation

$$dt \longrightarrow ds = \frac{1}{r} dt$$

i.e.

$$dt = u^2 ds \qquad (2.5.152)$$

the corresponding action becomes

$$S = \int ds \left[2m \left(\frac{du}{ds}\right)^2 + Ze^2 - Eu^2 \right]$$

and corresponds to a harmonic oscillator.

Of course, a rigorous realization of these transformations for the path integral requires careful treatment of the time-sliced version. In particular, the discrete version of (2.5.152) is

$$\Delta t_j \equiv \varepsilon \longrightarrow 4\tilde{r}_j \Delta s_j \equiv 4\tilde{r}_j \delta_j$$

Path integrals in curved spaces, spacetime transformations and the Coulomb problem 271

and the measure in the path integral transforms according to

$$\prod_{j=1}^{N}\left(\frac{m}{2\pi i\varepsilon\hbar}\right)^{2}\prod_{j=1}^{N-1}(2\widetilde{u}_{j})^{4}d^{4}u_{j} = \frac{1}{(4r)^{2}}\prod_{j=1}^{N}\left(\frac{m}{2\pi i\hbar\delta_{j}}\right)^{2}\prod_{j=1}^{N-1}d^{4}u_{j}. \qquad (2.5.153)$$

Using again equation (2.5.75) which produces a factor $4r$, we obtain the following path-integral transformation (cf theorem 2.4, page 257):

$$K(x,x_{0};T) = \frac{1}{2\pi i\hbar}\int_{-\infty}^{\infty}dE\,e^{-iTE/\hbar}G(x,x_{0};E) \qquad (2.5.154)$$

$$G(x,x_{0};E) = i\int_{0}^{\infty}ds\,e^{4iZe^{2}s/\hbar}\widetilde{K}(u,u_{0};s) \qquad (2.5.155)$$

where the spacetime-transformed path integral $\widetilde{K}$ is given by

$$\widetilde{K}(u,u_{0};\mathcal{S}) = \frac{1}{4r}\int_{-\infty}^{\infty}d\xi\lim_{N\to\infty}\left(\frac{m}{2\pi i\hbar\delta}\right)^{2N}\prod_{j=1}^{N-1}\int d^{4}u_{j}\exp\left[\frac{i}{\hbar}\sum_{j=1}^{N}\left(\frac{m}{2\delta}\Delta^{2}u_{j}+4\delta E\widetilde{r}_{j}(\widetilde{u})\right)\right]$$

$$= \left(\frac{m\omega}{2\pi i\hbar\sin\omega\mathcal{S}}\right)^{2}\int_{-\infty}^{\infty}\frac{d\xi}{4r}\exp\left\{-\frac{m\omega}{2i\hbar}\left[(u^{2}+u_{0}^{2})\cot\omega\mathcal{S}-2\frac{uu_{0}}{\sin\omega\mathcal{S}}\right]\right\}. \qquad (2.5.156)$$

Here we have used once again the known solution for the harmonic oscillator with constant frequency $\omega=\sqrt{-8E/m}$. We can check that no quantum correction (i.e. terms proportional to the Planck constant $\hbar$) appears in the transformation procedure.

The ξ-integration and explicit derivation of the energy-dependent Green function according to relation (2.5.155) require rather tedious calculations which we skip, presenting only the result:

$$G(x,x_{0};E) = G(r,\theta,\phi,r_{0},\theta_{0},\phi_{0};E)$$

$$= \frac{1}{4\pi}\sum_{l=0}^{\infty}(2l+1)P_{l}(\cos\gamma)G_{l}(r,r_{0};E)$$

$$= \sum_{l=0}^{\infty}\sum_{n=-l}^{l}Y_{l}^{n*}(\theta',\phi')Y_{l}^{n}(\theta'',\phi'')G_{l}(r,r_{0};E) \qquad (2.5.157)$$

where the radial Green function $G_{l}(r,r_{0};E)$ is given by

$$G_{l}(r,r_{0};E) = \frac{2m\omega}{\hbar\sqrt{r_{0}r}}\int_{0}^{\infty}\frac{ds}{\sin(\omega s)}$$

$$\times\exp\left\{\frac{4iZe^{2}s}{\hbar}-\frac{m\omega}{2i\hbar}(r_{0}+r)\cot(\omega s)\right\}I_{2l+1}\left(\frac{m\omega\sqrt{r_{0}r}}{i\hbar\sin(\omega s)}\right)$$

$$= \frac{1}{r_{0}r}\sqrt{-\frac{m}{2E}}\frac{1}{(2l+1)!}\Gamma\left(1+l+\frac{iZe^{2}}{\hbar}\sqrt{\frac{m}{2E}}\right)$$

$$\times W_{\frac{Ze^{2}}{\hbar}\sqrt{-\frac{m}{2E}},l+\frac{1}{2}}\left(\sqrt{\frac{-8mE}{\hbar^{2}}}\max(r,r_{0})\right)$$

$$\times M_{\frac{Ze^{2}}{\hbar}\sqrt{-\frac{m}{2E}},l+\frac{1}{2}}\left(\sqrt{\frac{-8mE}{\hbar^{2}}}\min(r,r_{0})\right). \qquad (2.5.158)$$

Here $W_{\nu,\mu}(z)$ and $M_{\nu,\mu}(z)$ denote the Whittaker functions. From the poles of the Γ-function at $z = -n_r = 0, -1, -2, \ldots$ we can read the bound-state energy levels

$$E_N = -\frac{mZ^2 e^4}{2\hbar^2 N^2} \qquad (N = n_r + l + 1 = 1, 2, 3, \ldots) \tag{2.5.159}$$

which are well known from the operator approach (i.e. solution of the Schrödinger equation).

Tedious manipulations with special functions (Kleinert 1995, Grosche 1992) allow us to write (2.5.158) in the form of the decomposition (2.1.81) and, hence, to the result for the wavefunctions of the hydrogen atom:

$$\psi_{N,l,n}(r,\theta,\phi) = \frac{2}{N^2}\left[\frac{(N-l-1)!}{a^3(N+l)!}\right]^{\frac{1}{2}}$$

$$\times \exp\left\{-\frac{r}{aN}\right\}\left(\frac{2r}{aN}\right)^l L_{N-l-1}^{(2l+1)}\left(\frac{2r}{aN}\right) Y_l^n(\theta,\phi) \qquad \text{(discrete spectrum)} \tag{2.5.160}$$

$$\psi_{p,l,n}(r,\theta,\phi) = \frac{1}{\sqrt{2\pi}(2l+1)!r}\Gamma\left(1+l+\frac{i}{ap}\right)$$

$$\times \exp\left(\frac{\pi}{2ap}\right) M_{\frac{i}{ap},l+\frac{1}{2}}(-2ipr) Y_l^n(\theta,\phi) \qquad \text{(continuous spectrum).} \tag{2.5.161}$$

Of course, these wavefunctions form a complete set. The spectrum of the continuous states is, of course, given by

$$E_p = \frac{\hbar^2 p^2}{2m} \qquad (p > 0). \tag{2.5.162}$$

These results coincide with those obtained by the operator approach. The complete Feynman kernel, therefore, reads as

$$K(x,x_0;T) = \sum_{l=0}^{\infty}\sum_{N=1}^{\infty} e^{-iTE_N/\hbar}\psi_{N,l}^*(r_0,\phi_0)\psi_{N,l}(r,\phi)$$

$$+ \sum_{l=0}^{\infty}\int_0^{\infty} dp\, e^{-iTE_p/\hbar}\psi_{p,l}^*(r_0,\phi_0)\psi_{p,l}(r,\phi) \tag{2.5.163}$$

with the wavefunctions given in equations (2.5.160) and (2.5.161) and the energy spectrum given in equations (2.5.159) and (2.5.162), respectively.

As we have already stressed, the method of space and time transformations can be applied to the path-integral treatment of many other solvable potentials, see Grosche and Steiner (1998). One example we suggest to the reader as an exercise (problem 2.5.7, page 285).

2.5.5 Path integrals on group manifolds

In this subsection, we consider quantum-mechanical systems defined on the manifolds of compact simple Lie groups. In a sense, the group manifold and homogeneous spaces of a Lie group $\mathfrak{G}$ (i.e. the factor, or coset, spaces $\mathfrak{G}/\mathfrak{G}_0$ for some stability subgroup $\mathfrak{G}_0$, cf supplement IV, volume II) are the closest analogs of flat spaces. This similarity is based on the fact that any two points of a homogeneous space or group manifold are related by a group transformation (transitivity of homogeneous spaces). This is the generalization of the characteristic property of a flat space, where arbitrary pairs of points are related by

Path integrals in curved spaces, spacetime transformations and the Coulomb problem 273

ordinary transformations. One more property of homogeneous spaces and group manifolds is that group-invariant connections on them produce a *constant* scalar curvature which can also be considered as the simplest generalization of the case of zero curvature (flat spaces).

The existence of group transformations (transitivity) provides the possibility of explicitly calculating the path integrals for free systems on group manifolds and homogeneous spaces (Marinov and Terentyev 1978, 1979, Schulman 1981).

We restrict our consideration to group manifolds only and, moreover, after a general introduction, we shall discuss some details for the basic example of the $SU(2)$ group (further generalizations can be found in Landsman and Linden (1991), McMillan and Tsutsui (1995), Kunstatter (1992), Klauder and Skagerstam (1985), Tomé (1998) and in references therein). In section 2.6.3, we shall consider reasons for the possibility of exactly calculating path integrals from a more general point of view.

◇ **Metrical quantities on a Lie group**

A compact simple Lie group $\mathfrak{G}$ may be defined as a differentiable manifold $\mathcal{M}$ equipped with a group structure (see basic notions on Lie group theory in supplement IV, volume II). Elements of $\mathfrak{G}$ correspond to points on $\mathcal{M}$ and may be parametrized in terms of the real coordinates q^a, $a = 1, \ldots, d$ (d is the dimension of the group). As usual, the origin $q^a = 0$ determines the unity element of the group. The group structure is fixed if we know the composition function Φ which determines the group multiplication. Let q_1^a and q_2^a be the parameters of two elements $g_1 = g(q_1)$ and $g_2 = g(q_2)$ of $\mathfrak{G}$. Then the equations

$$q^a = \Phi^a(q_1, q_2) \qquad a = 1, \ldots, d \tag{2.5.164}$$

correspond to the product $g = g_1 g_2$ of these two elements. The left auxiliary coordinate-dependent matrix is defined as

$$\eta^a{}_b(q) = \left.\frac{\partial \Phi^a(q, q_1)}{\partial q_1^b}\right|_{q_1=0}. \tag{2.5.165}$$

The inverse of η comprises the components of the *Maurer–Cartan form* λ:

$$\eta^a{}_b \lambda^b{}_c = \delta^a_c. \tag{2.5.166}$$

Now we consider a matrix representation $\mathsf{U}(q)$ of $\mathfrak{G}$, satisfying

$$\left.\frac{\partial}{\partial q^a}\mathsf{U}(q)\right|_{q=0} = -\mathrm{i}\mathsf{T}_a \tag{2.5.167}$$

where T^a are the generators of the group in the chosen representation. The generators T^a satisfy the commutation relations

$$[\mathsf{T}^a, \mathsf{T}^b] = \mathrm{i} f_{ab}{}^c \mathsf{T}^c \tag{2.5.168}$$

and they are normalized to

$$\operatorname{Tr} \mathsf{T}_a \mathsf{T}_b = \tfrac{1}{2}\delta_{ab}. \tag{2.5.169}$$

The left auxiliary functions are related to the derivatives of U. This can be seen by differentiating

$$\mathsf{U}(\Phi(q_1, q_2))) = \mathsf{U}(q_1)\mathsf{U}(q_2) \tag{2.5.170}$$

with respect to q_2, at $q_2 = 0$. This yields

$$\eta^a{}_b \partial_a \mathsf{U} = \mathrm{i}\mathsf{U}\mathsf{T}_b \qquad \partial_a \equiv \frac{\partial}{\partial q^a} \tag{2.5.171}$$

or, in terms of λ,
$$\lambda^a{}_b T^a = i U^{-1} \partial_b U \tag{2.5.172}$$

(this equality justifies our previous claim that $\lambda^a{}_b$ are components of the Maurer–Cartan form, cf supplement IV, volume II). Equations (2.5.171) and (2.5.172) imply that

$$L_a = \eta^b{}_a \partial_b \tag{2.5.173}$$

are the infinitesimal generators of transformations via group multiplication from the right: $U \to UU'$. Note that λ (and therefore also η and L) are invariant under global transformations from the left: $U \to U'U$. The roles of left and right are reversed, when in (2.5.165) the derivation of Φ is taken with respect to the first argument.

The natural metric on $\mathcal{M}$ is expected to be invariant under both left and right global multiplication. If $\mathfrak{G}$ is a simple compact group, this bi-invariant metric is unique up to multiplication by a positive constant:

$$g_{ab} = \sum_c \lambda^c{}_a \lambda^c{}_b. \tag{2.5.174}$$

Our choice of multiplicative constant is motivated by the desire that λ (and not a multiple of λ) can be interpreted as a vielbein field. Then the inverse of the metric reads:

$$g^{ab} = \sum_c \eta^a{}_c \lambda^b{}_c. \tag{2.5.175}$$

Note that the definition (2.5.174) of the metric is independent of any representation. For a given representation U, satisfying (2.5.167)–(2.5.169), the metric can also be written as

$$g_{ab} = -2 \operatorname{Tr}[U^{-1}(\partial_a U) U^{-1}(\partial_b U)]. \tag{2.5.176}$$

For the purpose of quantization (cf section 2.5.1), we also need the Christoffel symbols on the group manifolds and their scalar curvature. Substitution of the metric (2.5.174) into the general expressions (2.5.9) and (2.5.8) yields the result:

$$\begin{aligned}\Gamma^a{}_{bc} &= \tfrac{1}{2}\eta^{ad}(\partial_b \lambda_{dc} + \partial_c \lambda_{db}) \\ &= \eta^{ad}(\partial_b \lambda_{dc} + \tfrac{1}{2} f^{dd_1 d_2} \lambda_{d_1 b} \lambda_{d_2 c})\end{aligned} \tag{2.5.177}$$

(the second equality follows from the Maurer–Cartan equation; see supplement IV, volume II),

$$R = \tfrac{1}{4} \sum_{a,b,c} f_{ab}{}^c f_{ab}{}^c. \tag{2.5.178}$$

◇ **Canonical quantization**

The quantization on a group manifold is carried out according to the general prescription outlined in section 2.5.1. The quantum Hamiltonian is defined as in (2.5.12). Taking into account that on a group manifold, g_{ab}, $\Gamma^c{}_{ab}$ and R are given by (2.5.175), (2.5.177) and (2.5.178), respectively, the quantum correction $\Delta V_{\text{Weyl}}(q)$ can be calculated to be

$$\Delta V_{\text{Weyl}} = \frac{\hbar^2}{8m} \sum_c (\partial_a \eta^b{}_c)(\partial_b \eta^a{}_c). \tag{2.5.179}$$

The Hermitian left-invariant generators (2.5.173) of right multiplications can be expressed via the momentum operators:

$$\widehat{L}_a = -\tfrac{1}{2}(\eta^b{}_a \widehat{p}_b + \widehat{p}_b \eta^b{}_a)$$
$$= -\eta^b{}_a \widehat{p}_b + \frac{i}{2}(\partial_b \eta^b{}_a). \qquad (2.5.180)$$

The quadratic Casimir operator $\widehat{L}^2$ proves to be

$$\widehat{L}^2 = g^{ab}\widehat{p}_a \widehat{p}_b - \sum_a [i(\eta^b{}_a \partial_b \eta^c{}_a + \eta^c{}_a \partial_b \eta^b{}_a)\widehat{p}_c - \tfrac{1}{2}\eta^b{}_a(\partial_b \partial_c \eta^c{}_a) - \tfrac{1}{4}(\partial_b \eta^b{}_a)(\partial_c \eta^c{}_a)]. \qquad (2.5.181)$$

Comparing this to equations (2.5.12) and (2.5.179), we find

$$\widehat{H}_{\text{kin}} = \frac{1}{2m}\widehat{L}^2 \qquad (2.5.182)$$

where $\widehat{H}_{\text{kin}} = \widehat{H} - V(\widehat{q})$, that is, the kinetic part $\widehat{H}_{\text{kin}}$ of the Hamiltonian operator (2.5.12) is just the (unique) quadratic Casimir operator of the group.

◇ **Path-integral quantization**

The propagator for a particle on a group manifold is given by the general formulae (2.5.17) (in terms of the phase-space path integral) or (2.5.19) (configuration-space path integral) after substituting the specific form (2.5.175) of the metric on a Lie group and the corresponding ΔV_{Weyl} (see (2.5.179) and (2.5.182)).

Sometimes, it is desirable to present the path-integral measure in (2.5.19) in a more traditional form: $\prod_{k=1}^{N-1} \sqrt{g(q_k)} d^d q_k$ (i.e. to remove the midpoint prescription in this measure). This is achieved by the Taylor expansion of $g(q_k)$ and $g(q_{k+1})$ around $\widetilde{q}_k$ up to the order ε, with the use of (2.5.25) and subsequent exponentiation. This transformation converts (2.5.19) into the following form:

$$K(q, t; q_0, t_0) = \lim_{N \to \infty} \left(\frac{m}{2\pi i \varepsilon \hbar}\right)^{Nd/2} \int \left(\prod_{k=1}^{N-1} \sqrt{g(q_k)}\, d^d q_k\right)$$
$$\times \exp\left\{\frac{i}{\hbar} \sum_{k=0}^{N-1} \left[\frac{m}{2\varepsilon} g_{ab}(q) \Delta q_k^a \Delta q_k^b - \varepsilon \Delta V - \varepsilon V\right]\right\} \qquad (2.5.183)$$

where

$$\Delta V = \frac{\hbar^2}{8} g^{ab} \partial_a \Gamma^c{}_{bc} + \Delta V_{\text{Weyl}}. \qquad (2.5.184)$$

On a group manifold, ΔV_{Weyl} has already been calculated in (2.5.179) so that the complete ΔV is given by

$$\Delta V = \frac{\hbar^2}{8m} \sum_c [(\partial_a \eta^b{}_c)(\partial_b \eta^a{}_c) - \eta^a{}_c \partial_a \partial_b \eta^b{}_c - g^{cd} \eta^a{}_c (\partial_a \eta^b{}_d) \eta^f{}_e \partial_b \lambda^e{}_f]. \qquad (2.5.185)$$

Hence the path-integral representation (2.5.19) of the transition amplitudes becomes

$$K(q, t; q_0, t_0) = \lim_{N \to \infty} \left(\frac{m}{2\pi i \varepsilon \hbar}\right)^{Nd/2} \int \left(\prod_{k=1}^{N-1} \det \lambda(q_k)\, d^n q_k\right)$$

$$\times \exp\left\{i\sum_{k=0}^{N-1}\left[\frac{m}{2\varepsilon}\lambda^{ca}\lambda_{cb}\xi_k^a\xi_k^b - \varepsilon V\right.\right.$$
$$\left.\left. - \frac{\varepsilon\hbar^2}{8m}((\partial_a\eta^{bc})(\partial_b\eta^{ac}) - \eta^{ab}\partial_a\partial_c\eta^{cb} - \eta^{ab}(\partial_a\eta^{cb})\eta^{de}\partial_c\lambda^{ed})\right]\right\}. \quad (2.5.186)$$

◇ **Example: the path integral on the group manifold of $SU(2)$**

In $SU(2)$ the generators are $\mathsf{T}^a = \sigma^a/2$ with σ^a being the Pauli matrices. They satisfy

$$\mathsf{T}^a\mathsf{T}^b = \frac{1}{4}\delta^{ab} + \frac{i}{2}\varepsilon^{abc}\mathsf{T}^c. \quad (2.5.187)$$

We want to present the path integral for $SU(2)$ in two different well-known representations, namely, the exponential one $\mathsf{U} = \exp\{-iq^a\mathsf{T}^a\}$; and the parametrization in terms of Euler angles (see also problem 2.5.8, page 285 where we suggest the reader considers one more parametrization of $SU(2)$).

Example 2.1 ($SU(2)$ in the exponential parametrization). First, we want to use the general exponential parametrization

$$\mathsf{U} = e^{-iq^a\mathsf{T}^a} = c - 2i\frac{q^a\mathsf{T}^a}{|q|}s \quad (2.5.188)$$

where $|q| = \sqrt{q^aq^a}$ and we have denoted, for brevity,

$$c = \cos(|q|/2) \qquad s = \sin(|q|/2). \quad (2.5.189)$$

In this case, the composition functions (2.5.164) are rather complicated, but we do not need to know them. The Maurer–Cartan forms λ can be calculated using (2.5.172), the latter formula being applicable since the parametrization (2.5.188) satisfies the condition (2.5.167):

$$\lambda^{ab} \equiv \lambda^a{}_c\delta^{cb} = \frac{2cs}{|q|}P^{ab} + \frac{q^aq^b}{q^2} + 2\frac{s^2}{q^2}\varepsilon^{abc}q^c \quad (2.5.190)$$

where the projector P^{ab} is defined by

$$P^{ab} = \delta^{ab} - \frac{q^aq^b}{q^2}. \quad (2.5.191)$$

This implies left auxiliary functions of the form:

$$\eta^{ab} \equiv \eta^a{}_c\delta^{cb} = \frac{c}{2s}|q|P^{ab} + \frac{q^aq^b}{q^2} - \frac{1}{2}\varepsilon^{abc}q^c. \quad (2.5.192)$$

Equations (2.5.190)–(2.5.192) can be used to calculate the metric and its inverse:

$$g_{ab} = \frac{4s^2}{q^2}P^{ab} + \frac{q^aq^b}{q^2} \quad (2.5.193)$$

$$g^{ab} = \frac{q^2}{4s^2}P^{ab} + \frac{q^aq^b}{q^2}. \quad (2.5.194)$$

The Christoffel symbols can be calculated by using (2.5.177):

$$\Gamma^c{}_{ab} = \left(\frac{1}{|q|} - \frac{2cs}{q^2}\right)P^{ab}\frac{q^c}{|q|} + \left(\frac{c}{2s} - \frac{1}{|q|}\right)\left(P^{ac}\frac{q^b}{|q|} + P^{bc}\frac{q^a}{|q|}\right) \quad (2.5.195)$$

and the curvature is given by (2.5.178):

$$R = \tfrac{1}{4}\varepsilon^{abc}\varepsilon^{abc} = \tfrac{3}{2}. \tag{2.5.196}$$

The quantum correction ΔV_{Weyl} in the Weyl-ordered Hamiltonian is

$$\Delta V_{\text{Weyl}} = \frac{+\hbar^2}{8m}\left(-\frac{3}{2} - \frac{c^2}{2s^2} - \frac{1}{s^2} + \frac{c}{2s^3}|q| + \frac{2}{q^2}\right) \tag{2.5.197}$$

and the total correction ΔV, defined by (2.5.185) for the path integral (2.5.183), acquires the form

$$\Delta V = \frac{\hbar^2}{8m}\left(-\frac{3}{2} - \frac{c^2}{2s^2} - \frac{5}{2s^2} + \frac{c}{s^3}|q| + \frac{4}{q^2}\right). \tag{2.5.198}$$

Thus the path integral on the $SU(2)$ group manifold takes the form

$$K(q,t;q_0,t_0) = \lim_{N\to\infty}\left(\frac{m}{2\pi i\varepsilon\hbar}\right)^{3N/2}\int\left(\prod_{k=1}^{N-1}g^{1/2}(q_k)\,d^3q_k\right)$$

$$\times\exp\left\{\frac{i}{\hbar}\sum_{k=0}^{N-1}\left[\frac{m}{2\varepsilon}g_{ab}(\tilde{q}_k)\Delta q_k^a\Delta q_k^a - \varepsilon V(\tilde{q}_k)\right.\right.$$

$$\left.\left. + \frac{\varepsilon\hbar^2}{8m}\left(\frac{3}{2} + \frac{c^2}{2s^2} + \frac{5}{2s^2} - \frac{c}{s^3}|q| + \frac{4}{q^2}\right)\right]\right\}. \tag{2.5.199}$$

The path integral in the standard exponential parametrization (2.5.188) of the group elements can hardly be straightforwardly calculated even for the free Hamiltonian with $V(q) = 0$, because of the complicated form of the effective potential ΔV. However, in the specific case of the $SU(2)$ group, we can use the fact that its manifold is the three-dimensional sphere and in this way considerably simplify the problem (Marinov and Terentyev 1978, 1979, Peak and Inomata 1969, Duru 1984). For this aim we shall parametrize the manifold in terms of Euler angles and use the path integral in polar coordinates (cf section 2.5.3).

Example 2.2 (path integral over $SU(2)$ manifold in terms of the Euler angles). In this example, we parametrize the points of the $SU(2)$ manifold, which is the three-dimensional unit sphere $u^a u^a = 1$ ($\boldsymbol{u} = (u^1,\ldots,u^4)$ is the four-dimensional vector), in terms of the Euler angles:

$$\begin{aligned}u^1 &= \cos(\theta/2)\cos((\phi+\psi)/2) & u^2 &= \cos(\theta/2)\sin((\phi+\psi)/2) \\ u^3 &= \sin(\theta/2)\cos((\phi-\psi)/2) & u^4 &= \sin(\theta/2)\sin((\phi-\psi)/2)\end{aligned} \tag{2.5.200}$$

where the ranges of the angles are:

$$0\leq\theta\leq\pi \qquad 0\leq\phi\leq 2\pi \qquad -2\pi\leq\psi\leq 2\pi.$$

The free Lagrangian for a particle moving on the $SU(2)$ unit sphere is now given as

$$L = \frac{m}{2}\dot{\boldsymbol{u}}^2 = \frac{m}{2}(\dot{u}_1^2 + \dot{u}_2^2 + \dot{u}_3^2 + \dot{u}_4^2) \tag{2.5.201}$$

and the time-sliced version of the action reads

$$S = \sum_k \frac{m}{2\varepsilon}(\boldsymbol{u}_k - \boldsymbol{u}_{k-1})^2 = \sum_k \frac{m}{\varepsilon}(1 - \cos\Theta_k) \tag{2.5.202}$$

with $\cos\Theta_k = \boldsymbol{u}_{k-1} \cdot \boldsymbol{u}_k$. Note that Θ_k is the angle that rotates the vector $\boldsymbol{u}_{k-1}$ onto $\boldsymbol{u}_k$ around an axis perpendicular to the plane defined by $\boldsymbol{u}_{k-1}$ and $\boldsymbol{u}_k$. To find the relation of Θ_k with the Euler angles of $\boldsymbol{u}_{k-1}$ and $\boldsymbol{u}_k$, we recall the correspondence between the vectors $\boldsymbol{u}$ and the $SU(2)$ rotation matrices U:

$$U = \begin{pmatrix} u^1 + iu^2 & -u^4 + iu^3 \\ u^4 + iu^3 & u^1 - iu^2 \end{pmatrix}$$
$$= \begin{pmatrix} \cos(\theta/2)\exp\{i(\phi+\psi)/2\} & i\sin(\theta/2)\exp\{i(\phi-\psi)/2\} \\ i\sin(\theta/2)\exp\{-i(\phi-\psi)/2\} & \cos(\theta/2)\exp\{-i(\phi+\psi)/2\} \end{pmatrix}. \quad (2.5.203)$$

It can be directly verified that the scalar product of two vectors $\boldsymbol{u}_k, \boldsymbol{u}_j$ is given by

$$\boldsymbol{u}_k \cdot \boldsymbol{u}_j = \tfrac{1}{2}\operatorname{Tr}(U_k U_j^{-1}).$$

The kernel for the infinitesimal time shift in the parametrization by Euler angles has a rather simple form:

$$K(\boldsymbol{u}_k, \boldsymbol{u}_{k-1}; \varepsilon) = \exp\left\{i\frac{m}{\varepsilon}(1-\cos\Theta_k) + i\frac{\varepsilon}{8m}\right\} \quad (2.5.204)$$

where the extra term $\varepsilon/(8m)$ in the action is the ordering correction ΔV in the 'polar' coordinate Θ. If we expand $K(\boldsymbol{u}_k, \boldsymbol{u}_{k-1}; \varepsilon)$ up to $\mathcal{O}(\varepsilon^2)$ for $\varepsilon \to 0$ by using the formula (Peak and Inomata 1969)

$$\exp\left\{\frac{\varepsilon}{\theta}\cos\delta\right\} \approx \frac{\varepsilon}{2\theta}\sum_{j=-\infty}^{\infty}\exp\left\{ij\delta + \frac{\theta}{\varepsilon} - \frac{\varepsilon}{2\theta}\left(j^2 - \frac{1}{4}\right)\right\}$$

we obtain

$$K(\boldsymbol{u}_k, \boldsymbol{u}_{k-1}; \varepsilon) = \frac{i\varepsilon}{2m}e^{i\varepsilon/(8m)}\sum_{j=-\infty}^{\infty}\exp\left\{-\frac{i\varepsilon}{2m}\left(j_k^2 - \frac{1}{4}\right) + ij_k\Theta_k\right\} \quad (2.5.205)$$

which can also be written, by changing the order of the summation, as follows:

$$K(\boldsymbol{u}_k, \boldsymbol{u}_{k-1}; \varepsilon) = \frac{i\varepsilon}{2m}e^{i\varepsilon/(8m)}\sum_{l_k=0}^{\infty}\left[\exp\left\{-\frac{i\varepsilon}{2m}\left(l_k^2 - \frac{1}{4}\right)\right\}\right.$$
$$\left. - \exp\left\{-\frac{i\varepsilon}{2m}\left((l_k+1)^2 - \frac{1}{4}\right)\right\}\right]\sum_{m_k=-l_k}^{l}e^{im_k\Theta_k}. \quad (2.5.206)$$

The formula (Gradshteyn and Ryzhik (1980), formula 1.223)

$$e^{ax} - e^{bx} = (a-b)xe^{(a+b)x/2}\prod_{s=1}^{\infty}\left[1 + \frac{(a-b)^2x^2}{4s^2\pi^2}\right]$$

allows us to write the difference of the two exponents in (2.5.206) as

$$\left[\exp\left\{-\frac{i\varepsilon}{2m}\left(l_k^2 - \frac{1}{4}\right)\right\} - \exp\left\{-\frac{i\varepsilon}{2m}\left((l_k+1)^2 - \frac{1}{4}\right)\right\}\right]$$
$$= (2l_k+1)\frac{i\varepsilon}{2m}\exp\left\{-\frac{i\varepsilon}{2m}\left(l_k+\frac{1}{2}\right)^2\right\}\prod_{s=1}^{\infty}\left[1 - \frac{(2l_k+1)\varepsilon^2}{16m^2s^2\pi^2}\right]$$
$$\xrightarrow[\varepsilon\to 0]{} (2l_k+1)\frac{i\varepsilon}{2m}\exp\left\{-\frac{i\varepsilon}{2m}\left(l_k+\frac{1}{2}\right)^2\right\}. \quad (2.5.207)$$

The summation over m_k in (2.5.206) can be achieved by using the addition theorem for $SU(2)$ representations (see, e.g., Vilenkin (1968)):

$$\sum_{m_k=-l_k}^{l_k} e^{im_k \Theta_k} = \sum_{m_k=-l_k}^{l_k} \sum_{n_k=-l_k}^{l_k} e^{im_k(\phi_{k-1}-\phi_k)} e^{in_k(\psi_{k-1}-\psi_k)} P_{m_k n_k}^{l_k}(\cos\theta_{k-1}) P_{n_k m_k}^{l_k}(\cos\theta_k) \quad (2.5.208)$$

($P_{m_k n_k}^{l_k}(\cos\theta)$ is the θ-dependent part of the elements of the rotation matrix). Inserting (2.5.207) and (2.5.208) into (2.5.206), we obtain for the short-time-interval kernel:

$$K(\boldsymbol{u}_k, \boldsymbol{u}_{k-1}; \varepsilon) = \left(\frac{i\varepsilon}{2m}\right)^2 \sum_{l_k=0}^{\infty} \sum_{m_k=-l_k}^{l_k} \sum_{n_k=-l_k}^{l_k} (2l_k + 1) \exp\left\{-\frac{i\varepsilon}{2m} l_k(l_k+1)\right\}$$
$$\times e^{im_k(\phi_{k-1}-\phi_k)} e^{in_k(\psi_{k-1}-\psi_k)} P_{m_k n_k}^{l_k}(\cos\theta_{k-1}) P_{n_k m_k}^{l_k}(\cos\theta_k). \quad (2.5.209)$$

We can now introduce the finite-time-interval kernel in the usual time-sliced path-integral form:

$$K(\boldsymbol{u}, t; \boldsymbol{u}_0, t_0) = \lim_{N\to\infty} \frac{1}{(4\pi)^2} \prod_{k=1}^{N+1} \left(\frac{2m}{i\varepsilon}\right)^2 \int \prod_{k=1}^{N} \left(\frac{1}{(4\pi)^2} \sin\theta_k \, d\theta_k \, d\phi_k \, d\psi_k\right) \prod_{k=1}^{N+1} K(\boldsymbol{u}_k, \boldsymbol{u}_{k-1}; \varepsilon) \quad (2.5.210)$$

where

$$d\Omega \equiv \frac{1}{(4\pi)^2} \sin\theta_k \, d\theta_k \, d\phi_k \, d\psi_k$$

is the invariant volume element of $SU(2)$. By virtue of the normalization relation for the matrix elements of $SU(2)$ representations, we have

$$\frac{1}{(4\pi)^2} \int_0^\pi d\theta_k \sin\theta_k \int_0^{2\pi} d\phi_k \int_{-2\pi}^{2\pi} d\psi_k \, e^{i(m_k\phi_k+n_k\psi_k)} e^{-i(m_{k+1}\phi_k+n_{k+1}\psi_k)} P_{m_{k+1} n_{k+1}}^{l_{k+1}}(\cos\theta_k) P_{n_k m_k}^{l_k}(\cos\theta_k)$$
$$= \frac{1}{2l_k + 1} \delta_{l_k l_{k+1}} \delta_{m_k m_{k+1}} \delta_{n_k n_{k+1}}$$

the integral (2.5.210) can be calculated and we end up with the result:

$$K(\boldsymbol{u}, t; \boldsymbol{u}_0, t_0) = \frac{1}{(4\pi)^2} \sum_{l=0}^{\infty} (2l_k + 1) \exp\left\{-\frac{i(t-t_0)}{2m} l(l+1)\right\}$$
$$\times \sum_{m=-l}^{l} \sum_{n=-l}^{l} e^{i(m\phi+n\psi)} e^{-i(m\phi_0+n\psi_0)} P_{nm}^{l}(\cos\theta) P_{mn}^{l}(\cos\theta_0) \quad (2.5.211)$$

which displays the properly normalized wavefunctions

$$\psi_{mn}^{L}(\theta, \phi, \psi) = \frac{\sqrt{2l+1}}{4\pi} e^{-im\phi} e^{-in\psi} P_{mn}^{l}(\cos\theta) \quad (2.5.212)$$

and the energy spectrum

$$E = \frac{1}{2m} l(l+1). \quad (2.5.213)$$

To obtain a more compact expression for the transition amplitude (2.5.211), we sum over n by using (2.5.208), and arrive at the expression

$$K(\boldsymbol{u}, t; \boldsymbol{u}_0, t_0) = \frac{1}{(4\pi)^2} \sum_{l=0}^{\infty} (2l+1) \exp\left\{-\frac{i(t-t_0)}{2m} l(l+1)\right\} \sum_{m=-l}^{l} e^{im\Theta_{f,i}} \quad (2.5.214)$$

where $\Theta_{f,i}$ is the angle that rotates the vector $\boldsymbol{u}_0$ onto the vector $\boldsymbol{u}$. We can further simplify (2.5.214) by calculating the sum over m using

$$\sum_{m=-l}^{l} e^{im\Theta_{f,i}} = \frac{\sin((l+1/2)\Theta_{f,i})}{\sin(\Theta_{f,i}/2)}$$

and obtain

$$K(\boldsymbol{u}, t; \boldsymbol{u}_0, t_0) = \frac{1}{(4\pi)^2} \sum_{l=0}^{\infty} (2l+1) \exp\left\{-\frac{i(t-t_0)}{2m} l(l+1)\right\} \sin((l+1/2)\Theta_{f,i}). \quad (2.5.215)$$

This expression for the transition amplitude can also be presented as the derivatives of the *Jacobi θ-functions* (Schulman 1981).

Note that a similar technique allows us to derive and calculate the path integrals on higher-dimensional spheres S^n which are homogeneous spaces for the orthogonal Lie groups: $S^n \sim SO(n+1)/SO(n)$.

◇ **Coherent state path integral on the $SU(2)$-manifold**

The $su(2)$ Lie algebra is defined by the following commutation relations:

$$[\widehat{J}_+, \widehat{J}_-] = 2\widehat{J}_3 \qquad [\widehat{J}_3, \widehat{J}_\pm] = \pm\widehat{J}_\pm \quad (2.5.216)$$

where $\widehat{J}_\pm \equiv \widehat{J}_1 \pm i\widehat{J}_2$. Its unitary irreducible representations are given by the finite-dimensional matrices

$$\begin{aligned}\widehat{J}_3|J, M\rangle &= M|J, M\rangle \\ \widehat{J}_\pm|J, M\rangle &= \sqrt{(J \mp M)(J \pm M + 1)}|J, M \pm 1\rangle \qquad (|M| \leq J).\end{aligned} \quad (2.5.217)$$

Let us define the *non-normalized* $|\xi)$ and *normalized* $|\xi\rangle$ *coherent states* on the group $SU(2)$:

$$|\xi) \equiv e^{\xi \widehat{J}_+}|J, -J\rangle \qquad |\xi\rangle \equiv \frac{1}{(\xi|\xi)^{1/2}}|\xi) \qquad \xi \in \mathbb{C} \quad (2.5.218)$$

where we have introduced $|J, -J\rangle$ as the fiducial vector (Perelomov 1986, Klauder and Skagerstam 1985). Explicitly,

$$|\xi) = \sum_{m=0}^{2J} \xi^m \binom{2J}{m}^{1/2} |J, -J+m\rangle \quad (2.5.219)$$

and with the norm $(\xi|\xi) = (1 + |\xi|^2)^{2J}$, giving the normalized states:

$$|\xi\rangle = \frac{1}{(1+|\xi|^2)^J} \sum_{m=0}^{2J} \xi^m \binom{2J}{m}^{1/2} |J, -J+m\rangle. \quad (2.5.220)$$

They satisfy

$$\langle \xi | \xi' \rangle = \frac{(1 + \xi^* \xi')^{2J}}{(1 + |\xi|^2)^J (1 + |\xi'|^2)^J} \tag{2.5.221}$$

$$\frac{2J+1}{\pi} \int \frac{d\xi^* d\xi}{(1+|\xi|^2)^2} |\xi\rangle\langle\xi| \equiv \int d\mu(\xi^*, \xi) |\xi\rangle\langle\xi| = \mathbb{1}_J \tag{2.5.222}$$

where

$$d\xi^* d\xi \equiv d\operatorname{Re}(\xi) d\operatorname{Im}(\xi) \tag{2.5.223}$$

and

$$\mathbb{1}_J \equiv \sum_{M=-J}^{J} |J, M\rangle\langle J, M| \tag{2.5.224}$$

is the identity operator in the $(2J+1)$-dimensional irreducible representation. The matrix elements of the generators are found to be

$$\langle \xi | \widehat{J}_3 | \xi' \rangle = -J \frac{1 - \xi^* \xi'}{1 + \xi^* \xi'} \langle \xi | \xi' \rangle$$

$$\langle \xi | \widehat{J}_+ | \xi' \rangle = J \frac{2\xi^*}{1 + \xi^* \xi'} \langle \xi | \xi' \rangle. \tag{2.5.225}$$

Armed with this machinery, we can now discuss the path-integral formula for the Hamiltonian

$$\widehat{H} = h_1 \widehat{J}_1 + h_2 \widehat{J}_2 + h_3 \widehat{J}_3 \in su(2). \tag{2.5.226}$$

Let us, for simplicity, consider only the trace formula

$$\mathcal{Z}(T) \equiv \operatorname{Tr} e^{-i\widehat{H}T} = \operatorname{Tr} \lim_{N \to \infty} (1 - i\epsilon \widehat{H})^N = \lim_{N \to \infty} \mathcal{Z}_N \tag{2.5.227}$$

where

$$\mathcal{Z}_N \equiv \prod_{j=1}^{N} \int d\mu(\xi_j^*, \xi_j) \exp[i\{-i \ln\langle \xi_j | \xi_{j-1} \rangle - \epsilon H(\xi_j^*, \xi_{j-1})\}] \tag{2.5.228}$$

with $d\mu(\xi_j^*, \xi_j)$ being given by (2.5.222) and

$$H(\xi_j^*, \xi_{j-1}) \equiv \frac{\langle \xi_j | H | \xi_{j-1} \rangle}{\langle \xi_j | \xi_{j-1} \rangle}. \tag{2.5.229}$$

Here, similarly to the derivation of the path integral in problem 2.2.1, page 190, we have repeatedly inserted the resolution of unity (2.5.222) into the second relation in (2.5.227). With the aid of (2.5.225), (2.5.228) is found to be

$$\mathcal{Z}_N = \prod_{j=1}^{N} \int d\mu(\xi_j^*, \xi_j) \exp\left[iJ \left\{ 2i \ln\left(\frac{1 + \xi_j^* \xi_j}{1 + \xi_j^* \xi_{j-1}}\right) + \epsilon h \frac{1 - \xi_j^* \xi_{j-1}}{1 + \xi_j^* \xi_{j-1}} \right\}\right]. \tag{2.5.230}$$

Under the sign of the trace in (2.5.227), we can always diagonalize the Hamiltonian by using the $SU(2)$ rotation

$$\mathsf{U}\widehat{H}\mathsf{U}^\dagger = h\widehat{J}_3 \qquad \mathsf{U} \in SU(2) \tag{2.5.231}$$

so that (2.5.227) is equivalent to the so-called *character formula* (see supplement IV, volume II):

$$\mathcal{Z}(T) = \text{Tr} \exp(-i h \widehat{J}_3 T) = \frac{\sin((J+1/2)hT)}{\sin(hT/2)}. \quad (2.5.232)$$

Using the experience from sections 2.2–2.5, the reader can easily generalize this construction and obtain the path-integral representation for an arbitrary matrix element (transition amplitude), $\langle \xi | \exp\{-i\widehat{H}T\} | \xi_0 \rangle$. The interesting property of the path integral is that its calculation by the stationary-phase (WKB) approximation gives an *exact* result. In section 2.6.3 we shall discuss this fact from a general point of view.

Many further details on the coherent states for Lie groups and the corresponding path integrals the reader may find in Perelomov (1986) and Klauder and Skagerstam (1985).

2.5.6 Problems

Problem 2.5.1. Prove the formula for the differentiation of a determinant:

$$\frac{\partial}{\partial A_{ab}} \det A = (\det A)(A^{-1})_{ab}$$

where $(A^{-1})_{ab}$ is the element of the inverse matrix A^{-1}.

Hint. Use the standard formulae for the determinant of an $N \times N$ matrix

$$\det A = \varepsilon_{i_1 i_2 \ldots i_N} A_{1 i_1} A_{2 i_2} \cdots A_{N i_N}$$

where $\varepsilon_{i_1 i_2 \ldots i_N}$ is the absolutely antisymmetric tensor, and for the elements of an inverse matrix.

Problem 2.5.2. Calculate the integrals (2.5.25), (2.5.26) and (2.5.27).

Hint. We present a proof of the identity (2.5.25), the proof of the others being similar. Let us consider the integral

$$I(g_{cd}) = \int dq \, \exp\left\{\frac{im}{2\varepsilon\hbar} \xi^c g_{cd} \xi^d\right\} = \left(\frac{2\pi i\varepsilon\hbar}{m}\right)^{d/2} \frac{1}{\det g} \quad (2.5.233)$$

with g_{cd} considered as free parameters. Differentiation with respect to one of the parameters gives, on the one hand,

$$\frac{\partial}{\partial g_{ab}} I(g_{cd}) = \frac{im}{2\varepsilon\hbar} \int dq \, \exp\left(\frac{im}{2\varepsilon\hbar} \xi^c g_{cd} \xi^d\right) \xi^a \xi^b \quad (2.5.234)$$

and, on the other hand,

$$\frac{\partial}{\partial g_{ab}} I(g_{cd}) = \left(\frac{2\pi i\varepsilon\hbar}{m}\right)^{d/2} \frac{\partial}{\partial g_{ab}} g^{-1/2}$$

$$= -\frac{1}{2}\left(\frac{2\pi i\varepsilon\hbar}{m}\right)^{d/2} g^{-1/2} g^{ab} = -\frac{1}{2} g^{ab} I(g_{cd}). \quad (2.5.235)$$

Here we have used the formula for the differentiation of determinants from the preceding problem and the conventional notation g^{ab} for the inverse metric:

$$g_{ab} g^{bc} = \delta_a{}^c.$$

Problem 2.5.3. Derive the energy-dependent Green function (resolvent) for a free particle.

Hint. The corresponding energy-dependent Green function is given by (with $d = 1$)

$$G^{(1)}(x, x_0; E) = i \int_0^\infty K(x, x_0; T) e^{iET/\hbar} dT = \sqrt{-\frac{m}{2E}} \exp\left\{i\frac{|x - x_0|}{\hbar}\sqrt{2mE}\right\}.$$

For the d-dimensional case, we obtain

$$G^{(d)}(x, x_0; E) = 2i \left(\frac{m}{2\pi i\hbar}\right)^{d/2} \left(\frac{m}{2E}|x - x_0|^2\right)^{\frac{1}{2}(1-d/2)} K_{1-d/2}\left(-i\frac{|x - x_0|^2}{\hbar}\sqrt{2mE}\right)$$

$$= i\left(\frac{m}{2\hbar}\right)^{d/2} \left(\frac{m\pi^2}{2E}|x - x_0|^2\right)^{\frac{1}{2}(1-d/2)} H^{(1)}_{1-d/2}\left(\frac{|x - x_0|}{\hbar}\sqrt{2mE}\right)$$

where use has been made of the integral representation (see Gradshteyn and Ryzhik (1980)):

$$\int_0^\infty dt\, t^{\nu-1} \exp\left(-\frac{a}{4t} - pt\right) = 2\left(\frac{a}{4p}\right)^{\frac{\nu}{2}} K_\nu(\sqrt{ap})$$

and $K_\nu(z) = (i\pi/2) e^{i\pi\nu/2} H^{(1)}_\nu(iz)$, $K_{\pm\frac{1}{2}}(z) = \sqrt{\pi/2z}\, e^{-z}$. Here $H^{(1)}_\nu$ and K_ν are the Hankel and the modified Hankel functions.

Problem 2.5.4. Determine the energy-dependent Green function for the harmonic oscillator.

Hint. Use the integral representation (Gradshteyn and Ryzhik 1980):

$$\int_0^\infty \coth^{2\nu}\frac{x}{2} \exp\left[-\frac{a_1 + a_2}{2} t \cosh x\right] I_{2\mu}(t\sqrt{a_1 a_2} \sinh x)\, dx$$

$$= \frac{\Gamma(\frac{1}{2} + \mu - \nu)}{t\sqrt{a_1 a_2}\,\Gamma(1 + 2\mu)} W_{\nu,\mu}(a_1 t) M_{\nu,\mu}(a_2 t) \qquad (2.5.236)$$

where $a_1 > a_2$, $\text{Re}(\frac{1}{2} + \mu - \nu) > 0$; $W_{\nu,\mu}(z)$ and $M_{\nu,\mu}(z)$ denote the Whittaker functions. Now we should re-express

$$e^x = \sqrt{\pi x/2}[I_{\frac{1}{2}}(x) + I_{-\frac{1}{2}}(x)]$$

apply equation (2.5.236) for $\nu = E/2\hbar\omega$ and use the relations between the Whittaker and the parabolic cylinder functions U_α to obtain:

$$G(x, x_0; E) = -\frac{1}{2}\sqrt{\frac{m}{\pi\hbar\omega}} \Gamma\left(\frac{1}{2} - \frac{E}{\hbar\omega}\right) U_{-\frac{1}{2} + \frac{E}{\hbar\omega}}\left[\sqrt{\frac{2m\omega}{\hbar}}(x_0 + x + |x - x_0|)\right]$$

$$\times U_{-\frac{1}{2} + \frac{E}{\hbar\omega}}\left[-\sqrt{\frac{2m\omega}{\hbar}}(x_0 + x - |x - x_0|)\right].$$

Problem 2.5.5. Construct the path integral in polar coordinates for a particle in an isotropic potential $V(r)$ (in d-dimensional space) starting from the Hamiltonian written in polar coordinate representation.

Hint. The Hamiltonian for the particle reads

$$\widehat{H} = \left[-\frac{\hbar^2}{2m} \Delta_{\text{LB}} + V(r) \right] \quad (2.5.237)$$

where the Laplacian is written in polar coordinates

$$\Delta_{\text{LB}} = \frac{\partial^2}{\partial r^2} + \frac{d-1}{r} \frac{\partial}{\partial r} + \frac{1}{r^2} L^2_{(d)} \quad (2.5.238)$$

with the d-dimensional Legendre operator

$$L^2_{(d)} = \left[\frac{\partial^2}{\partial \theta_1^2} + (d-2) \cot \theta_1 \frac{\partial}{\partial \theta_1} \right] + \frac{1}{\sin^2 \theta_1} \left[\frac{\partial^2}{\partial \theta_2^2} + (d-3) \cot \theta_2 \frac{\partial}{\partial \theta_2} \right] + \cdots$$

$$+ \frac{1}{\sin^2 \theta_1 \cdots \sin^2 \theta_{d-3}} \left[\frac{\partial^2}{\partial \theta_{d-2}^2} + \cot \theta_{d-2} \frac{\partial}{\partial \theta_{d-2}} \right]$$

$$+ \frac{1}{\sin^2 \theta_1 \cdots \sin^2 \theta_{d-2}} \frac{\partial^2}{\partial \phi^2}. \quad (2.5.239)$$

The Legendre operator apparently contains non-commutative terms but an introduction of the Hermitian momenta

$$\widehat{p}_r = -i\hbar \left(\frac{\partial}{\partial r} + \frac{d-1}{2r} \right)$$

$$\widehat{p}_{\theta_\nu} = -i\hbar \left(\frac{\partial}{\partial \theta_\nu} + \frac{d-1-\nu}{2} \cot \theta_\nu \right) \quad (2.5.240)$$

$$\widehat{p}_\phi = -i\hbar \frac{\partial}{\partial \phi}$$

in the spirit of section 2.5.1 (cf (2.5.11)), casts Hamiltonian (2.5.237) into the form

$$H(p_r, r, \{p_\theta, \theta\}) = \frac{p_r^2}{2m} + \frac{1}{2mr^2} \left[p_{\theta_1}^2 + \frac{1}{\sin^2 \theta_1} p_{\theta_2}^2 + \cdots + \frac{1}{\sin^2 \theta_1 \cdots \sin^2 \theta_{d-2}} p_\phi^2 \right] + V(r) + \Delta V(r, \{\theta\}) \quad (2.5.241)$$

with

$$\Delta V(r, \{\theta\}) = -\frac{\hbar^2}{8mr^2} \left[1 + \frac{1}{\sin^2 \theta_1} + \cdots + \frac{1}{\sin^2 \theta_1 \cdots \sin^2 \theta_{d-2}} \right]. \quad (2.5.242)$$

As can be seen, Hamiltonian (2.5.241) does not contain products of non-commuting operators. Hence, the path integral for the corresponding evolution operator can be constructed straightforwardly, without discrete approximation (equivalent to operator-ordering) ambiguities. The result, of course, is equivalent to that in section 2.5.3.

Problem 2.5.6. Derive expression (2.5.118) for the propagator of the radial harmonic oscillator starting from expression (2.2.77) for the propagator of a one-dimensional oscillator.

Hint. Represent the propagator of a d-dimensional oscillator as a product of d one-dimensional propagators and use formula (2.5.88) to convert the product into (2.5.118).

Problem 2.5.7. Using the Duru–Kleinert method of space and time transformations in the framework of the path-integral formalism, find the propagator for a quantum-mechanical particle (in one-dimensional space) moving in the Morse potential

$$V(x) = C[e^{-2ax} - 2e^{-ax}]$$

where C and a are positive constants, $x \in \mathbb{R}$.

Hint. Perform the transformation

$$x = F(q) = -\frac{2}{a}\ln\left(\frac{qa}{\sqrt{2}}\right) \qquad q \in (0, \infty)$$

$$dt = f(x)\,ds \qquad f(x) = (F'(q))^2.$$

Then the new potential term in (2.5.40) becomes

$$W - EF'^2 = \frac{1}{2}a^2 C q^2 - \frac{1}{2q^2}\left(\frac{8E}{a^2} + \frac{1}{4}\right) - A\sqrt{8}.$$

Thus the problem reduces to consideration of the isotropic oscillator with frequency $\omega = a\sqrt{C}$ and effective angular momentum

$$l = -\frac{1}{2} + \sqrt{-\frac{8E}{a^2}}.$$

The next steps are quite similar to the case of the Coulomb potential.

Problem 2.5.8. Construct the path integral on the $SU(2)$ group manifold using the following parametrization of the group elements:

$$U = x^0 + ix^a \sigma^a \qquad x^0 = \sqrt{1 - x^a x^a} \qquad a = 1, 2, 3, \ldots \qquad (2.5.243)$$

(σ^a are the Pauli matrices).

Hint. Note that this parametrization does not fulfil condition (2.5.167). Therefore the results of section 2.5.5 cannot be applied straightforwardly so the reader should start from the very beginning. The composition function reads:

$$\Phi^a(x_1, x_2) = x_1^0 x_2^a + x_1^a x_2^0 - \varepsilon^{abc} x_1^b x_2^c \qquad (2.5.244)$$

which implies for the left auxiliary functions:

$$\eta^{ab} = x^0 \delta^{ab} + \varepsilon^{abc} x^c. \qquad (2.5.245)$$

For the Maurer–Cartan forms, we find that (the reader should be careful about the normalization of the generators of $su(2)$)

$$\lambda^{ab} = x^0 \delta^{ab} + \frac{x^a x^b}{x^0} - \varepsilon^{abc} x^c. \qquad (2.5.246)$$

The addition to the potential term now takes the form:

$$\Delta V = \frac{\hbar^2}{8m}\left(5\frac{x^a x^a}{(x^0)^2} - 3\right)$$

and the path integral reads as

$$K(x,t;x_0,t_0) = \lim_{N\to\infty} \left(\frac{m}{2\pi i\varepsilon\hbar}\right)^{3N/2} \int \left(\prod_{k=1}^{N-1} g^{1/2}(x_k)\, d^n x_k\right)$$

$$\times \exp\left\{i \sum_{k=0}^{N-1} \left[\frac{m}{2\varepsilon} g_{ab}(\widetilde{x}_k)\Delta x_k^a \Delta x_k^b - \varepsilon V(\widetilde{x}_k) - \frac{\varepsilon\hbar^2}{8m}\left(5\frac{x^a x^a}{(x^0)^2} - 3\right)\right]\right\}.$$

2.6 Path integrals over anticommuting variables for fermions and generalizations

We now turn to the problem of extending the path-integral formalism to cover theories containing both *bosons* and *fermions*. Although we have not so far used these terms (fermions and bosons) so far, since they pertain to the physics of many-body systems which we shall discuss in the next chapters, we assume that the reader is familiar with these basic notions of quantum mechanics and statistical physics. For the aims of the present section, it is enough to remember that the distinctive peculiarity of a fermionic system is that its many-body wavefunction is *antisymmetric*. As a result, the corresponding creation and annihilation operators of fermionic states satisfy *anticommutation* relations. This leads to a rather unusual, at first sight, representation for the corresponding evolution operator in terms of a 'path' integral over *anticommuting* variables, the latter being the generators of the so-called *Grassmann algebra*.

As a development of this idea of using unusual variables, we shall discuss in the subsequent part of this section a further generalization of path-integral construction, based on more general algebras than the Grassmann algebra. These more general 'path integrals' have not yet found any physical applications and the main reason for presenting them is to illustrate the power and to stress, once again, the main points of constructing path integrals using operator symbols.

The last subsection is devoted to a general analysis of the conditions under which the WKB approximation for a path integral gives an *exact result*. This topic is placed in the present section because the analysis essentially uses, together with other tools, the Grassmann variables.

2.6.1 Path integrals over anticommuting (Grassmann) variables for fermionic systems

The space of states of a fermionic system is constructed with the help of creation and annihilation operators satisfying the anticommutation relations

$$\{\widehat{\alpha}_j, \widehat{\alpha}_k^\dagger\} = \delta_{jk} \quad (2.6.1)$$

$$\{\widehat{\alpha}_j, \widehat{\alpha}_k\} = \{\widehat{\alpha}_j^\dagger, \widehat{\alpha}_k^\dagger\} = 0 \quad (2.6.2)$$

$$j,k = 1,\ldots,n$$

where $\{\widehat{A}, \widehat{B}\}$ denotes the *anticommutator* of two arbitrary operators $\widehat{A}$ and $\widehat{B}$:

$$\{\widehat{A}, \widehat{B}\} \stackrel{\text{def}}{\equiv} AB + BA. \quad (2.6.3)$$

Relations (2.6.1) and (2.6.2), being the direct generalization of the corresponding relations for bosonic creation and annihilation operators, are consistent with the assumption that $\widehat{\alpha}^\dagger$ is the Hermitian conjugate of $\widehat{\alpha}$. The standard introduction of self-adjoint coordinate and momentum operators starting from creation and annihilation operators via the formulae (cf (2.1.47))

$$\widehat{x}_j = \sqrt{\frac{\hbar}{2m\omega}}(\widehat{\alpha}_j + \widehat{\alpha}_j^\dagger)$$

$$\widehat{p}_j = i\sqrt{\frac{\hbar m\omega}{2}}(\widehat{\alpha}_j^\dagger - \widehat{\alpha}_j)$$

leads to the relations

$$\{\widehat{x}_j, \widehat{p}_k\} = 0$$

$$\{\widehat{x}_j, \widehat{x}_k\} = \frac{\hbar}{m\omega}\delta_{jk} \qquad \{\widehat{p}_j, \widehat{p}_k\} = \hbar m\omega \delta_{jk}$$

$$j, k = 1, \ldots, n.$$

These relations do not have the form of canonical commutation relations even up to the substitution $[\cdot, \cdot] \longrightarrow \{\cdot, \cdot\}$ and do not play such an important role for fermions as their counterpart in the case of bosons. Thus, as basic relations for fermion systems, we shall use (2.6.1) and (2.6.2).

◇ **The Hilbert space generated by the anticommuting creation and annihilation operators**

First of all, we wish to construct a complete basis for the states on which the fermionic creation and annihilation operators act. Note that for any given j, we have, from (2.6.2),

$$\widehat{\alpha}_j^2 = (\widehat{\alpha}_j^\dagger)^2 = 0. \tag{2.6.4}$$

These constraints imply that there exists a 'ket' vector $|0\rangle$ annihilated by all $\widehat{\alpha}_j$:

$$\widehat{\alpha}_j |0\rangle = 0 \qquad \text{for any } j = 1, \ldots, n \tag{2.6.5}$$

and the corresponding 'bra' vector $\langle 0|$ annihilated by all $\widehat{\alpha}_j^\dagger$:

$$\langle 0|\widehat{\alpha}_j^\dagger = 0 \qquad \text{for any } j = 1, \ldots, n. \tag{2.6.6}$$

Indeed, we can take

$$|0\rangle \sim \left(\prod_{k=1}^n \widehat{\alpha}_k\right)|f\rangle \qquad \langle 0| \sim \langle g|\left(\prod_{k=1}^n \widehat{\alpha}_k^\dagger\right)$$

where $|f\rangle$ and $\langle g|$ are some vectors for which these expressions do not vanish (note that if such vectors do not exist, the operators $(\prod_{k=1}^n \widehat{\alpha}_k)$ and $(\prod_{k=1}^n \widehat{\alpha}_k^\dagger)$ vanish identically, which we assume not to be the case). These states satisfy (2.6.5) and (2.6.6) by virtue of (2.6.4). We assume that the states $\langle 0|$ and $|0\rangle$ are unique and normalized, so that

$$\langle 0|0\rangle = 1 \tag{2.6.7}$$

(in general, a few such states may exist, labelled by some quantum numbers, but in this section, for simplicity, we omit this possibility, the generalization being quite straightforward).

Note that, because of the full symmetry of the relation $\{\widehat{\alpha}, \widehat{\alpha}^\dagger\} = 1$, the role of creation operator may be assigned equivalently either to $\widehat{\alpha}^\dagger$ or to $\widehat{\alpha}$.

A complete basis for the states of this system is provided by $|0\rangle$ and the states

$$|j, k, \ldots\rangle \stackrel{\text{def}}{\equiv} \widehat{\alpha}_j^\dagger \widehat{\alpha}_k^\dagger \cdots |0\rangle \tag{2.6.8}$$

with any number of *different* operators $\widehat{\alpha}_l^\dagger$ acting on $|0\rangle$. Note that the vectors $|j, k, \ldots\rangle$ are automatically antisymmetric in the indices $j, k, \ldots$ as should be the case for fermions. Similarly, we may define a complete dual basis, consisting of $\langle 0|$ and the states (also antisymmetric in the corresponding indices)

$$\langle j, k, \ldots | \stackrel{\text{def}}{\equiv} \langle 0| \cdots \widehat{\alpha}_j \widehat{\alpha}_k. \tag{2.6.9}$$

The action of operators $\widehat{\alpha}_m, \widehat{\alpha}_p^\dagger$ on these vectors is completely defined by the anticommutation relations (2.6.1) and (2.6.2). Using these relations together with (2.6.5)–(2.6.7), we can easily derive

$$\langle m, p, \ldots | j, k, \ldots\rangle = \langle 0| \cdots \widehat{\alpha}_m \widehat{\alpha}_p \widehat{\alpha}_j^\dagger \widehat{\alpha}_k^\dagger \cdots |0\rangle$$

$$= \begin{cases} 0 & \text{if the set of values } m, p, \ldots \\ & \text{does not coincide with } j, k, \ldots, \\ (-1)^{\sigma[P_{m,p,\ldots}^{j,k,\ldots}]} & \text{if the sets of values coincide with each other.} \end{cases} \tag{2.6.10}$$

Here $\sigma[P_{m,p,\ldots}^{j,k,\ldots}]$ denotes the number of transpositions necessary to convert the sequence of values $m, p, \ldots$ into the sequence $j, k, \ldots$.

In order to use the standard techniques of calculations in quantum mechanics, we would like to be able to rewrite sums over intermediate states like (2.6.8) as integrals over *eigenstates* of $\widehat{\alpha}_j$ or $\widehat{\alpha}_k^\dagger$. However, it is not possible for these operators to have eigenvalues (other than zero) in the usual sense. Suppose we try to find a state $|\xi; \alpha\rangle$ that satisfies

$$\widehat{\alpha}_j |\xi; \alpha\rangle = \xi_j |\xi; \alpha\rangle \qquad \text{(for all } j\text{)}. \tag{2.6.11}$$

From relation (2.6.2), we see that

$$\{\xi_j, \xi_k\} \equiv \xi_j \xi_k + \xi_k \xi_j = 0 \qquad \text{for all } j, k = 1, \ldots, n. \tag{2.6.12}$$

which is impossible for numbers. However, we can consider them as elements of an algebra and examine whether they are of some use for the description of fermionic states. From (2.6.12), we see that these new 'variables' are nilpotent elements, i.e.

$$\xi_j^2 = 0 \qquad \text{for any } j. \tag{2.6.13}$$

It is clear that an analogous consideration of left-sided eigenstates of the operator $\widehat{\alpha}^\dagger$:

$$\langle \xi; \alpha^\dagger | \widehat{\alpha}_j^\dagger = \langle \xi; \alpha^\dagger | \bar{\xi}_j$$

introduces the conjugate 'variables' $\bar{\xi}_j$ with the defining relations:

$$\{\bar{\xi}_j, \bar{\xi}_k\} \equiv \bar{\xi}_j \bar{\xi}_k + \bar{\xi}_k \bar{\xi}_j = 0 \tag{2.6.14}$$

$$\{\bar{\xi}_j, \xi_k\} \equiv \bar{\xi}_j \xi_k + \xi_k \bar{\xi}_j = 0 \tag{2.6.15}$$

$$\text{for all } j, k = 1, \ldots, n.$$

As a result, the general element g of the algebra $\mathcal{G}_n$, generated by $\xi_j, \bar{\xi}_k, j, k = 1, \ldots, n$ and called the *Grassmann algebra*, has the form of the finite sum

$$g = c_0 + \sum_j c_j \xi_j + \sum_m c^m \bar{\xi}_m + \sum_{m,j} c_j^m \xi_j \bar{\xi}_m + \cdots$$

$$+ \sum_{\substack{m_1 < m_2 < \cdots < m_n \\ j_1 < j_2 < \cdots < j_n}} c_{j_1 j_2 \cdots j_n}^{m_1 m_2 \cdots m_n} \xi_{j_1} \xi_{j_2} \cdots \xi_{j_n} \bar{\xi}_{m_1} \bar{\xi}_{m_2} \cdots \bar{\xi}_{m_n} \tag{2.6.16}$$

where the coefficients $c_{j_1 j_2 \ldots}^{m_1 m_2 \ldots}$ are arbitrary complex numbers.

Let us postulate that the Grassmann 'variables' $\xi_j, \bar{\xi}_k$ *anticommute* with the fermionic creation and annihilation operators:

$$\{\xi_j, \widehat{\alpha}_k\} = \{\xi_j, \widehat{\alpha}_k^\dagger\} = \{\bar{\xi}_j, \widehat{\alpha}_k\} = \{\bar{\xi}_j, \widehat{\alpha}_k^\dagger\} = 0 \qquad \text{for all } j, k = 1, \ldots, n. \tag{2.6.17}$$

We can now construct the eigenstates $|\xi; \alpha\rangle$ satisfying (2.6.11) (Schwinger 1970) (see, also e.g., Montonen (1974))

$$|\xi; \alpha\rangle = \exp\left\{-\sum_{j=1}^n \xi_j \widehat{\alpha}_j^\dagger\right\} |0\rangle \tag{2.6.18}$$

with the exponential defined as usual by its power series expansion (see problem 2.6.1, page 315). The left-sided eigenstates of $\widehat{\alpha}_j^\dagger$ have quite the same form:

$$\langle \xi; \alpha^\dagger| = \langle 0| \exp\left\{\sum_{j=1}^n \bar{\xi}_j \widehat{\alpha}_j\right\}. \tag{2.6.19}$$

One can also define the left-sided eigenvector of α:

$$\langle \zeta; \alpha| = \langle 0|\widehat{\alpha}_n \widehat{\alpha}_{n-1} \cdots \widehat{\alpha}_1 \exp\left\{\sum_{i=1}^n \zeta \widehat{\alpha}_i^\dagger\right\} \tag{2.6.20}$$

and the right-sided eigenvector of $\alpha^\dagger$:

$$|\zeta; \alpha^\dagger\rangle = \exp\left\{-\sum_{i=1}^n \bar{\zeta}_i \widehat{\alpha}_i\right\} \widehat{\alpha}_1^\dagger \cdots \widehat{\alpha}_{n-1}^\dagger \widehat{\alpha}_n^\dagger |0\rangle. \tag{2.6.21}$$

These vectors have the following scalar products (problem 2.6.2, page 315):

$$\begin{aligned}
\langle \zeta; \alpha|\xi; \alpha\rangle &= (-1)^{[(n+1)/2]} \prod_{j=1}^n (\zeta_j - \xi_j) \\
\langle \xi; \alpha^\dagger|\zeta; \alpha^\dagger\rangle &= (-1)^{[(n+1)/2]} \prod_{j=1}^n (\bar{\zeta}_j - \bar{\xi}_j) \\
\langle \zeta; \alpha|\zeta; \alpha^\dagger\rangle &= \exp\left\{-\sum_{j=1}^n \bar{\zeta}_j \zeta_j\right\} \\
\langle \xi; \alpha^\dagger|\xi; \alpha\rangle &= \exp\left\{\sum_{j=1}^n \bar{\xi}_j \xi_j\right\}
\end{aligned} \tag{2.6.22}$$

(here $[(n+1)/2]$ is the integer part of $(n+1)/2$).

◇ The Bargmann–Fock representation of the anticommuting operators

As we learned in section 2.3, there exists a convenient representation of creation and annihilation operators; namely the representation in the space of anti-holomorphic functions $\mathcal{F}^2$ on the corresponding classical phase space (with complex coordinates). This representation is useful for solving many problems and, in particular, for constructing path integrals in quantum mechanics.

Having at our disposal an analog of the complex coordinates of the classical phase space, i.e. the Grassmann variables $\xi_j, \bar{\xi}_k$, we may try to construct an analog of the anti-holomorphic representation for the fermionic creation and annihilation operators.

Let us start with a discussion of a system with one degree of freedom. In this case, a general element of the Grassmann algebra generated by $\xi, \bar{\xi}$ reads as

$$f(\xi, \bar{\xi}) = f_{00} + f_{01}\xi + f_{10}\bar{\xi} + f_{11}\xi\bar{\xi} \qquad (2.6.23)$$

where $f_{00}, f_{01}, f_{10}, f_{11}$ are arbitrary complex numbers. The elements of the form

$$f(\bar{\xi}) = f_0 + f_1 \bar{\xi} \qquad (2.6.24)$$

are considered as an analog of anti-holomorphic functions. Recall that in the case $n = 1$ (one degree of freedom) the operators $\widehat{\alpha}, \widehat{\alpha}^\dagger$ satisfying (2.6.1) can be represented by the 2×2 matrices

$$\widehat{\alpha} = \begin{pmatrix} 0 & 0 \\ 1 & 0 \end{pmatrix} \qquad \widehat{\alpha}^\dagger = \begin{pmatrix} 0 & 1 \\ 0 & 0 \end{pmatrix} \qquad (2.6.25)$$

i.e. the representation is two dimensional (this is a direct consequence of the nilpotency: $\widehat{\alpha}^2 = (\widehat{\alpha}^\dagger)^2 = 0$). The space of 'functions' (2.6.24) is also two-dimensional (because they are defined by two numbers, f_0 and f_1) and we shall use them for the new representation of the fermionic system.

As we know, in the case of bosonic oscillator operators the annihilation operator is represented by the derivative $\partial_{\bar{z}}$ (cf (2.3.41)). Thus we have to define an analog of differential calculus in Grassmann algebras. Of course, in a non-commutative algebra, we cannot define derivatives by a *straightforward* extension of the usual definition of a derivative as the limit of some finite difference. Instead, we shall understand the appropriate generalization of the notion of derivatives $\frac{\partial}{\partial \xi}, \frac{\partial}{\partial \bar{\xi}}$ as a map on the Grassmann algebra

$$\frac{\partial}{\partial \xi}, \frac{\partial}{\partial \bar{\xi}} : \mathcal{G} \to \mathcal{G}$$

possessing some properties to be clarified which allows us to consider this mapping as a reasonable generalization of derivatives. Note that a more refined mathematical treatment of the differential and integral calculi in Grassmann algebras (more precisely, on Grassmann manifolds), the reader can find in Berezin (1987). Another approach, which reveals an impressive analogy with the standard theory of holomorphic functions, is presented in Vladimirov and Volovich (1984).

In our simple one-dimensional case, it is natural to define the derivative as follows:

$$\frac{\partial}{\partial \bar{\xi}} f(\bar{\xi}) = \frac{\partial}{\partial \bar{\xi}}(f_0 + f_1 \bar{\xi}) \stackrel{\text{def}}{\equiv} f_1. \qquad (2.6.26)$$

Now, if we define the action of the operators on the set of anti-holomorphic functions in a manner similar to the bosonic case:

$$\widehat{\alpha}^\dagger f(\bar{\xi}) = \bar{\xi} f(\bar{\xi}) \qquad (2.6.27)$$

$$\widehat{\alpha} f(\bar{\xi}) = \frac{\partial}{\partial \bar{\xi}} f(\bar{\xi}) \qquad (2.6.28)$$

the reader can easily check that the one-dimensional relation for the fermionic oscillator operators, i.e.

$$\{\widehat{\alpha}, \widehat{\alpha}^\dagger\} = 1 \qquad (2.6.29)$$

(cf (2.6.1)) is satisfied indeed.

◇ The integral over Grassmann variables and scalar product in the Bargmann–Fock representation space

Our next task is to introduce a scalar product into the space of Grassmann anti-holomorphic functions, such that the operators (2.6.28) and (2.6.29) are conjugate to each other. We can do this using an appropriate definition of the 'integral' in the Grassmann algebra. Clearly, again, as in the case of the derivatives, this integral can not be constructed via a limiting procedure starting from a type of Darboux sum. Instead, we have to define a *linear functional* on the algebra:

$$\langle\!\langle \cdot \rangle\!\rangle : f \in \mathcal{G} \longrightarrow \langle\!\langle f \rangle\!\rangle \in \mathbb{C}$$

$$\langle\!\langle c_1 f_1 + c_2 f_2 \rangle\!\rangle = c_1 \langle\!\langle f_1 \rangle\!\rangle + c_2 \langle\!\langle f_2 \rangle\!\rangle \qquad (2.6.30)$$

$$c_1, c_2 \in \mathbb{C} \qquad f_1, f_1 \in \mathcal{G}$$

which puts some complex number in correspondence to any element of the algebra $\mathcal{G}$ and satisfies the property of linearity (2.6.30). Another property which we require to be satisfied by this functional is the following:

$$\left\langle\!\!\left\langle \frac{\partial}{\partial \xi} f \right\rangle\!\!\right\rangle = \left\langle\!\!\left\langle \frac{\partial}{\partial \bar{\xi}} f \right\rangle\!\!\right\rangle = 0. \qquad (2.6.31)$$

In fact, this is an analog of the Stokes formula for normal integrals. Indeed, the functional we seek is an analog of the integral

$$\int_{-\infty}^{\infty} dx\, F(x)$$

defined on the set of ordinary functions $F(x)$ and this normal integral definitely satisfies relations of the type (2.6.31) (i.e. $\int_{-\infty}^{\infty} dx\, \partial_x F(x) = 0$) for any integrable function $F(x)$ (due to the Stokes formula).

To construct the functional $\langle\!\langle \cdot \rangle\!\rangle$ explicitly, it is convenient to introduce, at first, the notion of anticommuting Grassmann 'differentials', $d\xi$, $d\bar{\xi}$:

$$\{d\xi, d\bar{\xi}\} = 0. \qquad (2.6.32)$$

Then introducing the standard symbol $\int$ for the 'integral' (i.e. for the linear functional $\langle\!\langle \cdot \rangle\!\rangle$) even in the Grassmann case, we define

$$\int d\xi\, c = c \int d\xi \left(\frac{\partial}{\partial \xi}\xi\right) = 0 \qquad c \in \mathbb{C}. \qquad (2.6.33)$$

Similarly,

$$\int d\bar{\xi}\, c = 0. \qquad (2.6.34)$$

Thus the integral for a constant Grassmann 'function' $f(\bar{\xi}, \xi) = c$ is equal to zero. Besides a constant, only linear 'functions' exist among holomorphic or anti-holomorphic elements of the form (2.6.24). Thus if we want to have non-trivial (non-zero) integral, we must put

$$\int d\xi\, \xi = 1 \qquad (2.6.35)$$

$$\int d\bar{\xi}\, \bar{\xi} = 1 \qquad (2.6.36)$$

(the normalization is, of course, a matter of convenience).

292   Path integrals in quantum mechanics

The multiple integral is understood as a repeated integral:

$$\int d\bar{\xi}\, d\xi\, f(\xi,\bar{\xi}) = \int d\bar{\xi}\left[\int d\xi\, f(\xi,\bar{\xi})\right]. \tag{2.6.37}$$

Rules (2.6.31)–(2.6.37), accompanied by the agreement that any Grassmann differential *anticommutes* with any variable, completely define the integral (linear functional) of any function (arbitrary element of the Grassmann algebra) (2.6.23):

$$\int d\bar{\xi}\, d\xi\, f(\xi,\bar{\xi}) = f_{00}\int d\bar{\xi}\, d\xi + f_{01}\int d\bar{\xi}\left[\int d\xi\, \xi\right] - f_{10}\int d\bar{\xi}\,\bar{\xi}\left[\int d\xi\right]$$
$$+ f_{11}\int d\bar{\xi}\left[\int d\xi\, \xi\right]\bar{\xi} = f_{11}. \tag{2.6.38}$$

This type of integral was initially introduced by Berezin (1966) and is called a *Berezin integral*.

Now we are ready to introduce the scalar product in the space of Grassmann anti-holomorphic functions (2.6.24)

$$\langle f|g\rangle \stackrel{\text{def}}{\equiv} \int d\bar{\xi}\, d\xi\, e^{-\bar{\xi}\xi} f^\dagger(\bar{\xi}) g(\bar{\xi}) \tag{2.6.39}$$

defining in this way the Hilbert space $\mathcal{F}^2_{\mathcal{G}}$. In (2.6.39), we understand that

$$f^\dagger(\bar{\xi}) = f_0^* + f_1^*\xi. \tag{2.6.40}$$

The reader can easily check that this scalar product is positively definite, and that the monomials

$$\psi_0 = 1 \qquad \psi_1 = \bar{\xi} \tag{2.6.41}$$

are orthonormalized (problem 2.6.4, page 315). The conjugacy of the operators (2.6.27) and (2.6.28) can be checked directly or derived from the fact that in the basis (2.6.41) they are equivalent to the matrices (2.6.25) because

$$\widehat{\alpha}^\dagger\psi_0 = \psi_1 \qquad \widehat{\alpha}^\dagger\psi_1 = 0 \qquad \widehat{\alpha}\psi_0 = 0 \qquad \widehat{\alpha}\psi_1 = \psi_0. \tag{2.6.42}$$

The Berezin integral shares many properties of normal integrals, in particular

$$\int d\bar{\xi}\, d\xi\, f(\xi+\eta,\bar{\xi}+\bar{\eta}) = \int d\bar{\xi}\, d\xi\, f(\xi,\bar{\xi}) \tag{2.6.43}$$

(see problem 2.6.5, page 316) where a further pair $\eta,\bar{\eta}$ of Grassmann variables, anticommuting with $\xi,\bar{\xi}$, has appeared. We postulate that any new variable, its corresponding Grassmann differential and its derivative *anticommute* with all other Grassmann variables, differentials and derivatives.

◇ The Gaussian integral over Grassmann variables

Let us apply these rules to calculate the Gaussian integral in the Grassmann case

$$\int d\bar{\xi}\, d\xi\, e^{\bar{\xi} A\xi + \bar{\xi}\eta + \bar{\eta}\xi} \tag{2.6.44}$$

where A is a complex number. Using (2.6.43) to make the substitution

$$\bar{\xi} \to \bar{\xi} - A^{-1}\bar{\eta} \qquad \xi \to \xi - A^{-1}\eta \tag{2.6.45}$$

and then expanding the exponent, we find

$$\int d\bar{\xi}\, d\xi\, e^{\bar{\xi}A\xi+\bar{\xi}\eta+\bar{\eta}\xi} = e^{-\bar{\eta}A^{-1}\eta} \int d\bar{\xi}\, d\xi\, e^{\bar{\xi}A\xi}$$
$$= -A e^{-\bar{\eta}A^{-1}\eta}. \qquad (2.6.46)$$

Note that formula (2.6.46) reads quite similarly to the case of the ordinary Gaussian integral for commuting variables, except for the pre-exponential factor A which has *positive* power, in contrast to the standard Gauss formula (2.1.73) where it appears in the denominator.

◇ **Generalization of the differential and integral calculi to the case of n degrees of freedom**

All the formulae can easily be generalized to n degrees of freedom. To this aim, we introduce $2n$ Grassmann variables

$$\xi_1, \ldots, \xi_n; \bar{\xi}_1, \ldots, \bar{\xi}_n \qquad (2.6.47)$$

together with their differentials and derivatives, with the postulate that all the quantities *anticommute* with each other, except pairs of variables and derivatives with the same indices:

$$\{\xi_j, \xi_k\} = \{\xi_j, \bar{\xi}_k\} = \{\bar{\xi}_j, \bar{\xi}_k\} = 0$$
$$\{d\xi_j, d\xi_k\} = \{d\xi_j, d\bar{\xi}_k\} = \{d\bar{\xi}_j, d\bar{\xi}_k\} = 0$$
$$\{\xi_j, d\xi_k\} = \{\bar{\xi}_j, d\bar{\xi}_k\} = \{\xi_j, d\bar{\xi}_k\} = \{\bar{\xi}_j, d\xi_k\} = 0$$
$$\{\partial_{\xi_j}, \partial_{\xi_k}\} = \{\partial_{\bar{\xi}_j}, \partial_{\bar{\xi}_k}\} = \{\partial_{\xi_j}, \partial_{\bar{\xi}_k}\} = 0 \qquad (2.6.48)$$
$$\{\partial_{\xi_j}, d\xi_k\} = \{\partial_{\bar{\xi}_j}, d\bar{\xi}_k\} = \{\partial_{\bar{\xi}_j}, d\xi_k\} = \{\partial_{\xi_j}, d\bar{\xi}_k\} = 0$$
$$\{\partial_{\xi_j}, \bar{\xi}_k\} = \{\partial_{\bar{\xi}_j}, \xi_k\} = 0$$
$$\{\partial_{\xi_j}, \xi_k\} = \{\partial_{\bar{\xi}_j}, \bar{\xi}_k\} = \delta_{jk}.$$

These commutation relations allow us to calculate the Grassmann–Berezin integral and the action of a derivative on any element of the Grassmann algebra $\mathcal{G}^{2n}$. Due to the linearity, it is enough to consider a monomial $\bar{\xi}_{j_1} \cdots \bar{\xi}_{j_l} \xi_{k_1} \cdots \xi_{k_m}$. Then the integral is given by

$$\int d\bar{\xi}_{j_r}\, \bar{\xi}_{j_1} \cdots \bar{\xi}_{j_r} \cdots \bar{\xi}_{j_l} \xi_{k_1} \cdots \xi_{k_m} = (-1)^{r-1} \bar{\xi}_{j_1} \cdots \underbrace{\bar{\xi}_{j_r}}_{\text{absent}} \cdots \bar{\xi}_{j_l} \xi_{k_1} \cdots \xi_{k_m} \qquad (2.6.49)$$

$$\int d\xi_{k_q}\, \bar{\xi}_{j_1} \cdots \bar{\xi}_{j_l} \xi_{k_1} \cdots \xi_{k_q} \cdots \xi_{k_m} = (-1)^{l+q-1} \bar{\xi}_{j_1} \cdots \bar{\xi}_{j_l} \xi_{k_1} \cdots \underbrace{\xi_{k_q}}_{\text{absent}} \cdots \xi_{k_m} \qquad (2.6.50)$$

where the underbraces indicate the absent factors in the monomial on the right-hand sides of the relations. As we have already discussed, a multiple integral is understood to be the repeated one

$$\int d\bar{\xi}_{j_1} \cdots d\bar{\xi}_{j_l}\, d\xi_{k_1} \cdots d\xi_{k_m}\, \bar{\xi}_{j_1} \cdots \bar{\xi}_{j_l} \xi_{k_1} \cdots \xi_{k_m}$$
$$= \int d\bar{\xi}_{j_1} \cdots d\bar{\xi}_{j_l}\, d\xi_{k_1} \cdots d\xi_{k_{m-1}} \left[\int d\xi_{k_m}\, \bar{\xi}_{j_1} \cdots \bar{\xi}_{j_l} \xi_{k_1} \cdots \xi_{k_m} \right]. \qquad (2.6.51)$$

Grassmann derivatives act quite similarly to the integrals:

$$\frac{\partial}{\partial \bar{\xi}_{j_r}} \bar{\xi}_{j_1} \cdots \bar{\xi}_{j_r} \cdots \bar{\xi}_{j_l} \xi_{k_1} \cdots \xi_{k_m} = (-1)^{r-1} \bar{\xi}_{j_1} \cdots \underbrace{\bar{\xi}_{j_r}}_{\text{absent}} \cdots \bar{\xi}_{j_l} \xi_{k_1} \cdots \xi_{k_m} \qquad (2.6.52)$$

$$\frac{\partial}{\partial \xi_{j_q}} \bar{\xi}_{j_1} \cdots \bar{\xi}_{j_l} \xi_{k_1} \cdots \xi_{k_q} \cdots \xi_{k_m} = (-1)^{l+q-1} \bar{\xi}_{j_1} \cdots \bar{\xi}_{j_l} \xi_{k_1} \cdots \underbrace{\xi_{k_q}}_{\text{absent}} \cdots \xi_{k_m}. \quad (2.6.53)$$

Note that in these formulae the derivatives are supposed to act on the monomial *from the left*, i.e. we must move the variable (corresponding to a derivative) to the left before dropping it. We could also use derivatives which act *from the right*:

$$\left(\frac{\partial}{\partial \bar{\xi}_{j_r}}\right)_R \bar{\xi}_{j_1} \cdots \bar{\xi}_{j_r} \cdots \bar{\xi}_{j_l} \xi_{k_1} \cdots \xi_{k_m} \equiv \bar{\xi}_{j_1} \cdots \bar{\xi}_{j_r} \cdots \bar{\xi}_{j_l} \xi_{k_1} \cdots \xi_{k_m} \frac{\overleftarrow{\partial}}{\partial \bar{\xi}_{j_r}}$$

$$= (-1)^{m+l-r} \bar{\xi}_{j_1} \cdots \underbrace{\bar{\xi}_{j_r}}_{\text{absent}} \cdots \bar{\xi}_{j_l} \xi_{k_1} \cdots \xi_{k_m} \quad (2.6.54)$$

$$\left(\frac{\partial}{\partial \xi_{j_q}}\right)_R \bar{\xi}_{j_1} \cdots \bar{\xi}_{j_l} \xi_{k_1} \cdots \xi_{k_q} \cdots \xi_{k_m} \equiv \bar{\xi}_{j_1} \cdots \bar{\xi}_{j_l} \xi_{k_1} \cdots \xi_{k_q} \cdots \xi_{k_m} \frac{\overleftarrow{\partial}}{\partial \xi_{j_q}}$$

$$= (-1)^{m-q} \bar{\xi}_{j_1} \cdots \bar{\xi}_{j_l} \xi_{k_1} \cdots \underbrace{\xi_{k_q}}_{\text{absent}} \cdots \xi_{k_m}. \quad (2.6.55)$$

We shall not, however, use the right derivatives in this chapter and, therefore, we shall not specifically indicate that all the Grassmann derivatives in this chapter are *left-sided* ones. In the next chapter, however, we shall need the right-hand derivatives as well.

The space of states $\mathcal{F}^2_{\mathcal{G}^{2n}}$ consists of anti-holomorphic functions $f(\bar{\xi}_1, \ldots, \bar{\xi}_n)$ and has the dimension 2^n. The fermionic oscillator operators with the defining relations (2.6.1) and (2.6.2) act in this space as follows:

$$\widehat{\alpha}_j f(\bar{\xi}_1, \ldots, \bar{\xi}_n) = \frac{\partial}{\partial \bar{\xi}_j} f(\bar{\xi}_1, \ldots, \bar{\xi}_n) \quad (2.6.56)$$

$$\widehat{\alpha}_j^\dagger f(\bar{\xi}_1, \ldots, \bar{\xi}_n) = \bar{\xi}_j f(\bar{\xi}_1, \ldots, \bar{\xi}_n) \quad (2.6.57)$$

and they are conjugate with respect to the scalar product

$$\langle f_1 | f_2 \rangle \stackrel{\text{def}}{\equiv} \int d\bar{\xi}_1 \cdots d\bar{\xi}_n \, d\xi_n \cdots d\xi_1 \exp\left\{-\sum_{j=1}^n \bar{\xi}_j \xi_j\right\} f_1^\dagger(\bar{\xi}_1, \ldots, \bar{\xi}_n) f_2(\bar{\xi}_1, \ldots, \bar{\xi}_n). \quad (2.6.58)$$

Here the conjugation in the Grassmann algebra is defined as follows:

$$(C\bar{\xi}_{j_1} \cdots \bar{\xi}_{j_l})^\dagger = C^* \xi_{j_l} \cdots \xi_{j_1} \quad (2.6.59)$$

the conjugation for differentials and derivatives being defined similarly. Note that the 'integration measure' (product of differentials) in (2.6.58) is defined in a self-conjugate manner and that the integral on the right-hand side of (2.6.58) for an arbitrary function $f(\bar{\xi}_1, \ldots, \bar{\xi}_n; \xi_1, \ldots, \xi_n)$ is

$$\int d\bar{\xi}_1 \cdots d\bar{\xi}_n \, d\xi_n \cdots d\xi_1 \, f(\bar{\xi}_1, \ldots, \bar{\xi}_n; \xi_1, \ldots, \xi_n) = f_{1,\ldots,n;n,\ldots,1} \quad (2.6.60)$$

where $f_{1,\ldots,n;n,\ldots,1}$ are the coefficients at $\xi_1 \cdots \xi_n \bar{\xi}_n \cdots \bar{\xi}_1$ in the expansion of the function $f(\bar{\xi}_1, \ldots, \xi_n)$ over the basic monomials (cf (2.6.16)).

The Gaussian integral

$$\int d\bar{\xi}_1 \cdots d\bar{\xi}_n \, d\xi_n \cdots d\xi_1 \, \exp\left\{\sum_{j,k=1}^{n}(\bar{\xi}_j A_{jk}\xi_k) + \sum_{j=1}^{n}(\bar{\xi}_j\zeta_j + \bar{\zeta}_j\xi_j)\right\} \qquad (2.6.61)$$

can be calculated by using a shift of variables as in (2.6.45) reducing it to

$$\exp\left\{-\sum_{j,k=1}^{n}(\bar{\zeta}_j(A^{-1})_{jk}\zeta_k)\right\} \in, d\bar{\xi}_1 \cdots d\bar{\xi}_n \, d\xi_n \cdots d\xi_1 \, \exp\left\{\sum_{j,k=1}^{n}(\bar{\xi}_j A_{jk}\xi_k)\right\}. \qquad (2.6.62)$$

The remaining integral is found with the help of formula (2.6.60):

$$\int d\bar{\xi}_1 \cdots d\bar{\xi}_n \, d\xi_n \cdots d\xi_1 \, \exp\left\{\sum_{j,k=1}^{n}(\bar{\xi}_j A_{jk}\xi_k)\right\} = (-1)^n \det \mathsf{A} \qquad (2.6.63)$$

(see the problem 2.6.6, page 316). Thus the final formula for the Gaussian integral in the Grassmann case reads as

$$\int d\bar{\xi}_1 \cdots d\bar{\xi}_n \, d\xi_n \cdots d\xi_1 \, \exp\left\{\sum_{j,k=1}^{n}(\bar{\xi}_j A_{jk}\xi_k) + \sum_{j=1}^{n}(\bar{\xi}_j\zeta_j + \bar{\zeta}_j\xi_j)\right\}$$

$$= (-1)^n \det \mathsf{A} \exp\left\{-\sum_{j,k=1}^{n}\bar{\zeta}_j(A^{-1})_{jk}\zeta_k\right\}. \qquad (2.6.64)$$

The main distinction from the bosonic case (i.e. a Gaussian integral with *commuting* variables) is that the determinant in (2.6.64) proves to be in the numerator (cf the bosonic formula (1.1.85) where it appears in the denominator). Note that the exponent in (2.6.64) can also be obtained as a solution of the 'extremal equations':

$$\frac{\partial}{\partial \bar{\xi}_m}\left(\sum_{j,k=1}^{n}(\bar{\xi}_j A_{jk}\xi_k) + \sum_{j=1}^{n}(\bar{\xi}_j\eta_j + \bar{\zeta}_j\xi_j)\right) = 0 \qquad (2.6.65)$$

$$\frac{\partial}{\partial \xi_m}\left(\sum_{j,k=1}^{n}(\bar{\xi}_j A_{jk}\xi_k) + \sum_{j=1}^{n}(\bar{\xi}_j\eta_j + \bar{\zeta}_j\xi_j)\right) = 0 \qquad (2.6.66)$$

$$m = 1, \ldots, n.$$

This is a general property of Gaussian-like integrals.

A similar calculation gives a useful formula for the square root of the determinant:

$$\int d\xi_1 \cdots d\xi_n \, \exp\left\{\sum_{j,k=1}^{n}(\xi_j A_{jk}\xi_k)\right\} = \sqrt{\det \mathsf{A}}. \qquad (2.6.67)$$

The monomials

$$\psi_{j_1,\ldots,j_r} = \bar{\xi}_{j_1} \cdots \bar{\xi}_{j_r} \qquad (j_1 < j_2 < \cdots < j_r) \qquad (2.6.68)$$

are orthonormalized and form a basis in the state space $\mathcal{F}^2_{\mathcal{G}^{2n}}$.

Note that the products $\prod_{j=1}^{n}(\zeta_j - \xi_j)$ and $\prod_{j=1}^{n}(\bar{\zeta}_j - \bar{\xi}_j)$ have a property analogous to the δ-function (problem 2.6.3, page 315):

$$\int d\xi_n \cdots d\xi_1 \left(\prod_{j=1}^{n}(\zeta_j - \xi_j) \right) f(\xi) = f(\zeta)$$
$$\int d\bar{\zeta}_n \cdots d\bar{\zeta}_1 \left(\prod_{j=1}^{n}(\bar{\xi}_j - \bar{\zeta}_j) \right) f(\bar{\xi}) = f(\bar{\xi}). \tag{2.6.69}$$

◇ **Symbols, operator kernels and path integrals in the case of fermionic (anticommuting) operators**

Now we are going to construct a path-integral representation for operators in the fermionic Hilbert space. This will be defined as the appropriate limit of a multiple integral over Grassmann variables. Of course, the reader may be confused by the very term '*path* integral'. Indeed, we can hardly imagine some real path in a space (which is usually called *superspace*) described by Grassmann variables serving as a type of coordinate. (Perhaps this is possible, at least to some extent, within the so-called 'point' approach to the definition of a superspace (Vladimirov and Volovich 1984), but up to now nobody has succeeded in such an interpretation.) Recall, however, that even in the bosonic case with the usual c-number coordinates of the phase space, the trajectories supporting the phase-space path integral are *discontinuous* functions so that the very notion of trajectories becomes doubtful (as it should in quantum mechanics). In fact, path integrals in quantum mechanics provide us with a convenient method for constructing a symbol or kernel of an evolution operator for *finite* time shifts, using its representation as a composition of an infinite number of *infinitesimal* evolution operators. This latter interpretation can be freely extended to the fermionic (Grassmann) case. One more remark about the Grassmann–Berezin path integrals is in order: if we consider a fermionic system with one or a finite number of degrees of freedom, path-integral techniques seems to be excessive because, in fact, all calculations can be reduced to manipulations with finite-dimensional matrices. However, in systems with an infinite number of degrees of freedom (quantum field theory), the path-integral approach proves to be very fruitful. In particular, this concerns most realistic models, which include both fermionic and bosonic degrees of freedom, where the path-integral approach provides a unified and powerful calculation method (see the next chapter).

Thus we proceed to the construction of the Grassmann–Berezin path integral and start by constructing the normal symbol and integral kernel in the space $\mathcal{F}_\mathcal{G}^2$ (one degree of freedom). The most general operator in this space is defined by the expression

$$\widehat{A} = A_{00} + A_{10}\widehat{\alpha}^\dagger + A_{01}\widehat{\alpha} + A_{11}\widehat{\alpha}^\dagger\widehat{\alpha} \tag{2.6.70}$$

where $A_{00}, A_{10}, A_{01}, A_{11}$ are complex numbers. Similarly to the bosonic case, we can associate with this operator two functions on the Grassmann algebra: the normal symbol

$$A(\bar{\xi}, \xi) = A_{00} + A_{10}\bar{\xi} + A_{01}\xi + A_{11}\bar{\xi}\xi \tag{2.6.71}$$

and the integral kernel

$$K_A(\bar{\xi}, \xi) = K_{00} + K_{10}\bar{\xi} + K_{01}\xi + K_{11}\bar{\xi}\xi \tag{2.6.72}$$

where the complex numbers $K_{00}, K_{10}, K_{01}, K_{11}$ are the matrix elements of the operator $\widehat{A}$ in the basis ψ_0, ψ_1 (cf (2.6.41)):

$$K_{jk} = \langle \psi_j | \widehat{A} | \psi_k \rangle. \tag{2.6.73}$$

Now the action of an operator $\widehat{A}$ in the space $\mathcal{F}_\mathcal{G}^2$ can be defined by its kernel:

$$(\widehat{A}f)(\bar{z}) = \int d\bar{\zeta}\, d\zeta\, e^{-\bar{z}\zeta} K_A(\bar{z}, \zeta) f(\bar{\zeta}) \tag{2.6.74}$$

and the product $\widehat{A}_1 \widehat{A}_2$ of two operators corresponds to the convolution $K_{A_1} * K_{A_2}$ of their kernels:

$$(K_{A_1} * K_{A_2})(\bar{\xi}, \xi) = \int d\bar{\zeta}\, d\zeta\, e^{-\bar{\zeta}\zeta} K_{A_1}(\bar{\xi}, \zeta) K_{A_2}(\bar{\zeta}, \xi). \quad (2.6.75)$$

Recall that the new Grassmann variables (in this case $\bar{\zeta}, \zeta$) anticommute with all other variables, as well as with the differentials and derivatives.

The relation between the normal symbol $A(\bar{\xi}, \xi)$ and the kernel K_A of an operator $\widehat{A}$ can be written in the form

$$K_A(\bar{\xi}, \xi) = e^{\bar{\xi}\xi} A(\bar{\xi}, \xi). \quad (2.6.76)$$

To prove this formula, it is enough to find the relation between the coefficients A_{jk} and K_{jk} in (2.6.70), (2.6.71) and (2.6.72), by use of (2.6.73):

$$A_{00} = K_{00} \qquad A_{01} = K_{01} \qquad A_{10} = K_{10} \qquad A_{11} = K_{11} - K_{00}. \quad (2.6.77)$$

In the same way as in the case of one degree of freedom, any operator

$$\widehat{A} = \sum_{r,q=1}^{n} \sum_{\substack{j_1 < \cdots < j_r \\ k_1 < \cdots < k_q}} A_{j_1,\ldots,j_r;k_1,\ldots,k_q} \widehat{a}^\dagger_{j_1} \cdots \widehat{a}^\dagger_{j_r} \widehat{a}_{k_1} \cdots \widehat{a}_{k_q} \quad (2.6.78)$$

can be defined either by its normal symbol

$$A(\bar{\xi}, \xi) = \sum_{r,q=1}^{n} \sum_{\substack{j_1 < \cdots < j_r \\ k_1 < \cdots < k_q}} A_{j_1,\ldots,j_r;k_1,\ldots,k_q} \bar{\xi}_{j_1} \cdots \bar{\xi}_{j_r} \xi_{k_1} \cdots \xi_{k_q} \quad (2.6.79)$$

or by its integral kernel

$$K_A(\bar{\xi}, \xi) = \sum_{r,q=1}^{n} \sum_{\substack{j_1 < \cdots < j_r \\ k_1 < \cdots < k_q}} (K_A)_{j_1,\ldots,j_r;k_1,\ldots,k_q} \bar{\xi}_{j_1} \cdots \bar{\xi}_{j_r} \xi_{k_1} \cdots \xi_{k_q} \quad (2.6.80)$$

where

$$(K_A)_{j_1,\ldots,j_r;k_1,\ldots,k_q} = \langle \psi_{j_1,\ldots,j_r} | \widehat{A} | \psi_{k_1,\ldots,k_q} \rangle. \quad (2.6.81)$$

The kernel and the normal symbol are related as follows:

$$K_A(\bar{\xi}, \xi) = \exp\left\{ \sum_{j=1}^{n} \bar{\xi}_j \xi_j \right\} A(\bar{\xi}, \xi) \quad (2.6.82)$$

(problem 2.6.8, page 316). The action of an operator in the space $\mathcal{F}^2_{\mathcal{G}^{2n}}$ is defined by the expression

$$(\widehat{A} f)(\bar{\xi}) = \int d\bar{\zeta}_1 \cdots d\bar{\zeta}_n\, d\zeta_n \cdots d\zeta_1\, \exp\left\{ \sum_{j=1}^{n} \bar{\zeta}_j \zeta_j \right\} K_A(\bar{\xi}, \zeta) f(\bar{\zeta}) \quad (2.6.83)$$

and the product of two operators corresponds to the convolution of the kernels:

$$(K_{A_1} * K_{A_2})(\bar{\xi}, \xi) = \int d\bar{\zeta}_1 \cdots d\bar{\zeta}_n\, d\zeta_n \cdots d\zeta_1\, \exp\left\{ \sum_{j=1}^{n} \bar{\zeta}_j \zeta_j \right\} K_{A_1}(\bar{\xi}, \zeta) K_{A_2}(\bar{\zeta}, \xi). \quad (2.6.84)$$

We invite the reader to compare these formulae with the corresponding expressions in the bosonic case for operator kernels and normal symbols and to convince him/herself that all the formulae have the same form. This implies, in turn, that an expression for a symbol or kernel of an evolution operator in terms of path integrals (understood as the limit of finite-dimensional multiple time-sliced integrals) also has the same form. Indeed, the whole derivation of the path-integral representation is based, in fact, on the formulae for the star-product of symbols or convolution of kernels and the relation between kernels and symbols. Since these formulae are identical in *form* in the bosonic and fermionic cases, we can immediately write the representation for the kernel of the evolution operator of a fermionic system with a Hamiltonian $\widehat{H}(\widehat{\alpha}^\dagger, \widehat{\alpha}; t)$ in the form (for details, see Borisov *et al* (1976) and Faddeev (1976)):

$$U(\bar{\xi}, \xi; t, t_0) = \int \prod_{\tau, j} d\bar{\xi}_j(\tau) \, d\xi_j(\tau) \, \exp \left\{ \frac{1}{2} \sum_j (\bar{\xi}_j(t)\xi_j(t) + \bar{\xi}_j(t_0)\xi_j(t_0)) \right.$$
$$\left. + i \int_{t_0}^{t} d\tau \left[\frac{1}{2i} \sum_j (\bar{\xi}_j \dot{\xi}_j - \dot{\bar{\xi}}_j \xi_j) - H(\bar{\xi}(\tau), \xi(\tau); \tau) \right] \right\} \quad (2.6.85)$$

where we assume the following boundary conditions (which are usual for formulae obtained via normal symbols):

$$\bar{\xi}_j(t) = \bar{\xi}_j \qquad \xi_j(t_0) = \xi_j \qquad j = 1, \ldots, n. \quad (2.6.86)$$

Here we integrate over an *infinite-dimensional* Grassmann algebra with the independent generators $\bar{\xi}_j(\tau), \xi_j(\tau), j = 1, \ldots, n$ for every $\tau \in [t_0, t]$. We suggest the reader explicitly repeats the steps of this path-integral derivation following the procedure outlined for the bosonic case in the preceding sections of this chapter.

Another useful exercise (cf problem 2.6.10, page 316) in order to obtain experience in handling Grassmann variables is to derive path integral (2.6.85) using the completeness relation in the space $\mathcal{F}^2_{\mathcal{G}^{2n}}$ similarly to the way considered for the bosonic case in problem 2.2.1, page 190.

Strictly speaking, we should write the evolution kernel in terms of the ratio of two path integrals (with a zero Hamiltonian in the denominator), as discussed in section 2.3 for the bosonic case (cf (2.3.77)) so that the overall factor in front of the integral becomes unessential. We do not go into detail here because, as we have already mentioned, the Grassmann–Berezin path integral finds a practical application in quantum field theories (for finite-dimensional fermionic systems, the operator method is more convenient). But in field theories, such peculiarities prove to be hidden by the renormalization procedure (we shall consider this in the next chapter). For the same reason, we do not explicitly indicate the shift in time variables (which, of course, have the same form as in the bosonic case, cf (2.3.122)). From the point of view of perturbation theory, the shifts again result in the rule that, after expanding the exponent in (2.6.85), we must drop the terms corresponding to the normal ordering of the Hamiltonian (cf (2.3.129) and the explanation following this condition). In quantum field theory, such a contraction anyway would result in a divergent expression and require renormalization. Thus in systems with an infinite number of degrees of freedom we may freely drop the shifts of the time variable transferring all the subtleties to the correct renormalization procedure.

2.6.2 Path integrals with generalized Grassmann variables

In the simplest case of one degree of freedom, the path integral over the Grassmann variables describes a quantum-mechanical system with two possible states. It is natural to generalize this to quantum-mechanical systems with an arbitrary but finite number of states. An operator description of such systems is realized with the help of the so-called *q-deformed oscillator algebra*, the parameter q being the root of unity: $q^p = 1$, for some integer p (for an introduction into q-deformed algebras and, in particular,

q-oscillator algebras see, e.g., Chaichian and Demichev (1996) and references therein). We shall consider even integers p, so that $q^{(k+1)} = -1$ for $k = p/2 - 1$. The case $q^2 = -1$ corresponds to the usual fermionic oscillator. Of course, generalization to field theory requires the more complicated algebra of a system of q-oscillators and its precise physical meaning depends on the form chosen for the commutation relations of the different q-oscillators (bosonic excitations on a lattice with a finite number of states at a given site or system with fractional statistics). In this subsection, we consider only one degree of freedom. Our aim is to present possible Bargmann–Fock representations of q-oscillators with a finite Fock space and to construct the corresponding q-deformed path integral. It is interesting that even in the simplest Grassmann-like case $q^2 = -1$, there are different forms of the Bargmann–Fock representation and the path integrals depend on the way in which the Planck constant $\hbar$ enters the commutation relation.

◇ **q-deformed oscillators (the case of a root-of-unity deformation parameter q)**

The initial commutation relations for q-oscillator operators are:

$$\widehat{\alpha}\widehat{\alpha}^\dagger - q\widehat{\alpha}^\dagger\widehat{\alpha} = q^{-\widehat{N}} \qquad \widehat{\alpha}^\dagger\widehat{\alpha} = [\widehat{N}]_q \equiv \frac{q^{\widehat{N}} - q^{-\widehat{N}}}{q - q^{-1}} \tag{2.6.87}$$

where $q^{k+1} = -1$ for some integer k,

$$[x]_q \stackrel{\text{def}}{\equiv} \frac{q^x - q^{-x}}{q - q^{-1}} \tag{2.6.88}$$

and $\widehat{\alpha}^\dagger$ is Hermitian conjugated to $\widehat{\alpha}$. To construct the Bargmann–Fock representation, we introduce the operators $\widehat{\beta}$ and $\widehat{\beta}^*$:

$$\widehat{\beta} = q^{\widehat{N}/2}\widehat{\alpha} \qquad \widehat{\beta}^* = \widehat{\alpha}^\dagger q^{\widehat{N}/2}$$

with the commutation relations

$$\widehat{\beta}\widehat{\beta}^* - q^2\widehat{\beta}^*\widehat{\beta} = 1 \tag{2.6.89}$$
$$\widehat{\beta}\widehat{\beta}^* - \widehat{\beta}^*\widehat{\beta} = q^{2\widehat{N}}. \tag{2.6.90}$$

For a root-of-unity parameter q, the operator $\widehat{\beta}^*$ is not Hermitian conjugate to $\widehat{\beta}$, but instead

$$\widehat{\beta}^\dagger = \widehat{\beta}^* q^{-\widehat{N}} \qquad (\widehat{\beta}^*)^\dagger = q^{-\widehat{N}}\widehat{\beta}. \tag{2.6.91}$$

The Fock space representation of the operators is:

$$\widehat{\beta}|n\rangle = q^{n-1}\sqrt{[n]_q}|n-1\rangle$$
$$\widehat{\beta}^*|n\rangle = \sqrt{[n+1]_q}|n+1\rangle \tag{2.6.92}$$
$$\widehat{N}|n\rangle = n|n\rangle.$$

Since $[k+1]_q = 0$, this Fock space is k dimensional.

There are two possibilities for an explicit realization of (2.6.92). One can construct the Bargmann–Fock representation in the space of anti-holomorphic functions with the basis

$$\psi_n = \frac{\bar{z}^n}{\sqrt{[n]_q!}}$$

with either *commuting* $\bar{z}, z$ variables or *non-commuting* ones. In both cases the variables must satisfy the condition of nilpotence, $z^{k+1} = \bar{z}^{k+1} = 0$, to provide the finiteness of the Fock space.

◇ The path integral for the Grassmann-like $q^2 = -1$ case

We call this case Grassmann-*like* because, although operators $\widehat{\beta}$ and $\widehat{\beta}^*$ satisfy anticommutation relations, we use either commuting variables or variables with non-Grassmann algebras on different time slices to construct the path integrals.

Physically these two possibilities correspond to the different forms of the commutation relations (2.6.89) and (2.6.90), after recovering the Planck constant $\hbar$. For the $q^2 = -1$ case and in the representation (2.6.92), both commutation relations (2.6.89) and (2.6.90) become the same:

$$\widehat{\beta}\widehat{\beta}^* + \widehat{\beta}^*\widehat{\beta} = 1. \tag{2.6.93}$$

In (2.6.89), (2.6.90) and (2.6.93) we have purposely dropped an explicit indication of the Planck constant. The point is that this can be recovered in two different ways. One way is to present (2.6.93) in the form

$$[\widehat{\beta}, \widehat{\beta}^*] = \hbar(1 - 2\gamma \widehat{\beta}^*\widehat{\beta}) \tag{2.6.94}$$

where γ is a dimensionful constant, such that $\hbar\gamma = 1$. This commutation relation corresponds in the limit $\hbar \to 0$ to a curved phase-space dynamics, with the Poisson bracket

$$\{z, \bar{z}\}_P = i(1 - 2\gamma\bar{z}z) \tag{2.6.95}$$

where $z, \bar{z}$ are the classical counterparts of $\widehat{\beta}, \widehat{\beta}^*$. The evolution operator corresponding to such quantization must be expressed with the help of q-path integrals with commuting variables.

The key observation for the development of path-integral representations is that the commutation relation written in the form (2.6.89) does not lead unavoidably to Bargmann–Fock variables with the commutation relation $z\bar{z} = q^2\bar{z}z$, but only to the commutation relation

$$\bar{\partial}\bar{z} - q^2\bar{z}\bar{\partial} = 1. \tag{2.6.96}$$

Thus, for the curved phase space (2.6.94), we can choose the *commuting* complex variables $z\bar{z} = \bar{z}z$, but with the following non-trivial commutation relations:

$$\bar{\partial}\bar{z} + \bar{z}\bar{\partial} = 1 \qquad d\bar{z}\bar{\partial} = -\bar{\partial}d\bar{z} \qquad \bar{z}d\bar{z} = -(d\bar{z})\bar{z} \tag{2.6.97}$$

and their conjugated counterparts. The creation $\widehat{\beta}^* = \bar{z}$ and annihilation $\widehat{\beta} = \bar{\partial}$ operators act in the Bargmann–Fock space with the basis $\{\psi_0 = 1, \psi_1 = \bar{z}\}$ and the scalar product

$$\int d\bar{z}\, dz\, e^{\bar{z}z} \bar{\psi}_n \psi_m = \delta_{nm}$$

where the integral is defined by the usual Berezin rules:

$$\int d\bar{z}\,\bar{z} = \int dz\, z = 1 \qquad \int d\bar{z} = \int dz = 0.$$

This method of path-integral construction is analogous to that in the case of the usual Grassmann path integral which we considered in the preceding subsection. As a result, the evolution operator kernel takes the form

$$U(t - t_0) = \int \left(\prod_\tau \frac{d\bar{z}(\tau)\, dz(\tau)}{1 - 2\bar{z}(\tau)z(\tau)}\right) \exp\left\{\bar{z}(t)z(t) - \int_{t_0}^t d\tau\, (\bar{z}(\tau)\dot{z}(\tau) + iH(\bar{z}(\tau)z(\tau)))\right\}.$$

Note that the integral measure corresponds to the form of the non-trivial Poisson bracket (2.6.95).

If, after recovering the Planck constant, the commutation relation (2.6.93) takes the form

$$\widehat{\beta}\widehat{\beta}^* + \widehat{\beta}^*\widehat{\beta} = \hbar$$

then it corresponds to the Grassmann phase space (in the limit $\hbar \to 0$), with anticommuting variables. In this case, the Bargmann–Fock variables seem to be identical to the usual Grassmann path integral. But there is one subtlety. In the usual construction, not only $z(t_i)$ and $\bar{z}(t_j)$ anticommute for any time slices t_i, t_j, but the same variables on different time slices also anticommute, e.g., $z(t_i)z(t_j) + z(t_j)z(t_i) = 0$. Such commutation relations cannot be generalized to other roots of unity of the parameter q. So we introduce another commutation relation on different time slices, valid for all $i \neq j$:

$$\begin{aligned} z(t_i)\bar{z}(t_j) &= -\bar{z}(t_j)z(t_i) & z^2(t_i) &= \bar{z}^2(t_i) = 0 \\ \bar{z}(t_i)\bar{z}(t_j) &= \bar{z}(t_j)\bar{z}(t_i) & z(t_i)z(t_j) &= z(t_j)z(t_i). \end{aligned} \quad (2.6.98)$$

We can check that, for such commutation relations, all the ingredients of path-integral construction (scalar product measure, relation between normal symbol and kernel of operators etc) remain the same as in the usual Grassmann path integral. Therefore, the path-integral representation of the evolution operator also has exactly the same form.

◇ The case $q^3 = -1$

The commutation relations (2.6.98) can easily be generalized to other values of the deformation parameter, for example, for $q^3 = -1$ we have

$$\begin{aligned} z(t_i)\bar{z}(t_j) &= q^2\bar{z}(t_j)z(t_i) & z^3(t_i) &= \bar{z}^3(t_i) = 0 \\ \bar{z}(t_i)\bar{z}(t_j) &= \bar{z}(t_j)\bar{z}(t_i) & z(t_i)z(t_j) &= z(t_j)z(t_i). \end{aligned} \quad (2.6.99)$$

In this case, the Bargmann–Fock representation is defined on the space of functions with the basis

$$\psi_0 = 1 \qquad \psi_1 = \bar{z} \qquad \psi_2 = \bar{z}^2 \quad (2.6.100)$$

which is orthonormal with respect to the scalar product

$$\int d\bar{z}\, dz\, \psi_n^\dagger(\bar{z})\mu(\bar{z}z)\psi_m(\bar{z}) = \delta_{nm} \quad (2.6.101)$$

where

$$\mu(\bar{z}z) = 1 + q^2\bar{z}z + q^2\bar{z}^2 z^2.$$

Here the integral is defined by the natural generalization of the Berezin rules for the Grassmann case:

$$\int d\bar{z}\, dz\, z^n \bar{z}^m = \delta_{nk}\delta_{mk}[k]_q! \qquad n, m = 0, 1, 2. \quad (2.6.102)$$

As a result of nilpotence, a general Hamiltonian has the form

$$\widehat{H} = \omega(u\widehat{\beta}^*\widehat{\beta} + v(\widehat{\beta}^*)^2\widehat{\beta}^2) \quad (2.6.103)$$

where the constants u, v are restricted by the hermiticity condition $\widehat{H}^\dagger = \widehat{H}$ (cf (2.6.91)) and can take three pairs of values: (i) $u = 1, v = -q$; (ii) $u = 1, v = 1 - 2q$; iii) $u = 0, v = -q^2$.

As usual, the action of any operator $\widehat{A}$ in the Bargmann–Fock Hilbert space can be represented with the help of its kernel K_A:

$$(\widehat{A}f)(\bar{z}_1) = \int d\bar{z}_2\, dz_2\, K_A(\bar{z}_1, z_2) f(\bar{z}_2) \qquad (2.6.104)$$

where

$$K_A(\bar{z}_1, z_2) = \sum_{m,n=0}^{2} K_{mn} \bar{z}_1^m z_2^n. \qquad (2.6.105)$$

Here a further pair of q-commuting coordinates is introduced, the commutation relations for different pairs being defined by (2.6.99). Now we can express K_{mn} through the scalar product

$$K_{mn} = \langle \psi_m | A | \psi_n \rangle \qquad (2.6.106)$$

and find a kernel of any operator by direct calculation. We consider the usual Schrödinger equation

$$i\hbar \frac{d}{dt} \psi(\bar{z}, t) = \widehat{H}(\widehat{\beta}^*, \widehat{\beta}) \psi(\bar{z}, t) \qquad (2.6.107)$$

with Hamiltonian (2.6.103). The integral kernel for the infinitesimal operator

$$\widehat{U} \approx 1 - \frac{i}{\hbar} \widehat{H} \varepsilon$$

takes the form

$$U(\bar{z}, z; \varepsilon) \approx g(\bar{z}z) \exp\left\{-\frac{i}{\hbar} H_{\text{eff}} \varepsilon\right\} \qquad (2.6.108)$$

where $g(\bar{z}z)$ is the kernel of the identity operator

$$g(\bar{z}z) = \sum_n \psi_n(\bar{z}) \psi_n^\dagger(\bar{z}) = 1 + \bar{z}z + \bar{z}^2 z^2$$

and the effective Hamiltonian H_{eff} is defined by the relation

$$H_{\text{eff}}(\bar{z}z) = g^{-1}(\bar{z}z) H(\bar{z}z) \qquad (2.6.109)$$

where $H(\bar{z}z)$ is the kernel of the initial Hamiltonian. Using nilpotence we obtain from (2.6.109):

$$H_{\text{eff}}(\bar{z}z) = \omega(u\bar{z}z + q(u + v - qu)\bar{z}^2 z^2).$$

Now we can write the convolution of N infinitesimal evolution operator kernels:

$$U(\bar{z}_N z_{N-1}) * U(\bar{z}_{N-1} z_{N-2}) * \ldots * U(\bar{z}_1 z_0)$$
$$= \int d\bar{z}_{N-1}\, dz_{N-1} \ldots d\bar{z}_1\, dz_1\, \mu(\bar{z}_{N-1} z_{N-1}) \ldots \mu(\bar{z}_1 z_1)$$
$$\times g(\bar{z}_N z_{N-1}) \ldots g(\bar{z}_1 z_0) e^{-iH_{\text{eff}}(\bar{z}_N z_{N-1})\varepsilon} \ldots e^{-iH_{\text{eff}}(\bar{z}_1 z_0)\varepsilon}. \qquad (2.6.110)$$

Due to nilpotence, it is possible to convert the functions g and μ into exponentials and in the continuous limit $\varepsilon \to 0$, we obtain the path integral for the case under consideration (i.e. $q^3 = -1$):

$$U(t - t_0) = \int \left(\prod_\tau d\bar{z}(\tau)\, dz(\tau)\, (1 + q\bar{z}(\tau)z(\tau)) \right) (1 + \bar{z}(t)z(t) + \bar{z}^2(t)z^2(t))$$
$$\times \exp\left\{ -\int_{t_0}^t d\tau\, [(1 + (1+2q)\bar{z}z)\bar{z}(\tau)\dot{z}(\tau) + iH_{\text{eff}}(\bar{z}(\tau)z(\tau))] \right\}. \qquad (2.6.111)$$

The kernels U in equation (2.6.111) correspond to the product of unitary operators up to the $(\varepsilon)^2$-terms, with the subsequent limit $\varepsilon \to 0$. Therefore, these kernels seem to be the kernels of unitary evolution operators. However, the integrands of these path integrals do not have the form $\exp\{iS/\hbar\}$, where S is a real functional. This is quite unusual and prevents us from an ordinary interpretation of the path integral. The obvious reason for this lies with the commutation relations (2.6.99) for the z and $\bar{z}$ coordinates which contain the complex parameter q.

Let us consider another possibility for $q^3 = -1$. Namely, we now introduce *commuting* variables, with the commutation relations (2.6.96) for variables and derivatives in analogy with the Grassmann-like case. The basis of the Bargmann–Fock representation has the same form (2.6.100) as in the case of non-commuting variables and is orthonormal with respect to the scalar product (2.6.101) with the measure $\mu = 1 + \bar{z}z + \bar{z}^2 z^2$, the integral being defined by (2.6.102) (but now with commuting variables). The derivation of the path integral is essentially the same as in the case of non-commuting variables and the result is similar to (2.6.111), the effective Hamiltonian again being defined by the general formula (2.6.109), but now for all Hermitian Hamiltonians of the form (2.6.103), H_{eff} proves to be a real function: (i) $H_{\text{eff}} = \omega \bar{z}z$ for $u = 1$, $v = -q$; (ii) $H_{\text{eff}} = \omega(\bar{z}z - \bar{z}^2 z^2)$ for $u = 1$, $v = 1 - 2q$; (iii) $H_{\text{eff}} = \omega \bar{z}^2 z^2$ for $u = 0$, $v = -q^2$.

Thus in the case of commuting variables $\bar{z}$ and z, the integrand of the path integral has the usual form $\exp\{iS/\hbar\}$ with the *real* functional S.

For definiteness, we present the path integral for Hamiltonian (2.6.109) with $u = 1$ and $v = -q$:

$$U(t - t_0) = \int \left[\prod_\tau \frac{d\bar{z}(\tau)dz(\tau)}{\hbar}\left(1 + 2\frac{\bar{z}(\tau)z(\tau)}{\hbar} + 3\frac{\bar{z}^2(\tau)z^2(\tau)}{\hbar^2}\right)\right]$$
$$\times \left(1 + \frac{\bar{z}(t)z(t)}{\hbar} + \frac{\bar{z}^2(t)z^2(t)}{\hbar^2}\right)$$
$$\times \exp\left\{-\frac{1}{\hbar}\int_{t_0}^t d\tau\left[\left(1 + \frac{\bar{z}z}{\hbar}\right)\bar{z}(\tau)\dot{z}(\tau) + i\omega\bar{z}(\tau)z(\tau)\right]\right\}. \quad (2.6.112)$$

We can see that there is simply no semiclassical approximation corresponding to the limit $\hbar \to 0$ (since the terms in the measure and the action diverge). This is an expected result because, as is well known, the semiclassical approximation for the spinlike systems considered here corresponds to the limit $j \to \infty$, where j is the spin of the system (Berezin 1975). Systems with fixed spins (numbers of states) have no (quasi)classical limit, as is once again confirmed by the expression (2.6.112).

The fact that creation and annihilation operators $\widehat{\beta}^*$, $\widehat{\beta}$ are not Hermitian conjugate (see (2.6.91)) is quite unusual and technically inconvenient (but since they are non-conjugate only up to a purely complex phase, this does not cause any real problems). We can try to use the Hermitian conjugate operators $\widehat{\alpha}^\dagger$, $\widehat{\alpha}$ with the commutation relations (2.6.87) for path-integral construction, noting that relation (2.6.89) is, in fact, a polynomial one due to nilpotence. In particular, in the $q^3 = -1$ case and in the Fock space representation, relation (2.6.89) is equivalent to the commutation relation

$$\widehat{\alpha}\widehat{\alpha}^\dagger + (\widehat{\alpha}^\dagger)^2 \widehat{\alpha}^2 = 1. \quad (2.6.113)$$

This commutation relation looks like a natural generalization of the usual relation (2.6.93) for fermionic operators. We can show that, in general, the algebra of operators

$$\widehat{\alpha}\widehat{\alpha}^\dagger + (\widehat{\alpha}^\dagger)^n \widehat{\alpha}^n = 1$$

has an $(n + 1)$-dimensional Fock representation (Chaichian and Demichev 1996). Using a specific differential calculus, we can construct the Bargmann–Fock representation for (2.6.113) and the corresponding path integral (see problem 2.6.11).

304 *Path integrals in quantum mechanics*

◇ **Higher roots case**

Since there are no universal functions in the case of the root-of-unity q-parameter which would connect operator kernels and normal symbols, there are no universal expressions for the path integrals. However, for any concrete value of the deformation parameter, the path integrals can be derived in the way considered earlier. For example, for the next value, $q^4 = -1$, and for non-commuting variables, we have

$$U(t - t_0) = \int \left(\prod_\tau d\bar{z}(\tau) \, dz(\tau) \, [1 + (1+i)\bar{z}(\tau)z(\tau) \right.$$
$$\left. + (2c - 1)\bar{z}^2(\tau)z^2(\tau) + i(1 - c)\bar{z}^3(\tau)z^3(\tau)] \right)$$
$$\times \exp\{\bar{z}(t)z(t) + (c - i/2)\bar{z}^2(t)z^2(t) + (2c - i/3)\bar{z}^3(t)z^3(t)\}$$
$$\times \exp\left\{ -\int_{t_0}^{t} d\tau \, [(1 - (1/2 + ic)\bar{z}z - 3(2c - i/2)\bar{z}^2 z^2)\bar{z}(\tau)\dot{z}(\tau) + iH_{\text{eff}}(\bar{z}(\tau)z(\tau))] \right\}$$
(2.6.114)

where, for brevity, $\hbar = 1$, $c \equiv 2^{-1/4} = 1/\sqrt{[2]_q}$ and H_{eff} is defined again by relation (2.6.109) but with the following explicit form of the kernel $g(\bar{z}z)$:

$$g(\bar{z}z) = 1 + \bar{z}z + c\bar{z}^2 z^2 + c\bar{z}^3 z^3.$$

Analogous expressions for q-path integrals with a different choice of variables can be derived for algebras which correspond to Fock spaces of an arbitrary dimension (i.e. for a q satisfying the condition $q^p = 1$ with an arbitrary integer p).

We note that the essential distinction of all *generalized* Grassmann path integrals is that they have non-trivial integral measures and non-Gaussian integrands.

A q-deformed path integral can also be derived for *real* values of the parameter q entering the modified commutation relations for creation and annihilation operators (Chaichian and Demichev 1994, 1996).

2.6.3 Localization techniques for the calculation of a certain class of path integrals

As we have already learned, there are only a few cases where an exact solution for path integrals can be obtained, most of them being different variations of Gaussian integrals (or integrals which can be reduced to the Gaussian form via an appropriate change of the spacetime variables). This situation has prompted the development and wide use of different approximate methods, e.g., perturbation theory in coupling constants or the semiclassical WKB approximation. We know that for a Gaussian path integral the semiclassical approximation gives an exact result. Then a natural question arises about the existence of a more general class of systems for which the semiclassical approximation would also give *an exact result*, that is, one for which the path integral is *localized* around the classical solution (more precisely, around extremal points, not only minima).

The idea of calculating path integrals by localizing them to sums or finite-dimensional integrals was instigated by the theorem by Duistermaat and Heckman (1982, 1983) dealing with finite phase-space integrals in classical mechanics. They proved that the partition function of a classical system expressed as an integral over the phase space is localized to critical points of the Hamiltonian function, under certain geometrical conditions. A natural generalization was to formulate these ideas for infinite-dimensional path integrals (Atiyah 1985, Blau *et al* 1990, Witten 1992, Niemi and Tirkkonen 1994). Instead of the critical points of the Hamiltonian, the path integral would be localized to the solutions of classical equations of motion determined by the extrema of the action.

Here we present, in short, the essence of these ideas dropping almost all (rather involved) proofs and the many beautiful geometrical constructions behind the general properties of systems with localizing integrals (for details, see the previously cited literature). The aim of this subsection is to give the reader a very preliminary idea about the subject.

◇ A short glance at finite-dimensional symplectic geometry

We start by recalling the basic notions of the symplectic geometry of a classical phase space Γ (see, e.g., Guillemin and Sternberg (1984)). The phase space Γ is described by a generic coordinate system z^a ($a = 1, \ldots, 2n = \dim(\Gamma)$). In these coordinates, the fundamental Poisson bracket is

$$\{z^a, z^b\} = \omega^{ab}(z) \tag{2.6.115}$$

and the inverse matrix ω_{ab}

$$\omega^{ac}\omega_{cb} = \delta^a_b \tag{2.6.116}$$

determines the components of the *symplectic two-form* on the phase space Γ,

$$\omega = \tfrac{1}{2}\omega_{ab}\, dz^a \wedge dz^b. \tag{2.6.117}$$

This symplectic two-form is closed:

$$d\omega = 0 \tag{2.6.118}$$

or, in components,

$$\partial_a \omega_{bc} + \partial_b \omega_{ca} + \partial_c \omega_{ab} = 0 \tag{2.6.119}$$

which is equivalent to the *Jacobi identity* for the Poisson bracket (2.6.115).

From (2.6.118), we conclude that ω can be represented locally as an exterior derivative of a one-form $\vartheta_a(z)$:

$$\omega = d\vartheta = \partial_a \vartheta_b\, dz^a \wedge dz^b \tag{2.6.120}$$

and smooth, real-valued functions $\psi(z)$ on Γ define diffeomorphisms that leave ω invariant: if we introduce a change of variables $z^a \to \tilde{z}^a$, such that

$$\vartheta_a\, dz^a = \vartheta \xrightarrow{\psi} \vartheta + d\psi = (\vartheta_a + \partial_a \psi)\, dz^a = \tilde{\vartheta}_a d\tilde{z}^a \tag{2.6.121}$$

we conclude from $d^2 = 0$ that ω remains intact,

$$\omega \xrightarrow{\psi} \tilde{\omega} \equiv \omega. \tag{2.6.122}$$

The change of variables (2.6.121) determines a *canonical transformation* on Γ and ψ is the generating function of this transformation. Indeed, Darboux's theorem states that locally in a neighborhood on Γ, we can always introduce a change of variables $z^a \to p_a, q^a$, such that ω becomes

$$\omega = dp_a \wedge dq^a \tag{2.6.123}$$

where p_a and q^a are canonical momentum and position variables on Γ. In these variables the symplectic one-form becomes

$$\vartheta = p_a\, dq^a \tag{2.6.124}$$

306   *Path integrals in quantum mechanics*

and (2.6.121) becomes
$$p_a dq^a = \vartheta \xrightarrow{\psi} \vartheta + d\psi = \tilde{\vartheta} = P_a dQ^a. \tag{2.6.125}$$

Consequently,
$$p_a dq^a - P_a dQ^a = d\psi \tag{2.6.126}$$

where both p_a, q^a and P_a, Q^a are canonical momentum and coordinate variables on Γ. This is the standard form of a canonical transformation determined by the generating functional ψ.

The exterior products of ω determine closed $2k$-forms on Γ. The $2n$-form (recall that $\dim(\Gamma) = 2n$)
$$\omega^n = \omega \wedge \cdots \wedge \omega \quad (n \text{ times}) \tag{2.6.127}$$

defines a natural volume element on Γ which is invariant under the canonical transformations (2.6.121). This is the *Liouville measure* and in the local Darboux coordinates (2.6.123), it becomes the familiar one:
$$\left\{\frac{1}{n!}(-)^{n(n-1)/2}\right\} \omega^n = dp_1 \wedge \cdots \wedge dp_n \wedge dq^1 \wedge \cdots \wedge dq^n. \tag{2.6.128}$$

Smooth, real-valued functions F on Γ are called classical observables. The symplectic two-form associates with the exterior derivative dF of a classical observable F a *Hamiltonian vector field* $\mathcal{X}_F$ by
$$\mathcal{X}_F^a = \omega^{ab} \partial_b F. \tag{2.6.129}$$

The Poisson bracket of two classical observables F and G can be expressed in terms of the corresponding vector fields:
$$\{F, G\} = \omega^{ab} \partial_a F \partial_b G = \mathcal{X}_F^a \partial_a G = \omega_{ab} \mathcal{X}_F^a \mathcal{X}_G^b = \omega(\mathcal{X}_F, \mathcal{X}_G). \tag{2.6.130}$$

This determines the internal multiplication i_F of the one-form dG by the vector field $\mathcal{X}_F$:
$$i_F dG = \mathcal{X}_F^a \partial_a G. \tag{2.6.131}$$

More generally, internal multiplication i_F by a vector field $\mathcal{X}_F$ is a nilpotent operation,
$$i_F^2 = 0 \tag{2.6.132}$$

which is defined on the exterior algebra Λ of the phase space Γ that maps the space Λ_k of k-forms to the space Λ_{k-1} of $(k-1)$-forms. Using d and i_F, we introduce the *equivariant exterior derivative*
$$d_F \stackrel{\text{def}}{\equiv} d + i_F \tag{2.6.133}$$

defined on the exterior algebra Λ of Γ. Since d maps the subspace Λ_k of k-forms onto the subspace Λ_{k+1} of $(k+1)$-forms, (2.6.133) does not preserve the form degree but maps an even form onto an odd form, and vice versa. Hence it can be viewed as a *supersymmetry operator* (cf supplement VI, volume II). The corresponding supersymmetry algebra closes to the *Lie derivative* $\mathcal{L}_F$ along the Hamiltonian vector field $\mathcal{X}_F$,
$$d_F^2 = d i_F + i_F d = \mathcal{L}_F \tag{2.6.134}$$

and the Poisson bracket (2.6.130) coincides with the Lie derivative of G along the Hamiltonian vector field $\mathcal{X}_F$:
$$\{F, G\} = \omega^{ab} \partial_a F \partial_b G = \mathcal{X}_F^a \partial_a G \equiv \mathcal{L}_F G. \tag{2.6.135}$$

The linear space of $\xi \in \Lambda$ which is annihilated by the Lie derivative (2.6.134),
$$\mathcal{L}_F \xi = 0 \tag{2.6.136}$$

determines an invariant subspace Λ_{inv} of the exterior algebra Λ which is mapped onto itself by the equivariant exterior derivative d_F. The restriction of d_F on this subspace is nilpotent, $d_F^2 = 0$, hence d_F determines a conventional exterior derivative in this subspace.

◇ **The Duistermaat–Heckman theorem**

Let us consider an oscillatory integral of the kind

$$I(t) \equiv \left(\frac{t}{2}\right)^n \int_{\mathcal{M}} d\mu \, e^{itf} \qquad (2.6.137)$$

over a $(2n)$-dimensional manifold $\mathcal{M}$ with an integration measure $d\mu$. If $\mathcal{M}$ is a Riemannian manifold, under rather mild hypotheses, namely that the function f is a *Morse function*, i.e. that the Hessian matrix of the function f is non-singular in all critical points of f:

$$\det \operatorname{Hess}_P(f) \neq 0 \qquad \text{if } \partial_a f(P) = 0, \; a = 1, \ldots, 2n.$$

We recall that the Hessian of a function $f(x)$, defined on a d-dimensional space is the matrix

$$\operatorname{Hess}(f(x)) \stackrel{\text{def}}{\equiv} \frac{\partial^2 f}{\partial x^a \partial x^b} \qquad a, b = 1, \ldots, d.$$

Then it is possible to show (see, e.g., Guillemin and Sternberg (1984)) that, for large values of the parameter t, we have

$$I(t) = \sum_P c_P e^{itf(P)} + \mathcal{O}(t^{-1}) \qquad (2.6.138)$$

where the sum ranges over all critical points of f and the coefficients are given in terms of the determinant of the Gaussian fluctuations of f around the critical points:

$$c_P = \exp[i\tfrac{1}{4} \operatorname{sgn} \operatorname{Hess}_P(f)][\det \operatorname{Hess}_P(f)]^{-\frac{1}{2}}. \qquad (2.6.139)$$

Here the signature $\operatorname{sgn} \mathsf{A}$ of a symmetric real-valued non-singular matrix A is defined as the number of its positive eigenvalues minus the number of its negative eigenvalues.

Of course, the remainder term $\mathcal{O}(t^{-1})$ vanishes identically if $\mathcal{M}$ is the linear manifold $\mathbb{R}^{2n}$ with the standard integration measure $\sigma = dx_1 \cdots dx_{2n}$ and the function f has a quadratic form (i.e. for Gaussian integrals): $f = \tfrac{1}{2}\mathsf{Q}x \cdot x - \xi \cdot x$, Q being any symmetric real-valued $(2n)$-dimensional non-singular matrix. In this case the only critical point of f is $x_0 = \mathsf{Q}^{-1}\xi$ and $\operatorname{Hess}_{x_0}(f) = \mathsf{Q}$, so that (2.6.138) with $\mathcal{O}(t^{-1}) \equiv 0$ simply gives the formula for a Gaussian integral.

The Duistermaat–Heckman theorem establishes the conditions under which an integral of the form (2.6.137) can be *exactly* evaluated in the stationary-phase approximation (all the remainder terms $\mathcal{O}(t^{-1})$ in (2.6.138) vanish).

Theorem 2.5 (Duistermaat–Heckman). Consider an integral of the form (2.6.137)

$$\mathcal{Z} = \int_{\mathcal{M}} \omega^n \frac{1}{n!} e^{itH} \qquad (2.6.140)$$

(a characteristic functional for a classical system in which the dynamics is governed by the Hamiltonian H), where the integration measure is defined by the volume form ω^n (cf (2.6.127)). If

- the Hamiltonian vector field $\chi_H = \omega^{kj} \partial_j H \partial_k$ is a Killing vector field with respect to some metric g: $\mathcal{L}_H g = 0$; and
- χ_H is the fundamental vector associated with an element of the Lie algebra $\mathfrak{g}$ of a *compact* Lie group $\mathfrak{G}$ acting on the manifold $\mathcal{M}$

then the corrections to the stationary-phase approximation (cf (2.6.138)) vanish:

$$\mathcal{Z} = \sum_P \frac{e^{it H(P)}}{(-it)^n D(P)}$$

where P are the critical points of the Hamiltonian (i.e. $dH(P) = 0$) and $D(P)$ is the product of the non-zero eigenvalues of the Hessian, Hess $H = \det(\partial_i \partial_j H)$, at the critical points.

◇ Path integrals in a phase space with an arbitrary symplectic form and their interpretation in terms of the symplectic geometry of a loop space

In the case of an arbitrary non-constant symplectic form $\omega_{ab}(z)$ and, hence, non-constant right-hand side of the Poisson brackets (2.6.115), we do not have at our disposal all the necessary ingredients to construct path integrals from the operator formulation of quantum mechanics. For example, we do not have the *explicit* form of a complete basis of states for the algebra which appears after the quantization (substitution $\{\cdot,\cdot\} \to (i/\hbar)[\cdot,\cdot]$), or explicit formulae for the star-product of the operator symbols. Such formulae can be found only after specifying an explicit form of $\omega_{ab}(z)$. Therefore, we shall not derive, but just *postulate* the form of the path integral in the phase space with an arbitrary symplectic form using, as a guiding principle, the general covariance and correspondence with the standard constant symplectic form (2.6.123). These requirements imply the following path-integral representation for the generating functional:

$$\mathcal{Z} = \int \mathcal{D}z^a \prod_\tau \sqrt{\det(\omega_{ab})} \exp\left\{i \int_0^t d\tau \, [\vartheta_a \dot{z}^a - H(z)]\right\} \qquad (2.6.141)$$

(recall that $\omega^n = \det(\omega_{ab}) \, d^{2n}z$ is the volume form and that the one-form ϑ has in the Darboux coordinates the canonical form (2.6.124); for brevity, we put the Planck constant equal to unity, $\hbar = 1$). We shall argue that, if the Hamiltonian $H(z)$ satisfies a certain condition which we shall specify in the following, the path integral (2.6.141) can be evaluated exactly in the sense that it reduces to an ordinary integral over the classical phase space Γ.

In this subsection, we shall consider periodic boundary conditions for the trajectories: $z^a(0) = z^a(T)$ (i.e. we restrict our consideration only to the trace of the evolution operator). Then the appropriate geometrical interpretation of (2.6.141) is, in terms of symplectic geometry, in a *canonical loop space* $L\Gamma$ over the classical phase space Γ. The symplectic geometry of $L\Gamma$ is constructed from the symplectic geometry of Γ.

The loop space $L\Gamma$ is parametrized by the time evolution $z^a \to z^a(\tau)$, with the periodic boundary conditions $z^a(0) = z^a(t)$. The exterior derivative in $L\Gamma$ is obtained by lifting the exterior derivative of the phase space Γ,

$$d = \int_0^t d\tau \, dz^a(\tau) \frac{\delta}{\delta z^a(\tau)} \equiv dz^a \frac{\delta}{\delta z^a} \qquad (2.6.142)$$

where $dz^a(\tau)$ denotes a basis of loop-space one-forms, obtained by lifting a basis of one-forms in the phase space Γ to the loop space.

The loop-space symplectic geometry is determined by a loop-space symplectic two-form

$$\Omega = \int d\tau \, d\tau' \, \tfrac{1}{2} \Omega_{ab}(\tau, \tau') \, dz^a(\tau) \wedge dz^b(\tau'). \qquad (2.6.143)$$

This is a closed two-form in the loop space,

$$d\Omega = 0 \qquad (2.6.144)$$

or, in local coordinates $z^a(\tau)$,

$$\frac{\delta}{\delta z^a}\Omega_{bc} + \frac{\delta}{\delta z^b}\Omega_{ca} + \frac{\delta}{\delta z^c}\Omega_{ab} = 0. \qquad (2.6.145)$$

Hence, we can locally represent Ω as an exterior derivative of a loop-space one-form,

$$\Omega = d\Theta \qquad (2.6.146)$$

where

$$\Theta = \int d\tau\, \Theta_a(\tau)\, dz^a(\tau). \qquad (2.6.147)$$

We shall assume that (2.6.143) is non-degenerate, i.e. the matrix $\Omega_{ab}(\tau, \tau')$ can be inverted in the loop space. Examples of such non-degenerate two-forms are obtained by lifting the symplectic two-forms $\omega_{ab}(z)$ from the original phase space Γ to the loop space:

$$\Omega_{ab}(\tau, \tau') = \omega_{ab}[z(\tau)]\delta(\tau - \tau'). \qquad (2.6.148)$$

Similarly, other quantities can be lifted from the original phase space to the loop space.

In particular, we define the loop-space canonical transformations as those loop-space changes of variables that leave Ω invariant. These transformations have the form

$$\Theta \xrightarrow{\Psi} \tilde{\Theta} = \Theta + d\Psi \qquad (2.6.149)$$

with $\Psi[z(\tau)]$ being the generating functional of the canonical transformation.

The exterior products of Ω determine canonically invariant closed forms on $L\Gamma$, and the top form yields a natural volume element, the loop-space Liouville measure. We are particularly interested in the corresponding integrals which are of the form

$$\mathcal{Z} = \int \mathcal{D}z^a \sqrt{\det(\Omega_{ab})}\, \exp\{iS_B\} \qquad (2.6.150)$$

$S_B(z)$ being a loop-space observable, i.e. a functional on $L\Gamma$. If we specify (2.6.148) and identify S_B with the action in (2.6.141), we can interpret (2.6.141) as an example of such a loop-space integral.

We exponentiate the determinant in (2.6.150) using the *anticommuting* variables $c^a(t)$,

$$\mathcal{Z} = \int \mathcal{D}z^a \mathcal{D}c^a\, \exp\{iS_B + ic^b \Omega_{bd} c^d\} = \int \mathcal{D}z^a \mathcal{D}c^a\, \exp\{iS_B + iS_F\} \qquad (2.6.151)$$

(cf (2.6.67)). This integral is invariant under a type of supersymmetry transformation (cf supplement VI, volume II): if $\mathcal{X}_S^a$ is the loop-space Hamiltonian vector field determined by the functional S_B,

$$\frac{\delta S_B}{\delta z^a} = \Omega_{ab}\mathcal{X}_S^b \qquad (2.6.152)$$

the supersymmetry transformation is defined as follows:

$$d_S z^a = c^a \qquad (2.6.153)$$
$$d_S c^a = -\mathcal{X}_S^a. \qquad (2.6.154)$$

Let us interpret $c^a(t)$ as the loop-space one-forms $dz^a(t) \sim c^a(t)$, and introduce the loop-space equivariant exterior derivative

$$d_S = d + i_S. \qquad (2.6.155)$$

Here i_S denotes the contraction along the Hamiltonian vector field $\mathcal{X}_S^a$ (cf (2.6.131))

$$i_S = \mathcal{X}_S^a i_a \qquad (2.6.156)$$

and $i_a(t)$ is the basis of loop-space contractions which is dual to $c^a(t)$:

$$i_a(t)c^b(t') = \delta_a^b(t-t'). \qquad (2.6.157)$$

Again, (2.6.155) fails to be nilpotent (cf (2.6.134)) and its square determines the loop-space Lie derivative along $\mathcal{X}_S^a$,

$$d_S^2 = di_S + i_S d = \mathcal{L}_S. \qquad (2.6.158)$$

The action in (2.6.151) is a linear combination of a loop-space zero-form (S_B) and a two-form (S_F). The supersymmetry (2.6.153), (2.6.154) means that it is equivariantly closed in the loop space, that is

$$d_S(S_B + S_F) = 0. \qquad (2.6.159)$$

Hence the action can be represented *locally* as an equivariant exterior derivative of a one-form $\hat{\Theta}$,

$$S_B + S_F = (d + i_S)\hat{\Theta} = \hat{\Theta}_a \mathcal{X}_S^a + c^a \Omega_{ab} c^b \qquad (2.6.160)$$

and the supersymmetry (2.6.159) implies that

$$d_S^2 \hat{\Theta} = (di_S + i_S d)\hat{\Theta} = \mathcal{L}_S \hat{\Theta} = 0 \qquad (2.6.161)$$

so that $\hat{\Theta}$ is in the subspace where d_S is nilpotent.

◇ **Infinite-dimensional (loop-space) generalization of the Duistermaat–Heckman integration formula**

The supersymmetry (2.6.159) can be used to derive a loop-space generalization of the Duistermaat–Heckman integration formula:

$$\mathcal{Z} = \int \mathcal{D}z^a \sqrt{\det(\Omega)} e^{iS_B} = \sum_{\delta S_B = 0} \frac{\sqrt{\det(\Omega)}}{\sqrt{\det(\delta^2 S_B)}} e^{iS_B}. \qquad (2.6.162)$$

Here the sum on the right-hand side is taken over all critical points of the action S_B, i.e. over the zeros of the Hamiltonian vector field $\mathcal{X}_S^a$. The derivation of (2.6.162) assumes that the loop space admits a Riemannian structure with a *globally defined* loop-space metric tensor $G_{ab}(z; \tau, \tau')$ which is Lie-derived by the vector field $\mathcal{X}_S^a$,

$$\mathcal{L}_S G = 0 \qquad (2.6.163)$$

or, in component form,

$$\partial_a \mathcal{X}_S^b G_{bc} + \partial_c \mathcal{X}_S^b G_{ba} + \mathcal{X}_S^b \partial_b G_{ac} = 0. \qquad (2.6.164)$$

For a *compact* phase space, the corresponding condition would mean that the canonical flow generated by the Hamiltonian vector field $\mathcal{X}_S^a$ corresponds to the global action of a circle $S^{(1)} \sim U(1)$ on the phase space Γ. We assume that this is also the relevant case in the loop space. The circle $S^{(1)}$ is parametrized by a continuous parameter s, i.e. $z^a \to z^a(s)$ with $z^a(1) = z^a(0)$. Thus

$$\mathcal{X}_S^a(z[\tau]) = \left.\frac{\partial z^a(\tau; s)}{\partial s}\right|_{s=0} \qquad (2.6.165)$$

and if we also assume that we have selected the coordinates $z^a(t)$ so that the flow parameter τ shifts the loop (time) parameter $\tau \to \tau + s$, we obtain

$$\mathcal{X}_S^a(z[\tau]) = \frac{\partial z^a(\tau; s)}{\partial s}\bigg|_{s=0} = \frac{dz^a(\tau)}{d\tau} \equiv \dot{z}^a. \qquad (2.6.166)$$

The relation (2.6.152) then simplifies to

$$\frac{\delta S_B}{\delta z^a} = \Omega_{ab}(z)\dot{z}^b \qquad (2.6.167)$$

and the supersymmetry transformation (2.6.153), (2.6.154) becomes

$$d_{\dot{z}} z^a = c^a \qquad (2.6.168)$$
$$d_{\dot{z}} c^a = \dot{z}^a. \qquad (2.6.169)$$

Here

$$d_S \to d_{\dot{z}} = d + i_{\dot{z}} \qquad (2.6.170)$$

is the equivariant exterior derivative along the $S^{(1)}$-vector field $\mathcal{X}_S^a \to \dot{z}^a$, and the corresponding Lie derivative is simply

$$\mathcal{L}_{\dot{z}} = di_{\dot{z}} + i_{\dot{z}}d \sim \frac{d}{d\tau}. \qquad (2.6.171)$$

◇ Sketch of a proof of the generalized Duistermaat–Heckman integration formula (in loop space) for Hamiltonians which generate a circle action

We want to evaluate the path integral (2.6.141) for a Hamiltonian H that generates the global action of $S^{(1)} \sim U(1)$ on the classical phase space Γ. We shall first evaluate this path integral by interpreting it in the space of loops that are defined in the original phase space Γ. The relevant loop-space equivariant exterior derivative has the functional form

$$d + i_{\dot{z}} + i_H. \qquad (2.6.172)$$

As in (2.6.151), we introduce the anticommuting variables c^a and write (2.6.141) as

$$\mathcal{Z} = \int \mathcal{D}z^a\, \mathcal{D}c^a\, \exp\left\{i\int_0^t d\tau\, [\vartheta_a \dot{z}^a - H + \tfrac{1}{2} c^a \omega_{ab} c^b]\right\}. \qquad (2.6.173)$$

The loop-space Hamiltonian vector field that corresponds to the bosonic part of the action is

$$\mathcal{X}_S^a = \dot{z}^a - \omega^{ab}\partial_b H \qquad (2.6.174)$$

and we identify $c^a(t)$ as the basis of one-forms on this loop space. The corresponding loop-space equivariant exterior derivative then has the form (2.6.155):

$$d_S = d + i_S = c^a \partial_a + \mathcal{X}_S^a i_a = d + i_{\dot{z}} + i_H \qquad (2.6.175)$$

where i_a denotes again the basis for loop-space interior multiplication which is dual to the c^a, as in (2.6.157):

$$i_a(\tau) c^b(\tau') = \delta_a^b(\tau - \tau'). \qquad (2.6.176)$$

In order to evaluate (2.6.173) using the supersymmetry determined by (2.6.175), we introduce, at first, the following generalization of the path integral (2.6.173):

$$\mathcal{Z}_\xi = \int \mathcal{D}z^a \, \mathcal{D}c^a \, \exp\left\{i \int_0^t \delta\tau \, [\vartheta_b \dot{z}^b - H + \tfrac{1}{2} c^b \omega_{bd} c^d + d_S \xi]\right\}. \qquad (2.6.177)$$

Here ξ is an arbitrary one-form on the loop space. Using the supersymmetry determined by (2.6.175), we find that if we introduce a 'small' variation

$$\xi \to \xi + \delta\xi \qquad (2.6.178)$$

where $\delta\xi$ is a homotopically trivial element in the subspace defined by the condition

$$\mathcal{L}_S \delta\xi = 0 \qquad (2.6.179)$$

the path integral (2.6.177) is invariant under this variation:

$$\mathcal{Z}_\xi = \mathcal{Z}_{\xi + \delta\xi}. \qquad (2.6.180)$$

In particular, if ξ itself is a homotopically trivial element in the subspace

$$\mathcal{L}_S \xi = 0 \qquad (2.6.181)$$

we conclude that the path integral (2.6.177) is independent of ξ and coincides with the original path integral (2.6.141). The idea is then to evaluate (2.6.173) and (2.6.177) by selecting ξ in (2.6.177) properly, so that the path integral simplifies to the extent that it can be evaluated exactly.

Since the Hamiltonian H in (2.6.173) and (2.6.177) generates a global action of $S^{(1)}$, we conclude that the phase space Γ admits a Riemannian structure with a metric tensor g_{ab} which is Lie-derived by the Hamiltonian vector field $\mathcal{X}_H^a$:

$$\mathcal{L}_H g = 0 \qquad (2.6.182)$$

or, in component form,

$$\partial_a \mathcal{X}_H^c g_{cb} + \partial_b \mathcal{X}_H^c g_{ca} + \mathcal{X}_H^c \partial_c g_{ab} = 0. \qquad (2.6.183)$$

Locally, such a metric tensor always exists in domains where H does not have any critical points. To achieve this, it is sufficient to introduce local Darboux coordinates, so that the Hamiltonian coincides with one of the coordinates, say $H \sim p_1$. For g_{ab} we can select, e.g., $g_{ab} \sim \delta_{ab}$. However, since we require (2.6.182) to be valid *globally* on Γ, for a compact phase space, this is equivalent to the requirement that H generates the global action of $S^{(1)}$. We can construct such a metric tensor from an *arbitrary* metric tensor on Γ, by averaging it over the circle $S^{(1)} \sim U(1)$. Obviously, this metric tensor is not unique. For example, if g_{ab} satisfies the condition (2.6.182), the following one-parameter generalization of g_{ab} also satisfies (2.6.182):

$$g_{ab} \to g_{ab} + \mu \cdot g_{ac} \mathcal{X}_H^c \mathcal{X}_H^d g_{db}. \qquad (2.6.184)$$

If we select g_{ab} so that it satisfies (2.6.182), the following one-parameter family of functionals is in the subspace (2.6.181):

$$\xi_\lambda = \frac{\lambda}{2} g_{ab} \mathcal{X}_S^a c^b. \qquad (2.6.185)$$

If a variation of the parameter λ indeed determines a homotopically trivial variation (2.6.180), the corresponding path integral (2.6.177) is independent of λ and, for $\lambda \to 0$, it reduces to the original path integral (2.6.173). Consequently, (2.6.177) and (2.6.185) coincide with (2.6.173) for all values of λ, and the evaluation of (2.6.177) with ξ defined by (2.6.185) in the $\lambda \to \infty$ limit gives the path-integral version (2.6.162) of the Duistermaat–Heckman integration formula (see Keski-Vakkuri *et al* (1991) and Niemi and Tirkkonen (1994)).

◇ Hamiltonians which are generic functions of H

The localization formula for the evaluation of the path integral (2.6.141) proves to be valid (Niemi and Tirkkonen 1994) for a Hamiltonian which is quite a general (almost arbitrary, see later) function $P(H)$ of an observable H that generates the action of $S^{(1)}$ on the phase space Γ,

$$\mathcal{Z} = \int \mathcal{D}z^a \, \mathcal{D}c^a \, \exp\left\{i \int_0^t d\tau \, [\vartheta_b \dot{z}^b - P(H) + \tfrac{1}{2}c^b \omega_{bd} c^d]\right\}. \tag{2.6.186}$$

In order to evaluate (2.6.186), we first consider the quantity

$$\exp\left\{-i \int P(H)\right\}. \tag{2.6.187}$$

In fact, the generalization to a function of H is valid under the assumption that there exists another function $\phi(\xi)$, so that we can write (2.6.187) as a Gaussian path-integral transformation of $\phi(\xi)$:

$$\exp\left\{-i \int P(H)\right\} = \int \mathcal{D}\xi \, \exp\left\{i \int d\tau \, [\tfrac{1}{2}\xi^2 - \phi(\xi) H]\right\}. \tag{2.6.188}$$

Note that *locally* such a function $\phi(\xi)$ can always be constructed, but there might be obstructions to constructing $\phi(\xi)$ globally.

◇ Examples of the application of the localization formula and comparison with known results

Let us consider now the quantum mechanics of a spinning particle, described by the Hamiltonian

$$\hat{H} = \boldsymbol{B} \hat{\boldsymbol{J}} \tag{2.6.189}$$

where the spin operators $\hat{J}_j$ ($j = 1, 2, 3$) satisfy the usual commutation relations ($\hbar = 1$):

$$[\hat{J}_i, \hat{J}_j] = i\epsilon_{ijk} \hat{J}_k. \tag{2.6.190}$$

Thus we consider the quantization of spin, i.e. the path integral defined on the two-dimensional sphere S^2 (co-adjoint orbit of $SU(2)$). Choosing the magnetic field along the third axis, the Hamiltonian becomes a function of the generator J_3 of $SU(2)$: $H \sim J_3$.

In the spin-j representation of $SU(2)$, the canonical realization of J_3 on the Riemann sphere S^2 is (Berezin 1975)

$$H \sim J_3 = j \frac{1 - zz^*}{1 + zz^*} \tag{2.6.191}$$

and the corresponding symplectic structure is determined by the two-form

$$\omega = \frac{1}{2}\omega_{ab} c^a c^b = \frac{2ij}{(1 + z\bar{z})^2} c^z c^{z^*}. \tag{2.6.192}$$

Calculations with the help of the localization formula reduce the corresponding path integral to the finite-dimensional one, the integrand being a function of the combination $zz^* + c\bar{c}$ only:

$$\mathcal{Z} = \frac{i}{\pi t} \int dz \, dz^* \, dc \, d\bar{c} \, F(zz^* + c\bar{c}) \tag{2.6.193}$$

where the function $F(y)$ is

$$F(y) = \frac{\frac{t}{2}\frac{1-y}{1+y}}{\sin\left[\frac{t}{2}\left(\frac{1-y}{1+y}\right)\right]} \exp\left\{-ijt\left(\frac{1-y}{1+y}\right)\right\}. \tag{2.6.194}$$

The integral (2.6.193) can be evaluated using the so-called *Parisi–Sourlas* integration formula (recall that c, $\bar{c}$ are Grassmann variables):

$$\frac{1}{\pi}\int d^2x\, d\theta\, d\bar\theta\, F(x^2 + \theta\bar\theta) = \int_0^\infty du\, \frac{dF(u)}{du} = F(\infty) - F(0) \tag{2.6.195}$$

and the result is

$$\mathcal{Z} = \frac{\sin(tj)}{\sin(\frac{1}{2}t)}. \tag{2.6.196}$$

We refer to Keski-Vakkuri *et al* (1991) for a discussion of the reasons for the necessity of making the so-called Weyl shift $j \to j + \frac{1}{2}$, in order to obtain the correct result for the $SU(2)$ group:

$$\mathcal{Z} = \frac{\sin(t[j + \frac{1}{2}])}{\sin(\frac{1}{2}t)}. \tag{2.6.197}$$

We only note that the general explanation can be traced back to the way in which the path integral has been regularized (discretization).

This method can also be applied to evaluate the path integral for the Hamiltonian $H \sim J_3^2$ in the spin-j representation of $SU(2)$:

$$H \sim J^2 = j^2\left(\frac{1-zz^*}{1+zz^*}\right)^2. \tag{2.6.198}$$

In this case, the corresponding integration formula gives the following finite-dimensional integral:

$$\mathcal{Z} = \frac{i}{\sqrt{4\pi it}}\int_{-\infty}^\infty d\phi\, \frac{1}{\phi}\int dz\, d\bar z\, dc\, d\bar c\, F_\phi(zz^* + c\bar c) \tag{2.6.199}$$

where

$$F_\phi(y) = \frac{\frac{t\phi}{2}\frac{1-y}{1+y}}{\sin\left[\frac{t\phi}{2}\left(\frac{1-y}{1+y}\right)\right]} \exp\left\{\frac{i}{4}t\phi^2 - ijt\phi\left(\frac{1-y}{1+y}\right)\right\} \tag{2.6.200}$$

and we have redefined

$$c^a \to \sqrt{\frac{i}{\phi}}c^a. \tag{2.6.201}$$

It is possible to evaluate this integral using the Parisi–Sourlas integration formula (2.6.195) and, introducing the Weyl shift $j \to j + \frac{1}{2}$, we obtain

$$\mathcal{Z} = \sqrt{\frac{t}{4\pi i}}\int_{-\infty}^\infty d\phi\, e^{\frac{i}{4}t\phi^2}\frac{\sin[(j+\frac{1}{2})t\phi]}{\sin(\frac{1}{2}t\phi)} = \sum_{m=-j}^j \sqrt{\frac{t}{4\pi i}}\int_{-\infty}^\infty d\phi\, e^{-it\phi m}e^{\frac{i}{4}t\phi^2}$$

$$= \sum_{m=-j}^j e^{-itm^2} = \text{Tr}(e^{-itH^2}). \tag{2.6.202}$$

The two last equalities confirm the correctness of the functional Duistermaat–Heckman formula for the trace of the evolution operator.

2.6.4 Problems

Problem 2.6.1. Verify that the state

$$|\xi; \alpha\rangle = \exp\left\{-\sum_{j=1}^{n} \xi_j \widehat{\alpha}_j^\dagger\right\}|0\rangle$$

is the eigenstate of the fermionic annihilation operators $\widehat{\alpha}_j$.

Hint. Use the relations (2.6.1) together with the fact that the $\xi_j \widehat{\alpha}_j^\dagger$ commute with each other and have zero square.

Problem 2.6.2. Calculate the scalar products of the eigenstates given by (2.6.18), (2.6.19), (2.6.20) and (2.6.21) and prove that the result is given by (2.6.22).

Hint. Use the basic properties (2.6.5) and (2.6.6) of the vacuum vectors and the nilpotence (2.6.13) of the Grassmann variables. For example:

$$\langle \xi; \alpha^\dagger | \zeta; \alpha^\dagger \rangle = \langle 0| \prod_{j=1}^{n}(1 + (\bar{\xi}_j - \bar{\zeta}_j)\widehat{\alpha}_j) \prod_{k=1}^{n} \widehat{\alpha}_k^\dagger |0\rangle$$

$$= \langle 0| \prod_{j=1}^{n}((\bar{\xi}_j - \bar{\zeta}_j)\widehat{\alpha}_j) \prod_{k=1}^{n} \widehat{\alpha}_k^\dagger |0\rangle$$

$$= \prod_{j=1}^{n}(\bar{\xi}_j - \bar{\zeta}_j).$$

Other scalar products can be calculated similarly.

Problem 2.6.3. Prove that the product $\prod_{j=1}^{n}(\zeta_j - \xi_j)$ satisfies the equality (2.6.69) and hence in the space of functions of Grassmann variables it can be used similarly to the Dirac δ-function.

Hint. Use the identity

$$\left[\prod_{j=1}^{n}(\zeta_j - \xi_j)\right] f(\xi) = \left[\prod_{j=1}^{n}(\zeta_j - \xi_j)\right] f(\zeta - (\zeta - \xi))$$

$$= \left[\prod_{j=1}^{n}(\zeta_j - \xi_j)\right] f(\zeta)$$

(the second equality is fulfilled due to the nilpotence property (2.6.13) of the Grassmann variables).

Problem 2.6.4. Verify that the scalar product (2.6.39) is positively defined and that the monomials (2.6.41) are orthonormalized.

Hint.

$$\langle \psi_0 | \psi_0 \rangle = \int d\bar{\xi}\, d\xi\, e^{-\bar{\xi}\xi} = \int d\bar{\xi}\, d\xi\, (1 - \bar{\xi}\xi) = 1.$$

The scalar products for other pairs of functions can be calculated similarly and the positive definiteness follows from this fact.

316   *Path integrals in quantum mechanics*

Problem 2.6.5. Prove the 'translational invariance' (2.6.43) of the Berezin integral.

Hint. Check straightforwardly that, for an arbitrary Grassmann element f, the right- and left-hand sides of (2.6.43) are equal.

Problem 2.6.6. Prove the formula (2.6.63) for the Grassmann–Berezin Gauss-like integral.

Hint. Use formula (2.6.60).

Problem 2.6.7. Prove the transformation rule for the Grassmann measure under the change of variables:

$$\xi_i \to \zeta_i = \sum_j A_{ij}\xi_j \qquad (i, j = 1, \ldots, n)$$

$$\prod_{i=1}^n d\zeta_i = (\det A)^{-1} \prod_{i=1}^n d\xi_i. \qquad (2.6.203)$$

Hint. Show that rule (2.6.203) gives the correct result for the integration of an arbitrary function of Grassmann variables.

Problem 2.6.8. Prove relation (2.6.82) between the normal symbol and integral kernel of a fermionic operator for an arbitrary number of degrees of freedom.

Hint. Use as a hint the discussion of the fermionic system with one degree of freedom (cf (2.6.77)).

Problem 2.6.9. Prove the completeness for the eigenvectors $|\xi; \alpha\rangle$, $|\zeta; \alpha^\dagger\rangle$, $\langle\xi; \alpha^\dagger|$, $\langle\zeta; \alpha|$ (defined by (2.6.18), (2.6.19), (2.6.20) and (2.6.21)):

$$(-1)^n \int |\xi; \alpha\rangle\, d\xi_n \cdots d\xi_1\, \langle\xi; \alpha^\dagger| = \mathbb{I}$$

$$\int |\zeta; \alpha^\dagger\rangle\, d\zeta_n \cdots d\zeta_1\, \langle\zeta; \alpha| = \mathbb{I}. \qquad (2.6.204)$$

Hint. Represent an arbitrary vector $|f\rangle$ as a expansion over the eigenvectors $|\xi; \alpha\rangle$:

$$|f\rangle = \int |\xi; \alpha\rangle\, d\xi_n \cdots d\xi_1\, f(\xi)$$

and find that the 'coefficients' $f(\xi)$ of the expansion are given by the equality:

$$f(\xi) = (-1)^n \langle\xi; \alpha^\dagger|f\rangle.$$

Substitution of this expression for $f(\xi)$ into the expansion immediately gives the first required completeness relation. The second relation is proved similarly.

Problem 2.6.10. Derive the path-integral representation for the transition amplitude $\langle\xi; \alpha^\dagger, t|\xi_0; \alpha, t_0\rangle$ in the case of a fermionic system with a normally ordered Hamiltonian $\widehat{H}(\widehat{a}^\dagger, \widehat{a})$, using the short-time expansion together with the completeness relation (2.6.204) and acting similarly to the method suggested in problem 2.2.1, page 190 for the bosonic case.

Hint. Let us start from the case of one fermionic degree of freedom. The matrix element of the infinitesimal evolution operator has the form

$$\langle \xi'; \alpha^\dagger | e^{-i\varepsilon \widehat{H}/\hbar} | \xi; \alpha \rangle \approx \int d\zeta \, e^{-\bar{\zeta}(\xi'-\xi)} e^{-i\varepsilon H(\bar{\zeta},\xi)/\hbar}. \qquad (2.6.205)$$

The completeness relation (2.6.204) allows us to write the time-sliced approximation for the fermionic transition amplitude as follows:

$$\langle \xi'; \alpha^\dagger, t | e^{-i(t-t_0)\widehat{H}/\hbar} | \xi; \alpha, t_0 \rangle = \langle \xi'; \alpha^\dagger | (e^{-i\varepsilon \widehat{H}/\hbar})^N | \xi; \alpha \rangle$$

$$= \int \cdots \int \langle \xi', t | e^{-i\varepsilon \widehat{H}/\hbar} | \zeta_N, t_N \rangle \, d\zeta_N \, \langle \zeta_N, t_N | e^{-i\varepsilon \widehat{H}/\hbar} | \zeta_{N-1}, t_{N-1} \rangle \, d\zeta_{N-1}$$

$$\cdots d\zeta_1 \, \langle \zeta_1, t_1 | e^{-i\varepsilon \widehat{H}/\hbar} | \xi, t_0 \rangle$$

(here $|\xi_i, t_i\rangle$, $\langle \zeta_i, t_i|$ are eigenvectors of $\hat{\alpha}$ and $\langle \xi_i, t_i|$, $|\zeta_i, t_i\rangle$ are eigenvectors of $\hat{\alpha}^\dagger$). Using the expression (2.6.205), reordering the factors and passing to the continuum time limit, we obtain the path integral (2.6.85). The generalization to the case of n degrees of freedom is straightforward.

Problem 2.6.11. Construct the path-integral representation for operator satisfying the defining relations (2.6.113), by using the appropriate generalization of Grassmann-like differential and integral calculi.

Hint. Introduce the following unusual 'differential' operators $\mathcal{D}$, $\bar{\mathcal{D}}$, with the properties:

$$\bar{\mathcal{D}}\bar{z} + \bar{z}^2 \bar{\mathcal{D}}^2 = 1 \qquad \mathcal{D}z + z^2 \mathcal{D}^2 = 1 \qquad \bar{z}^3 = z^3 = 1. \qquad (2.6.206)$$

Consider, for definiteness, the commuting variables $z\bar{z} = \bar{z}z$ (we can consider other possibilities, e.g., anticommuting z and $\bar{z}$). The basis of functions (2.6.100) is orthonormal with respect to the scalar product (2.6.101), with the function $\mu = 1 + \bar{z}z + \bar{z}^2 z^2$, the integral again being defined by (2.6.102). The operators $\bar{z}$ and $\bar{\mathcal{D}}$ are Hermitian conjugate with respect to this scalar product. We can check that the normal symbols $A_N(\bar{z}z)$ of operators A and their kernels $\mathcal{A}(\bar{z}z)$ are related with the help of the same function $\mu(\bar{z}z)$ that defines the scalar product, the kernel being defined in the usual way:

$$A\psi(\bar{z}) = \int d\bar{z}' \, dz' \, \mathcal{A}(\bar{z}, z') \psi(\bar{z}').$$

Thus, using nilpotence and the usual procedure for path-integral derivation, we obtain

$$U(t - t_0) = \int \left(\prod_\tau d\bar{z}(\tau) \, dz(\tau) \, (1 + 2\bar{z}(\tau)z(\tau) + 3\bar{z}^2(\tau)z^2(\tau)) \right) \mu(\bar{z}(t)z(t))$$

$$\times \exp\left\{ -\int_{t_0}^t d\tau \, [(1 + \bar{z}z)\bar{z}(\tau)\dot{z}(\tau) + iH_{\text{eff}}(\bar{z}(\tau)z(\tau))] \right\}.$$

Appendices

A General pattern of different ways of construction and applications of path integrals

In order to give the reader an idea about the different approaches to the construction of path integrals and their main applications, we present schematically the general pattern in figure 2.5 (page 319). Obviously, it is impossible to reflect in a single figure the many different path-integral applications and thus this diagram may serve only for a preliminary orientation in the subject and for visualizing the links which exist among various topics discussed in this monograph.

B Proof of the inequality used for the study of the spectra of Hamiltonians

This appendix contains the proof of the second condition (2.1.110) which is used for proving the discreteness of the spectrum of Hamiltonians in which the potential energy satisfies the conditions (2.1.105). To this aim, rewrite the Cauchy–Schwarz–Bunyakowskii inequality (2.1.114) in the form

$$\psi^2(x_t) \le \int_{-\infty}^{\infty} dx_0\, \psi_0^2(x_0) \frac{\exp\left\{-\frac{(x_t-x_0)^2}{t}\right\}}{\sqrt{\pi t}}$$
$$\times \int_{\mathcal{C}\{0;x_t,t\}} d_W x(s)\, \exp\left\{-2\int_0^t ds\, V(x(s))\right\}. \tag{B.1}$$

The set $\mathcal{C}$ of continuous functions can be separated into two non-intersecting subsets $\mathcal{C}_1$ and $\mathcal{C}_2$ ($\mathcal{C} = \mathcal{C}_1 \cup \mathcal{C}_2$, $\mathcal{C}_1 \cap \mathcal{C}_2 = 0$), so that

$$\exp\left\{-2\int_0^t ds\, V(x(s))\right\} < \varepsilon \qquad \forall x(s) \in \mathcal{C}_1. \tag{B.2}$$

This condition is fulfilled if

$$|x(s)| > M(\varepsilon) \tag{B.3}$$

(due to (2.1.105)), for a suitable function $M(\varepsilon)$. We also imply that $x_t > M(\varepsilon)$; recall that inequality (2.1.110) is expected to be correct for $x_t > X(\varepsilon)$ with a suitably chosen function $X(\varepsilon)$.

Thus the second factor in (B.1) can be estimated as follows:

$$\int_{\mathcal{C}\{0;x_t,t\}} d_W x(s)\, \exp\left\{-2\int_0^t ds\, V(x(s))\right\}$$
$$< \varepsilon + \int_{\mathcal{C}_2\{0;x_t,t\}} d_W x(s)\, \exp\left\{-2\int_0^t ds\, V(x(s))\right\} \tag{B.4}$$

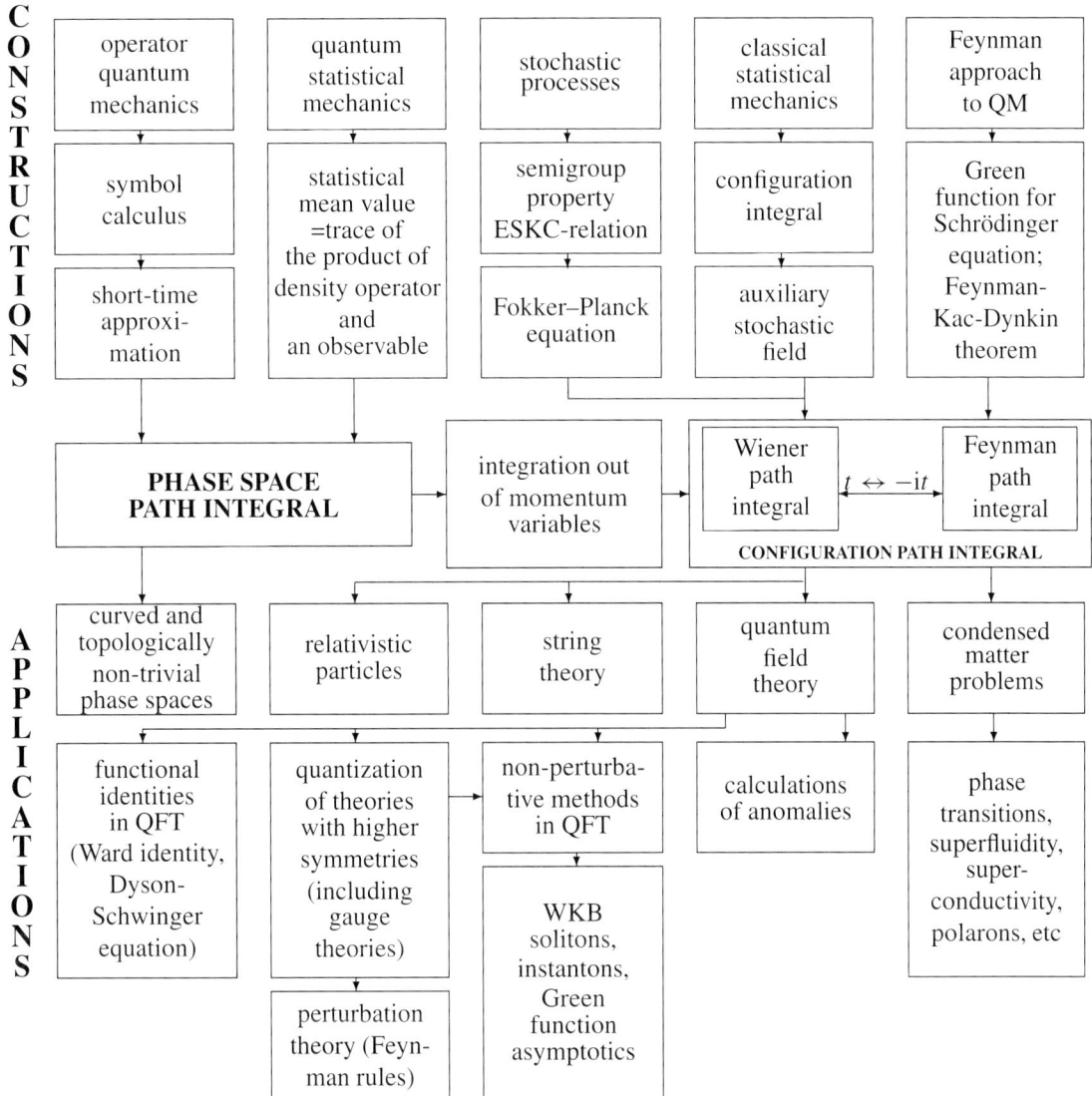

Figure 2.5. General pattern of different ways of construction and applications of path integrals.

or, taking into account the inequality (2.1.115), as

$$\int_{\mathcal{C}\{0;x_t,t\}} d_W x(s) \exp\left\{-2\int_0^t ds\, V(x(s))\right\} < \varepsilon + \int_{\mathcal{C}_2\{0;x_t,t\}} d_W x(s). \tag{B.5}$$

The next step is to show that the second term of (B.5) can be made arbitrarily small for sufficiently large x_t and restricted t.

First, we estimate this term by the inequality

$$\int_{\mathcal{C}_2\{0;x_t,t\}} d_W x(s) < \int_{\mathcal{C}_3\{0;x_t,t\}} d_W x(s) \tag{B.6}$$

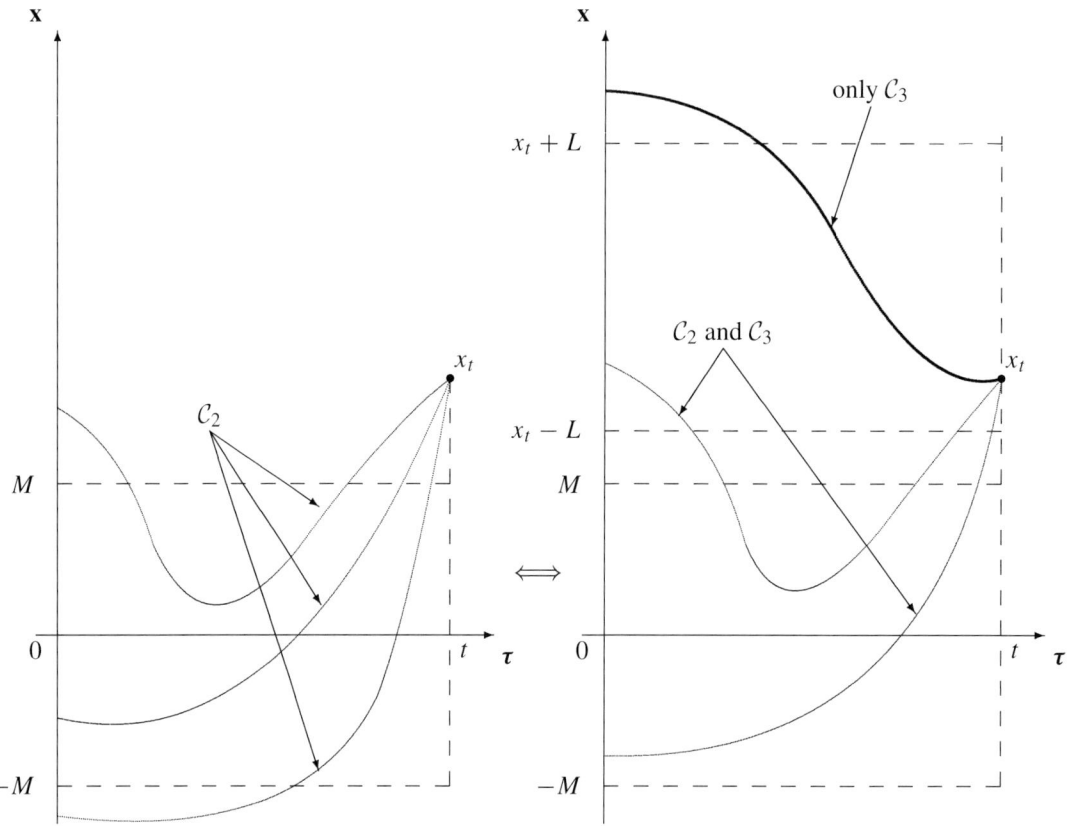

Figure 2.6. Illustration of the inclusion $C_2 \subset C_3$: all trajectories from C_2 must drop inside the interval $[-M, M]$ and hence belong to C_3, as shown on the right; the sample trajectory which belongs only to C_3 is depicted by the bold curve.

where the subset C_3 is defined by the condition

$$C_3 = \left\{ x(s) \middle| \sup_{0 \le s \le t} |x(s) - x_t| > L \right\}. \tag{B.7}$$

The inequality (B.6) follows from the positivity of the Wiener measure and the inclusion $C_2 \subset C_3$. Indeed, we can choose x_t so that

$$x_t - L > M$$

and this provides (see figure 2.6) that any trajectory, belonging to the set C_2 (i.e. violating the inequality (B.3)), satisfies also the defining condition (B.7) for the set C_3.

The right-hand side of (B.6) can be substituted by the path integral over the set of trajectories terminating at the origin, due to the spatial homogeneity of Wiener processes:

$$\int_{C_3\{0;x_t,t\}} d_W x(s) = \int_{C_4\{0;x_t=0,t\}} d_W x(s) \tag{B.8}$$

where

$$C_4 = \left\{x(s) \Big| \sup_{0 \leq s \leq t} |x(s)| > L\right\}.$$

It is clear that

$$1 = \int_{C\{0; x_t=0,t\}} d_W x(s) = \int_{C_4\{0; x_t=0,t\}} d_W x(s) + \int_{C_5\{0; x_t=0,t\}} d_W x(s) \tag{B.9}$$

where

$$C_5 = \left\{x(s) \Big| \sup_{0 \leq s \leq t} |x(s)| < L\right\}$$

because $C = C_4 \cup C_5$.

Thus, the problem is now reduced to estimating (from below) the following Wiener integral:

$$\int_{C_5\{0; x_t=0,t\}} d_W x(s) = \int_{-L}^{L} dx_0 \int_{C_6\{x_0, 0; x_t=0,t\}} d_W x(s)$$
$$= \int_{-L}^{L} dx_0 \, \widetilde{K}(0, t|x_0, 0) \tag{B.10}$$

where

$$C_6 = \left\{x(s) \Big| x(0) = x_0, x(t) = 0, \sup_{0 \leq s \leq t} |x(s)| < L\right\}$$

so that $\widetilde{K}(0, t|x_0, 0)$ is the fundamental solution of the diffusion equation with the modified boundary conditions

$$\frac{\partial \widetilde{K}}{\partial t} = \frac{1}{4} \frac{\partial^2 \widetilde{K}}{\partial x^2}$$
$$\widetilde{K}|_{x=L} = \widetilde{K}|_{x=-L} = 0. \tag{B.11}$$

This equation can be solved by the well-known *method of images* (cf problem 1.2.15, page 117), with the result

$$\widetilde{K}(0, t|x_0, 0) = \frac{\exp\left\{-\frac{x_0^2}{t}\right\}}{\sqrt{\pi t}}\left[1 - \sum_{n=1}^{\infty}(-1)^n \left(\exp\left\{-\frac{4Ln}{t}(Ln + x_0)\right\} + \exp\left\{-\frac{4Ln}{t}(Ln - x_0)\right\}\right)\right]. \tag{B.12}$$

Using the obvious properties of series with alternating signs and with monotonically decreasing terms, we obtain the estimation

$$\widetilde{K}(0, t|x_0, 0) > \frac{\exp\left\{-\frac{x_0^2}{t}\right\}}{\sqrt{\pi t}}\left(1 - \exp\left\{-\frac{4L}{t}(L + x_0)\right\} - \exp\left\{-\frac{4L}{t}(L - x_0)\right\}\right)$$
$$> \frac{\exp\left\{-\frac{x_0^2}{t}\right\}}{\sqrt{\pi t}}\left(1 - 2\exp\left\{-\frac{4L}{t}(L - |x_0|)\right\}\right)$$

which, together with (B.10), gives

$$\int_{C_5\{0;x_t=0,t\}} d_W x(s) > \int_{-L}^{L} dx_0 \frac{\exp\left\{-\frac{x_0^2}{t}\right\}}{\sqrt{\pi t}} - 2\int_{-L}^{L} dx_0 \frac{\exp\left\{-\frac{(2L-|x_0|)^2}{t}\right\}}{\sqrt{\pi t}}$$

$$= 1 - \int_{|\xi|>L} d\xi \frac{\exp\left\{-\frac{\xi^2}{t}\right\}}{\sqrt{\pi t}} - 2\left(\int_{-2L}^{-L} d\xi \frac{\exp\left\{-\frac{\xi^2}{t}\right\}}{\sqrt{\pi t}} + \int_{L}^{2L} d\xi \frac{\exp\left\{-\frac{\xi^2}{t}\right\}}{\sqrt{\pi t}}\right)$$

or

$$\int_{C_5\{0;x_t=0,t\}} d_W x(s) > 1 - 3\int_{|\xi|>L} d\xi \frac{\exp\left\{-\frac{\xi^2}{t}\right\}}{\sqrt{\pi t}}.$$

Taking into account (B.9) this means, in turn, that

$$\int_{C_4\{0;x_t=0,t\}} d_W x(s) < 3\int_{|\xi|>L} d\xi \frac{\exp\left\{-\frac{\xi^2}{t}\right\}}{\sqrt{\pi t}} \quad \text{(B.13)}$$

and, using (B.4), (B.7) and (B.8), we derive

$$\int_{C\{0;x_t,t\}} d_W x(s) \exp\left\{-2\int_0^t ds\, V(x(s))\right\} < \varepsilon + \varepsilon'(L) \xrightarrow[|x_t|\to\infty]{} 0 \quad \text{(B.14)}$$

(because at $|x_t| \to \infty$, we may take L arbitrarily large).

Finally, with the help of the inequalities (B.1) and (B.14), we can conclude that

$$\psi^2(x_t) < \int_{-\infty}^{\infty} dx_0\, \psi_0^2(x_0) \frac{\exp\left\{-\frac{(x_t-x_0)^2}{t}\right\}}{\sqrt{\pi t}} (\varepsilon + \varepsilon') \xrightarrow[|x_t|\to\infty]{} 0 \quad \text{(B.15)}$$

i.e. $\psi(x_t)$ uniformly tends to zero at large values of x_t and the second condition (2.1.110) has been proved.

C Proof of lemma 2.1 used to derive the Bohr–Sommerfeld quantization condition

The proof of the lemma starts from the inequality

$$\exp\left\{-\sum_{i=1}^n V(x_i)\frac{t}{n}\right\} \leq \frac{1}{t}\sum_{i=1}^n \exp\{-tV(x_i)\}\frac{t}{n} \quad \text{(C.1)}$$

which, in turn, is the particular case of the following proposition.

Proposition 2.2 (Jensen inequality). Let $\phi(x)$ be a convex function (so that $\phi''(x) > 0$) defined on the interval $[a,b] \in \mathbb{R}$ and $x_i, i=1,\ldots,n$ be some points in the interval: $x_i \in [a,b]$. Then

$$\sum_{i=1}^n \alpha_i \phi(x_i) \geq \phi\left(\sum_{i=1}^n \alpha_i x_i\right) \quad \text{(C.2)}$$

where

$$\sum_{i=1}^n \alpha_i = 1 \qquad \alpha_i \in \mathbb{R}_+ \text{ (positive numbers).} \quad \text{(C.3)}$$

Proof of the Jensen inequality. It is clear that $\tilde{x} \stackrel{\text{def}}{=} \sum_{i=1}^{n} \alpha_i x_i$ belongs to the interval $[a, b]$. Consider the Taylor series with the remainder term

$$\phi(x_i) = \phi(\tilde{x}) + (x_i - \tilde{x})\phi'(\tilde{x}) + \tfrac{1}{2}(x_i - \tilde{x})^2 \phi''(\xi_i) \qquad i = 1, \ldots, n$$

where $\xi_i \in [a, b]$. Since the function ϕ is convex, we have $\phi''(x) > 0$. Multiplying each of these series by α_i and summing them up over i, we obtain the required inequality

$$\sum_{i=1}^{n} \alpha_i \phi(x_i) \geq \phi(\tilde{x}).$$

□

Corollary 2.2. In the particular case $\alpha_i = 1/n$, $\phi(x) = e^{-x}$, we introduce $a_i = e^{-x_i}$ and the Jensen inequality becomes equivalent to the relation:

$$(a_1 a_2 \cdots a_n)^{1/n} \leq \frac{1}{n}(a_1 + a_2 + \cdots + a_n) \qquad a_i \in \mathbb{R}_+. \tag{C.4}$$

Corollary 2.3. The set of the positive numbers $\alpha = \{\alpha_i\}$ can be considered as a probability distribution due to condition (C.3), so that the Jensen inequality can be written in the more general form

$$\phi(\langle x \rangle) \leq \langle \phi(x) \rangle \tag{C.5}$$

where $\langle \cdot \rangle$ denotes the mean value with respect to the distribution $\alpha = \{\alpha_i\}$.

Relation (C.1) follows from (C.4) if we put

$$a_i = e^{-tV(x_i)}.$$

Since the inequality (C.1) is correct for arbitrary n, we can take the limit $n \to \infty$ with the result

$$\exp\left\{-\frac{1}{t}\int_0^t ds\, V(x(s))\right\} \leq \frac{1}{t}\int_0^t ds\, \exp\{-tV(x(s))\} \tag{C.6}$$

and since the latter is correct for arbitrary $x(s)$, we can integrate it with the Wiener measure to get

$$K_B(x, t|x, 0) = \int_{C\{x,0;x,t\}} d_W x(s)\, \exp\left\{-\int_0^t ds\, V(x(s))\right\}$$

$$\leq \frac{1}{t}\int_{C\{x,0;x,t\}} d_W x(s) \int_0^t ds\, \exp\{-tV(x(s))\}. \tag{C.7}$$

Changing the order of integration and using the ESKC relation, we can derive from (C.7) the inequality

$$\int_{-\infty}^{\infty} dx\, K_B(x, t|x, 0) \leq \frac{1}{t}\int_{-\infty}^{\infty} d\xi \int_0^t ds\, \exp\{-tV(\xi)\}\frac{1}{\sqrt{\pi t}}$$

$$= \frac{1}{\sqrt{\pi t}}\int_{-\infty}^{\infty} d\xi\, \exp\{-tV(\xi)\} \tag{C.8}$$

(problem 2.1.8). This relation is correct for arbitrary t. Now we shall prove that for small values of t, there exists the inequality with the opposite sign. Thus, actually, (C.8) is an *equality* and this proves the lemma.

Let us make the substitution:

$$x(s) \longrightarrow x + x(s) \qquad x = x(t) = x(0)$$

in the Wiener integral

$$\int_{\mathcal{C}\{x,0;x,t\}} d_W x(s) \exp\left\{-\int_0^t ds\, V(x(s))\right\} = \int_{\mathcal{C}\{0,0;0,t\}} d_W x(s) \exp\left\{-\int_0^t ds\, V(x + x(s))\right\}. \quad (C.9)$$

The integrand of the latter path integral can be rewritten with the help of the step-function θ via the *Stieltjes* integral

$$\exp\left\{-\int_0^t ds\, V(x + x(s))\right\} = \int_0^\infty e^{-u} du\, \theta\left(u - \int_0^t ds\, V(x + x(s))\right) \quad (C.10)$$

due to the well-known relation

$$\frac{d}{du}\theta(u - a) = \delta(u - a). \quad (C.11)$$

Using (C.10), we can write (changing the order of the integration)

$$\int_{\mathcal{C}\{0,0;0,t\}} d_W x(s) \exp\left\{-\int_0^t ds\, V(x(s))\right\}$$

$$= \int_0^\infty e^{-u} du \left[\int_{\mathcal{C}\{0,0;0,t\}} d_W x(s)\, \theta\left(u - \int_0^t ds\, V(x + x(s))\right)\right]$$

$$\geq \int_0^\infty e^{-u} du \left[\int_{\mathcal{C}_\delta\{0,0;0,t\}} d_W x(s)\, \theta\left(u - \int_0^t ds\, V(x + x(s))\right)\right] \quad (C.12)$$

where $\mathcal{C}_\delta\{0, 0; 0, t\}$ is a subset of $\mathcal{C}\{0, 0; 0, t\}$ such that

$$|x(s)| < \delta.$$

The inequality in (C.12) follows from the positivity of the Wiener measure and the positive semi-definiteness of the step-function θ. With the help of the inequality

$$\sup_{|x(s)| < \delta} \int_0^t ds\, V(x + x(s)) \leq t \max_{|y| < \delta} V(x + y) \quad (C.13)$$

and the characteristic property of the θ-function, we can make the estimation (C.12) stronger:

$$\int_{\mathcal{C}_\delta\{0,0;0,t\}} d_W x(s)\, \theta\left(u - \int_0^t ds\, V(x + x(s))\right) \geq \theta\left(u - t \max_{|y| < \delta} V(x + y)\right) \int_{\mathcal{C}_\delta\{0,0;0,t\}} d_W x(s). \quad (C.14)$$

Thus, (C.9) gives

$$\int_{-\infty}^\infty dx\, K_B(x, t|x, 0) \geq \int_0^\infty e^{-u} du\, \text{Meas}\left\{x : t \max_{|y| < \delta} V(x + y) < u\right\} \int_{\mathcal{C}_\delta\{0,0;0,t\}} d_W x(s)$$

$$= \int_0^\infty e^{-ut} du\, \text{Meas}\left\{x : \max_{|y| < \delta} V(x + y) < u\right\} \int_{\mathcal{C}_\delta\{0,0;0,t\}} d_W x(s). \quad (C.15)$$

Here we have used (cf (2.1.131))

$$\int_{-\infty}^{\infty} dx\, \theta\left(u - t \max_{|y|<\delta} V(x+y)\right) = \text{Meas}\left\{x : t \max_{|y|<\delta} V(x+y) < u\right\}$$

and the substitution $u \to ut$.

We want to derive asymptotic behaviour for the right-hand side of (C.15) as $t \to 0$. The path integral can be estimated as follows (the influence of the boundary condition $|x(s)| < \delta$ becomes negligible at very small t because the particle does not have time to reach the boundaries, cf also the discussion in the preceding appendix, inequality (B.13)):

$$\left.\int_{\mathcal{C}_\delta\{0,t;0,0\}} d_W x(s)\right|_{t\to 0} \simeq \int_{\mathcal{C}\{0,t;0,0\}} d_W x(s) = \frac{1}{\sqrt{\pi t}}. \qquad (C.16)$$

The first factor in (C.15), at small values of t, can be estimated using conditions (2) and (3) on the potential energy (see (2.1.130)). The point is that the asymptotics in t of the integrals

$$f(t) \stackrel{\text{def}}{\equiv} \int_0^\infty du\, e^{-tu} F(u)$$

do not depend at small t on the actual form of the function $F(u)$, provided that the latter has the fixed asymptotics in u:

$$F(u)|_{u\to\infty} \sim Cu^\alpha \qquad \alpha > 0.$$

In fact, this statement is a particular variant of the Tauberian theorem (see appendix D). In the case under consideration, the function $F(u)$ can be written as

$$F(u) = Cu^\alpha + \varepsilon(u)u^\alpha$$

where $\varepsilon(u) \xrightarrow[u\to\infty]{} 0$. Hence,

$$f(t) = C \int_0^\infty du\, e^{-tu} u^\alpha + \int_0^U du\, e^{-tu} \varepsilon(u) u^\alpha + \int_U^\infty du\, e^{-tu} \varepsilon(u) u^\alpha.$$

Here U is chosen so that

$$\varepsilon(u) < \delta \qquad \text{iff } u > U.$$

Using the integral representation for the Γ-function:

$$\Gamma(\beta) = \int_0^\infty du\, e^{-u} u^{\beta-1} \qquad (C.17)$$

we obtain

$$f(t)|_{t\to 0} \approx \frac{C\Gamma(\alpha+1)}{t^{\alpha+1}}$$

because

$$\int_U^\infty du\, e^{-tu} \varepsilon(u) u^\alpha < \frac{\delta\Gamma(\alpha+1)}{t^{\alpha+1}}$$

and

$$\int_0^U du\, e^{-tu} \varepsilon(u) u^\alpha \xrightarrow[t\to 0]{} \text{constant}.$$

This result, together with conditions (2.1.130), allows us to write

$$\int_0^\infty e^{-ut} du \operatorname{Meas}\left\{x : \max_{|y|<\delta} V(x+y) < u\right\}\bigg|_{t\to 0} \simeq \int_0^\infty e^{-ut} du \operatorname{Meas}\{x : V(x) < u\}$$
$$= \int_{-\infty}^\infty dx\, e^{-tV(x)}. \qquad (C.18)$$

Thus, we have obtained the inequality

$$\int_{-\infty}^\infty dx\, K(x,t|x,0)\bigg|_{t\approx 0} \geq \frac{1}{\sqrt{\pi t}} \int_{-\infty}^\infty dx\, e^{-tV(x)}$$

and comparison with (C.8) gives the statement of lemma 2.1. $\square$

D Tauberian theorem

Theorem 2.6 (Tauber). Let a function $F(u)$ satisfy the following asymptotic condition

$$\int_0^\infty du\, e^{-tu} F(u)\bigg|_{t\to 0} \simeq Ct^{-\gamma} \qquad \gamma > 0,\ C, \gamma \in \mathbb{R}. \qquad (D.1)$$

Then the asymptotic behaviour of the integral

$$\int_0^R du\, F(u)\bigg|_{R\to\infty} \qquad (D.2)$$

at $R \to \infty$ is uniquely defined.

Sketch of proof of the Tauber theorem. Since

$$\int_0^\infty du\, e^{-tu}(e^{-tu})^n u^{\gamma-1} = \frac{\Gamma(\gamma)}{(n+1)} \qquad n \in \mathbb{Z}$$

and since the relation (D.1) implies

$$\int_0^\infty du\, e^{-tu}(e^{-tu})^n F(u)\bigg|_{t\to 0} \simeq Ct^{-\gamma}(n+1)^{-\gamma}$$

there exist the following asymptotic relations

$$\int_0^\infty du\, e^{-tu}(e^{-tu})^n F(u)\bigg|_{t\to 0} \simeq \frac{C}{t^\gamma \Gamma(\gamma)} \int_0^\infty du\, e^{-tu}(e^{-u})^n u^{\gamma-1}$$

or, summing them with the coefficients a_n, we obtain the relation

$$\int_0^\infty du\, e^{-tu} \sum_n a_n (e^{-tu})^n F(u)\bigg|_{t\to 0} \simeq \frac{C}{t^\gamma \Gamma(\gamma)} \int_0^\infty du\, e^{-tu} \sum_n a_n (e^{-u})^n u^{\gamma-1}. \qquad (D.3)$$

We can show that there exists such a choice of coefficients a_n that

$$\sum_n a_n (e^{-tu})^n = \begin{cases} 0 & u > \dfrac{1}{t} \\ e^{tu} & u < \dfrac{1}{t}. \end{cases}$$

With this choice, (D.3) gives the following asymptotics:

$$\int_0^{1/t} du\, F(u) \bigg|_{t\to 0} \simeq \frac{C}{t^\gamma \Gamma(\gamma)} \int_0^1 du\, u^{\gamma-1} = \frac{C}{t^\gamma \Gamma(\gamma+1)}$$

and this proves the statement of the Tauberian theorem. □

Bibliography

Abramowitz M and Stegun I A 1965 *Handbook of Mathematical Functions with Formulae, Graphs, and Mathematical Tables* (New York: Dover)
Albeverio S A and Høegh-Krohn R J 1976 *Mathematical Theory of Feynman Path Integrals (Lecture Notes in Mathematics 523)* (Berlin: Springer)
Albeverio S A *et al* 1979 *Feynman Path Integrals (Lecture Notes in Physics 106)* (Berlin: Springer)
Atiyah M F 1985 *Asterisque* **131** 43
Balazs N L and Jennings B K 1984 *Phys. Rep.* **104** 347
Barber M N and Ninham B W 1970 *Random and Restricted Walks* (New York: Gordon and Breach)
Barut A O and Rączka R 1977 *Theory of Group Representations and Applications* (Warsaw: Polish Scientific)
Beilinson A A 1959 *Dokl. Akad. Nauk SSSR* **128** 123
Belavin A *et al* 1975 *Phys. Lett.* **59B** 85
Berezin F A 1966 *The Method of Second Quantization* (New York: Academic)
——1971 *Teor. Mat. Fiz.* **6** 194
——1974 *Izv. Akad. Nauk SSSR Ser. Mat.* **38** 1116
——1975 *Commun. Math. Phys.* **40** 153
——1981 *Sov. Phys.–Usp.* **23** 409
——1987 *Introduction to Superanalysis* ed A A Kirillov and D Leites (Dordrecht: Reidel)
Berry M V and Mount K E 1972 *Rep. Prog. Phys.* **35** 315
Berry M V and Tabor M 1977 *J. Phys. A: Math. Gen.* **10** 371
Billingsley P 1979 *Probability and Measure* (New York: Wiley)
Blau M *et al* 1990 *Phys. Lett.* B **246** 92
Bogoliubov N N 1954 *Dokl. Akad. Nauk SSSR* **99** 225
Böhm M and Junker G 1987 *J. Math. Phys.* **28** 1978
Borel E 1909 *Rend. Circ. Mat. Palermo* **47** 247
Borisov N V *et al* 1976 *Sov. J. Theor. Math. Phys.* **29** 906
Born M and Wolf E 1959 *Principles of Optics* (New York: Pergamon)
Breiman L 1968 *Probability* (Reading, MA: Addison-Wesley)
Brush S G 1961 *Rev. Mod. Phys.* **33** 79
Cartier P and DeWitt-Morette C 1996 *Path integrals from meV to MeV* ed V S Yarunin and M A Smondyrev (Dubna: JINR)
Chaichian M and Demichev A 1994 *Phys. Lett.* B **320** 273
——1996 *Introduction to Quantum Groups* (Singapore: World Scientific)
Chaichian M and Hagedorn R 1998 *Symmetries in Quantum Mechanics. From Angular Momentum to Supersymmetry* (Bristol: IOP Publishing)
Chandrasekhar S 1943 *Rev. Mod. Phys.* **15** 1
Chernoff P 1968 *J. Funct. Anal.* **2** 238
Chung K L and Williams R J 1983 *Introduction to Stochastic Integration* (Boston, MA: Birkhauser)
Courant R and Hilbert D 1953 *Methods in Mathematical Physics* vols 1 and 2 (New York: Interscience)
Das A 1993 *Field Theory. A Path Integral Approach* (Singapore: World Scientific)
Daubechies I and Klauder J R 1985 *Phys. Rev. Lett.* **52** 1161

——1985 *J. Math. Phys.* **26** 2239
Davydov A S 1976 *Quantum Mechanics* (Oxford: Pergamon)
de Almeida O 1988 *Hamiltonian Systems: Chaos and Quantization* (Cambridge: Cambridge University Press)
de Gennes P G 1969 *Rep. Prog. Phys.* **32** 187
De Witt B S 1957 *Rev.Mod.Phys.* **29** 377
——1967 *Phys. Rev.* **160** 1113
——1967 *Phys. Rev.* **162** 1195
——1967 *Phys. Rev.* **162** 1239
DeWitt-Morette C *et al* 1979 *Phys. Rep.* **50** 255
Dirac P A M 1933 *Phys. Z. Sowjetunion* **3** 64
——1947 *The Principles of Quantum Mechanics* (New York: Oxford University Press)
Dittrich W and Reuter M 1992 *Classical and Quantum Dynamics. From Classical Paths to Path Integrals* (Berlin: Springer)
Doob J L 1953 *Stochastic Processes* (New York: Wiley)
Dowker J S 1972 *J. Phys. A: Math. Gen.* **5** 936
Duistermaat J H and Heckman G J 1982 *Invent. Math.* **69** 259
——1983 *Invent. Math.* **72** 153
Duru I H 1984 *Phys. Rev.* D **30** 2121
Duru I H and Kleinert H 1979 *Phys. Lett.* **84B** 185
——1982 *Forsch. Phys.* **30** 401
Dynkin E B 1955 *Dokl. Akad. Nauk SSSR* **104** 321
Eckhardt B *et al* 1992 *Phys. Rev.* **A45** 3531
Eckhardt B and Wintgen D 1990 *J. Phys. B: At. Mol. Opt. Phys.* **23** 355
Edwards S F and Peierls R E 1954 *Proc. R. Soc.* A **224** 24
Eichmann G 1971 *J. Opt. Soc. Am.* **61** 161
Einstein A 1905 *Ann. Phys., Lpz.* **17** 549
——1906 *Ann. Phys., Lpz.* **19** 371
——1926 *Investigations on the Theory of the Brownian Movement* (London: Methuen) (republished 1956 *Investigations on the Theory of the Brownian Movement* ed (with notes) R Frth (New York: Dover))
Elliot R J 1982 *Stochastic Calculus and Applications* (Berlin: Springer)
Evgrafov M A 1970 *Sov. Math. Dokl.* **11** 474
Faddeev L D 1976 *Methods in Field Theory (Proc. Les Houches 1975)* ed R Balian and J Zinn-Justin (Amsterdam: North-Holland)
Faddeev L D and Popov V N 1967 *Phys. Lett.* B **25** 29
Feller W 1951 *An Introduction to Probability Theory and Its Applications* vol 1 (New York: Wiley)
——1961 *An Introduction to Probability Theory and Its Applications* vol 2 (New York: Wiley)
Feynman R P 1942 The principle of least action in quantum mechanics *PhD Thesis* (Princeton, NJ: Princeton University) unpublished
——1948 *Rev. Mod. Phys.* **20** 367 (reprinted in 1958 *Selected Papers on Quantum Electrodynamics* ed J Schwinger (New York: Dover))
——1950 *Phys. Rev.* **80** 440
——1951 *Phys. Rev.* **84** 108
——1953 *Phys. Rev.* **90** 1116
——1953 *Phys. Rev.* **90** 1291
——1953 *Phys. Rev.* **90** 1301
——1954 *Phys. Rev.* **94** 262
——1955 *Phys. Rev.* **97** 660
——1963 *Acta Phys. Polon.* **24** 697
——1972 *Statistical Mechanics: A Set of Lectures* (Reading, MA: Benjamin)
Feynman R P and Hibbs A R 1965 *Quantum Mechanics and Path Integrals* (New York: McGraw-Hill)
Foch V A 1965 *Diffraction and Propagation Problems* (New York: Pergamon)
Fradkin E S 1954 *Dokl. Akad. Nauk SSSR* **98** 47

Fradkin E S and Tyutin I V 1969 *Phys. Lett.* B **30** 562
Fujikawa K 1979 *Phys. Rev. Lett.* **42** 1195
Garrod G 1966 *Rev. Mod. Phys.* **38** 483
Gelfand I M and Minlos R A 1954 *Dokl. Akad. Nauk SSSR* **97** 209
Gelfand I M and Yaglom A M 1960 *J. Math. Phys.* **1** 48
Gihman I I and Skorohod A W 1972 *Stochastic Differential Equations* (Berlin: Springer)
Glashow S L 1961 *Nucl. Phys.* **22** 579
Glimm J and Jaffe A 1987 *Quantum Physics: a Functional Integral Point of View* (New York: Springer)
Gnedenko B V 1968 *The Theory of Probability* (New York: Chelsea)
Gomez-Reino C and Liñares J 1987 *J. Opt. Soc. Am.* **4** 1337
Gradshteyn I S and Ryzhik I M 1980 *Table of Integrals, Series and Products* (New York: Academic)
Grosche C 1992 An introduction into the Feynman path integral *Preprint* NTZ-29-92
——1996 *Path Integrals, Hyperbolic Spaces and Selberg Trace Formulae* (Singapore: World Scientific)
Grosche C and Steiner F 1998 *Handbook of Feynman Path Integrals* (Berlin: Springer)
Gross D J and Wilczek F 1973 *Phys. Rev. Lett.* **30** 1343
——1973 *Phys. Rev.* D **8** 3633
Guillemin V and Sternberg S 1984 *Symplectic Techniques in Physics* (Cambridge: Cambridge University Press)
Gutzwiller M C 1990 *Chaos in Classical and Quantum Mechanics* (Berlin: Springer)
Hida J 1980 *Brownian Motion* (Berlin: Springer)
Hunt K L C and Ross J 1981 *J. Chem. Phys.* **75** 976
Ikeda N and Watanabe S 1981 *Stochastic Differential Equations and Diffusion Processes* (Amsterdam: North-Holland)
Ilyin V A and Poznyak E G 1982 *Fundamentals of Mathematical Analysis* vols 1 and 2 (Moscow: Mir)
Isham C I 1989 *Modern Differential Geometry for Physicists* (Singapore: World Scientific)
Ito K 1951 *Mem. Am. Math. Sci.* **4**
Itzykson C and Zuber J -B 1980 *Quantum Field Theory* (New York: McGraw-Hill)
Jahnke E and Emde F 1965 *Tables of Functions with Formulae and Curves* (New York: Dover)
Jauch J M 1968 *Foundations of Quantum Mechanics* (Reading, MA: Addison-Wesley)
Jeffreys H 1962 *Asymptotic Approximations* (London: Oxford University Press)
Kac M 1949 *Trans. Am. Math. Soc.* **65** 1
——1959 *Probability and Related Topics in Physical Sciences* (New York: Interscience)
Keller J 1958 *Ann. Phys.* **4** 180
Keski-Vakkuri E *et al* 1991 *Phys. Rev.* D **44** 3899
Khalatnikov I M 1952 *Dokl. Akad. Nauk SSSR* **87** 538
——1955 *Sov. Phys.–JETP* **28** 633
Khandekar D C and Lawande S V 1986 *Phys. Rep.* **137** 115
Kikuchi R 1954 *Phys. Rev.* **96** 563
——1955 *Phys. Rev.* **99** 1684
Klauder J R and Skagerstam Bo-S (ed) 1985 *Coherent States: Applications to Physics and Mathematical Physics* (Singapore: World Scientific)
Kleinert H 1995 *Path Integrals in Quantum Mechanics* (Singapore: World Scientific)
Kobayashi S and Nomizu K 1969 *Foundations of Differential Geometry* vols 1 and 2 (New York: Interscience)
Kolmogorov A N 1938 *Usp. Mat. Nauk* **5** 5
——1956 *Foundations of the Theory of Probability* (New York: Chelsea)
Korn G A and Korn T M 1968 *Mathematical Handbook for Scientists and Engineers* (New York: McGraw-Hill)
Kunstatter G 1992 *Class. Quantum Grav.* **9** 1469
Kuo H -H 1975 *Gaussian Measures in Banach Spaces (Lecture Notes in Mathematics 463)* (Berlin: Springer)
Landau L D and Lifshitz E M 1981 *Quantum Mechanics* (New York: Pergamon)
Landsman N P and Linden N 1991 *Nucl. Phys.* B **365** 121
Langouche F *et al* 1982 *Functional Integration and Semiclassical Expansion* (Dordrecht: Reidel)
Levit S and Smilansky U 1977 *Ann. Phys.* **103** 198
Lichtenberg A J and Lieberman M A 1983 *Regular and Stochastic Motion* (New York: Springer)

Mandelbrot B B 1977 *Fractals* (San Francisco, CA: Freeman)
——1982 *The Fractal Geometry of Nature* (San Francisco, CA: Freeman)
Mandelstam S 1968 *Phys. Rev.* **175** 1580
——1968 *Phys. Rev.* **175** 1604
Marinov M S and Terentyev M V 1978 *Sov. J. Nucl. Phys.* **28** 729
——1979 *Fortsch. Phys.* **27** 511
Maslov V P and Fedoriuk M V 1982 *Semiclassical Approximation in Quantum Mechanics* (Dordrecht: Reidel)
Matthews P T and Salam A 1954 *Nuovo Cimento* **12** 563
McMillan D and Tsutsui I 1995 *Ann. Phys.* **237** 269
Mensky M B 1993 *Continuous Quantum Measurements and Path Integrals* (Bristol: IOP Publishing)
Mizrahi M M 1976 *J. Math. Phys.* **17** 556
Montonen C 1974 *Nuovo Cimento* A **19** 69
Morse M 1973 *Variational Analysis* (New York: Wiley)
Nakamura K 1993 *Quantum Chaos: a New Paradigm of Nonlinear Dynamics* (Cambridge: Cambridge University Press)
Nelson E 1964 *J. Math. Phys.* **5** 332
Neveu A 1977 *Rep. Prog. Phys.* **40** 599
Niemi A J and Tirkkonen O 1994 *Ann. Phys.* **235** 318
Paley R and Wiener N 1934 *Fourier Transforms in the Complex Domain* (New York: American Mathematical Society)
Papadopoulos G J 1978 *Path Integrals* ed G J Papadopoulos and J T Devreese (New York: Plenum)
Pauli W 1973 *Pauli Lectures on Physics* vol 6 (Cambridge, MA: MIT Press) ch 7
Peak D and Inomata A 1969 *J. Math. Phys.* **10** 1422
Perelomov A 1986 *Generalized Coherent States and their Applications* (Berlin: Springer)
Politzer H D 1973 *Phys. Rev. Lett.* **30** 1346
Polyakov A 1975 *Phys. Lett.* **59B** 82
Popov V N 1983 *Functional Integrals in Quantum Field Theory and Statistical Physics* (Dordrecht: Reidel)
Ranfagni A *et al* 1990 *Trajectories and Rays: The Path-Summation in Quantum Mechanics and Optics* (Singapore: World Scientific)
Ray D 1954 *Trans. Am. Math. Soc.* **77** 299
Reed M and Simon B 1972 *Methods of Modern Mathematical Physics, Vol 1: Functional Analysis* (New York: Academic)
——1975 *Methods of Modern Mathematical Physics, Vol 2: Fourier Analysis, Self-Adjointness* (New York: Academic)
Rivers R J 1987 *Path Integral Methods in Quantum Field Theory* (Cambridge: Cambridge University Press)
Roepstorff G 1996 *Path Integral Approach to Quantum Physics. An Introduction* (Berlin: Springer)
Rosenblatt M 1962 *Random Processes* (New York: Oxford University Press)
Sagan H 1992 *Introduction to the Calculus of Variations* (New York: Dover)
Salam A 1968 *Elementary Particle Physics (Nobel Symp. No. 8)* ed N Svartholm (Stockholm: Almqvist and Wilsell)
Schiff L I 1955 *Quantum Mechanics* (New York: McGraw-Hill)
Schulman L S 1968 *Phys. Rev.* **176** 1558
——1981 *Techniques and Applications of Path Integration* (New York: Wiley)
Schwinger J 1970 *Quantum Kinematics and Dynamics* (New York: Benjamin)
Shilov G E and Gurevich B L 1966 *Integral, Measure and Derivative: a Unified Approach* (Englewood Cliffs, NJ: Prentice-Hall)
Simon B 1974 *The $P(\varphi)_2$ Euclidean (Quantum) Field Theory* (Princeton, NJ: Princeton University Press)
——1979 *Functional Integration and Quantum Physics* (New York: Academic)
Solymar L and Cooke D J 1981 *Volume Holography and Volume Gratings* (London: Academic)
Symanzik K 1954 *Z. Naturforsch.* a **9** 809
Taylor J R 1972 *Scattering Theory. The Quantum Theory of Nonrelativistic Scattering* (New York: Wiley)
ter Haar D 1954 *Phys. Rev.* **95** 895
——1971 *Elements of Hamiltonian Mechanics* (New York: Pergamon)
't Hooft G 1971 *Nucl. Phys.* B **33** 173

——1971 *Nucl. Phys.* B **35** 167
Tomé W 1998 *Path Integrals on Group Manifolds* (Singapore: World Scientific)
Tricomi F G 1957 *Integral Equations* (London: Blackie)
Trotter H F 1959 *Proc. Am. Math. Soc.* **10** 545.
Truman A 1978 *J. Math. Phys.* **19** 1742
Uhlenbeck G E and Ornstein L S 1930 *Phys. Rev.* **36** 823
Vilenkin N Ya 1968 *Special Functions and the Theory of Group Representations (Transl. Math. Monographs)* (New York: American Mathematical Society)
Visconti A 1992 *Introductory Differential Geometry for Physicists* (Singapore: World Scientific)
Vladimirov V S and Volovich I V 1984 *Theor. Math. Phys.* **59** 317
Volterra V 1965 *Theory of Functionals and of Integral and Integrodifferential Equations* (New York: McGraw-Hill)
von Neumann J 1955 *Mathematical Foundations of Quantum Mechanics* (Princeton, NJ: Princeton University Press)
Voros A and Grammaticos B 1979 *Ann. Phys.* **123** 359
Weinberg S 1967 *Phys. Rev. Lett.* **19** 1264
Widder D V 1971 *An Introduction to Transform Theory* (New York: Academic)
Wiegel F W 1975 *Phys. Rep.* **16** 57
——1983 *Phys. Rep.* **95** 283
——1986 *Introduction to Path-Integral Methods in Physics and Polymer Science* (Singapore: World Scientific)
Wiener N 1921 *Proc. Natl Acad. Sci., USA* **7** 253
——1921 *Proc. Natl Acad. Sci., USA* **7** 294
——1923 *J. Math. Phys. Sci.* **2** 132
——1924 *Proc. London Math. Soc.* **22** 454
——1930 *Acta Math.* **55** 117
Wilson K 1974 *Phys. Rev.* D **10** 2445
Witten E 1992 *J. Geom. Phys.* **9** 303
Wybourn B G 1974 *Classical Groups for Physicists* (New York: Wiley)

Index

action functional, 140
annihilation operators, 132
anti-normal symbols, 207
anticommutation relations, 286
anticommutator, 286
anticommuting variables, 286
asymptotic series, 171
asymptotic states, 222

Baker–Campbell–Hausdorff formula, 155
Bargmann–Fock realization of CCR, 206
Berezin integral, 292
Bernoullian random walk, 49
Bloch equation, 65, 72, 76
Bohr–Sommerfeld quantization condition, 123, 145, 146, 176, 180
Borel set, 5
bra-vector, 127
Brownian bridge, 82
Brownian motion, 12, 13
    and fractal theory, 28
    discrete version, 13
    in field of non-conservative force, 115
    independence of increments, 23
    of interacting particles, 67
        under external forces, 68
    under an arbitrary external force, 66
    under an external harmonic force, 64, 84
    with absorption, 73
Brownian particle, 6, 13
    drift velocity, 49
    time to reach a point, 118
    under an external force, 76
    with inertia, 71

canonical commutation relations (CCRs), 132
canonical loop space, 308
canonical transformations, 305
Cauchy–Schwarz–Bunyakowskii inequality, 143

caustics, 176
change of variables in path integrals, 45
characteristic function, 101
characteristic functional, 101
chronological ordering, 74
coherent-state path integrals, 200, 218
coherent states, 207
    normalized, 218
      on the group $SU(2)$, 280
      overcompleteness, 219
commutator of operators, 129
compound event, 22
conditional probability, 15
conjugate points, 174, 175
continuity equation, 13
continuous integral, 1
contraction operator, 144
contravariant symbol, 220
coordinate representation, 129
correspondence principle, 201
Coulomb problem, 122
covariance, 103
creation operators, 132
curvature, 246
    scalar, 246
cyclotron frequency, 196

deformed oscillator algebra, 298
delta-functional (δ-functional), 71
diffusion constant, 13
diffusion equation, 13
    inhomogeneous, 20
    solution, 18, 19
discrete random walks, 108
discrete-time approximation, 37
double-well potential, 91
driven harmonic oscillator, 104
    classical (stochastic), 104

334   Index

transition probability, 107
quantum
  propagator, 198
Duistermaat–Heckman theorem, 307
  loop-space generalization, 310
Duru–Kleinert method, 267

effective action, 247
equivariant exterior derivative, 306, 309
ESKC (semigroup) relation, 21
Euler–Lagrange equations, 79, 80, 141
evolution operator, 125
  as a ratio of path integrals, 213
  normal symbol, 217
  Weyl symbol, 212
excluded volume problem and the Feynman–Kac formula, 110

Fermat principle, 188
Feynman–Kac formula, 65, 73, 137
  in quantum mechanics, 123
  proof for the Bloch equation, 76
finite-difference operators, 95
fluctuation factor, 80, 82
focal points, 176
Fokker–Planck equation, 61, 111
Fourier decomposition, 95, 100
  for Brownian trajectories, 95
  independence of coefficients, 117
free Hamiltonian, 222
functional derivatives, 86
functional integral, 1
functional space, 26
functionals
  characteristic, 31
  generating, 101
  integrable, 34
  simple, 33
    Cauchy sequence, 34
fundamental solution, 20

gauge invariance, 186
gauge transformations, 186
gauge-fixing conditions
  Coulomb, 195
Gaussian distributions, 59
Gaussian integral, 18
  complex, 135

  Grassmann case, 295
  multidimensional, 38
Gelfand–Yaglom method, 43, 87, 168, 263
generalized eigenfunctions, 130
generating function, 101
generating functional, 101
Grassmann algebra, 286, 288
  infinite-dimensional, 298
Grassmann variables, 289
Green functions, 20
  of the stationary Schrödinger equation, 177
ground state, 132

Hamiltonian vector field, 306
harmonic oscillator, 103
  quantum, 131
  with time-dependent frequency, 168
Hermite function, 132
Hermite polynomial, 132
Hilbert–Schmidt theorem, 142
hopping-path approximation, 91
hopping-path solution, 92

index of a bilinear functional, 175
instantons, 91
integral kernel of an operator, 78, 103
  convolution, 134
integral over histories, 1
integral over trajectories, 1
interaction representation, 223
invariant torus, 178
Ito stochastic integral, 63, 120, 183
  Stokes formula, 120

ket-vector, 127
Kolmogorov second equation, 61
Kolmogorov's theorem, 36
Kustaanheimo–Stiefel transformation, 245, 266, 269

Lagrangian, 141
Langevin equation, 61, 62
Laplace–Beltrami operator, 246
large fluctuations, 91
lattice derivative, 95
Legendre transformation, 180
Lie derivative, 306
Liouville measure, 306
Lyapunov exponent, 181

Markov chain, 13
Markov process, 23
Maslov–Morse index, 174, 181
Maurer–Cartan form, 273
mean value, 58
measure, 5
   Feynman (formal), 138
   Lebesgue, 5
   Wiener, 25
method of images, 231
method of square completion, 159
midpoint prescription, 47, 186, 210
mixed states, 127
mode expansion, 100
Morse function, 307
Morse theorem, 176

normal symbol, 206, 207

observable, 123
operator
   annihilation, 132
   compact, 142
   conjugate, 124
   creation, 132
   Hamiltonian, 125
   self-adjoint (Hermitian), 124
   symmetric, 124
operator ordering problem, 129, 183, 187, 200
operator spectrum, 124
   study by the path-integral technique, 141
ordering rules, 129
Ornstein–Uhlenbeck process, 63, 90
overcompleted basis, 207

Parisi–Sourlas integration formula, 314
partition function, 56
path integrals, 1
   and singular potentials, 267
   calculation by ESKC relation, 39
   change of variables
      via Fredholm equation, 45
      via Volterra equation, 46
   coherent state PI on $SU(2)$ group, 281
   discrete-time (time-sliced) approximation, 36
   Feynman, 122, 137
   in phase space, 122, 139, 155
   in terms of coherent states, 218
   Wiener, 25
   with constraints, 122
   with topological constraints, 122
path length, 188
periodic orbit theory, 154, 181
periodic orbits, 176, 178
perturbation expansion, 152
phase-space path integral, 122
physical-optical disturbance, 187
Poincaré map, 181
Poisson brackets, 129
Poisson distribution, 50
Poisson formula, 232
Poisson stochastic process, 51
postulates of quantum mechanics, 123
probability amplitude, 124, 125
probability density, 57
probability distribution, 19, 56
   initial, 14
   normal (Gaussian), 19
probability space, 56
propagator (transition amplitude), 135
   for a particle in a box, 232
   for a particle in a curved space, 248
   for a particle in a linear potential, 197
   for a particle in a magnetic field, 195
   for a particle on a circle, 240
      and α-quantization, 241
   for a particle on a half-line, 236
   for a short time interval, 140
   for a torus-like phase space, 242
   for the driven oscillator, 198
   interrelation in different coordinate systems, 257
   radial part for a free particle, 265
   radial part for the harmonic oscillator, 265
pure states, 127
px-symbol, 202

quadratic approximation, 86, 87
quantization
   canonical, 128
quantum chaos, 181
quantum fluctuations, 162
quasi-geometric optics, 187
quasi-periodic boundary conditions, 230, 241

radial path integrals, 258, 260

random field, 102
random force, 62
random function, 57
random variable, 56
random walk model, 108
renormalization in quantum field theory, 212
resolvent of a Hamiltonian, 177
Riemann ζ-function, 100
Riemann–Lebesgue lemma, 170

S-matrix, 222
saddle-point approximation, 170
scalar curvature, 246
scattering operator, 222, 223
   adiabatic, 224
Schrödinger equation, 125
   stationary, 133
Schwarz test functions, 130
Schwinger variational equation, 1
semiclassical approximation, 80, 144
semigroup property, 21
source functions, 102
spacetime
   Euclidean, 139
   Minkowski, 139
spacetime transformations in path integrals, 253
star-product (star-operation), 201
state vector, 123
stationary Schrödinger equation, 126
stationary state, 126
stationary-phase approximation, 170
steepest descent method, 170
stochastic (random) field, 57, 102
   Gaussian, 102
stochastic chain, 57
stochastic equations, 61
stochastic function, 57
stochastic integral, 63
stochastic process, 17, 23, 57
   Gaussian (normal), 58
   Markov, 58
   stationary, 58
   white noise, 59
   Wiener, 59
stochastic sequence, 57
summation by parts, 96

superposition principle, 126
superselection rules, 126
superselection sectors, 127
superspace, 296
supersymmetry operator, 306
symbol of an operator, 200
   px-symbol, 202
   xp-symbol, 201
   contravariant, 220
   normal, 207
   Weyl, 202
symplectic two-form, 305

Tauberian theorem, 149, 326
time-ordering operator, 126
time-slicing, 28
topological term, 234
transition amplitude (propagator), 135
transition matrix, 14
transition probability, 14
Trotter product formula, 156, 157

uncertainty principle
   and path integrals, 159, 216

vacuum state, 132
Van Vleck–Pauli–Morette determinant, 173
variational methods, 80
volume quantization condition, 243

wavefunction, 127
Weyl symbol, 202, 203
Wick theorem, 225
Wiener measure, 25
   conditional, 25
   unconditional (full, absolute), 25, 43
Wiener path integral, 1, 25
Wiener process, 24
   its derivative (white noise), 113
Wiener theorem, 29
   analog for phase-space path integrals, 214
   and differential operators in path integrals, 100
WKB approximation, 169, 172

xp-symbol, 201